genetics

a human perspective

genetics

a human perspective

linda r. maxson
pennsylvania state university

charles h. daugherty
victoria university of wellington

 Wm. C. Brown Publishers

Book Team

Editor *Kevin Kane*
Developmental Editor *Megan Johnson*
Production Editor *Anne E. Scroggin*
Designer *Eric Engelby*
Art Editor *Margaret Rose Buhr*
Photo Editor *Lori Gockel*
Permissions Editor *Gail Wheatley*
Visuals Processor *Joyce E. Watters*

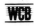 **Wm. C. Brown Publishers**

President *G. Franklin Lewis*
Vice President, Publisher *George Wm. Bergquist*
Vice President, Operations and Production *Beverly Kolz*
National Sales Manager *Virginia S. Moffat*
Group Sales Manager *Vincent R. Di Blasi*
Vice President, Editor in Chief *Edward G. Jaffe*
Marketing Manager *Paul Ducham*
Advertising Manager *Amy Schmitz*
Managing Editor, Production *Colleen A. Yonda*
Manager of Visuals and Design *Faye M. Schilling*
Production Editorial Manager *Julie A. Kennedy*
Production Editorial Manager *Ann Fuerste*
Publishing Services Manager *Karen J. Slaght*

WCB Group

President and Chief Executive Officer *Mark C. Falb*
Chairman of the Board *Wm. C. Brown*

Calligraphy by David C. Lansdon

Library of Congress Catalog Card Number: 91–71084

ISBN 0–697–09918–0 (Paper)
      0–697–14554–9 (Case)

Printed in the United States of America by Wm. C. Brown Publishers,
2460 Kerper Boulevard, Dubuque, IA 52001

10 9 8 7 6 5 4 3 2 1

contents

preface

No science is more central to contemporary life than genetics. Human health, reproduction, and food increasingly rely on our understanding of the hereditary mechanism of humans and other species. Furthermore, the hereditary mechanism shared by all living organisms, including humans, reflects the evolutionary relatedness of all life. Understanding the meaning of being a human requires knowledge of our biological origins, which genetics clearly provides.

Genetics so permeates human life that knowledge of genetic mechanisms and of their many influences on contemporary society is indispensable for living an informed life. Our goal in writing this book is to provide this essential genetic information—the genetics that humans need to know.

This book is written primarily for college and university students of all backgrounds and majors. The most important task is to convey in some detail the structure and functioning of the genetic apparatus. In each chapter, we attempt to relate basic genetic facts to important aspects of everyday life, and also to key issues of social and ethical responsibilities raised by advancing genetic knowledge.

The range of issues related to genetics is so broad that we have necessarily been highly selective in our choices of social and ethical topics. However, instructors should have no difficulty adding to our coverage by reading both the scientific literature and the mass media, where genetic issues receive almost daily treatment.

When we wrote our first edition less than ten years ago, the revolution in DNA technologies was in its early stages. In the 1980s, DNA technologies penetrated every aspect of biological science. The most significant change in the present edition is the greatly expanded coverage of these technologies and their many uses. Chapter 10, "Genetic Engineering: A Miracle Made Reality," introduces the technologies at a level that should be comprehensible to the non-science undergraduate and demonstrates the range of their uses. In other chapters we have included new information derived from DNA technologies that support and expand genetic knowledge from more traditional approaches.

Rapidly advancing genetic knowledge also underlies a second theme of the book—that new knowledge sometimes solves old problems, but equally often presents us with new dilemmas. The use of DNA probes to identify carriers of harmful genes only complicates the lives of such persons and their families, for example, when science cannot also offer a cure.

A third new dimension of this edition is the increased coverage of genetic issues in human history. Scholarship in the 1980s paid increased attention to the role of natural selection in European conquest, and to the role of physicians and geneticists in the abuses of genetics in Germany, Britain, and the United States in the first half of the twentieth century. While our coverage of these topics is necessarily limited, we hope students will benefit from exposure to the relationship of genetics to some of the central events of this and earlier centuries.

Two of the important teaching aids of this text are significantly expanded in this edition. *Boxed readings* cover topics ranging from biological weapons to selfish genes, the use of fetuses in research, the genetics of wheat, and species extinction. We hope the diversity of these topics may spur interested students to seek more information on their own. We have attempted to assist their search by the inclusion of expanded lists of *Suggested readings* that provide original sources for these topics.

Acknowledgments

We especially thank Richard D. Maxson, Kevin R. Maxson, Marie Daugherty, and Brendan Daugherty for patience, good humor, and support throughout the writing of this and previous editions. We thank Jennifer M. Hay for research and technical assistance and Dr. Ann Bell, Dr. A. P. Blair, Dr. G. K. Chambers, and Dr. G. K. Rickards for helpful commentary on specific sections of the text.

The following reviewers provided invaluable assistance in their reviews of this and previous editions of the book:

Ann A. Lopez
San Jose City College

George A. Hudock
Indiana University

Ronald W. Wilson
Michigan State University

genetics

a human perspective

Introduction

Humans and alleles

(a) Gregor Mendel (1822–1884) discovered the principles of genetics through meticulous experiments on inheritance in peas.
(b) Charles Darwin (1809–1882) proposed the theory of evolution by means of natural selection and spent the remainder of his life gathering and analyzing data to test his theory. (c) Francis H. C. Crick (b. 1916) and James D. Watson (b. 1928) discovered the double helical structure of DNA.
(d) Stanley N. Cohen (b. 1935) and (e) Herbert W. Boyer (b. 1936) developed techniques for moving segments of DNA-genes-between organisms, thus founding the genetic engineering revolution.

(a and b) © William Ferguson; (c) J. D. Watson, "The Double Helix," Atheneum, New York, 1968; (d) Courtesy of Stanley N. Cohen; (e) VCSF/David Powers.

(a)

(b)

(c)

(d)

(e)

Genetics is the study of inheritance. Why are we each like our parents? Why do most organisms strongly resemble their parents? Why do a few organisms differ greatly from their parents?

Genetics thus involves a fundamental paradox: we are like our parents, but we are also different. Each person has many features that are similar to each parent, but also many features that resemble neither parent. Dark-haired parents usually have children with dark hair, but they often have blond children as well. Some seeds from the tallest plants give rise to short plants. Brown cows can produce white calves. How can this be?

The first geneticists were people who observed that "like begets like" and used that basic observation to improve their lives. About ten thousand years ago, and perhaps even one hundred thousand years ago, humans began to domesticate animals such as dogs and sheep. By controlling the breeding of those individuals with the most useful traits—docility, affection, strong mothering instincts, good meat quality—humans used the natural mechanisms

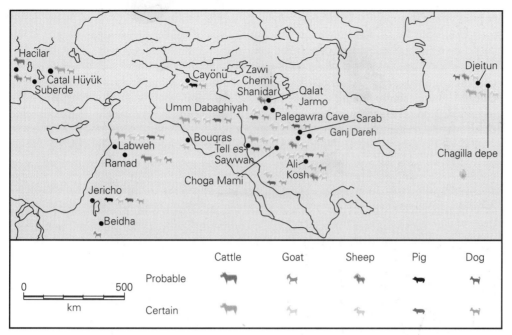

Figure 1.2

(a) Sites where animal remains show early domestication of animals, about ten thousand years ago. Changes associated with selective breeding for muzzle length in skulls of (b) the European Wolf, *Canis lupus;* domestic dogs, *Canis familiaris;* (c) a mutt; (d) a Great Dane; and (e) a Pekingese.

(a–e) From Juliet Clutton-Brock, *Domesticated Animals From Early Times.* Copyright © 1981 University of Texas Press. Used by courtesy of the Natural History Museum, London.

(a)

(b)   (c)

(d)   (e)

of inheritance to improve their lives. Control of inherited traits led to improved food supplies, sources of skins for clothing and other uses, and beasts of burden and companionship. Docile cattle gave rise to more docile cattle, and control of the lineage remained with humans.

At about the same time, plants that produced edible grains, nuts, and fruits came into cultivation. Seeds from the largest, healthiest plants were saved to produce the next generation of increasingly productive plants. Predictable, improved food supplies allowed cities to arise around permanent plantations. Control of inherited traits of wild animals and plants thus formed an essential ingredient in the development of human civilization.

While plant breeding and animal breeding have been practiced successfully for at least as long as humans have recorded their history, an understanding of the mechanisms of inheritance—genetic mechanisms—was almost nonexistent until about a century ago. In 1866, an Austrian monk named **Gregor Mendel** (figure 1.1) published the first paper accurately describing aspects of the hereditary mechanism ("Experiments with Plant Hybrids"). These studies were mostly ignored by the scientific community until 1900, when the importance of Mendel's work was independently recognized by three biologists, Hugo de Vries in Holland, Carl Correns in Germany, and Erich von Tschermak in Austria. Since 1900, genetic knowledge has expanded with increasing rapidity. Genetics is a science that is truly a child of the twentieth century.

Thus, although humans have successfully practiced plant breeding and animal breeding for ten thousand years (and perhaps much longer), theoretical knowledge of principles of inheritance has existed for only 1% of that time: the past one hundred years. What is unknown about genetic mechanisms still dwarfs what we know. If you are now twenty years old, you can expect to live a half century longer. During the rest of your life genetic knowledge will increase immensely. The doubling time of scientific knowledge is now about ten years. What is known today will be only a small fraction of what is known fifty years from now. Furthermore, during the next fifty years genetic technology will be applied (and *is being applied*) to hundreds of problems that affect your daily life.

How rapidly is new genetic knowledge changing the lives of humans? Humans now use knowledge of genetic mechanisms for purposes unimaginable only a few decades ago: to cure disease, to improve crop yield, to create pest-resistant crops, to produce self-fertilizing crops, to produce organisms that can clean up oil spills, to conserve rare species, to insert genetic characteristics of one species into another species, and even to produce species that have never before existed on the face of the earth. The limits of the use of genetic knowledge, for good or bad, are the limits of the human imagination. Knowledge can indeed be power.

The Hereditary Mechanism

Long before Mendel's experiments, some scientists believed that tiny particles carrying hereditary information were somehow transmitted from parent to offspring. Mendel proved that such particles exist and, furthermore, that for some specific characteristics each individual receives one hereditary particle from each parent. These hereditary particles are now called **genes.** Mendel showed that flower color in garden peas, for example, was governed by two genes, one from each parent. Each organism possesses thousands of genes, half of which are contributed by each parent. Humans are estimated to have about 50,000–100,000 genes. **Genetics** *is the study of the structure and function of genes and of the transmission of genes between generations.*

One of the most important scientific discoveries of the twentieth century was the establishment of the physical nature of the gene. In a classic paper published in 1953, James Watson and Francis Crick (figure 1.1) proposed a structure for a complex molecule called **deoxyribonucleic acid**—usually called **DNA.** They suggested that DNA was a long double helix, like a twisted ladder (figure 1.3). Each of the uprights is a chain of alternating sugar molecules and phosphate molecules. The two uprights are linked together by steps of nitrogenous bases, much like the rungs of a ladder. Since the dramatic publication of the **Watson-Crick model,** further experimentation has confirmed the double helical nature of most DNA. DNA and RNA (a very similar molecule) constitute the genetic material of all living organisms.

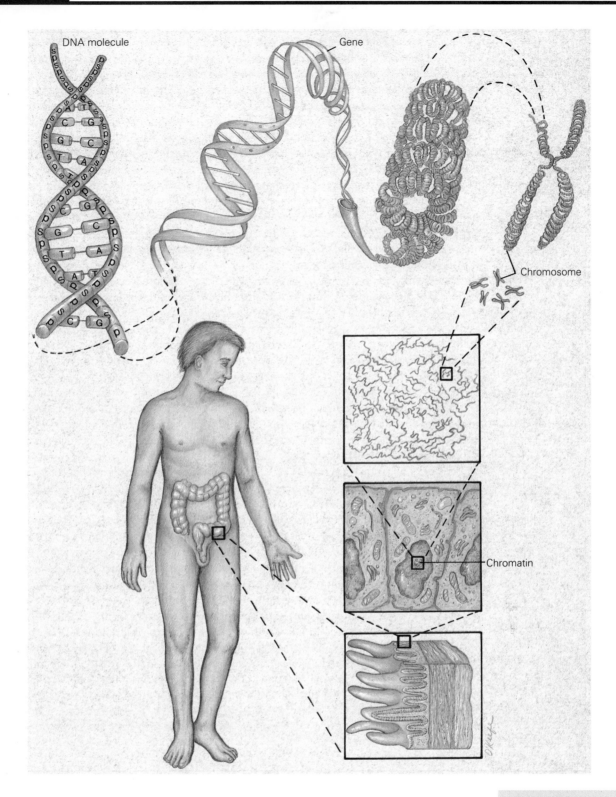

DNA molecule

Gene

Chromosome

Chromatin

Figure 1.3

DNA in humans. DNA is
found in chromosomes that
are located in the nuclei of
cells in almost all tissues of
humans.

Watson and Crick also proposed that the sequence of nitrogenous bases along the DNA molecule (the rungs of the ladder) carries the enormous amount of genetic information necessary for life. Their proposal turned out to be correct, and DNA is now often called an **informational macromolecule**—that is, DNA is a very long molecule that contains information encoded in its structure. The structure of DNA thus encodes the genetic message. A *gene* is a segment of DNA, and a DNA molecule may have a sequence of many genes along its length. The information encoded in each gene is used to direct the production or synthesis of other types of molecules. The interaction of genes with the external environment and with the molecular products of other genes produces a living organism.

In addition to describing the mechanism by which DNA encodes the information of life, the Watson-Crick model also predicted other key features of the genetic mechanism: **replication,** a method for accurate self-reproduction of the genetic material; and **mutation,** the mechanism of genetic change. In conjunction with other available knowledge of reproduction, the Watson-Crick model also explained how genetic information is passed between generations, *transmission genetics*.

The enormous amount of information encoded in one hundred thousand or more genes makes apparent the precise control that the genes exert on the organism—its development, structure, and function. This recognition has in turn altered the science of genetics. The study of transmission of genes is no longer the only goal of geneticists. Many geneticists now study gene function and gene regulation: How is the information in genes used to influence and regulate various aspects of the life of the organism? How are the genes themselves turned on and off?

One aspect of transmission genetics has important implications for all of biology. Some genes are transmitted to succeeding generations far more frequently than other genes, which may be transmitted less efficiently. Genes that increase an individual's capacity to survive and reproduce will spread throughout a species, while other genes may disappear.

Differential survival and reproduction of individuals as a result of different genetic characteristics is called **natural selection.** In 1859, **Charles Darwin** (figure 1.1) proposed that natural selection was the mechanism responsible for evolution (*On the Origin of Species*). The discovery of the principle of natural selection by Darwin is one of the greatest achievements in the history of science. Darwin stressed that all species produce more young than survive, and that within each species is a vast reservoir of heritable variation. Moreover, conditions on earth are always changing. Through time, natural selection acts on the variation present in populations, changing populations, species, and their genes. Some species change into new species; some do not change fast enough and become extinct. Some species may appear not to change externally, but through time the genes of all species undergo some change. Darwin called this process "descent with modification." Today we call it **evolution.**

A genetic mechanism is fundamental to the evolutionary process. Traits favorable to survival and reproduction can increase in a population only if they are heritable. Furthermore, evolution can occur only when genetic variation exists. Although Darwin did not understand the mechanism of inheritance (Mendel's work was never seen by Darwin), he did understand that a mechanism of inheritance was required for the process of natural selection. Perhaps the most precise and simple definition of evolution is a genetic definition: Evolution equals change in genes; or Evolution equals "descent with modification" of genes.

The Importance of Genetics

Knowledge of the hereditary mechanism is essential information for any informed individual. To a large extent, the development of civilization has coincided with the capacity of humans to control their basic needs such as food and shelter. The enormous expansion of the human population in this century is directly related to genetic knowledge that allowed vast improvement in productivity of domesticated plant species used for food—rice, wheat, and corn, in particular. Genetic knowledge has also been a key component of the revolution in health and medical care in this century.

In the early 1970s, **Herbert W. Boyer** and **Stanley N. Cohen** (figure 1.1) developed techniques for directly altering the genetic material of an organism. Called **bioengineering,** these techniques allow segments of DNA (genes) to be moved to different locations within the long DNA molecule, or even to be inserted into the DNA of another species. Individuals can thus acquire new genes and new genetic traits.

Bioengineering is already changing human life by playing a major role in meeting fundamental human needs. Bioengineering is increasing food, fiber, and wood production, and combatting many diseases. In less than twenty years, bioengineering has become a worldwide industry involving tens of billions of dollars. Inevitably, rapid social changes and technological advances have social costs. Society is now confronting complex social and ethical questions arising from bioengineering. To deal with these exciting but difficult changes, we all need a basic understanding of genetics and the many ways in which the science of genetics affects our daily lives.

Health

About 3–5% of the world population—200 million people—are estimated to be afflicted by serious genetic disease (figure 1.4). Millions more have genetic anomalies that cause relatively few problems in their lives. Knowledge of the genetic basis of diseases such as phenylketonuria (PKU) has already allowed treatment. Other conditions such as Down's syndrome are clearly identified as genetic conditions though cures have not been developed. However, for both PKU and Down's syndrome, as well as many other diseases, genetic testing and counseling can prevent recurrence of certain genetic conditions.

Advances in medical genetics have also been accompanied by new problems. The wide use of antibiotics such as penicillin has led to the evolution of strains of disease-causing organisms resistant to these drugs. Thus, recent technologies have actually helped produce new organisms (chapter 10) that are less susceptible to treatment.

Bioengineering offers the hope of creating more effective antibiotics. Already, bioengineering techniques are being used to produce chemicals for medical use. For example, synthetic human growth hormone is being used to treat some forms of dwarfism. The limits of use of such chemicals have not yet been established: Should synthetic growth hormone be used on people of normal height to make them potentially better basketball players, or soldiers, or more intimidating business people? Experiments are now underway to treat or cure genetic diseases by introducing normal genes into affected people. What are the limits to such technologies? Should we be developing techniques that will allow future generations to produce "designer people," engineered to have just the "right" genes? What are the "right" genes, anyway?

Such technologies challenge human ethical codes. Many people fear that changing human genes is "playing God" and should thus be forbidden. Others argue that gene therapy is no different in principle from more conventional therapies and that withholding such treatment from the ill would be immoral. Even the ability to diagnose genetic disease presents difficult ethical choices. Should the results of screening for genetic diseases be made available to employers, insurance companies, or a future spouse?

Reproduction

Widely applied contemporary reproductive technologies such as artificial insemination and *in vitro* fertilization are not strictly genetic techniques. They do, however, raise numerous ethical questions and point to the widely held concern that identical copies of particular persons can be "mass produced," or **cloned.** Although cloning of humans may not be possible for some decades, it will certainly be subject to very close political and ethical scrutiny.

In the meantime, ethical aspects of present technologies present difficult choices. *In vitro* fertilization (*IVF*) is increasingly used to treat infertility, but it is expensive, inefficient, and appropriate only for a relatively uncommon type of infertility. Should the treatment be limited only to the affluent, who can afford it? Might the money spent on *IVF* be used more responsibly to prevent or treat more common types of infertility? *IVF* results in many excess frozen embryos. Should experimentation be allowed on these embryos?

Artificial insemination, often using anonymously donated sperm, has been used since the early nineteenth century for couples with male-related infertility. Should a woman be

Figure 1.4

Examples of serious genetic diseases. (a) Folksinger Woody Guthrie (1912–1967) before onset of symptoms of Huntington's disease, a dominantly inherited degenerative disease, and (b) after effects of the disease became obvious, (c) red blood cells of a normal black person and (d) misshapen (sickled) red cells of a black person affected with sickle cell anemia. Bioengineering techniques allow direct genetic screening to identify people who carry genes causing these diseases, but some people carrying these genes lead perfectly normal lives. Should employers have the right to test potential employees to see if they carry these genes and *might* be affected with these diseases?

(a) The Bettmann Archive; (b) Woody Guthrie Publications, Inc.; (c) R. Becker/Custom Medical Stock Photo; (d) © Science VU/Visuals Unlimited.

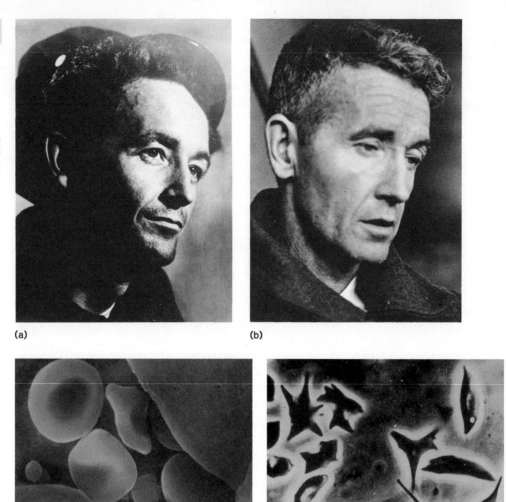

(a)   (b)

(c)   (d)

allowed to conceive using sperm that her husband placed in a sperm bank before his death? How should donors to such banks be chosen? For what diseases, including genetic diseases, should the sperm be screened?

Agriculture

Human life depends on thousands of other species. Like most animals, humans use plants and animals for food. Additionally, we rely upon other species for countless, often highly imaginative, purposes: shelter, clothing, medicinal and recreational drugs, companionship, transportation, musical instruments, decoration, scientific experimentation, communication, and recreation.

Throughout human history, the ingenuity of plant and animal breeders has led to the development of an enormous variety of domesticated breeds (figure 1.5). The ancient Chinese, for example, selected for tiny Pekingese dogs, the Romans had both lap dogs and fighting dogs, Egyptians produced hunting dogs much like modern greyhounds, and heavier hunting dogs were bred in eastern Asia.

Since World War II, plant breeding has reached its culmination in programs of the Green Revolution, which have developed new, highly productive varieties of important grain crops for use in third world countries. These programs have been scientifically successful—grain production has increased greatly. They have often had important social consequences, as well, which are widely debated. Ancient social systems have changed to meet new agricultural practices, but hunger and famine remain. Bioengineering will allow continued and probably dramatic increases in agricultural productivity. Ensuring that larger food supplies leads to a reduction in world hunger will involve all of us, not only geneticists and farmers.

Figure 1.6

Logging of a rain forest in South America. Extinction of the plant species occurring in such forests means that unique genetic resources of unknown value are lost forever.

© Science VU/Visuals Unlimited.

Conservation

Genetic variation is a fundamental natural resource. Genetic differences exist both between different species and within the same species (between individuals, populations, and geographical races). Unique genetic types are often the basis for major advances in agriculturally important species. Genes for such important traits as disease resistance and increased productivity are identified from species and individuals in nature, and then introduced to domestic strains by hybridization. Natural genetic diversity is so important for supporting the development of new varieties of grain crops that gene banks containing hundreds of thousands of specimens of wild rice, wheat, and maize have been established in some tropical countries.

Wild plant species are often the source of medically important drugs, as well. About 75% of Americans regularly take medications derived from plants, and the active ingredients in about 25% of all prescription pharmaceuticals are derived from plants. A variety of marine animals—snails, sea urchins, fish—are now being surveyed in the hope of identifying new drugs to fight cancer.

Despite the practical importance of wild plants and animals, however, fewer than 1% of the 10–30 million living species have been examined for their medicinal, agricultural, and other uses. Destruction of tropical forests and other reservoirs of biological diversity is proceeding so rapidly that many species are becoming extinct before they can be named by scientists and their importance identified (figure 1.6).

Nature **conservation** is thus increasingly recognized as an important genetic issue. The World Conservation Strategy, produced by the International Union for the Conservation of Nature, the United Nations, and the World Wildlife Fund, states as one of its three major goals: "*Preserve genetic diversity.*" This recognizes that genes constitute the primary biological resource whose conservation is essential to human survival.

Genetics and Evolution

The examples in the preceding sections give some idea of the extraordinary diversity of applications of genetic knowledge to practical aspects of human life. Genetics also plays a central role in larger issues of human experience.

Genetics was recognized by Darwin as the key to understanding evolution. Although Darwin's concepts of genetics are now known to be wrong, Darwin understood clearly that evolutionary change is the *inevitable consequence* of natural selection acting on an hereditary system. Genetics thus provides one of the foundations for the theory of evolution.

Evolution, in turn, is the central concept in biology. The famous biologist Theodosius Dobzhansky observed that "Nothing in biology makes sense except in the light of evolution." All aspects of all living organisms are influenced to some extent by their genes. And, because all organisms trace their origins back to a common ancestor several billion years ago, the studies of evolution and genetics merge with philosophy and help us understand ourselves and our place in the universe.

The Evolutionary Process

How does evolution occur? The Darwinian mechanism is surprisingly simple and requires only three basic parts:

1. *Heritable variation.* The genetic mechanism that is not perfect, but produces offspring that are sometimes slightly different from their parent(s);
2. *Limited environmental opportunities.* There are insufficient resources for all individuals to survive and reproduce, leading to *competition* for necessary resources such as food and shelter;
3. *Changing environments.* This condition is easily met, as environments have continually changed throughout the long history of the earth. As conditions change, different individuals are best able to compete and survive to reproduce.

When organisms compete for resources like food, some are more able than others to obtain what they need (figure 1.7). When wolves hunt reindeer for a meal, beneficial traits for hunting success include large size, strong jaws, intelligence, cooperative ability, and endurance for the chase. From the reindeer's perspective, similar traits—large size, intelligence, endurance, the ability to cooperate in defense—enhance survival.

Individuals with heritable traits that allow them to survive will pass those traits along to their offspring, while traits that are detrimental to survival will not be transmitted to future generations. Thus, genes beneficial to survival and reproduction are *naturally selected* and increase in frequency in future generations.

The process of natural selection is virtually identical to a farmer selecting seeds from the largest plants for next year's crops or selectively breeding for larger cattle or smarter sheep dogs. The only difference is that natural selection occurs without human intervention. One definition of natural selection is *differential reproduction,* meaning that some individuals produce more offspring than others, and thus their genes and traits allowing improved survival become proportionally more abundant in future generations.

The struggle for existence began with the first organisms on earth and continues today. Environments have continually changed throughout history, and thus natural selection has favored different genes in different times and places. In turn, this has led to a continually changing sequence of plants, animals, and other organisms that have inhabited the earth. The process of change in the biota of earth is known as *evolution*. In fact, change is the simplest definition of evolution.

The Results of Evolution

Fossils clearly show that life originated on earth over 3 billion years ago. Since then evolution has produced an astonishing array of species. In general, the numbers and types of living organisms have increased throughout evolutionary history, although occasional calamities such as impacts of asteroids may have sometimes caused massive extinctions.

As many as thirty million different species may survive on earth today. Genetic data indicate that there are three main evolutionary lineages: (1) Archaea (*archaebacteria*), primitive single-celled organisms found mainly in extremely unusual and hostile environments; (2) Bacteria (*true bacteria*), the many species of single-celled bacteria found in virtually every habitat in the world; and (3) Eucarya (*eukaryotes*), including all the species of larger, often familiar organisms: fungi, plants, animals, and single-celled protozoans such as amoebae. The diversity of life is briefly reviewed in Appendix A.

Human Evolution

Animals with backbones, **vertebrates,** have an exceptionally well-known evolutionary history because their bones can be preserved in the geological record as fossils. The first vertebrates were small fish, whose fossilized skeletons are found in rocks about 550 million years old. Descendants of these primitive fish diversified, and one lineage gave rise to an entirely new class of vertebrates, the *amphibians.* Further evolution produced the *reptiles,* the *birds,* and eventually the *mammals.*

The first *primate* appeared about fifty to sixty million years ago. Early primates were small-bodied, large-eyed, nocturnal tree dwellers whose evolutionary history was confined primarily to the tropical regions of South America, Africa, and Asia. Most primates are small monkey-like creatures, but one lineage consists of a few species of relatively large-bodied forms—the **Great Apes,** which include gorillas, chimpanzees, orangutans, and gibbons. About five million years ago, a group diverged from the common ancestor of the Great

Figure 1.8

"Lucy," the most complete early hominid fossil ever discovered. Lucy is about 3.5 million years old, lived in northern Africa and belongs to the earliest hominid species, *Australopithecus afarensis*. Discovery of this fossil helped confirm that the lineage leading to modern humans diverged from the Great Apes only about 5 million years ago and that the genus *Homo* arose less than 3 million years ago.

E. L. Simon, "Human Origins," *Science,* Vol. 245: 1343, 1989. Photo courtesy of Peter Schmidt.

Ape lineage and began to change rapidly in posture, size, and intelligence. This evolutionary lineage has thus existed for less than one-tenth of 1% of the total history of the earth, and only one species now survives: *Homo sapiens,* wise man.

In spite of the substantial differences in appearance between humans and chimpanzees, evidence of their recent common ancestry is revealed by their genes and gene products. The genes and proteins of humans and chimpanzees are over 99% identical in molecular sequence (see chapter 13). The genetic difference between humans and chimpanzees is less than that usually found between two species in the same genus, such as the coyote and the domestic dog.

Lucy, Hominids, and Humans

The extraordinary genetic similarity of humans and chimpanzees was first identified in the late 1960s. Previously, the evolutionary lineage of the Family Hominidae (hominids—humans, as distinct from the Great Apes) was thought to have originated between fifteen and forty million years ago. Molecular genetic data collected by biochemist Allan Wilson and anthropologist Vincent Sarich supported a very recent origin (3–5 million years) for hominids—but many scientists wanted more (non-molecular) proof.

Dramatic evidence was provided in 1974. Working in northern Ethiopia, Donald Johanson and Tim White discovered the oldest known hominid fossil. The extraordinary fossil was named **Lucy** after the song "Lucy in the Sky with Diamonds" (the discovery team celebrated by drinking beer and playing a Beatles' tape). Usually, only fragments of jaw or teeth are found, but this fossil skeleton was 40% complete (figure 1.8).

Lucy belongs to the species *Australopithecus afarensis,* and she is about 3.5 million years old. *Australopithecus* is the first hominid genus and a descendant of the Great Apes. Johanson and White believe that her species is probably the common ancestor of all other species of *Australopithecus* and *Homo.* Because Lucy is such a recent fossil (3.5 million years old rather than 10, 15, or more), her descendants must therefore be even more recent.

Johanson and White thus argue that the genus *Homo* is less than 3 million years old, agreeing closely with the date first suggested on the basis of molecular data in 1967 by Wilson and Sarich. More recent fossils and additional genetic data now substantiate this view.

Figure 1.9

The family tree of humans, as proposed by Donald Johanson and Tim White. Small variations on the branching pattern have been proposed by other scientists, but the general relationship of *Australopithecus* and *Homo* is now accepted by most scientists, as are the approximate divergence dates of the main lineages.

Other fossils show that several species of *Australopithecus* evolved and probably co-existed with the first human species (*Homo habilis*) in Africa 2–3 million years ago (figure 1.9). It is possible that the Australopithecines became extinct through competition with *Homo habilis*. Subsequently, *Homo habilis* gave rise to *Homo erectus,* which in turn led to *Homo sapiens,* the only surviving hominid.

Eve and Modern Humans

Studies of human mitochondrial DNA from throughout the world support the view that all modern humans are extremely closely related to one another. These studies from Allan Wilson's lab, published in 1987, show that present genetic variation can be traced back to a single female—called **Eve,** who lived about 200,000 years ago, probably in Africa. While populations of humans would have lived in Europe and Asia at that time, these data suggest that migrations of humans from Africa to Europe and Asia may have resulted in the African populations replacing those in Europe and Asia.

The significance of this date, 200,000 years, is that it represents only about ten thousand human generations. The differences among the various surviving human racial and ethnic groups are of extremely recent evolutionary origin. Thus, all humans share a very recent common ancestor—indeed all of us may trace our ancestry back to the same ten thousandth-great-great-grandmother, "Eve."

BOX 1.1

Disease, natural selection, and European conquest

In the early 1520s the Spanish conquistador Hernando Cortés and several hundred soldiers captured the Aztec capitol of Tenochtitlán (Mexico City), a city of 300,000 with an army of 100,000, after a siege of only a few weeks. The entire Aztec empire of several million people soon fell to the Spanish conquerors. Within the next decade, the conquistador Francisco Pizarro and a tiny Spanish army had conquered the mighty Incas, with a population and empire even larger than that of the Aztecs. How could the vastly outnumbered Spanish achieve their massive military success so quickly?

Disease, not superior weaponry, was the chief agent of Spanish conquest. No one knows how many Amerindians lived in North and South America when Columbus first landed in the Caribbean. Some estimates are as high as 100 million, but much smaller numbers— twenty million or less—are supported by most scholars. What is not in doubt, however, is that native populations precipitously declined, often by 90% or more, as a result of epidemics of disease brought by European invaders. The advanced civilizations of the Aztecs and Incas were thus brought to an end by natural selection of the strongest kind.

The first documented epidemic in the Americas occurred on the island of Hispaniola (now the countries of Haiti and the Dominican Republic) in 1518, where smallpox reduced the Indian population from perhaps one million to only one thousand within a year. Smallpox (or red plague, as it was called) had been a common viral disease throughout Eurasia for many centuries prior to 1500. Most people in Europe would have had the disease in childhood and recovered from it, although it was still a common cause of death. Thus, through a long history of natural selection, European populations had acquired some immunity to the smallpox virus.

Amerindians, on the other hand, had never been exposed to smallpox and the effects were devastating. Native peoples seemed almost to fall dead at the sight of a European, undoubtedly reinforcing a sense of natural superiority already felt by the Spanish. The few survivors were often debilitated and demoralized, unable to resist the Spanish armies.

Epidemics often spread faster than the invading armies. Smallpox rampaged through Tenochtitlán immediately prior to Cortés' victory. Subsequently, smallpox spread throughout South America ahead of Pizarro's army, making the Incas an easy conquest. Indian populations in North America were smaller than in South America, but similarly suffered from European diseases. Malaria and, later, yellow fever were other Old World diseases brought to the Americas that devastated native peoples. As recently as the nineteenth century, blankets infected with smallpox were used as agents of genocide, given to Amerindians in western North America and Australian Aboriginals to destroy native tribes and make way for expanding European populations.

Are Humans Still Evolving?

All the genes that occur within one species are called the **gene pool** for that species. Many genes occur in alternate forms called **alleles.** All humans have the same basic genes (involved in the production of eye color, hair color, blood type, and so on). Different individuals, however, may have different forms or alleles of these genes. An individual may have alleles producing blue eyes or alleles that result in brown eyes, alleles producing blood type A or alleles producing blood type AB, etc. Thus, a gene pool contains many different alleles for each gene.

Evolution refers to changing proportions of alleles within a gene pool, as well as the introduction of new alleles and the loss of others. During the past four hundred years, the allelic content of the human gene pool has continually changed. Since 1600, the total population of the earth has dramatically increased, doubling many times. By 1850, the world's population was about one billion; by 2050, it is likely to be at least ten billion, and perhaps even higher. However, not all populations have increased equally. Some have gone extinct,

Figure 1.10

Battle of Cholula, Mexico—
conquest by the Spanish in
1519.

Historical Picture Service.

Diseases governed many aspects of European conquest. Europeans themselves were not always immune. European influence in tropical America was limited by disease, ironically by malaria and yellow fever that were brought from the Old World but thrived in the American tropics. Black Africans, however, were genetically resistant to these diseases and were thus imported as slaves to work the plantations of tropical America. British expeditions of conquest in South America in the 1700s were thwarted by epidemics of yellow fever. The Panama Canal, linking the Atlantic and Pacific Oceans, was as much a medical as an engineering triumph, being completed only in the early 1900s after yellow fever was defeated.

The flow of disease was not entirely one way. Syphilis is thought to be a disease of American origin that Columbus' sailors acquired and brought back to Europe in 1493, where its victims included even a Pope in 1516. Syphilis subsequently accompanied European voyages of exploration along with smallpox, measles, and other Old World diseases. Sadly, this New World disease became yet another lethal weapon of natural selection and European conquest, killing tens of thousands of non-resistant native peoples throughout the Pacific, Australia, and New Zealand.

while others have increased enormously. Different racial and ethnic groups, differing in allelic proportions, have differentially contributed to the present human gene pool, and the flux is continuing.

Loss of alleles. The total genetic diversity of humans has decreased in recent history, as many racial and ethnic groups have become extinct. As Europeans explored the world and imposed their culture on the native peoples of North America, South America, Africa, Australia, and the Pacific Islands, many native tribes were greatly reduced in numbers or even exterminated as a result of European conquest.

When Captain Cook declared Australia to be the property of the King of England, for example, about three million Aborigines lived in Australia. Today, the population of Australia is about eighteen million, fewer than 100,000 of whom are Aborigines. In two hundred years, the proportion of alleles in Australia of Aboriginal origin has decreased from 100%

to less than 1%. Changes in the Americas were equally dramatic. Alleles of Amerindian origin now comprise a very small proportion of the gene pool(s) in North and South America (see box 1.1).

The demise of indigenous peoples throughout the world in the face of European conquest can be ascribed to many factors, including technological superiority in warfare and deliberate genocide. Recently, historians have emphasized the role of the "European armoury of diseases"—that is, diseases such as smallpox, influenza, and measles, which Europeans could tolerate as a result of centuries of exposure (i.e., natural selection) but which native populations had never encountered. In fact, it now seems certain that these diseases often travelled faster than invading Europeans. Thus, native populations in the Americas, Australia, and New Zealand succumbed quickly to diseases inadvertently introduced by the first wave of European explorers. Survivors, if any, of these epidemics were easily conquered by subsequent invading military expeditions.

Natural selection can thus be seen as an important agent assisting European expansionism in recent centuries. Native populations were reduced by disease and replaced by European peoples more resistant to these diseases. Intermarriage with invaders quickly blended the surviving natives into the gene pool of the new culture. The result was almost always the disproportionate reduction of the genetic contribution of native peoples to succeeding generations.

Increasing allele frequencies.   As some genetic groups decline or disappear, others gain proportionately. The expansion of European cultures throughout the world increased the frequencies of alleles common to Caucasians. India, China, and Japan also maintained large and increasing populations. Since World War II, developed countries have become aware of the dangers of overpopulation, and birth control has increasingly been used to limit population growth. Now, the developed nations are stabilizing at low rates of growth, while many of the developing countries are experiencing extremely rapid rates of growth. In the coming decades, developing nations will comprise an increasingly greater proportion of the population of the world. The doubling time of populations in developed countries is 50–200 years, but only 20–35 years for other countries. Thus, the allelic composition of the human gene pool will continue to change significantly.

Within a heterogeneous population such as the United States, individual groups also change rapidly as a proportion of the total population. The proportion of blacks in the population of the United States has increased throughout the twentieth century and is expected to continue increasing. The proportion of alleles contributed by persons of Hispanic ancestry is also increasing rapidly. These racial, ethnic, and hence genetic changes now occurring reflect a continuation of historical trends in population composition of the United States. Successive periods in American history have been marked by pulses of immigration by nationalities and ethnic groups experiencing political and economic oppression and hardship.

Many persons consider the uncontrolled increase of the world's population to be the single largest problem facing humanity. For that reason, birth control has been widely advocated, and efforts have been made to distribute birth-control information and technology to those groups with the highest birth rates—minorities and people of the developing nations. While there is no question of the seriousness of the population problem, many persons are suspicious of birth-control programs, particularly those who have been the objects of oppression in the past. Many blacks, both in African countries and the United States, view birth-control programs as racist. Many Jews remember the Holocaust and have similar fears. Most methods of birth control are opposed for religious reasons by the Roman Catholic Church and other religious groups. Resistance to population regulation favors continuing genetic change in the human gene pool.

Genetics and Ethics

Ethics is the study of standards of morality and human conduct. By contributing to our understanding of human origins, evolutionary relationships, and disease, genetics has helped to define standards of behavior. Contemporary geneticists, for example, universally accept that differences among various human groups are literally only skin deep. Because all humans share a very recent common ancestor, most of our genes (over 99%) are virtually identical.

Figure 1.11

Babi Yar in the Ukraine, USSR, is a monument to the September, 1941, massacre in Kiev where an estimated 200,000 humans were brutally killed. On September 29 and 30, 1941, over 33,000 Jews were shot at Babi Yar—a greater killing rate than when the gas chambers at Auschwitz were at full operation.

The superficial traits of appearance such as skin color, size, and body shape reflect differences between individuals in relatively few genes as well as the environments in which we are raised.

These limited differences offer no justification for discriminating among people of different races, sexes, or other arbitrarily defined groups, such as persons with particular diseases. Genetics confirms that each individual is different from all others, but those differences do not make any individual intrinsically better or worse than another. **Racism**—the stigmatization of an individual because of his or her race—is not an ethical view that can be supported by genetics. Neither is **sexism**—stigmatization because of an individual's sex.

Even if humans were very distantly related to one another and thus very different genetically (which they are not), genetic differences would not be a basis for conferring different worth on different people. *Genes themselves are not morally good or bad—just different*. Geneticists speak of genes as being beneficial or harmful, but only in terms of physical or mental health. Thus, beneficial genes may allow a person to live longer or free from disease, while harmful genes may cause disease, pain, or death. But individuals who carry genes harmful to health or genes conferring different skin or hair color are of equal moral worth to all other people.

We raise this topic at the beginning of this book because racist and sexist beliefs are (and long have been) deeply held by people in many societies. Genetic differences are frequently invoked entirely inappropriately to justify such beliefs. In the next section we discuss a particularly horrible example of racism claimed to have a scientific justification. In later chapters we deal with other aspects of race and sex as they relate to genetics and ethics.

Racism: An Example

It is difficult to underestimate the evil that has resulted from racism. In this century, the most appalling example is the rise to power of the National Socialist political party—the Nazis—in Germany in the 1930s. Led by Adolf Hitler, the Nazis established a racist regime that undertook systematic genocide of millions of Jews, gypsies, blacks, homosexuals, Slavs, and almost anyone the Nazis believed did not belong to a "master race" they called Aryan. Eventually, their aggression led to World War II and the deaths of 30–60 million people. Genocide on a massive scale has repeatedly occurred in this century, but the Nazis are distinguished by their systematic use of science, genetics, and scientists to justify and implement their policies.

In the United States, immigration policies early in this century excluded people of eastern European, Asian and other ethnic or racial origins for strictly racial reasons. By 1939, over thirty thousand mentally ill or criminally insane people had been involuntarily sterilized for genetic (= racist) reasons. Segregation of black people was strictly maintained in southern states and considered legal until the 1950s and 1960s. The U.S. Public Health Service conducted "experiments" that involved withholding medical treatment from black men with syphilis from the 1930s to the early 1970s.

The point is that the horrors of Nazi Germany occurred within the framework of widely held beliefs in western society, beliefs that often were claimed to be justified by "the laws of Nature." Nothing could be further from the truth. *Genetic knowledge provides no justification for racism, and never has.* Quite the contrary, in fact. Genes themselves are not inherently good or evil; genes do not make one person or race better than another; and even the concept of "race" has very little meaning when applied to humans (chapter 13). The shameful history of the misuse of genetics serves as a constant reminder of how ideas can be perverted to serve evil causes.

Furthermore, racism is not confined to history. In South Africa, laws guaranteeing privilege to one race while impoverishing another were rescinded only in 1991. In countries such as the United States and Great Britain where equality is guaranteed under the law, tensions among racial groups often lead to covert discrimination and violence.

The horrors of Nazi Germany do not guarantee that similar movements will not arise again. In the 1980s a political party in France, called the National Front, achieved a very large popular following by associating itself with ideas reminiscent of the Nazis, especially anti-Semitism. Jewish cemeteries, synagogues, and even corpses were desecrated. Some historians have supported the National Front with claims that Nazi concentration camps for the extermination of Jews and other groups never existed; such ideas, they insist, were propaganda invented by Jews themselves in order to gain sympathy! Such astonishing lies demonstrate the relationship of racism to truth.

A Challenge: Medical Ethics

A very different category of ethical problems is also related to contemporary genetics. The rapidly developing ability to test people for a wide variety of genes harmful to health raises complex questions of medical ethics that society is only slowly confronting. The range of problems is so great that most people are likely to be involved in difficult choices, often of life and death, for themselves or family members at some time during their lives—choices that often have no obvious correct or painless answers: Should I get married, or have a child, if I am likely to transmit a serious genetic defect to my child? Should a fetus carrying a disease-causing gene, for which there is no treatment, be aborted? Should a fetus *likely* to carry such a gene be aborted? Should a newborn child suffering from a serious genetic disease for whom there is no hope of a life free from pain be allowed to die? or be assisted to die? Should my employer have access to results of tests showing that I have a serious genetic disease? or AIDS? Should my family have access to results of my genetic tests, because they might carry the same harmful genes? even if they do not wish to know?

The many dilemmas posed by new genetic technologies are an important theme of this book. Genetics does not and cannot provide the answers to such questions. Genetics does, however, shape the questions that society must confront. Geneticists, scientists, lawyers, philosophers, and theologians alone cannot provide the answers. All of us must contribute if satisfactory answers are to be found.

Summary

1. Genetics is the study of inheritance, that is, the study of the structure, function, and transmission of genes. The modern study of genetics began in 1900 with the rediscovery of Mendel's work.
2. Genes are segments of deoxyribonucleic acid (DNA). DNA is an informational macromolecule that bears hereditary information encoded in its structure.
3. Evolution is defined most simply as "genetic change." Evolution is the inevitable consequence of natural selection acting on heritable variation.
4. The theory of evolution provides the foundation of modern biology. Genetics and evolution are interdependent sciences, and all other biological sciences draw upon them in understanding life on earth.

5. The rise of civilization has coincided with the development of human capacity to control the gene pool of domesticated plant and animal species. The most direct application of practical genetics has been to increase productivity of species used for food. Bioengineering means that productivity of agricultural systems should continue to increase significantly.

6. Genes that might be of value to humans are disappearing rapidly due to the extinction of many wild species, which have yet to be examined for their potential benefits. Conservation of genetic resources is now recognized as an important economic, as well as ethical, issue.

7. Serious genetic conditions affect 3–5% of all humans, and minor conditions probably affect most of us. Medical genetics has become an important health specialty in this century, and bioengineering is likely to enable treatment of many genetic conditions in the coming decades.

8. Among the results of the process of evolution are 10–30 million living species. Among the most recent species to arise are the Great Apes and the hominids. "Lucy" is a fossil *Australopithecus afarensis* who lived about 3.5 million years ago. Her species was ancestral to ours, *Homo sapiens*.

9. Genetic data indicate that "Eve," a woman who lived about ten thousand generations ago, was the ancestor of all living humans. Thus, all present human races and ethnic groups are very closely related.

10. Natural selection played an important role in human evolution, as it did in the evolution of all other species. Even recent human history has been strongly influenced by natural selection, as shown by the variable effects of diseases on different human populations.

11. Because genetics is a science that deals with fundamental issues of human identity, it has a close relationship to the philosophical study of ethics (the study of morality and human conduct). Genetic differences between individuals are not differences in individual worth, and thus they provide no basis for stigmatizing individuals. Nonetheless, history and the present are full of examples of racism and sexism, where people are stigmatized and treated unfairly because of their race or sex.

12. Existing and developing genetic technologies confront society with life-and-death choices that we have never before been required to make. An example is prenatal diagnosis for severe genetic diseases; parents who discover their fetus is affected must make a difficult choice to carry the child, or to abort it. Later, they may have to decide if they wish to risk having a second child that may be similarly affected.

Key Words and Phrases

allele	Eve	Gregor Mendel
bioengineering	evolution	mutation
Herbert W. Boyer	extinction	natural selection
cloning	gene	racism
Stanley N. Cohen	gene pool	replication
conservation	genetics	sexism
Charles Darwin	Great Apes	vertebrates
deoxyribonucleic acid (DNA)	informational macromolecule	Watson-Crick double helix model
ethics	Lucy	

Questions

1. Consider our statement that "for some specific characteristics each individual receives one hereditary particle from each parent." What particular types of characteristics might be directly governed by genes? Might some characteristics be governed only indirectly? not at all? Are some features, such as an arm or a leg, likely to be controlled by one or by many genes?

2. The significance of the chemical DNA in defining life on earth has been recognized only since the 1950s. While genetic factors unquestionably play an important role in human behavior, why is it incorrect to attribute either good or evil actions to the genes one possesses?

3. Look around you where you now sit. Identify products of biological origin. Name the species from which these products were derived that might have been subjected to artificial selection of genes to improve their useful characteristics. Name the species that you ate at your last meal that have been genetically modified.

4. Why might medical and genetic techniques that improve human health or lengthen life be considered controversial? Give examples.

5. Why do you suppose there was controversy over the evidence of a relatively recent common ancestry of humans and other primates?

6. Speculate about other simple characteristics (such as height) that might be or have been the object of natural selection in human evolution.

7. Truly important scientific research can often be identified only many decades after it was done. Do you think we are justified in listing Herbert W. Boyer and Stanley N. Cohen along with Mendel, Watson, and Crick as major figures in the history of genetics? Why?

8. We said earlier that Eve lived "only about ten thousand human generations" ago and was thus a "very recent common ancestor." To understand why we think that 200,000 years is *very* recent, find the answers to the following questions: How old is the earth? How old is life on earth? How long have vertebrates existed? Terrestrial vertebrates? Mammals? Primates? How long have humans lived in Europe? in the Americas?

9. Consider the various wars and other serious conflicts now underway in the world. Which of them have an important racial, ethnic, or religious component? (Remember that religions are often based in race or ethnic group.)

10. World food production, aided by genetic advances, has continued to increase at least as rapidly as population. There is more than enough food to provide for the world's entire population, but starvation and hunger persist and even increase. What are some of the reasons for this contradiction?

11. Name as many medical drugs as you can that are derived from plants.

For More Information

Attenborough, D. 1979. *Life on Earth*. William Collins Sons & Co. Ltd., London.

Bökönyi, S. 1976. "Development of early stock rearing in the Near East." *Nature,* 264:19–23.

Clutton-Brock, J. 1981. *Domesticated Animals from Early Times*. University of Texas Press, Austin.

Diamond, J. 1990. "A Pox Upon Our Genes." *Natural History,* 2:26–30.

"Evolution." 1978. A special issue of *Scientific American,* 239.

Gonick, L., and M. Wheelis. 1983. *The Cartoon Guide to Genetics*. Barnes and Noble Books, New York.

Gould, S. J. 1977. *Ever Since Darwin*. W. W. Norton & Co., New York.

———. 1983. *The Mismeasure of Man*. W. W. Norton & Co., New York.

Hobhouse, H. 1985. *Seeds of Change*. Harper & Row, New York.

———. 1989. *Forces of Change*. Sidgwick & Jackson, London.

Howard, J. 1982. *Darwin*. Oxford University Press, New York.

International Union for the Conservation of Nature—United Nations Environmental Programme—World Wildlife Fund. 1980. *World Conservation Strategy*. IUCN, Gland, Switzerland.

Johanson, D., and M. Edey, 1982. *Lucy: The Beginnings of Mankind.* Warner Books, Inc., New York.

Miller, J., and B. Van Loon. 1982. *Darwin for Beginners.* Pantheon Books, New York.

Müller-Hill, B. 1988. *Murderous Science. Elimination by Scientific Selection of Jews, Gypsies, and Others, Germany, 1933–1945.* Oxford University Press, New York. Translated by George R. Fraser.

Orel, V. 1984. *Mendel.* Oxford University Press, New York.

Proctor, R. N. 1988. *Racial Hygiene: Medicine under the Nazis.* Harvard University Press, Cambridge, Massachusetts.

Roberts, L. 1989. "To Test or Not to Test?" *Science,* 247:17–19.

Sarich, V. M., and A. C. Wilson. 1967. "Immunological Time Scale for Hominid Evolution." *Science,* 158:1200–1203.

Sayre, A. 1975. *Rosalind Franklin and DNA.* W. W. Norton & Co. Inc., New York. (An alternative view of Watson's view of events.)

Stepan, N. 1982. *The Idea of Race in Science. Great Britain 1800–1960.* Archon (Shoe String), Hamden, Connecticut.

Suzuki, D., and P. Knudtson. 1989. *Genetics.* Harvard University Press, Cambridge, MA.

Tierney, J., and L. Wright. 1988. "The Search for Adam and Eve." *Newsweek,* Jan. 11:46–52.

Volpe, E. P. 1985. *Understanding Evolution,* fifth edition. Wm. C. Brown Publishers, Dubuque, Iowa.

Watson, J. D. 1968. *The Double Helix.* Atheneum, New York.

Watson, J. D., and F. H. C. Crick. 1953. "Genetical Implications of the Structure of Deoxyribonucleic Acid." *Nature,* 171:964–67.

Zohary, D., and M. Hopf. 1988. *Domestication of Plants in the Old World.* Oxford University Press, New York.

Reproduction and Inheritance

Cellular basis of structure and growth

Figure 2.1

(a) Young embryo twelve days after fertilization.
(b) Human egg and sperm (magnification 400×); stages in the development of a human embryo—approximate age after ovulation
(c) twenty-eight days,
(d) thirty-two days,
(e) thirty-eight days,
(f) forty-two days, (g) forty-nine days. (h–k) Development continues until death.

(a–g) Dr. Landrum Shettles;
(h and i) Bettman Archives;
(j) Historical Picture Service.

(a) (b) (c)
(d) (e) (f)
(g) (h) (i)
(j) (k)

All living things are composed of cells. Each of us arose from a single cell through a complicated process of development (figure 2.1). This special cell, the **zygote,** results from the fertilization of an egg from the mother by a single sperm from the father. The cells of the developing **embryo** divide many times to produce, at eight weeks of age, a **fetus.** A fetus has essentially all the structures that the adult human will possess. Rudiments of the external sex organs are the last to develop, not appearing clearly until about sixteen weeks. At birth, thirty-eight weeks after conception, the newborn infant consists of approximately one trillion cells (10^{12}). Remarkably, the nuclear genetic content of almost all cells in the body is identical.

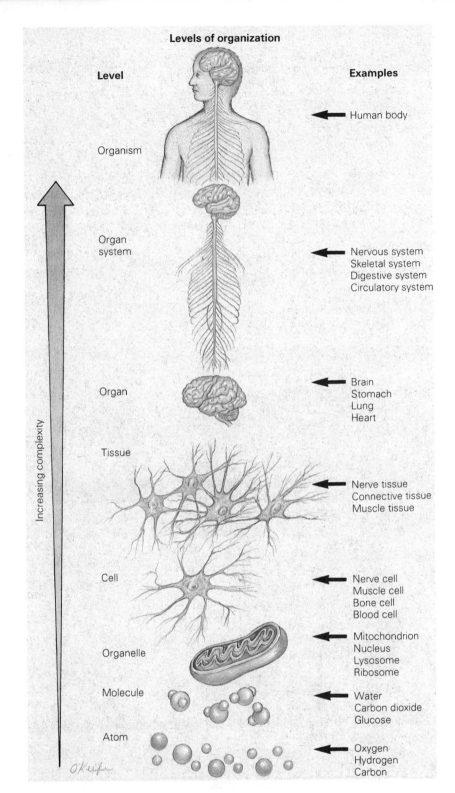

Levels of organization

Level	Examples
Organism	Human body
Organ system	Nervous system Skeletal system Digestive system Circulatory system
Organ	Brain Stomach Lung Heart
Tissue	Nerve tissue Connective tissue Muscle tissue
Cell	Nerve cell Muscle cell Bone cell Blood cell
Organelle	Mitochondrion Nucleus Lysosome Ribosome
Molecule	Water Carbon dioxide Glucose
Atom	Oxygen Hydrogen Carbon

Increasing complexity

Figure 2.2

The physical structure of a human being, like that of any organism, is complex, hierarchical, and highly organized.

From: John W. Hole, Jr., *Human Anatomy and Physiology,* 5th edition Copyright © 1990 Wm. C. Brown Publishers, Dubuque, Iowa. All Rights Reserved. Reprinted by permission.

Organization of the Body

The physical structure of a human being, like that of any organism, is complex and highly organized. The structure consists of a series of levels of organization and is thus *hierarchical* (figure 2.2).

The basic unit of all matter, both living and non-living, is the *atom*. One hundred five types of atoms (or elements) have been discovered, but living organisms require only about one-fourth of those. The elemental requirements vary between species, but humans, for example, need twenty-four elements. Even so, most of these twenty-four elements are not needed

Figure 2.3

Comparison of typical
prokaryotic cell and typical
eukaryotic animal cell:
(a) generalized diagram of a
bacterial cell; (b) generalized
representation of an animal
cell. Note the increased
complexity of the animal cell
with its numerous organelles
and internal membranes.
(Size not to scale.)

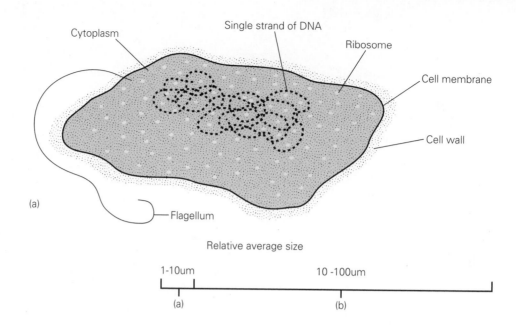

Relative average size

1-10um   10 -100um

(a)   (b)

in large quantities. Four fundamental elements—carbon, hydrogen, oxygen, and nitrogen—comprise 99% of the atoms in the body.

Atoms of the same or different types combine to form *molecules.* Molecules occur in an astonishing variety of sizes and shapes, but again, living organisms require specific amounts of only a few basic types. Over 60% of the molecules in the human body are water—a molecule composed of two hydrogen atoms and one oxygen atom.

The majority of the remaining molecules are *organic molecules,* whose structure and chemistry depend primarily on the carbon atom, and the chemistry of carbon is often considered to be the chemistry of life. There are four basic types of organic molecules—carbohydrates, lipids, proteins, and nucleic acids. Proteins and nucleic acids are macromolecules crucial to genetic function, which we examine in detail in chapter 8.

Molecules do not combine randomly to form living organisms. Instead, they occur in highly organized structures called **cells,** the fundamental unit of structure and function for life on earth. Two basic types of cells can be distinguished (figure 2.3).

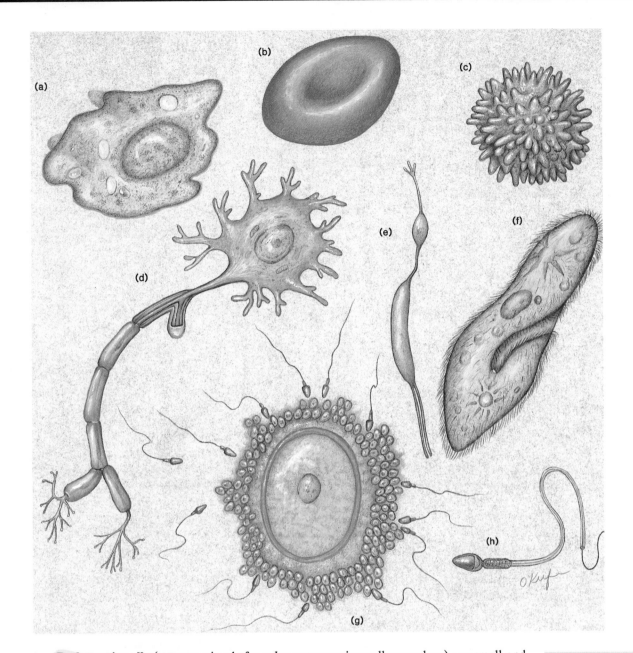

Figure 2.4

A small sample of the diversity of eukaryotic cell types: (a) an amoeba, (b) a mammalian red blood cell (when mature, these cells are unusual in totally lacking a nucleus), (c) morning glory pollen, (d) a nerve cell (with long projections—nerve cells are among the longest known cells), (e) a rod cell from the retina of the eye, (f) a paramecium, (g) a human ovum, (h) a human sperm.

Prokaryotic cells (*pro* meaning before; *karyon* meaning cell or nucleus) are small and simple in appearance, with few distinguishable features under the light microscope. With an electron microscope, DNA of prokaryotes appears as a long ring. No proteins have been found associated with prokaryotic DNA. For support, prokaryotic cells are surrounded by a polysaccharide (complex sugar) cell wall. They reproduce primarily by fission, or equal division, and not by sexual reproduction. On an average, prokaryotic cells vary from one to ten micrometers in diameter (1,000 micrometers = 1 millimeter). Bacteria and blue-green algae are examples of prokaryotes.

Eukaryotic cells (*eu* meaning true; *karyon* meaning cell or nucleus) usually have a volume one thousand times greater than prokaryotic cells. They vary from ten to over one hundred micrometers in diameter and are highly variable in shape (figure 2.4). Unlike prokaryotes, they have a very well-defined nucleus, and within the nucleus, DNA occurs in association with proteins. During cellular reproduction, the DNA-protein complex appears as many dark rods instead of as a single ring. The rods of DNA and protein are known as **chromosomes.** Many other structures are visible within the cell.

In multicellular organisms, cells must meet their own needs and those of the entire organism as well. Often, cells of the same type grow together or combine to form *tissues.* Tissues usually are identified by the same name as the cells that form them. For example, muscle tissue is composed of muscle cells. Tissues of several types are organized into larger units called *organs,* such as kidneys, lungs, or the brain. Organs are in turn grouped into *organ systems.* The nervous system consists of the brain, spinal cord, peripheral nerves, and all the sense organs, such as the eyes and ears.

Finally, of course, organ systems are combined to form the *organism.* A primary feature of a living organism is its high level of structural and functional organization. The science of genetics used to be considered primarily the study of *inheritance,* the transmission of characters from one generation to the next. Now, genetics is equally viewed as the mechanism of *control* of the extraordinary organization of the individual organism as well.

The Eukaryotic Cell

Cells come in many shapes and sizes (figure 2.4). Within each cell are structural and functional units called *organelles* ("little organs"), which carry out basic metabolic functions for the cells. Approximately two hundred types of cells have been described in humans, differing not only in size and shape, but also in the relative proportion of various types of organelles. This variety is not surprising because cells fulfill many different functions. Thus, cells may be spherical, flattened, or tubular. They may have a relatively smooth surface or possess numerous projections. Cells are usually too small to be seen without a microscope, but some nerve cells are more than a meter long, and an ostrich egg (which is only one cell) weighs several kilograms.

All cells conduct the same basic metabolic and reproductive processes common to all life; thus, all share certain common features. Biologists have learned much about cells by examining them under microscopes. Often, special stains are applied to the cells, which make particular organelles highly visible. Figure 2.5 shows the structure of a generalized eukaryotic cell.

The *cell membrane* encloses and supports the cell and its organelles. The cell membrane regulates the passage of molecules into and out of the cell. Prokaryotic, fungal, and plant cells possess an additional external supportive structure, the *cell wall.*

The nuclear membrane surrounds the largest organelle within the cell, the **nucleus.** The nucleus is usually visible through a microscope as a central darkened area, except when the cell is undergoing division. External to the nucleus is the **cytoplasm,** a viscous fluid containing those organelles responsible for the metabolic functions of the cell. The cytoplasm consists of water and many complex molecules, including sugars, lipids, and amino acids.

Several organelles within the cytoplasm are crucial to genetic function. A complex series of folds of the cell membrane within the cytoplasm is called the *endoplasmic reticulum.* Attached to some sections of the endoplasmic reticulum are sequences of small, spherical bodies called **ribosomes.** Ribosomes are the sites of the synthesis of proteins.

In many (including human) eukaryotic cells, a small spherical *centrosome* lies close to the nucleus. It contains two cylindrical structures oriented at right angles to each other, the *centrioles.* The centrioles play an important role in organizing the process of cell division.

Throughout the cytoplasm are numerous **mitochondria.** Mitochondria release the energy from carbohydrate molecules, thus supplying the cell with energy. They also contain their own genetic information, independent of the DNA within the nucleus. They are inherited only from the female parent. Plant (but not animal) cells often contain **chloroplasts.** Chloroplasts are the site of photosynthesis, where sunlight is used to produce carbohydrate molecules from carbon dioxide and water. (All human life is ultimately dependent upon plants' ability to capture energy and make food.) Like mitochondria, chloroplasts also contain their own DNA.

The Nucleus

Although some cytoplasmic organelles contain DNA, the nucleus is the primary site of genetic information within the cell. Both mitochondria and chloroplasts require the use of nuclear genetic information, as well as their own. Two structures are usually visible within the nucleus (figure 2.6): a very dark **nucleolus** ("little nucleus") and filamentous **chromatin.** Chromatin that stains lightly is called **euchromatin;** darkly staining chromatin is called **heterochromatin.**

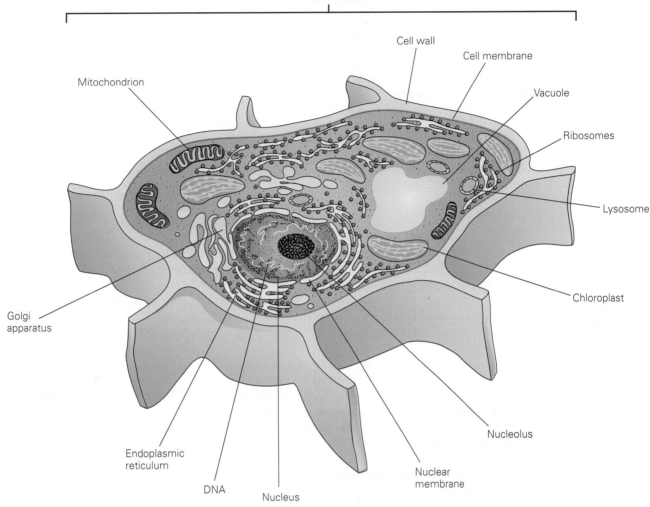

10 Micrometers

Cell wall

Cell membrane

Mitochondrion

Vacuole

Ribosomes

Lysosome

Golgi apparatus

Chloroplast

Endoplasmic reticulum

Nucleolus

DNA

Nucleus

Nuclear membrane

Chromatin is distributed throughout the nucleus and consists of two types of molecules: proteins and nucleic acids. **Histones** are the major type of nuclear protein. There are five types of histones, and each shows almost no variability in structure among the eukaryotes. Prokaryotes have no histones. Histones are usually bound tightly to the surface of the long chains of nucleic acids. Approximately twenty other types of proteins are found in the nucleus, also bound to the nucleic acids. All nuclear proteins appear to play a role in controlling the function of nucleic acids.

The nucleic acids in chromatin are of a particular type, deoxyribonucleic acid, or DNA. The genetic information that organizes and governs every aspect of the life of a cell is encoded in its DNA. DNA is the fundamental genetic material and provides the physical basis of inheritance.

For most of the life of the cell, DNA and its associated proteins occur as extremely long, diffuse threads only barely visible microscopically. At this time, the DNA-protein complex is called chromatin. During cell division, however, the chromatin becomes highly condensed and clearly visible for a brief time. These highly condensed lengths of chromatin are then called *chromosomes*.

Chromosomes

The appearance of chromosomes is governed by the coiling of the chromatin. When the chromatin is most tightly coiled, or condensed, the chromosomes are the shortest and most visible. This normally occurs during cell division, when the chromosomes appear as dense, compact rods (figures 2.7 and 2.8).

Figure 2.5

The structure of a typical eukaryotic plant cell.

Figure 2.6

This is a plant cell illustrating the nucleolus (Nu) and chromatin (Ch) within the nucleus as visualized under (a) a light microscope and (b) an electron microscope. Other organelles of note include the mitochondria (M) and vacuoles (V); ER is the endoplasmic reticulum and CW is the cell wall.

Cell ultrastructure by William A. Jenson and Roderic B. Park, © 1967 by Wadsworth Publishing Co., Inc., Belmont, California. Reprinted by permission of the publisher.

(a)

(b)

Chromosomes can be seen in living cells, but they are best studied in cells that have been killed (carefully, so that they are not damaged) and then treated with special stains (figure 2.8). Stains are simply chemicals that adhere to specific molecules and make them more visible. In fact, chromosome means "colored body," referring to the appearance of chromosomes when stained. Along the length of a chromosome, certain segments stain darkly because they are tightly coiled, giving the chromosome a characteristic black- and white-banded appearance.

Most chromosomes have a unique pattern of bands that allows identification of specific chromosomes. Several types of stains are used to visualize different bands. Giemsa is a common stain that seems to adhere to the tightly coiled chromosome segments and produces **G bands.** C-banding uses a modification of the Giemsa technique to darken the centromeric region of the chromosome. Most (but not all) chromosomes produce either G or C bands when stained appropriately.

The primary constriction that occurs somewhere along the length of the chromosome is the **centromere.** The centromere plays a crucial role in directing chromosome movement during cell division. Chromosomes are classified according to the location of the centromere (figure 2.7): **metacentric chromosomes** have a centromere in the middle of the chromosome; **submetacentric chromosomes** have a centromere slightly off-center; **acrocentric chromosomes** have the centromere almost at one end; and **telocentric chromosomes** have a terminal centromere. Some chromosomes have a second constriction (usually near one end) that pinches off a small segment known as a **chromosome satellite.**

The term **karyotype** refers to the total chromosome complement of an individual. A karyotype is also a formal arrangement of photographed chromosomes of a single cell from one individual, in order of decreasing size (figure 2.9). Among eukaryotes, each species possesses a unique karyotype. Members of a species usually have the same number of chromosomes. Exceptions are found in species such as bees, in which females have twice the number of chromosomes as males, and many species with varying numbers of tiny *supernumerary chromosomes.* Individual chromosomes are identified on the basis of length, location of the centromere, and the banding pattern. Each pair of chromosomes is given a number. Within the karyotype, groups of chromosomes are often recognized by shared characteristics, for example, large metacentrics, short acrocentrics, etc.

In many species, the karyotype of some individuals contains one pair of chromosomes that are *not* identical in appearance. These are the *sex chromosomes* (figure 2.9). In species that have sex chromosomes, members of that chromosome pair will appear identical in one sex, but different in the other sex. The sex chromosomes are always given the last (highest) number in the karyotypic designation. All chromosomes that are not sex chromosomes are called **autosomes.**

The number of chromosomes is not related to the apparent complexity or evolutionary position of a species. Humans have forty-six chromosomes—which, if not condensed, would stretch two meters long. Most species possess between ten and fifty chromosomes. One species of roundworm has only two chromosomes, and some ferns have over twelve hundred chromosomes.

(a)

Figure 2.7

(a) Human chromosomes as seen at mitotic metaphase. Each chromosome consists of two sister chromatids held together by a single centromere. (b) Classification of chromosomes based on gross morphology. For each type, the chromosome illustrated in (a) is nondividing; that in (b) has been arrested in metaphase of mitosis.

(a) Kleberg Cytogenetics Laboratory, Institute for Molecular Genetics, Baylor College of Medicine, Houston, Texas.

Metacentric	Submetacentric	Acrocentric	Telocentric	Chromosome with satellites

(b)

For many species the total number of chromosomes in an adult is known as the **diploid** or **2N number** for that species. In diploid species the chromosomes occur in pairs. One member of each pair originally came from the female parent and is referred to as the maternal chromosome; the other chromosome from the male parent is known as the paternal chromosome. Each member of a chromosome pair is said to be **homologous** to the other. This means that the chromosomes are similar in appearance and structure and possess the same sequence of genes along their length (figure 2.10).

Recall that a gene occupies a particular position, or **locus,** on a chromosome. A gene is a segment of DNA, which can take one of several different forms called **alleles.** Because most chromosomes occur in pairs, an individual possesses two alleles for most genes, one on the maternal chromosome and one on the paternal chromosome (exceptions will be discussed in later chapters). An individual with two copies of the same allele at a locus is **homozygous** at that locus, with two different alleles, **heterozygous.** This pair of alleles is the individual's **genotype.**

The number of pairs of chromosomes in a species is called the **haploid** or **N number.** The haploid number is one-half the diploid number. Each new individual receives a complete haploid *set* (one member of each pair of chromosomes) from each parent. The egg carries the maternal set of chromosomes, and the sperm carries the paternal set of chromosomes. At fertilization, the two haploid sets unite to form the diploid number that characterizes each individual.

On occasion genetic mistakes occur in the process of cell division. An individual who receives too many or too few chromosomes is **aneuploid** ("without the true number"). Such

Cellular Basis of Structure and Growth   **37**

Figure 2.8

Haploid karyotype of human male, as shown by diagrammatic representation according to Paris Conference nomenclature. The left chromatid represents the G-banding pattern seen in mid-metaphase, and the right chromatid shows the pattern in late prophase.

From J. J. Yunis, "High Resolution of Human Chromosomes," *Science*, Vol. 191:1268–1270, 26 March 1967. Copyright © 1976 by the American Association for the Advancement of Science. Reprinted by permission.

karyotypes are designated 2N + 1, 2N − 1, 2N + 2, and so on. Aneuploidy usually has serious medical consequences for humans.

Instead of having an excess or deficit of single chromosomes, some individuals possess entire extra sets of chromosomes. They are called **polyploid.** Polyploidy is fatal in humans. Specific names identify how many haploid sets a polyploid individual has: 3N = *triploid,* 4N = *tetraploid,* 5N = *pentaploid,* and so on. Sometimes the polyploid condition is not harmful. In fact, among plants, some polyploids are exceptionally healthy and vigorous. Agricultural geneticists have used this fact to select many of our common food crops. Cereal grains, such as wheat and rye, and many other commercial crops are polyploids.

Evolution of Chromosomes

Closely related species, such as members of the same genus, usually have similar but not necessarily identical numbers of chromosomes. For example, dogs, wolves, foxes, and coyotes (all members of the genus *Canis*) have seventy-eight chromosomes; humans (*Homo*) have forty-six chromosomes and the closely related chimpanzees (*Pan*) and gorillas (*Gorilla*) have forty-eight. The process of speciation is often accompanied by changes in chromosome number

(a)

(b)

Figure 2.9

Karyotypes of normal
(a) human male and
(b) human female.

(a and b) The Kleberg
Cytogenetics Laboratory,
Baylor College of Medicine,
Houston, Texas.

as a result of *fusion,* the union of two or more pairs of chromosomes, which reduces the diploid number. Human chromosome number 2 is homologous to two chromosomes in chimpanzees and gorillas, accounting for our lower chromosome number (figure 2.11). *Fission* is also common, in which a pair divides into two or more pairs. Thus, closely related species possess many of the same genes, but those genes may be organized into chromosomes very differently.

The genetic similarity of closely related species means that they are sometimes capable of **hybridizing,** or interbreeding. The mule is a hybrid between a male donkey (*Equus asinus,* $2N = 62$) and a female horse (*Equus caballus,* $2N = 64$), but like many hybrids, it is usually not fertile. New mules are made by continued hybridization, not by breeding mule to mule.

An Overview of Reproduction

An organism may use *sexual* or *asexual* processes to reproduce itself, and some species use both. Sexual reproduction involves the fusion of genetic material from two parents to produce an offspring, while in asexual reproduction a single parent produces offspring that are all genetically identical.

During sexual reproduction in most eukaryotes, a new organism is formed by the union of two reproductive cells called **gametes,** each containing one haploid set of chromosomes from one parent. The offspring thus begins life as a single cell with a complete diploid complement of chromosomes.

Gametes are formed within special reproductive organs, the *gonads.* The female gonad, the *ovary,* produces the female gamete or *ovum* (egg). The male gonad, the *testis,* produces *sperm.* Formation of haploid gametes by diploid parent cells within the gonad requires a type of cell division that is unique to reproductive or *germinal* tissues: **meiosis.** The distinguishing feature of meiosis is the reduction of the karyotype from a diploid to a haploid condition, and it occurs nowhere in the body except the gonads. In fact, meiosis is derived from a Greek word meaning "diminution." Meiosis is discussed in detail in the next chapter.

Genetic recombination also occurs in prokaryotes. It involves the transfer of genetic material between two bacteria, which may or may not undergo fission to produce new individuals at the time. Thus, the key element of sexual reproduction is the union of DNA from two individuals to produce new genetic combinations, rather than actual reproduction of the individual itself. Transfers and exchanges of DNA may also occur among viruses that infect a bacterial cell, and two, three, or more individual viruses may participate.

All nonreproductive cells of the body are called *somatic* cells (*soma* meaning body). Somatic cells and some gonadal cells of some species reproduce by **mitosis** (*mitos* meaning

Figure 2.10

A homologous pair of human chromosome·number 9, showing the heterozygous condition at two loci and the homozygous condition at one locus. The person with this pair of chromosomes would be a carrier of the recessive condition galactosemia, would have blood type AB, and would not have the dominant condition of nail-patella syndrome.

BOX 2.1

Chromosomal relationships of humans and Great Apes

Chimpanzees, gorillas, and orangutans all have a diploid number of 48 chromosomes. A fusion of one chromosome pair in a human ancestor, producing the large, submetacentric chromosome 2, reduced the diploid number in humans to 46. High resolution G-banding techniques now permit the identification of about one thousand bands in humans and in their Great Ape relatives. Comparative studies reveal virtually total homology of these one thousand bands.

Two major types of differences are found among the chromosomes of these species. First, variations in the amount of nongenic heterochromatin are common. Many chromosomes of the chimpanzee and gorilla, for example, have terminal bands, heterochromatic regions not present in the other two species. The Y chromosomes also show differences in heterochromatin that do not reflect differences in genic content. If these nongenic sequences are disregarded, many chromosomes are essentially identical among the various species: thirteen pairs for human and chimpanzee (chromosomes 3, 6, 7, 8, 10, 11, 13, 14, 19, 20, 21, 22, and XY); nine pairs for human and gorilla (3, 6, 11, 13, 19, 20, 21, 22, and XY); and eight pairs for human and orangutan (5, 6, 12, 13, 14, 19, 21, and 22).

The other type of chromosomal difference is structural rearrangement. (These will be explained in detail in chapter 11.) The most common, pericentric inversion, has occurred between chromosomes 4, 5, 9, 12, 15, and 16 of human and chimpanzee. A paracentric inversion can be seen in chromosome 7 of chimpanzee and gorilla, and in several instances both paracentric and pericentric inversions have occurred (chromosome 16 of human and gorilla; chromosomes 3 and 17 of human and orangutan). A reciprocal translocation can be seen between chromosomes 5 and 17 of the gorilla.

These comparisons show that essentially the same one thousand nonheterochromatic bands are present in these species. In other words, they all have the same genes. The primary differences are in the order of those genes along the chromosomes. These karyological differences agree with other genetic studies that have found extraordinarily high levels of genetic similarity among humans and Great Apes—over 99% identity in the amino-acid sequence of proteins among humans and chimpanzees, for example. Such findings suggest that structural rearrangements of chromosomes may be associated with regulatory alterations of the genes that produce the phenotypic differences between species.

Such high levels of genetic similarity also indicate an exceptionally close evolutionary relationship of humans to Great Apes. Only a few decades ago, the lines leading to humans and the Great Apes were believed to have diverged perhaps forty million years ago. The accumulating evidence of high genetic similarity, however, combined with the discoveries of key fossils such as Lucy, suggests that divergence of the human–chimpanzee lineages occurred far more recently, perhaps five million years ago, and divergence from the other Great Apes not too many million years earlier. In evolutionary time, humans and Great Apes are separated by only a few ticks of the clock.

thread—a reference to the appearance of the chromosomes). During mitosis, a single diploid cell divides equally to produce two diploid daughter cells. Mitosis thus generates new somatic cells for multicellular organisms, and it is also the mechanism of *asexual reproduction*. One-celled organisms and some multicellular organisms reproduce simply by mitotic fission or budding of cells. In contrast to sexual reproduction, asexual reproduction involves a single diploid parent giving rise to a diploid offspring. Asexual reproduction thus produces offspring that have nuclei that are genetically identical to—a **clone** of—their parents. A clear understanding of mitosis and the life cycle of the cell is essential to an understanding of genetics.

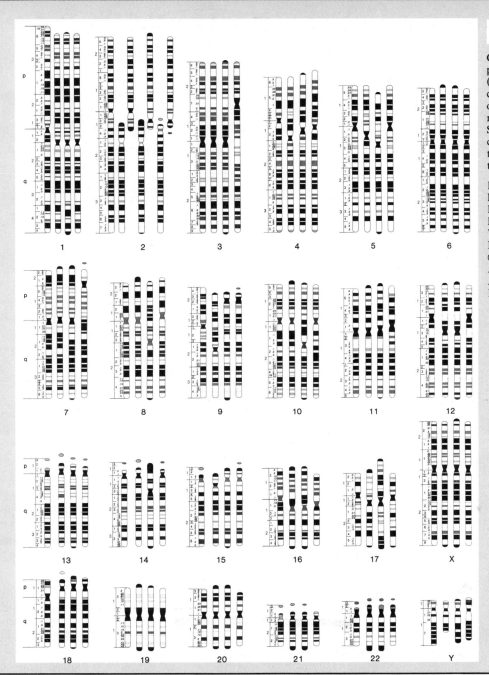

Figure 2.11

Comparison of late-prophase chromosomes of (left to right) human, chimpanzee, gorilla, and orangutan. Drawings represent the 1,000-band Stage. Note: Human chromosome 2 is a fusion of two chromosomes found in the Great Apes.

From J. J. Yunis and O. Prakash, "The Origin of Man: A Chromosomal Pictorial Legacy." *Science*, Vol. 215:1525–1530, 19 March 1982. Copyright © 1982 by the American Association for the Advancement of Science. Reprinted by permission.

The Mitotic Cell Cycle

The complete life history of a single cell, from its origin as one of two daughter cells to its own mitotic division, is called the *mitotic cell cycle* (figure 2.12). There are two major segments of the cell cycle: interphase and mitosis.

Interphase

For almost all its life, the cell nucleus is clearly visible, and the chromosomes appear as diffuse chromatin. This period of cell growth, called **interphase,** is a time of high metabolic

Figure 2.12

The cell cycle of the human white blood cell. Note that interphase is by far the longest part of the cycle and consists of three segments. The duration of the cell cycle phases varies with species, cell age, and type.

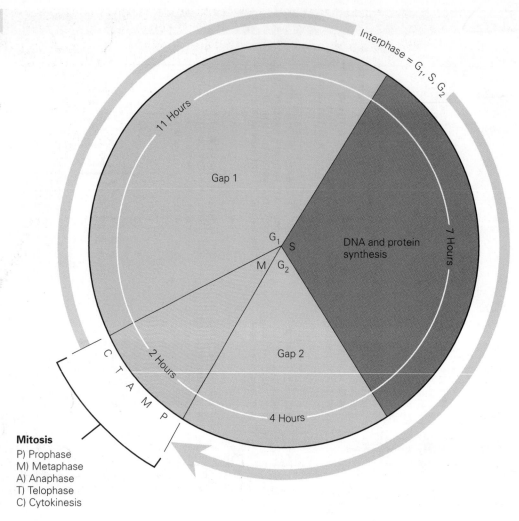

Mitosis
P) Prophase
M) Metaphase
A) Anaphase
T) Telophase
C) Cytokinesis

activity. The cell is in a period of vegetative growth. Some cells, such as nerve cells and adult muscle cells, never divide and spend their entire life in interphase. For most cells, however, interphase is a time of preparation for mitosis. The key genetic event of interphase is the precise replication of the chromosomes. Three distinct phases occur during interphase.

G_1

The *first gap,* or G_1 phase, is the longest period of the cell cycle. It was originally designated as a gap phase because nothing appears to be happening when the cell is viewed microscopically. However, recent studies show that two very different activities occur during G_1. First, the cells may prepare metabolically for the next phase of the cycle. G_1 may last only a few minutes (for example in many embryonic cells), but usually most cells require many hours to complete G_1. Alternatively cells may enter a quiescent (non-proliferating) phase, (G_0) whose duration may be of any length, even including the remainder of the individual's life, as in the case of nerve cells that never divide. Some cells maintained in the laboratory in tissue cultures also can remain in the G_0 state for very long periods. In the quiescent state, the cell neither grows nor duplicates its DNA.

S

The *synthesis,* or S phase, immediately follows G_1. All chromosomes are **replicated** (precisely duplicated) during this interval. An average human chromosome contains about forty millimeters of DNA and millions of nitrogenous bases in an exact sequence, all surrounded by histones and other proteins. Replication begins at many sites along each chromosome and moves in both directions along its length. Mistakes in replication are usually fatal to the daughter cells, so the process of replication must be precisely correct. Embryonic cells of

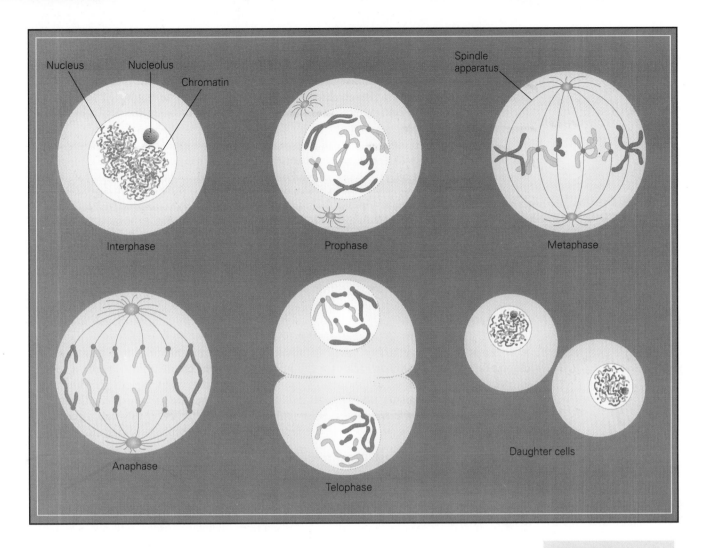

Nucleus Nucleolus
Chromatin
Spindle apparatus

Interphase

Prophase

Metaphase

Anaphase

Telophase

Daughter cells

Figure 2.13

Mitotic cell cycle for a species with diploid number of 6 (2N = 6). Note: Brown chromosomes are from one parent and white from the other.

Drosophila complete S in less than four minutes, but adult cells require about ten hours. The length of S not only varies between embryonic and adult cells, but also between species.

G_2

A *second gap,* or G_2, phase follows S. Like G_1, G_2 microscopically appears to be a time of cellular inactivity, but is a period of increased protein synthesis when the cell metabolically prepares for its shortest but most active phase, mitosis.

Mitosis

All cells require genetic information to control their activities. Mitosis ensures that each cell receives a complete set of chromosomes identical to the chromosomes of its parent. An extraordinary feature of multicellular organisms is that every cell in the body possesses all the genetic information required to control the activities of every other cell in the body. In an adult, about fifty million cells die every second and are immediately replaced by mitotic division of other cells.

Mitosis is a continuous sequence of events. For purposes of analysis, however, four stages have been defined on the basis of nuclear and, particularly, chromosomal behavior (figure 2.13).

Prophase

As mitosis begins, the chromatin begins to coil and condense. **Prophase** begins when the condensing chromosomes become microscopically visible. At this stage, each chromosome appears to be double stranded. Each of the dense strands of a chromosome is called a *chro-*

matid, and each chromatid is attached to its **sister chromatid** at the centromere. Although each chromosome consists of two sister chromatids, the structure is still considered to be just one chromosome as long as only one functional centromere exists. In other words, the number of chromosomes equals the number of centromeres. The existence of two chromatids at this point reflects the replication of each chromosome that has occurred during interphase. Except for the lack of its own centromere, each chromatid is in fact a complete chromosome and identical to its sister.

Also during prophase, the nucleoli and nuclear membrane disappear. In the cytoplasm a new structure forms—the **spindle apparatus.** The spindle, shaped like its name suggests, appears to consist of a series of lines or spindle fibers stretched between two poles. In animal cells, the centrioles separate to form the poles of the spindle. In plants, centrioles do not exist, but a spindle still forms. The spindle serves as a structure that organizes the movement of chromosomes during later phases of mitosis.

The key feature of mitosis is the precise movement of chromosomes, ensuring that each daughter cell receives a complete diploid set. The extreme condensing of chromosomes that characterises prophase may be essential for their efficient movement. If they attempted movement while elongated as chromatin, the result might be an enormous tangle.

Metaphase

The pace now quickens. As **metaphase** begins, spindle formation is completed. The chromosomes appear to be attached to the spindle fibers by the centromere of each pair of sister chromatids. Then, the chromosomes move to the *equatorial* or *metaphase plate,* a plane located midway between the centrioles (or spindle poles) and at right angles to the line connecting them. At the conclusion of metaphase, all chromosomes are lined up along the metaphase plate. Chromosomes are most condensed and most visible during metaphase. Therefore, karyotypes are normally studied from preparations made at metaphase of mitosis.

Anaphase

At the beginning of **anaphase,** each centromere divides. Each sister chromatid becomes a full-fledged chromosome with its own centromere. The spindle fibers contract, and the two daughter chromosomes (former sister chromatids) are drawn toward opposite poles of the cell. Because it appears to be pulled by its centromere, each chromosome now has a V, J, or I shape, depending upon whether it is metacentric, submetacentric, or telocentric. Imagine the complexity of twenty-five hundred chromosomes of one fern cell going their separate ways.

Anaphase is the shortest period of mitosis. The result of anaphase is the exactly equal division of chromosomes. The dividing cell now possesses two daughter groups of chromosomes, each with a complete diploid complement equivalent to that of the parent cell.

Telophase

The last stage of mitosis, **telophase,** begins when the nuclear membranes reappear and enclose each set of daughter chromosomes. Nucleoli also reappear. In animal cells, a furrow appears in the cell membrane that deepens until finally the dividing cell is pinched in two. The developing furrow, of course, also divides the cytoplasmic contents of the parent cell, but the division of cytoplasm does not appear to occur with the precision of nuclear and chromosomal division. In plants, a cell plate forms dividing the cytoplasm among the daughter cells. Mitochondria and, in plants, chloroplasts are also divided among the daughter cells. If a daughter cell does not receive any mitochondria or chloroplasts, it will die. Mitosis ends when the chromosomes de-condense and the daughter cells separate.

Control of the Cell Cycle

The result of the cell cycle is **proliferation,** or the increase in number of cells resulting from cell division. Proliferation is the primary agent of growth in multicelled organisms, and uncontrolled proliferation leads to **cancer.** Thus, understanding control of the cell cycle may be an important key to curing cancer.

Fortunately, the past decade has seen an enormous increase in understanding the system that governs the cell cycle. Although much remains to be learned, it now appears that two proteins regulate the cell cycle (figure 2.12). The first, appropriately named **cyclin,** is more or less continuously produced by the cells and thus regularly increases in concentration until it activates a second protein called *maturation promoting factor,* or **MPF.** In turn, active MPF initiates mitosis but also degrades cyclin rapidly. Once cyclin concentrations are reduced, MPF is inactivated, mitosis is completed, and the build-up of cyclin begins anew.

Obviously, many other factors must be involved, because many cells never divide, while others rapidly divide. The healing of a wound, for example, requires extremely rapid proliferation, perhaps of cells that have not divided in many years and will not divide again once healing is complete. But the cyclin-MPF system seems to be the first step to understanding this very complex process.

Equally important, the cyclin-MPF system appears to control the cell cycle in virtually all eukaryotes. Such a universal system demonstrates the common evolutionary origin of these organisms. It also has an important practical advantage: Scientists can learn about the cell cycle by studying species such as yeasts and sea urchins, which are easy to manipulate and study in the laboratory. Findings from these species then can be translated into knowledge of direct use in curing cancer and other human diseases.

Development

Plants, animals, and fungi all begin life as one-celled organisms, but then enter a multicellular stage for most of their lives. The process by which a single, unspecialized cell becomes a large, complex, and highly organized aggregation of specialized cells of many types is called *development.* Development consists of two processes—*growth* and *differentiation.*

Growth

Individual cells exist only within a small range of sizes. Organisms attain large size through mitotic division (proliferation) and the subsequent aggregation of many thousands of daughter cells. Such enlargement via mitosis is the primary mechanism of **growth.**

Proliferation is a continuous occurrence throughout the life of multicellular organisms, even after full size has been attained. Surface cells are continually sloughed off, so the cells of our skin or digestive tract, as well as blood cells and hair cells, grow throughout life. Some cells, however, do not: nerve cells and muscle cells stop dividing, once fully formed.

As with single cells, there are physical limits to the size of multicelled organisms. The largest terrestrial animals, now extinct, were not much larger than elephants. The largest animal ever, the blue whale, lives in the ocean, where water helps support the huge mass. The largest known organisms are the giant sequoias of California. Almost all the tissues of these trees are dead supportive tissue; only a small amount of the total mass of a tree is living, growing tissue.

Differentiation

Multicellular organisms do not consist only of one type of cell. **Differentiation** occurs when cells become specialized, both structurally and functionally, to meet particular needs of the entire organism. As development progresses, the organism not only becomes larger as a result of mitotic increase in the number of cells, but differentiation produces a series of highly specialized cells, then tissues, organs, and systems.

We noted earlier that all cells possess all the genetic information necessary to produce the entire organism. Each cell is thus potentially **totipotent**—"all powerful" genetically. A truly totipotent cell has the capacity to recreate the entire organism on the basis of the genetic information it contains. However, most cells use only a fraction of their genetic information.

A key aspect of development is precise control of specialization of cells. Each bit of genetic information guiding development is used only in a specific sequence, for a specific duration of time, and is then shut down or "turned off." At conception a new human consists of only one relatively nondescript cell; at birth a human consists of approximately one trillion cells, and an adult consists of perhaps seventy-five trillion cells. What distinguishes humans or any species from all others is the pattern of organization of specialized cells that occurs during development. Thus, *control* and *organization* of development have become fundamental aspects of the study of genetics.

One aspect of the control or limit to cell growth is **contact inhibition.** When normal cells are grown in a cell culture, they grow only in a single layer or monolayer. When the enlarging monolayer encounters a barrier such as the wall of the dish or another cell, it will grow around but not over that barrier. When the surface of the culture dish is covered with cells, growth stops. Thus, contact inhibition is an expression of orderly and limited cell growth.

Growth and healing in a live organism are equally orderly. For example, if a portion of the liver of a rat is removed surgically, the remaining liver cells grow to replace the removed section. Growth stops when the liver attains its normal size. This process can be repeated many times with the same result. Thus, normal liver cells are able to respond to environmental changes. The genetic information in liver cells directs the orderly replacement of missing tissue until normal size is reached, at which point growth stops.

The G_1 phase of the cell cycle may be of particular importance in the control of cell growth. Cells can persist for years or even decades in G_1 or G_0, but suddenly proceed to mitosis if required, for example, to heal a wound. What factors limit division for so long, and then change so quickly to initiate renewed proliferation? Furthermore, differentiation begins in G_1, suggesting that a genetic *switch* has been turned on, directing the cell into a new developmental pathway.

Cancerous cells not only proliferate uncontrollably, but they also exhibit defective differentiation and lack contact inhibition. Understanding G_1 thus may be a key to understanding the causes of cancer.

Summary

1. Eukaryotic cells are larger and more complex than prokaryotic cells. They possess many organelles and have a discrete, well-organized nucleus.
2. Within the nucleus are one or more nucleoli and many chromosomes. Eukaryotic chromosomes consist of segments of DNA bound to nuclear proteins, such as histones.
3. Chromosomes are highly organized. A centromere is always present, and G bands and C bands are often apparent with appropriate stains. A satellite may be present at the end of a chromosome.
4. In diploid species, chromosomes occur in homologous pairs. The two homologues are similar in size, location of the centromere, and banding patterns, and they possess the same sequence of genes on each (except for sex chromosomes).
5. A specific form of a gene is known as an allele. An allele is one of many possible molecular sequences that govern a particular character.
6. An individual almost always has two homologous genes governing a specific characteristic. A diploid individual can have two identical alleles or two different alleles for each gene. Within a species there may be dozens of different alleles for a given gene.
7. Chromosome number and structure is usually similar in closely related species. Often, speciation involves fusion or fission of pairs of chromosomes. Closely related species can often hybridize, but such hybrids are usually sterile.
8. Species may reproduce either sexually or asexually. Prokaryotes most commonly use asexual reproduction; eukaryotes, sexual.
9. The growth and division of one cell is called a mitotic cell cycle. Interphase is a prolonged period of vegetative growth and the replication of genetic material. Mitosis is a brief period of actual cell division in the cell cycle.
10. Development has two basic aspects—growth and differentiation. Growth is achieved primarily by division of existing cells to produce more cells, while differentiation involves the structural and functional specialization of individual cells.

I PMAT

Key Words and Phrases

acrocentric chromosomes

allele

anaphase

aneuploid

autosomes

cancer

cell

centromere

chloroplast

chromatin

chromosome

chromosome satellite

clone

contact inhibition

cyclin

cytoplasm

differentiation

diploid (2N) number

embryo

euchromatin

eukaryotic cell

fetus

G_0, G_1, G_2 phases

gametes

genotype

growth

haploid (N) number

heterochromatin

heterozygous

histones

homologous chromosomes

homozygous

hybridization

interphase

karyotype

locus

meiosis

metacentric chromosomes

metaphase

mitochondria

mitosis

MPF

nucleolus

nucleus

polyploid

prokaryotic cell

proliferation

prophase

replicate

ribosome

sister chromatids

S phase

spindle apparatus

submetacentric chromosomes

telocentric chromosomes

telophase

totipotent

zygote

Questions

1. The diploid number of donkeys is sixty-two and that of horses is sixty-four. What is the diploid number of a mule? Why are mules usually sterile?
2. Diagram mitosis in a species with 2N = 8, where two pairs of chromosomes are metacentric and two pairs are telocentric. Color paternal chromosomes solid.
3. How many chromosomes are present in prophase in a species with 2N = 30? How many in anaphase?
4. Diagram a pair of metacentric chromosomes at mitotic metaphase. Label two genetic loci on this chromosome pair by A and B. Place alleles A and a on genetic locus A and alleles B and b on genetic locus B.
5. Most treefrogs have a diploid number of twenty-four. How many chromosomes would be found in a triploid treefrog? How many in a tetraploid treefrog?
6. Identify the types of chromosomes (i.e., metacentric, telocentric, etc.) in figure 2.9. Do males and females have the same numbers of each type of chromosome?
7. How do the G_1 and S phases of the cell cycle differ?

For More Information

"The Cell." 1961. A special edition of *Scientific American*, 205.

Laskey, R. A., M. P. Fairman, and J. J. Blow. 1989. "S Phase of the Cell Cycle." *Science* 246: 609–614.

Marx, J. L. 1989. "The Cell Cycle Coming Under Control." *Science* 246: 252–255.

Mazia, D. 1974. "The Cell Cycle." *Scientific American*, 230: 54–69.

McIntosh, J. R., and M. P. Koonce. 1989. "Mitosis." *Science* 246: 622–628.

Murray, A. W., and M. W. Kirschner. 1989. "Dominoes and Clocks: The Union of Two Views of the Cell Cycle." *Science* 246: 614–621.

Pardee, A. B. 1989. "G_1 Events and Regulation of Cell Proliferation." *Science* 246: 603–608.

Chapter

Mendel, meiosis, and basic principles of inheritance

BOX 3.1

Gregor Mendel

Gregor Mendel was born on a farm in northern Moravia, part of modern Czechoslovakia near the Polish border. As a youth, he performed well in school, especially in natural history (biology) and physics. After a period of hardship as a teenager, he entered the Augustinian monastery at Brno at age twenty-one in order to (in his own words) "free [me] from the bitter struggle for existence." The monastery was administered by Abbot Napp, who for many years had pursued an interest in plant breeding and hybridization. Abbot Napp organized a course in hybridization, which was taught for the first time during the year that Mendel entered the monastery.

When he was twenty-nine Mendel was sent by the Abbot to the University of Vienna to study math and physics. Biologists had long sought unsuccessfully to understand mechanisms of heredity, and Abbot Napp was convinced that progress would result from the techniques of disciplines other than biology. Mendel studied at the University for two years before returning to the monastery. He then spent the next seven years growing and hybridizing common garden peas. Peas had been the subject of breeding studies since the 1820s. They offered many advantages for experimental study and were already under cultivation in the monastery.

Mendel reported the results of his breeding studies to the local natural history society in 1865, and his paper ("Experiments with Plant Hybrids") was published in their journal the following year. In 1868 Abbot Napp died and Mendel was elected to succeed him as Abbot, thus ending his scientific research. He died in 1884, his work still unknown to the scientific community.

Figure 3.1

(a) The first page of Mendel's manuscript "Versuche über Pflanzenhybriden" (Experiments in Plant-Hybridization) from 1865. (b) The first page of the reprint of "Versuche über Pflanzenhybriden" from 1866.

Courtesy of Anna Matalová, chief of the Mendelianum Musei Moraviae, Brno.

BOX 3.2

Genotype and phenotype

Geneticists distinguish between genes and their expressions. The actual genes of an individual are referred to as its **genotype;** the expressions of those genes are its **phenotype.** An individual has alleles that govern hair color; those specific alleles constitute the individual's genotype for hair color. The actual hair color that the individual exhibits is its phenotype (blond, brown, red, etc.).

Why is such a distinction necessary? Few things are ever as simple as they seem. The phenotype is determined not only by the genotype, but also by the interaction of the genotype with the **environment.** While the influence of the environment may be difficult to assess, *all phenotypic characteristics are determined by the joint effects of genotype and the environment.*

In the case of hair color, for example, how might an individual whose alleles encode information for brown hair actually possess a phenotype other than brown? In most environments, the genotype will be expressed as the phenotype brown hair, but if the environment includes long periods of intense sunlight (such as a spring trip to a Florida beach), the hair may become bleached and not at all brown. A deficiency of protein may result in the inability to produce hair at all, in which case baldness rather than brown hair is the phenotype. Furthermore, part of the environment for one gene is all other genes. That is, there is a genetic environment, as well as an external environment, in which all genes function. Thus, with increasing age, the effects of alleles encoding brown hair will be superceded by the effects of alleles governing the aging process, and gray hair results.

Phenotype should be used in the broadest possible sense. Phenotype commonly refers to visible characteristics, such as size, shape, and color of the parts of an organism—height, weight, eye color, length of limb, and so on. But other, less obvious characteristics also comprise the phenotype: behavioral traits (twitches, yawns, posture, smiles per hour, psychosis), physiology (heart rate, blood pressure, level of basal metabolism), and biochemical characteristics (cholesterol level, presence or absence of particular enzymes, blood type). In fact, any characteristic of an organism that can be described and measured is a phenotypic character.

In 1865 Gregor Mendel (box 3.1) presented his now classic paper describing the mechanisms of inheritance. In it, he demonstrated the particulate basis of heredity and laid the foundation of modern genetics. Mendel showed that physical traits of an organism corresponded to invisible *elementen* in the cells. More important, however, he showed that the genes (*elementen*) controlling these characters exist in pairs, and that only one member of each pair is incorporated into each germ cell during gametogenesis. As we will see, these concepts allowed Mendel to suggest a mechanism for explaining the genetic structure of hybrids, as well as to develop the laws of segregation and independent assortment.

During the subsequent "Golden Age of Cytology" (1880–1900), the discovery of the behavior of the chromosomes during meiosis provided a physical explanation for the segregation and independent assortment of genes described by Mendel in 1865. It is important to remember that Mendel's work appeared at a time when there was little understanding of cell structure and function, and nothing was known about chromosomes, mitosis, meiosis, or fertilization. Thus, Mendel's work was in a sense premature—the scientific community did not possess a framework within which his principles could be understood. Furthermore, Mendel used mathematical language and concepts to explain his findings, which probably helped to make his work relatively inaccessible to the biologists of his day.

Mendel's Principles

Mendel could not study genes directly. Instead he studied their physical expression, what we now call the **phenotype** (box 3.2). Mendel wanted to know how phenotypic characters such as color were transmitted between generations. Were individual characteristics transmitted unchanged from parent to offspring, or were they somehow altered or recreated between generations? Or, said another way, do hereditary particles exist, and, if so, what is their nature?

To answer such questions, Mendel conducted a series of experiments that are a model of scientific simplicity. His experiments succeeded because of their careful design, and because Mendel collected his data very systematically. Previous investigators had attempted to examine the big picture, collecting much data on inheritance of many characters over many generations. But Mendel categorized his data, focusing on the transmission of a few carefully chosen characters. Most importantly, Mendel counted. He simply tallied the number of offspring from different types of matings. His systematic work resulted in the first demonstration of principles of inheritance.

Mendel chose for his experiments the simple garden pea, *Pisum sativum* (figure 3.2). It was easy to cultivate domestically, and it produced one generation with many offspring per year. Many different varieties of peas had been developed and were available for study. Usually, peas are self-fertilizing (pollen from the anthers of one flower fertilize the stigma of that same flower by simply falling on the stigma) (figure 3.2); but when it was necessary to cross-fertilize, the stamens (or male, pollen-bearing organs) could be removed from one plant and used to pollinate another. Often, such manipulation was used to produce **reciprocal crosses,** so that females of one phenotype were mated with males of another, and vice versa. Reciprocal experiments can help detect the influence of sex in governing genetic traits.

Geneticists still select experimental organisms that can be easily cultured (or grown and bred under strict control), have short generation times, and produce many offspring. Laboratory rats and mice, *Drosophila* (fruit flies), *Neurospora* (bread mold), yeast, and the common gut bacterium, *Escherichia coli,* all meet those criteria, and they are the organisms that are best known genetically (figure 3.3).

Mendel studied seven different traits of pea plants (figure 3.4). Each plant showed one of two possible states or phenotypes for each trait. For example, stems were either long (about six to seven feet) or short (about one foot). Most importantly, the type of difference shown by each of the seven characters was **qualitative.** The two phenotypes occurred in discrete categories, and an individual plant could be easily and accurately assigned to one of the categories. Stems were long or short, never intermediate. Thus, Mendel could count the number of offspring from each cross possessing each phenotype.

The alternate type of variation is **quantitative** or continuous variation. Quantitative characters are those that occur on a continuous scale, such as height and weight (figure 3.12). Within a large population, almost all possible values of a character will be shown if the character is a quantitative variable. (Consider, for example, the range of weights shown by all the adult males in New York City.) In most instances, stem length of peas would be a quantitative variable. Although limited variation occurred within the short phenotype and within the long phenotype, the differences found by Mendel were so great that there was no overlap *between* the two phenotypes. Plants clearly fell into either the long category or the short category.

Mendel began his experiments with a set of **pure-breeding** pea plants. For each trait, he obtained seeds that always gave rise to plants with phenotypes identical to the parents. In other words, pure-breeding round seeds always grew into plants that produced more round seeds, and so on. Mendel then conducted a series of **monohybrid crosses.** For each of the seven characters he studied, he crossed pure-breeding parents of opposite phenotypes—the parental or P generation (figure 3.5). The experiments involved one trait at a time, hence were monohybrid. The descendants of the P generation are called the F1 or first filial (meaning child) generation. Descendants of the F1 generation are called the F2 generation, and so on.

For each of the seven monohybrid crosses (figure 3.4), similar results were obtained. All F1s in each cross resembled one parental type, and none resembled the other parental type. For example, all the offspring from the "round seed × wrinkled seed" cross possessed round seeds. In other words, one phenotype was not expressed in each of the F1 generations. The trait that was not expressed is thus **recessive** to the trait that was expressed, the **dominant** trait. This phenomenon is known as **Mendel's Law of Dominance.**

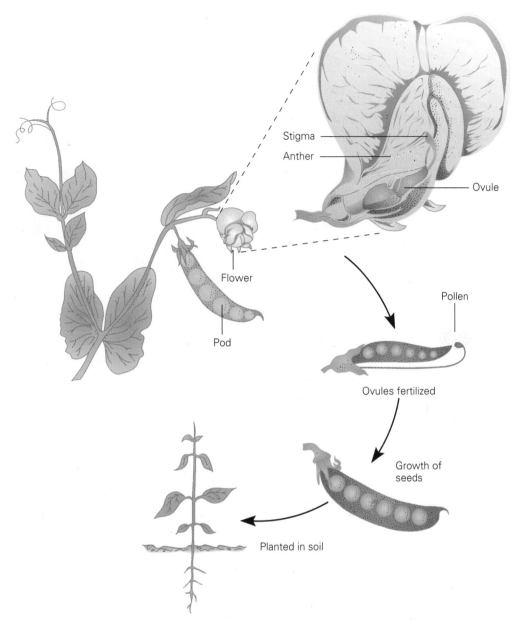

Figure 3.2

Anatomy of a pea plant showing the reproductive features that make artificial pollination possible.

Stigma

Anther

Ovule

Flower

Pollen

Pod

Ovules fertilized

Growth of seeds

Planted in soil

Further, the sex of the parent had no influence over which phenotype was absent and which expressed, because reciprocal crosses produced the same results. A round-seed female $\times$ a wrinkled-seed male produced a round-seed F1, as did a wrinkled-seed female $\times$ a round-seed male. This result is called **Mendel's Law of Parental Equivalence.**

When the F1s were crossed with each other, however, an extraordinary thing happened: the phenotype absent in the F1 generation reappeared in the F2. Furthermore, it always occurred in about one-fourth of the total progeny. Thus, round-seed F1 plants, when bred with one another or self-fertilized, always gave rise to $\frac{3}{4}$ round and $\frac{1}{4}$ wrinkled F2s. It was impossible to predict with certainty the phenotype of a particular individual, but collectively the F2s always occurred in a ratio of 3 round: 1 wrinkled. Furthermore, the 3:1 ratio held true for all seven traits.

Mendel's Results and Modern Concepts

Mendel explained his findings by postulating the existence of hereditary particles or factors (*elementen*) that are transmitted unchanged between generations. These factors control particular traits, but each factor is not necessarily expressed in each generation. A factor—what we call an *allele*—that is always expressed when present is called *dominant,* while one that can be suppressed or masked is called *recessive.*

(a)

(b)

(d)

Figure 3.3

Organisms commonly used in genetic research:
(a) *Neurospora* (bread mold), (b) *E. coli* infected with virus, (c) mouse,
(d) fruit flies, and
(e) *Cryptococcus* (yeast).

(a) © Omikron/Photo Researchers; (b) © Lee D. Simon/Photo Researchers; (e) © Martin M. Rotker/ Taurus Photos.

(c)

♀ ♂

(e)

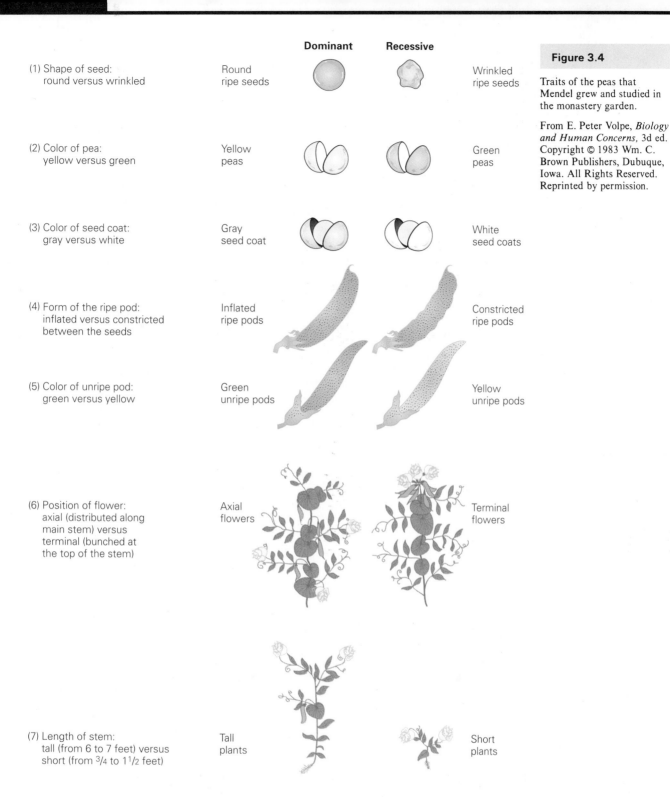

	Dominant	Recessive	
(1) Shape of seed: round versus wrinkled	Round ripe seeds		Wrinkled ripe seeds
(2) Color of pea: yellow versus green	Yellow peas		Green peas
(3) Color of seed coat: gray versus white	Gray seed coat		White seed coats
(4) Form of the ripe pod: inflated versus constricted between the seeds	Inflated ripe pods		Constricted ripe pods
(5) Color of unripe pod: green versus yellow	Green unripe pods		Yellow unripe pods
(6) Position of flower: axial (distributed along main stem) versus terminal (bunched at the top of the stem)	Axial flowers		Terminal flowers
(7) Length of stem: tall (from 6 to 7 feet) versus short (from 3/4 to 1 1/2 feet)	Tall plants		Short plants

Figure 3.4

Traits of the peas that Mendel grew and studied in the monastery garden.

From E. Peter Volpe, *Biology and Human Concerns,* 3d ed. Copyright © 1983 Wm. C. Brown Publishers, Dubuque, Iowa. All Rights Reserved. Reprinted by permission.

    Mendel's hypothesis is easily interpreted in terms of our understanding of DNA and homologous chromosomes introduced in chapter 2. Each phenotypic trait studied by Mendel is governed by two alleles that occur on a homologous pair of chromosomes. If the same allele is present on each homologous chromosome, the individual is said to be **homozygous** for that trait. Homozygous individuals—called *homozygotes*—are pure breeding for that trait, if they are self-fertilizing like pea plants. They can only transmit one type of allele to their offspring.

    On the other hand, an individual with two different alleles of a gene is said to be **heterozygous.** Heterozygous individuals—called *heterozygotes*—transmit one of two different

Mendel, Meiosis, and Basic Principles of Inheritance

Figure 3.5

Results of one of Mendel's monohybrid crosses. Tall plants are dominant to short plants.

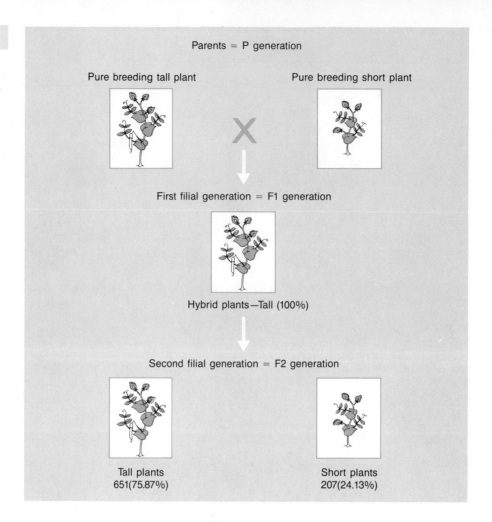

Parents = P generation

Pure breeding tall plant   Pure breeding short plant

First filial generation = F1 generation

Hybrid plants—Tall (100%)

Second filial generation = F2 generation

Tall plants
651(75.87%)

Short plants
207(24.13%)

types of alleles to their offspring (one allele per gamete). All of Mendel's F1 progeny were heterozygous for a particular trait; and in every case, one allele was dominant to the other and was the only one expressed (figure 3.5). Nonetheless, the recessive allele retained its integrity. It was not changed in the F1 and could be transmitted by the F1 to succeeding generations.

We will use a shorthand to label alleles. The alleles at a gene *locus* (position on a chromosome) will be designated by a letter or a set of a few letters. The first letter of abbreviations of dominant alleles is capitalized, while recessive alleles are indicated in lower case. Thus, the allele for round seeds is represented by R, while r is used for wrinkled seeds (box 3.3). Where there are more than two alleles per locus, subscripts and superscripts are often used. Genetic nomenclature can be complex, and we will explain other variations as necessary.

The existence of two alleles at a locus means that three genotypes are possible: RR (the dominant homozygote), Rr (the heterozygote), and rr (the recessive homozygote). However, because of dominance both RR and Rr exhibit the same phenotype (producing round seeds). These two genotypes cannot be told apart simply by appearance.

The results of Mendel's F1 crosses are most clearly represented by a visual device called a **Punnett square**—named after the first person to use it (figure 3.6). Each parent possesses both R and r alleles, and each offspring receives one allele from each parent. Which allele is received from each parent is a **random** event. In this context, random means that the chance that the progeny will receive one allele is identical to the chance it will receive the alternate allele. Transmission of an allele is analogous to the toss of a coin—heads for R, tails for r (Appendix B).

Genotypes can be formed in four ways. An offspring receiving one R from each parent has the homozygous dominant genotype. If it receives one r from each, it is homozygous

Figure 3.6

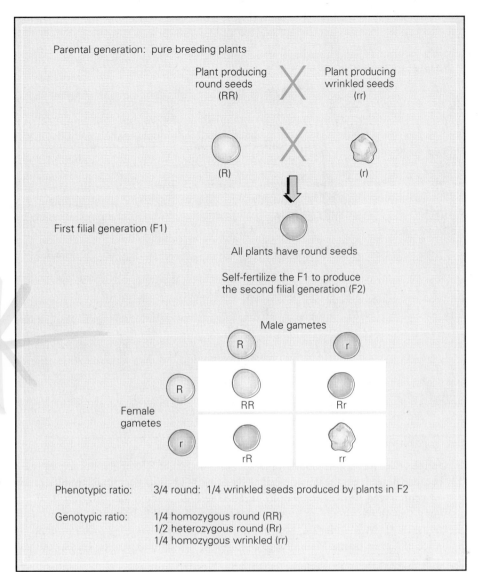

A Punnett square, used to determine the types and proportions of offspring of a cross between two F1 hybrid plants. Each of the F1 parents produces round seeds, but are heterozygous for the recessive wrinkled trait. The two types of gametes produced by each parent combine at random, giving rise to three genotypes and two phenotypes in the F2 generation.

recessive. Finally, it could receive one R from the mother and one r from the father (Rr), or vice versa (rR), but in both cases it is heterozygous and always represented as Rr. Each of the four possibilities is equally likely, that is, each has a 25% chance of occurrence, because the chance of receiving R from each parent is the same as that of receiving r. However, since three of the four genotypic possibilities produce a round-seed phenotype, the offspring of a mating of two heterozygotes occur in a 3:1 phenotypic ratio—exactly Mendel's result.

Mendel's data were so important and so compelling precisely because he expressed his results in terms of ratios. His 3:1 ratio expresses a general phenomenon of the genetic mechanism, and this 3:1 ratio has been duplicated in many experiments by other scientists and by students.

Furthermore, the fact that the recessive phenotype reappears in the F2 after being absent in the F1 argues strongly for the integrity of hereditary particles. Alleles are transmitted between generations without being changed, even though they may not always be expressed. Thus, Mendel's results provided the first strong evidence for the existence of discrete hereditary particles, transmitted unchanged from generation to generation.

Mendel's hypothesis to explain his data fulfills a key requirement of a good scientific hypothesis: it makes predictions that can be tested by further experiments, thus helping to validate the hypothesis. For example, in the F2 generation, Mendel postulated the existence of three different genotypes, RR, Rr, and rr. Individuals with the wrinkled phenotype should be rr, and if permitted to self-fertilize they should produce only more wrinkled individuals, rr, in the F3 generation. In other words, recessive homozygotes should breed true.

Mendel, Meiosis, and Basic Principles of Inheritance

Figure 3.7

The results of Mendel's Law of Segregation over many generations. Offspring of the F1 generation are either round or wrinkled. If self-fertilized, the homozygous wrinkled F2 plants will breed pure, producing only more wrinkled seeds in all succeeding generations. However, because the round-seeded F2 plants represent two different genotypes, two differing results occur if these plants are self-fertilized. Roughly $\frac{1}{3}$ of the F2 plants with round seeds are homozygous and will breed pure, producing only more plants with round seeds. The other $\frac{2}{3}$ of F2 plants with round seeds are heterozygous and thus continue to produce progeny in a ratio of $\frac{3}{4}$ round and $\frac{1}{4}$ wrinkled, just like their F1 parents.

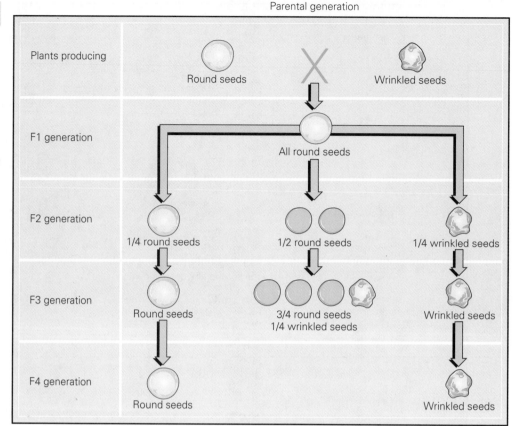

On the other hand, round individuals should be either RR or Rr, occurring in a ratio of 1:2 (because heterozygotes may be either Rr or rR). Thus, if permitted to self-fertilize, one-third of round-seed pea plants should breed true and produce only more round peas. The remaining two-thirds should produce both round and wrinkled progeny in a ratio of, again, 3:1 (figure 3.7). In Mendel's experiments using self-fertilizing F2s, he obtained just such results for all seven traits. Mendel's genetic model not only explained results of past experiments, but also predicted results of future experiments.

An additional experiment was performed by Mendel that verified the genotype of his F1 generation. Mendel predicted that all F1s were heterozygous. Therefore, the mating of the F1 hybrid with the recessive homozygous parental type should produce two types of progeny, Rr and rr (figure 3.8). No RR will occur, because the recessive homozygous parent can transmit only r alleles to the next generation. Furthermore, because the probability of a progeny receiving the R allele is equal to the probability of receiving the r allele, the two types of progeny should occur in approximately equal numbers.

This type of cross—mating a heterozygote to the recessive homozygote—is called a **testcross.** A testcross is a diagnostic test for the heterozygous genotype in the F1 generation. Mendel performed a series of testcrosses using the seven traits in peas. In every case, his data showed the expected 1:1 ratio of the two phenotypes.

Mendel's Law of Segregation

The reappearance of the recessive phenotype in the F2 generation led to **Mendel's Law of Segregation.** This principle of inheritance states that each member of a pair of alleles governing a character maintains its own integrity, and that at reproduction only one allele of a pair is transmitted to each gamete. Normally, the allele does not change between generations. Both types of alleles are incorporated into gametes, but only one per gamete. Which allele occurs in a particular gamete is a random event (i.e., impossible to predict), although in general each of the two types occurs in 50% of the gametes. As we shall see, meiosis is the process governing segregation of alleles into gametes.

Figure 3.8

A *testcross,* a mating of an heterozygote and a recessive homozygote. A testcross produces heterozygous and recessive homozygous offspring in approximately equal proportions. No dominant homozygotes are produced. Data are from Mendel's experiments.

Mendel's Law of Independent Assortment

Each of the seven traits Mendel examined demonstrated a 3:1 phenotypic ratio in the F2 generation. As a logical development of these experiments, Mendel conducted a series of **dihybrid crosses;** that is, crosses between strains breeding true for two characters. As an example, consider the cross between seeds differing for both seed shape and seed color. In addition to seeds showing round or wrinkled phenotypes, Mendel also used seeds that were either yellow (allele Y) or green (allele y) in color, where Y is dominant to y.

In the P generation, the double dominant homozygote, RRYY, was crossed with double recessive homozygote, rryy. Because each F1 received one dominant and one recessive gene for each character, all F1s expressed a round/yellow phenotype (genotype RrYy). Mendel then permitted the doubly heterozygous F1s to self-fertilize and produce the F2 generation.

Examination of these data for one trait at a time shows that the expected monohybrid ratios still occur (figure 3.10). For seed shape, the data show 423 smooth: 133 wrinkled, and for color, 416 yellow: 140 green, both very close to 3:1. Thus, the Law of Segregation applies independently to each trait in the dihybrid cross. Next, if the same 556 seeds are classified by both shape *and* color simultaneously, four distinct phenotypes are possible: round/yellow, round/green, wrinkled/yellow, and wrinkled/green. Mendel observed 315 seeds of round/yellow phenotype, 108 round/green, 101 wrinkled/yellow, and 32 wrinkled/green. If the number in each category is divided by the number in the smallest category (32), the result approximates the expected 9:3:3:1 phenotypic ratio of a *dihybrid cross* (figure 3.10).

The simplest explanation of a dihybrid cross is that the two characters, seed shape and seed color, do not interact in the formation of gametes. Thus, each F1 produces equal proportions of four types of gametes for the pollen and eggs, RY, Ry, rY, and ry. A Punnett square (figure 3.10) shows that nine different genetic combinations occur within the sixteen (4 × 4) possible ways of randomly combining the gametes. However, because of dominance, only four phenotypic classes are possible. These do not occur in equal proportions, but in a ratio of 9:3:3:1, almost exactly what Mendel observed.

This 9:3:3:1 dihybrid ratio can be obtained by using the laws of probability (Appendix B). For example, $\frac{3}{4}$ of all seeds are round, and $\frac{3}{4}$ of those are also yellow. Therefore, the proportion of seeds that are *both* round *and* yellow is $\frac{3}{4} \times \frac{3}{4}$, or $\frac{9}{16}$. The proportion that are round and green is $\frac{3}{4} \times \frac{1}{4}$, or $\frac{3}{16}$; wrinkled and yellow, $\frac{1}{4} \times \frac{3}{4}$, or $\frac{3}{16}$; and wrinkled and green, $\frac{1}{4} \times \frac{1}{4}$, or $\frac{1}{16}$—9:3:3:1. Thus, not only does the Law of Segregation apply to di-

BOX 3.3

Mendel's wrinkled peas

Wrinkled pea seeds are produced when the pea plant is homozygous recessive (rr) for a gene first described by Gregor Mendel in his classic studies of inheritance in pea plants. Now, over 100 years after Mendel published his studies, geneticists have studied in detail the gene responsible for this notable phenotypic effect. The genetic difference between the allele, R, and its recessive counterpart, r, is a small extra piece of DNA that interferes with the synthesis of an important enzyme needed for starch production. An rr plant is unable to synthesize the enyzme. This causes sugar and water to accumulate in the developing seeds. As the seeds mature, most of the water is lost, causing the seeds to wrinkle.

Starch is often added to frozen foods by food processors to help maintain food structure. Thus it is hoped that understanding how this gene influences starch production will enable scientists to genetically modify plant starches to improve the quality of prepackaged foods.

Figure 3.9

Each pea in the pod is the result of one pollen grain and one ovum. Thus, round (RR or Rr) and wrinkled (rr) peas can occur in a single pod.

© Cell Press.

hybrid crosses, but traits for characters controlled by different genes also assort during reproduction independently of each other. This phenomenon is now referred to as **Mendel's Law of Independent Assortment.**

Mendel demonstrated the generality of the Law of Independent Assortment by conducting experiments involving *trihybrid crosses*—hybridizing strains breeding true for three separate traits. In this case, the triply heterozygous F1s produced eight different gametic types, twenty-seven different genotypes, and eight phenotypes. The eight phenotypes occurred in a 27:9:9:9:3:3:3:1 ratio, just as Mendel's laws predict.

The Importance of Mendel's Laws

Be sure that you understand the distinction between Mendel's Law of Segregation and Mendel's Law of Independent Assortment. These principles describe the ways in which genes and alleles occur in the gametes of an individual.

The Law of Segregation refers to the distribution of alleles of one pair of homologous genes among gametes. Because a single diploid individual has two alleles for one gene, a heterozygote produces two different types of gametes with respect to that gene, and those two types of gametes are produced in equal numbers. A homozygote produces only one type of gamete.

The Law of Independent Assortment refers to the distribution of non-homologous chromosomes to gametes (genes on the same chromosome may be an exception; see the following discussion of linkage). Genes on non-homologous chromosomes occur in gametes independently of each other. In the case of Mendel's experiment, knowledge that a gamete contains allele R gives no information as to whether that gamete contains a Y allele or a y allele. In other words, alleles on separate chromosomes occur in gametes (assort) randomly with respect to one another.

Figure 3.10

Mendel's Law of Independent Assortment, shown by a dihybrid cross of genes controlling seed shape and seed color. The Punnett square shows the nine different genotypes and their proportions. Because of dominance, only four phenotypes occur. The table at the bottom shows Mendel's actual data.

Parental generation

Plant with round/yellow seeds   X   Plant with wrinkled/green seeds

F1 generation

All plants produce round/yellow seeds

Gametes of male parent in F1

	RY	Ry	rY	ry
RY	RRYY	RRYy	RrYY	RrYy
Ry	RRYy	RRyy	RrYy	Rryy
rY	RrYY	RrYy	rrYY	rrYy
ry	RrYy	Rryy	rrYy	rryy

Gametes of female parent in F1

F2 generation:
423 plants producing round seeds	RR and Rr	¾	
133 plants producing wrinkled seeds	rr	¼	
416 plants producing yellow seeds	YY and Yy	¾	
140 plants producing green seeds	yy	¼	

315 round/yellow seeds	R– Y–	$9/16$	[¾ × ¾]
108 round/green seeds	R– yy	$3/16$	[¾ × ¼]
101 wrinkled/yellow seeds	rr Y–	$3/16$	[¼ × ¾]
32 wrinkled/green seeds	rr yy	$1/16$	[¼ × ¼]

In combination, these two laws demonstrate how immense amounts of genetic variation are produced in every generation, simply by the random mixing of alleles of a single gene and the mixing of genes over the entire genome. Remember that humans have about one hundred thousand genes. The number of possible genetic combinations that may occur in a gamete from one individual is immense: if a person is heterozygous at only one gene on each pair of chromosomes, he or she will be capable of producing 2^{23} or over eight million different types of gametes. A new individual is produced, of course, by the union of two gametes from parents who probably possess a far greater amount of genetic variation than in this example, essentially guaranteeing that every human fertilization produces a unique combination of alleles. Thus, these laws explain much of the variation that is observed among individuals of a species.

Mendel's laws also show that a large component of randomness exists in the genetic mechanism. It is impossible to predict which combinations of alleles will occur in a particular gamete and, if fertilized, whether the combination of the two gametes will be viable. Clearly, some of the billions of possible combinations must be extremely vigorous and fertile, while some must be lethal.

Further Considerations of Mendel's Laws

Mendel's laws apply to all sexually reproducing diploid organisms. Many traits of many organisms have been examined for agreement with the predictions of these laws. In many cases, data have not agreed completely with the laws as Mendel stated them. Such cases have led to modification or elaboration of the principles as first stated by Mendel. Here, we mention some of those modifications, which will be discussed further in later chapters.

TABLE 3.1

Examples of allelic diversity in a few organisms

Organism	Trait	Locus	Number of alleles
Human	ABO blood group	ABO	4 (common)
Human	HLA antigens*	HLA-A	23 (at least)
		HLA-B	47 (at least)
Fruit fly	Eye color	White	12
Cat	Coat color	C-full color	5
Rabbit	Coat color	Color	4
Red clover	Self-sterility	Sterility	200 (at least)
Snapdragon	Flower color	Solid color	9

*See chapter 4.

Multiple Alleles and Codominance

Mendel experimented with seven genes, each having two different alleles, one dominant to the other. Nature, of course, is not designed for the convenience of scientific experimentation, and it is hardly surprising that many variations on this pattern occur.

Although a diploid individual can have only two alleles of one particular gene, within a population more than two alleles commonly occur at many loci. A particularly well-known example of **multiple allelism** is the ABO blood group locus of humans, which has four common alleles and several other extremely rare ones. The ABO locus will be discussed in detail in chapter 4. Many other genes in humans and other species are known to have multiple alleles (table 3.1), and many genes are known to have several dozen alleles.

With so many alleles commonly occurring, dominance relationships are also often variable. In many cases, both alleles are expressed in a heterozygote, and neither suppresses the other. In such a case, the alleles are said to be **codominant** with respect to each other. Codominance is a common type of expression of alleles at loci governing enzyme production. Some alleles at the ABO blood group locus exhibit codominance in their relationships to other alleles, while other alleles exhibit dominance or recessiveness.

Quantitative Inheritance

Prior to the discovery of the particulate nature of inheritance, one genetic theory was that of **blending inheritance.** This theory claimed that hereditary particles from both parents combine in the progeny. The particles then lose their integrity and are permanently changed. Evidence for blending came from quantitatively varying characters, such as height and weight, where offspring are often intermediate with respect to their parents. Such events may in fact be explained by variations on Mendelian principles.

One of the discoverers of Mendel's work, Carl Correns, conducted experiments similar to Mendel's, but examining flower color in four-o'clocks (figure 3.11). Upon hybridizing true-breeding red and white flowering parents, he discovered that all F1s had pink flowers. The F2 plants resulting from self-fertilizing the F1s produced red, pink, and white flowers in the ratio 1:2:1. The pink flowered plants were all heterozygotes. Thus, each genotype was represented by a unique phenotype, explaining the F2 ratio.

The pink phenotype resulted because the heterozygote did not express red pigment as strongly as did the homozygote. When the expression of a character is reduced in the heterozygote as compared to the homozygote, the alleles exhibit **partial** or **incomplete dominance.** Incomplete dominance thus produces F1 hybrids that appear intermediate to the parental phenotypes. However, the segregation of alleles in the F2 generation demonstrates clearly that the genes themselves remain unchanged between generations, retaining their integrity in accordance with Mendelian principles, but contrary to ideas of blending inheritance.

With incomplete dominance, phenotypes still may occur in discrete categories—red, pink, and white, for example. However, if the gene controlling flower color in four-o'clocks had many alleles, instead of just two as in Correns' experiment, and if each allele gave rise to

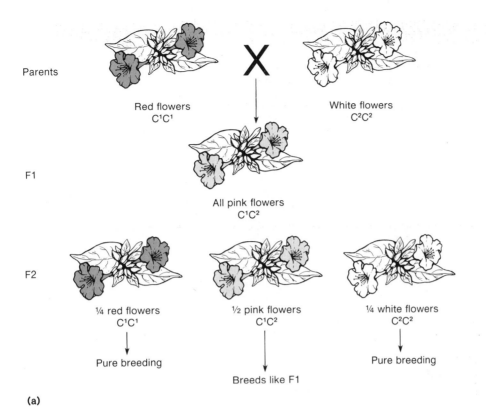

Parents

Red flowers
C^1C^1

X

White flowers
C^2C^2

F1

All pink flowers
C^1C^2

F2

¼ red flowers
C^1C^1

½ pink flowers
C^1C^2

¼ white flowers
C^2C^2

Pure breeding

Breeds like F1

Pure breeding

(a)

Figure 3.11

(a) Incomplete or partial dominance among the alleles for color in four-o'clocks. Because each genotype is represented by a unique phenotype, Mendel's Law of Segregation produces a 1:2:1 ratio of phenotypes. (b) An early attempt to explain "blending inheritance" by Francis Galton, an eminent nineteenth century scientist, in a letter to his equally famous cousin, Charles Darwin, in 1875. Are these explanations consistent with the example shown in (a)?

Source: *Clinical Genetics: Problems in Diagnosis and Counseling*, edited by A. M. Willey, T. P. Carter, S. Kelly, and I. H. Porter. Copyright © 1982 by Academic Press, Inc., New York, NY.

42, RUTLAND GATE *Dec.* 19 75

MY DEAR DARWIN, The explanation of what you propose does not seem to me in any way different on my theory, to what it would be in any theory of organic units. It would be this:

Let us deal with a single quality, for clearness of explanation, and suppose that in some particular plant or animal and in some particular structure, the hybrid between white and black forms was exactly intermediate, viz: grey—thenceforward forever. Then a bit of the tinted structure under the microscope would have a form which might be drawn as in a diagram, as follows:—

White Form.   Black Form.

whereas in the hybrid it would be either that some cells were white and others black, and nearly the same proportion of each, as in (1) giving *on the whole* when less highly magnified a

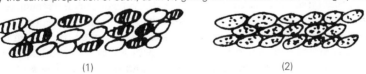

(1)   (2)

uniform grey tint,—or else as in (2) in which *each cell* had a uniform grey tint.

(b)

differing shades of pink, then it is possible to imagine a population of different individuals exhibiting the entire spectrum of pink, grading imperceptibly from red to white.

Yet another genetic phenomenon also may explain continuous variation: many characters are controlled not by one but by many genes. This type of inheritance is called **polygenic inheritance.** If many different genes contribute to a character such as height, the value of that character will depend upon the way in which those genes interact. Their expressions

Number of individuals in each rank	1	0	0	1	5	7	7	22	25	26	27	17	11	17	4	4	1
Height in feet and inches	4:10	4:11	5:0	5:1	5:2	5:3	5:4	5:5	5:6	5:7	5:8	5:9	5:10	5:11	6:0	6:1	6:2

Figure 3.12

Frequency distribution of heights of male students at Connecticut Agricultural College.

Library of Congress.

may be cumulative, or they may interact in more complex ways. Most phenotypic characters are the result of the interactions of many, perhaps dozens to hundreds of different genes. The result of these interactions is the enormous amount of variation shown by most phenotypic traits (figure 3.12).

Linkage: A Modification of the Law of Independent Assortment

Genes do not exist independently in the nucleus. Genes occur sequentially on chromosomes, and most chromosomes have hundreds or thousands of genes. Genes located on the same chromosome are *linked* and are transmitted as a group to gametes. Linkage groups may be broken during meiosis, but the existence of **linkage** means that the Law of Independent Assortment does not always apply to transmission of linked genes.

Independent assortment occurs when genes are located on non-homologous chromosomes, or when genes on the same chromosome are relatively distant from each other. Genes close to one another on the same chromosome do *not* show the 9:3:3:1 ratio of the F2 generation of a dihybrid cross, because certain allelic combinations occur only rarely. Linkage always alters the phenotypic ratios from those expected if genes assort independently. Mendel was either exceptionally lucky or highly selective in the seven traits he reported examining in peas. Although several of the characters he reported were linked, they were far enough apart that recombination during meiosis caused them to assort independently. Therefore, Mendel obtained the appropriate ratios in all dihybrid and trihybrid crosses.

Mutations

Mendel's laws demonstrate the constancy of genetic factors between generations. Genes are transmitted from parent to progeny unaltered—most of the time. In fact, **mutations** or changes in the genetic material occur regularly. The probability of a particular gene changing in one cell cycle (meiosis or mitosis) is less than one in a million in most cases. However, because there are so many genes, and because cell division occurs continuously, mutation is a common occurrence. Many birth defects, for example, result from gene mutations.

Mutation was not a major concern for Mendel because it would have affected only a tiny fraction of the individuals and traits he examined. Thus mutation would not be expected to alter his ratios significantly. However, the phenomenon of mutation is of great importance to modern genetics, medicine, and agriculture. We discuss the causes of mutation in chapter 11.

Meiosis: The Cellular Basis of Mendel's Laws

Shortly after the discovery and acceptance of Mendel's work in the early 1900s, it was noticed that the behavior of chromosomes during sexual reproduction corresponded to the behavior of the hereditary factors postulated by Mendel. In chapter 2 we showed how somatic cells divide mitotically to produce daughter cells that have a chromosome complement identical to the parent cell. The germinal cells residing in the gonads, however, divide to give a very different result.

Meiosis is a sequence of two divisions of a diploid germinal cell to produce haploid cells. The continued development and differentiation of the haploid cells to produce gametes is called **gametogenesis.**

Meiosis is the cellular basis for the process of sexual reproduction, and it has three crucial results. First, the number of chromosomes is reduced by one-half during meiosis from the diploid to the haploid state. This serves to maintain a constant number of chromosomes in all diploid members of a species. Because sexual reproduction involves the union of cells from two individuals, the absence of a reduction in chromosome number during the formation of gametes would produce individuals with immense numbers of chromosomes in only a few generations.

Second, halving of the chromosome number is not a random process. The reduction to the haploid state involves the orderly assortment of one member of each homologous pair into daughter cells. This ensures that each gamete contains one of each type of chromosome and, hence, one copy of each gene. Without such balance, most gametes would not produce viable progeny. However, the separation of members of any homologous pair of chromosomes occurs randomly with respect to all other pairs. Thus, some chromosomes of maternal and some of paternal origin occur together in individual gametes. This phenomenon is the basis for independent assortment observed by Mendel and explains much of the genetic variation that occurs between individuals.

Third, during meiosis, chromosomes routinely exchange homologous segments by a process called **crossing-over,** causing genetic **recombination.** Crossing-over brings together entirely new combinations of alleles on a chromosome. The chromosomes that an individual passes on to its offspring via its gametes are almost never identical to the chromosomes received from either of its parents. The genes of each parent are recombined on chromosomes during meiosis to form a new combination transmitted to the next generation.

Meiosis occurs only during gametogenesis of sexually reproducing animals. In many respects it resembles mitosis, but in key aspects it is very different. It consists of two separate divisions, called meiosis I and meiosis II. Each division has the same four phases as occur in mitosis, but each is labelled with a Roman numeral to distinguish the phases of the two divisions—for example, prophase I, metaphase I, and so on. We present here an outline of meiosis as it occurs in a human male. The process we describe is a general model for meiosis in diploid organisms, although there are many variations on this pattern. Meiosis in females is a slightly modified version of male meiosis, and we will discuss meiosis in females in chapter 6.

Meiosis I

Interphase prior to the first meiotic division is a time of vegetative growth and replication of the chromosomes, just as in mitosis. When the chromosomes first become visible at the beginning of meiosis, each consists of a pair of sister chromatids, held together by a common centromere. The first meiotic division separates members of each homologous pair into two daughter cells, but each chromosome remains in the duplicated condition. Recall that the number of chromosomes equals the number of centromeres. Because meiosis I reduces the number of centromeres in each of the daughter cells to the haploid number, it is considered a **reduction division.**

Prophase I (figures 3.13a and 3.15a)

The first phase of meiosis is the most complex of the entire cycle. Prophase I is a period of high chromosome activity, with several distinct segments. It begins with the appearance of the chromosomes as long, single filaments. Their duplicated condition is not yet obvious under the microscope.

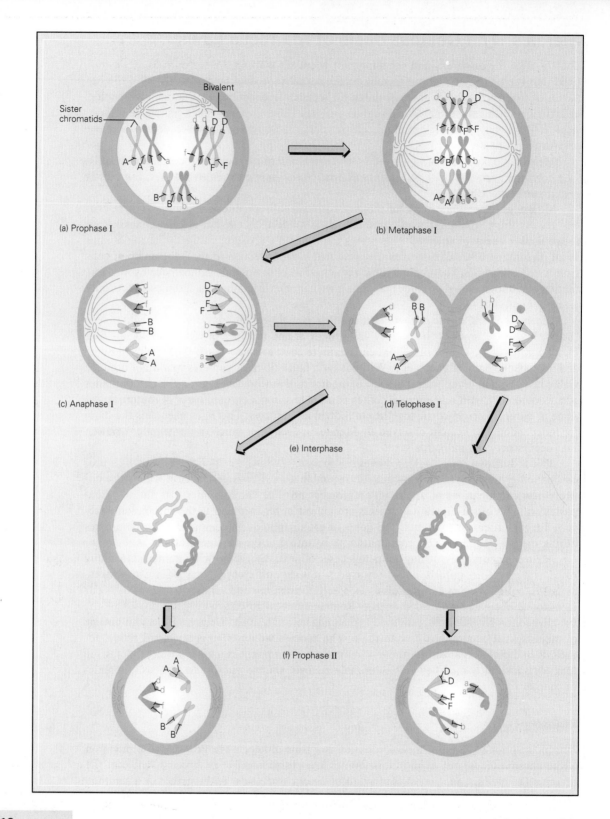

Figure 3.13

Meiosis in a male with 2N = 6: no crossover. Yellow chromosomes come from one parent; brown from the other parent.

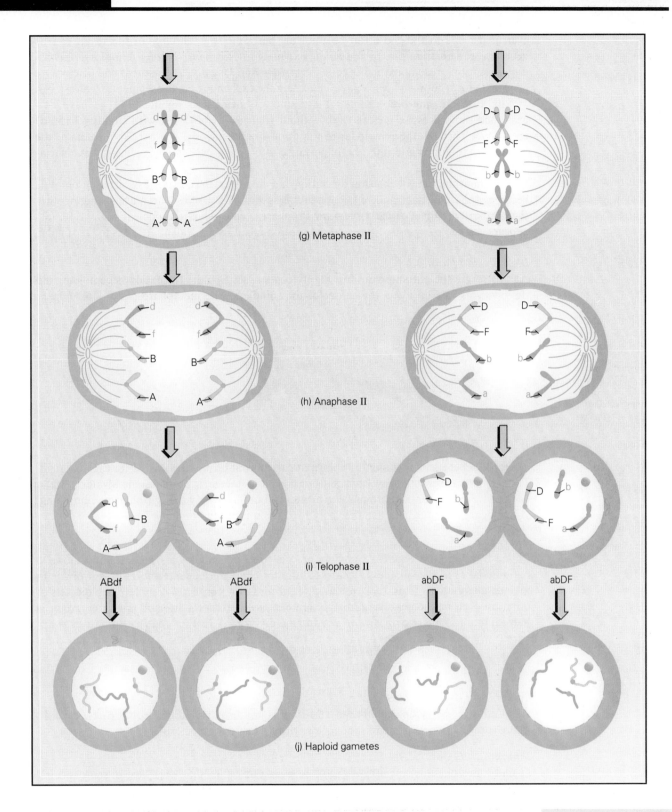

(g) Metaphase II

(h) Anaphase II

(i) Telophase II

ABdf   ABdf   abDF   abDF

(j) Haploid gametes

Next, the members of a homologous pair line up tightly side by side, a process called **synapsis** (meaning juncture). Synapsis is a key feature that distinguishes meiosis from mitosis. The chromosomes progressively condense throughout prophase I, and as they do, a number of attachments between homologues become visible. Because of the X-shaped appearance of these attachments, they are called **chiasmata.** Formation of chiasmata is a crucial event, because they are the mechanism of crossing-over. At a chiasma, chromosomes break and then usually re-attach to another broken end. If a chromosome heals to its own broken end or that of its sister chromatid, it remains identical to its original condition. How-

Figure 3.13

(continued)

ever, if it heals to the end of an homologous but *non-sister* chromatid, the result can be **genetic recombination** due to the *equal exchange of segments among homologous chromosomes*. Such recombination is called **crossing-over.** Usually, at least two and often more chiasmata form between homologues, indicating the exchange of many segments (figure 3.14).

Because crossing-over changes the results of meiosis so significantly, we have diagrammed the process of meiosis both without (figures 3.13a-j) and with (figures 3.15a-d) crossovers. Through the remainder of the discussion of meiosis, compare the results of the two figures to appreciate fully the importance of crossing-over in generating genetic variation.

At about the time of chiasma formation and crossover, the existence of sister chromatids becomes apparent. Each pair of homologues is seen to possess four chromatids, and together the duplicated pair is called a **bivalent.** Prophase I ends when the nuclear membrane disappears, a spindle begins to form, and the centromere of each bivalent attaches to the spindle.

Metaphase I (figure 3.13b)

At this stage, the chromosomes are highly condensed. Spindle formation is complete, and the bivalents line up along the center of the spindle at the metaphase plate. Homologues

appear to repel each other and the chiasmata are now at terminal positions for each bivalent. For example, this would leave two chiasmata on a metacentric chromosome bivalent, one at each end. An acrocentric chromosome bivalent would have only a single chiasma remaining. Homologues are thus held together only by their chiasmata.

Anaphase I (figures 3.13c and 3.15b)

Pulled by their centromeres, members of each pair of homologues separate and move to opposite poles of the spindle. In contrast to mitotic anaphase, centromeres do not divide in meiotic anaphase I. Thus, the two daughter groups now consist of complete haploid sets of duplicated chromosomes. Because the number of centromeres in each daughter group is halved, the first meiotic division is a reduction division. A further contrast to mitosis is apparent when considering the condition of the two sister chromatids in each duplicated chromosome. Because of crossing-over, these two sister chromatids may no longer be identical (figure 3.15b).

Telophase I (figure 3.13d)

As in mitosis, the cytoplasm and its constituents are also divided between the new cells. In many species, a nuclear membrane forms within each daughter cell.

Interphase (figure 3.13e)

Usually, interphase is only a brief pause between the two meiotic divisions. In some species, both telophase I and interphase may be skipped entirely, and meiosis II ensues directly from anaphase I. There is never any replication of chromosomes during this interval.

Meiosis II

Because the genetic material was duplicated once prior to initiation of meiosis, there must be two divisions to achieve haploid gametes whose chromosomes are not duplicated. The second division involves division of the centromeres and the separation of the two sister chro-

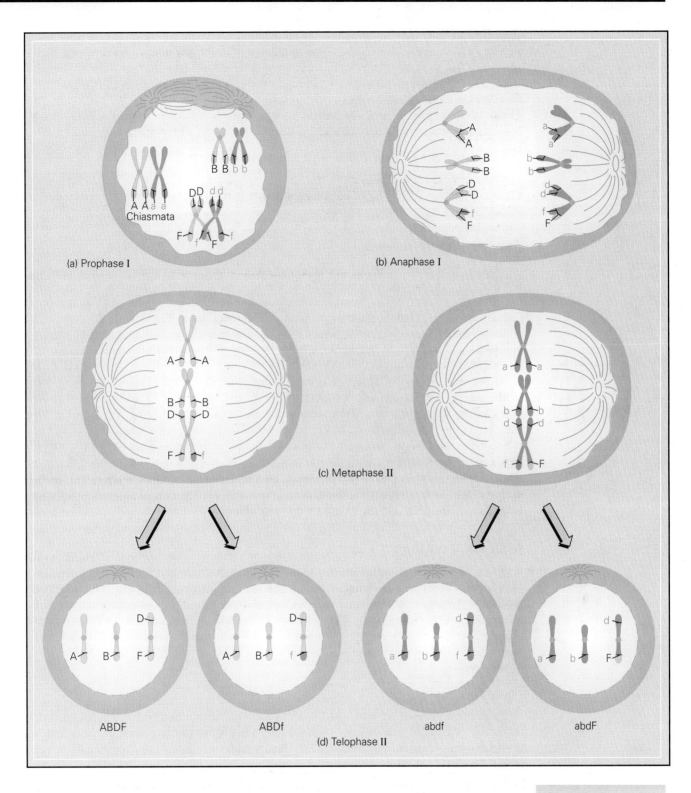

(a) Prophase I

Chiasmata

(b) Anaphase I

(c) Metaphase II

(d) Telophase II

ABDF

ABDf

abdf

abdF

Figure 3.15

Critical stages of meiosis in a male with 2N = 6; one crossover.

matids. Bear in mind, as pointed out previously, that the sister chromatids may not be identical at this point, if crossing-over has occurred. Meiosis II is less complex than meiosis I and similar in many respects to mitosis.

Prophase II (figure 3.13f)

The nuclear membrane (if present) disappears. The chromosomes, which had become relatively diffuse late in meiosis I, once again begin condensing. Unlike prophase I, there is no pairing of chromosomes because the duplicated chromosomes have no homologous partners with which to pair. A spindle begins to form.

Metaphase II (figures 3.13g and 3.15c)

The still duplicated chromosomes move to the equator of the spindle, each attached by its single centromere.

Anaphase II (figure 3.13h)

The centromere divides, and now two new and single chromosomes exist. Each new set of haploid daughter chromosomes moves toward opposite poles of the spindle.

Telophase II (figures 3.13i and 3.15d)

A nuclear membrane forms around each daughter set, and the chromosomes once again become diffuse. The contents of the cytoplasm are divided, and meiosis is complete (figure 3.13j).

The Results of Meiosis

Mitosis involves one division that produces two daughter cells, each chromosomally identical to the parent. In meiosis, each parent cell undergoes two divisions to produce four daughter cells in males (one functional gamete in females), each containing a potentially unique haploid set. Because of independent assortment of chromosomes in anaphase I, the number of potential combinations of chromosomes in a gamete depends upon the number of pairs of chromosomes, the haploid number, n. The number of different possible combinations is 2^n. For humans, that number is 2^{23}, or 8,388,608.

Even that amount of variation is small when compared to that due to crossing-over. While independent assortment produces different combinations of chromosomes, crossing-over changes the allelic combinations that occur within each chromosome. These two factors thus assure an immense amount of variation within the gametes of each individual and between individuals. With the single exception of monozygotic twins (discussed in chapter 4), the meiotic mechanism virtually guarantees that no two humans will ever have the same genetic constitution. A geneticist views each human as a unique result of the interaction of a unique set of environmental influences and a unique combination of alleles and chromosomes. Each of us is a genetic experiment that will be tried only once.

Meiosis and Mendel's Laws

Chromosome behavior during meiosis is the cellular basis for the Law of Segregation and the Law of Independent Assortment. Mendel's hereditary factors, genes, are located on chromosomes. Each gamete randomly receives one member of each homologous pair of chromosomes, ensuring segregation of alleles and also providing for independent assortment of unlinked genes. Crossing-over provides a measure of independent assortment even for linked genes, although *how much* independence is related to the proximity of the linked genes on a chromosome.

In retrospect, it is extraordinary indeed that Mendel discovered the Law of Segregation and the Law of Independent Assortment with no knowledge of the cellular mechanism underlying them. The combination of Mendel's inheritance experiments and cytological examination of meiosis provided the evidence that quickly led to acceptance of these laws as the foundation of modern genetic theory. Thus, modern genetics dates from the first years of the twentieth century, when Mendel's work was "discovered" and linked to the newly understood behavior of chromosomes.

Summary

1. In 1865 Mendel described the basic laws of particulate inheritance in diploids. His laws combined with the cytology of meiosis are the foundation of modern genetics.
2. The phenotype is the expression of the combined effects of the genotype and environmental influences.
3. All phenotypic characters exhibit either quantitative or qualitative variations. Mendel studied qualitatively variable characters in the garden pea.

4. Mendel's Law of Dominance describes the interaction of alleles at a locus. Dominant alleles are always expressed, while recessive alleles are expressed only when homozygous.

5. Mendel's Law of Parental Equivalence states that the results of a cross do not depend upon which parent has the genetic characters or alleles being studied.

6. Mendel's Law of Segregation states that genes governing a phenotypic character occur in pairs and that one member of each pair is transmitted to each progeny. The two alleles that each individual has are transmitted to the next generation in approximately equal proportions.

7. Mendel's Law of Independent Assortment states that unlinked genes are distributed among gametes independently with respect to each other. This law does not apply to genes that are closely linked.

8. The primary results of meiosis are the production of haploid daughter cells that will become gametes and the generation of huge amounts of genetic variation.

9. Partial dominance and polygenic inheritance are mechanisms that can account for quantitative variation.

10. Linkage alters the gametic ratios obtained in crosses involving more than one character.

11. Meiosis is the cytological process that provides the physical basis of Mendel's laws. Meiosis consists of one replication of the chromosomes, followed by two divisions.

12. Independent assortment of chromosomes during meiosis I distributes random mixtures of parental chromosomes, but exactly one of each homologous pair, to daughter cells. Crossing-over produces new combinations of alleles within one chromosome.

Key Words and Phrases

bivalent

blending inheritance

chiasma

codominance

crossing-over

dihybrid cross

dominant

environment

gametogenesis

genotype

heterozygous

homozygous

incomplete dominance

independent assortment

linkage

meiosis

Mendel's Law of Dominance

Mendel's Law of Independent Assortment

Mendel's Law of Parental Equivalence

Mendel's Law of Segregation

monohybrid cross

multiple alleles

mutation

partial dominance

phenotype

polygenic inheritance

Punnett square

pure-breeding

qualitative variation

quantitative variation

random

recessive

reciprocal cross

recombination

reduction division

synapsis

testcross

Questions

1. Assume that in Asian yaks a gene on chromosome one controls appearance. At that locus, allele C for cuteness is dominant to c for ugliness. On a second chromosome is the locus controlling hair color: allele P for pink hair is dominant to allele p for blond hair. With respect to these two characters answer the following:
   a. If an individual possesses the genotype CcPP, how many different types of gametes can it produce? What are they?
   b. What are the possible genotypes of offspring of a mating between this individual (genotype CcPP) and an individual of genotype CcPp? ccpp?
   c. What phenotypes will result from these different matings?

2. About 70% of Americans get a bitter taste from the drug phenylthiocarbamide (PTC). The ability to taste this compound is dominant to the inability to taste it, and this is an autosomal trait. Albinism is an autosomal recessive trait. A normally pigmented woman who is a nontaster has a father who is an albino taster. She marries an albino taster whose mother is an albino nontaster. Assuming that the genetic factors for albinism and the ability to taste PTC assort independently, what types of offspring can this couple produce, and what is the probability of each type?

3. An individual possesses the genotype AaBbCcDdEE. All the genes are unlinked except DdEE, which are so closely linked that no crossover occurs. How many types of different gametes can this individual produce? What if there was no linkage at all?

4. How many ways are discussed in this chapter that can cause genetic variation? What are they?

5. Name some phenotypic characters of humans that are obviously caused by the effects of both genes and the environment. What are those environmental factors?

6. In addition to the ABO blood group, many others have been identified in humans. One such group is the MN group, controlled by two codominant alleles, M and N, at one locus. What genotypes and phenotypes would be produced in crosses involving the following phenotypes: M × N; M × MN; N × N; and MN × MN?

7. Could a child of type N result from the mating of M × MN?

8. In a cell with 2N = 14, how many chromatids are present at meiotic prophase I? At meiotic prophase II? At mitotic metaphase?

9. In a cell with 2N = 14, how many chromosomes are present during meiotic anaphase I? During meiotic metaphase II? During mitotic anaphase?

10. Re-examine figure 3.7. If all offspring in the F3 generation are only allowed to self-fertilize for ten more generations, will the proportion of heterozygotes in the experimental population increase, decrease, or remain the same in succeeding generations?

For More Information

Burns, G. W., and P. J. Bottino. 1989. *The Science of Genetics,* Sixth edition. Macmillan Publishing Co., New York.

Dunn, L. C. 1965. *A Short History of Genetics.* McGraw-Hill, New York.

"Gregor Mendel's Autobiography." Reprinted in English in *The Journal of Heredity,* 1954. Volume 45, pp. 231–34.

Fincham, J. R. S. 1990. "Mendel—Now Down to the Molecular Level." *Nature,* 343:208–209. (Molecular explanation for wrinkled seed shape in Mendel's peas.)

John, B., and K. R. Lewis. 1984. *The Meiotic Mechanism.* Carolina Biological Reader #65. Carolina Biological Supply Co., North Carolina.

Klug, W. S., and M. R. Cummings. 1986. *Concepts of Genetics,* second edition. Scott, Foresman & Company, Glenview, Illinois.

Novitski, E. 1982. "The Formation of the Germ Cells." *Human Genetics,* second edition (chapter 3). Macmillan Publishing Co., New York.

AP/Wide World
Photos.

Mendelian genetics of humans

Figure 4.1

Some Mendelian traits in humans: (a) polydactyly (extra digits on hands and feet); (b) hereditary ptosis (drooping eyelid, due to muscle degeneration); (c) progeria (premature aging and greatly reduced life expectancy); (d) Ellis van Creveld syndrome (a type of dwarfism common in the Pennsylvania Amish, resulting in shortened limbs, malformed hearts, and six fingers on each hand).

(a) Lester V. Bergman and Associates, Inc.; (b) Ira A. Abrahamson, Jr., M. D., "Opthamologic Manifestations of Medical and Neurologic Diseases," 1972, by Medcom, Inc. Reprinted by permission of Medcom, Inc.; (c and d) Victor McKusick.

(a)

(b)

(c)

(d)

Mendel analyzed the proportions of offspring of various types of matings to deduce fundamental laws of genetics. He chose a species that produced many offspring, that had a short generation time, and whose matings he could carefully control. Most genetic experimentation is still conducted on organisms meeting these criteria.

Pedigree Analysis

Many examples of Mendelian inheritance have now been discovered in humans (figure 4.1), but not by using the techniques of Mendel. In fact, it is hard to imagine a species more difficult to examine experimentally than humans. Humans produce a relatively small number of offspring, they have one of the longest generation times of any species, and ethical considerations prohibit controlled matings. Therefore, discovery of traits exhibiting Mendelian patterns of inheritance has depended upon a methodology not used by Mendel—*pedigree analysis.*

A **pedigree** is a diagrammatic representation of a genealogy, a set of familial relationships. The word *pedigree* is derived from the Latin words *pes* and *grus,* meaning literally "foot of the crane," a reference to the lines in the pedigree specifying particular relationships.

Figure 4.2

An example of the common symbols used in pedigree analysis. Circles represent females, squares represent males, and diamonds represent persons of unknown or unspecified sex. Each row represents a single generation and is labelled with a Roman numeral. Within a generation, each individual is identified with an Arabic number. When persons do not need to be indicated individually, a number is inserted within a symbol to identify more than one person, as the three persons indicated in generation III, numbers 12–14. Horizontal lines connect reproductive partners (for example, II-7 and II-8), and vertical lines descending from this horizontal line lead to all their offspring (for example, individuals 8–14 in generation III). Persons exhibiting a particular phenotypic trait or affected by a specific condition are indicated by a yellow symbol. This hypothetical pedigree portrays a trait caused by an autosomal recessive allele that is expressed in individuals II-3, II-4, II-6, III-2, III-4, III-6, IV-2, and IV-7. The person first bringing the condition to medical attention, the index case or proband, is indicated with an arrow (II-4).

Figure 4.2 shows a hypothetical pedigree that illustrates the shorthand symbolism used by geneticists.

Pedigrees are usually constructed with reference to a particular phenotypic (potentially genetic) trait. Squares represent males, circles represent females, and diamonds represent offspring of unknown or unspecified sex. The symbols are solid for affected individuals, open or of different color, for individuals who are normal (for a particular trait). The affected individual who first called attention to the pedigree is known as the *proband* or **index case** and is indicated by an arrow. Mated pairs are directly connected by a horizontal line. A vertical line descending from the marriage line connects the progeny (brothers and sisters, *siblings,* or members of a *sibship*) to the reproductive couple. Offspring are listed in order of birth from left to right. Each individual within a pedigree is identified with a unique number: a Roman numeral indicates the generation (row) and an Arabic numeral indicates the position in the row. In figure 4.2, for example, the lineage begins with female I-1, who is married to male I-2. Their offspring are II-1, II-3, II-4, II-6, and II-7.

Relatedness

Individuals who share genes derived from a common ancestor are said to be related, or **consanguineous** (meaning of the same blood). In an evolutionary sense, all humans are descended from a few common ancestors that gave rise to our species, such as *Eve,* the female hypothesized to have lived about two hundred thousand years ago and from whom all surviving human groups are descended (see chapter 1). This means that all humans are genetically identical at most (well over half) of their loci, even members of different races. Differences between individuals are the result of differences at a very small fraction of the genome (see chapter 13). In general, the members of any species are identical at most of their loci.

Genetically, however, time is measured on a smaller scale, and consanguinity, or relatedness, usually refers to the presence in two or more individuals of particular alleles that are **identical by descent,** that is, copies of a gene or genes that are identical because they were passed on from a recent common ancestor. Geneticists usually trace genetic identity no further than five or ten generations.

Figure 4.3

Some common coefficients of
relationship (r) and examples
of how they are determined
for (a) parent-child, (b) full
siblings, (c) half siblings, and
(d) grandparent-grandchild.
Note that the coefficient
decreases by 1/2 for each
step in the chain of
relationship, and that, as in
(b), two pathways of
relatedness may connect two
individuals.

The proportion of alleles that are identical by descent is a measure of relatedness between individuals, called the **coefficient of relationship,** r (figure 4.3). Recall that as a result of meiosis, each parent transmits $\frac{1}{2}$ of their genes (and $\frac{1}{2}$ of their chromosomes) to each offspring. The proportion of shared alleles between parent and offspring, r, is thus $\frac{1}{2}$ (figure 4.3a). (Unless there is reason to believe otherwise, all matings with which a pedigree begins are believed to be random, thus having an initial relatedness of 0.)

Full siblings, sibs with the same parents, normally share some alleles derived from the mother, and some derived from the father. On average, each should share $\frac{1}{2}$ of the genes received from each parent. Thus, a pair of siblings share $\frac{1}{2}$ of the $\frac{1}{2}$—or $\frac{1}{4}$—of alleles derived from the mother, and similarly $\frac{1}{4}$ of the alleles derived from the father. Their total relatedness is the sum of the probability that they share alleles from their mother ($\frac{1}{4}$) and the probability that they share alleles from their father ($\frac{1}{4}$) or $\frac{1}{4} + \frac{1}{4} = \frac{1}{2}$ (figure 4.3b). Note the relatedness of parent to offspring is the same as the relatedness of sibling to sibling (that is a parent and child have 50% of their genes in common and any two siblings, on average, also have 50% of their genes identical), but a different set of alleles is shared in each case (figures 4.3a and 4.3b). The farther away (genetically) two people are from one another, the fewer genes they share in common, the smaller the coefficient of relationship (figure 4.3).

In general, the coefficient of relationship can be determined by analysis of the pathway of relationships shown in a pedigree. Because the degree of relatedness decreases by $\frac{1}{2}$ with each step (usually a generation) in a pedigree, the probability of a particular allele being shared between two individuals is $(\frac{1}{2})^n$, where n is the number of steps in the pathway of relationships connecting two individuals.

Where the relationship in question involves only one ancestor and a direct descendant—for example, parent-child (figure 4.3a)—only one pathway of relatedness exists for sharing alleles, and the degree of relationship is simply $(\frac{1}{2})^1 = \frac{1}{2}$ for that single pathway. The relationship of half-siblings (figure 4.3c) also involves only one pathway, in this case with two steps; the relatedness of half-siblings is thus $(\frac{1}{2})^2 = \frac{1}{4}$. For many other relationships, genes are shared through two or more ancestors, as in the case of full siblings (note there are two pathways of two steps each in figure 4.3b). As a result of **inbreeding**—the mating of related individuals—it is possible for relatives to share alleles through three or more pathways of relationship. The calculation of relatedness involves the sum of relationships along all possible pathways of shared descent.

Thus, the coefficient of relationship of an individual to almost all others is 0. Genetically identical individuals, on the other hand, have a relatedness of 1. Among humans, only **monozygotic** (or *identical*) **twins** are genetically identical. They arise as the result of a single zygote splitting very early during development, giving rise to two individuals of identical genotype. The phenotypic similarities of monozygotic twins are extraordinary, but it should be remembered that they are usually raised in similar environments (including the same womb for the first nine months of life), so a part of the phenotypic similarity is due to environmental factors. Monozygotic twins are always of the same sex.

Nine-banded armadillos are extraordinary because they always give birth to identical quadruplets, the result of two sequential fissions of one zygote. The genetic identity of the offspring have made this species a valuable laboratory organism for genetic and medical studies. Among invertebrates and plants, many species reproduce asexually, often producing thousands or millions of genetically identical copies—**clones**—of themselves.

Twins may also arise as a result of the fertilization of two separate eggs by two separate sperm. These are **dizygotic** (or *fraternal*) **twins.** Dizygotic twins are no more closely related genetically than are normal full siblings—$r = \frac{1}{2}$.

Consanguineous Matings

Most *deleterious* (harmful) recessive genes occur at a low frequency among humans. Because the expression of recessive alleles requires that they occur in a homozygous condition, relatively few individuals expressing the trait actually occur. For example, consider a locus with a normal allele D and a recessive allele d (which is lethal when homozygous). If one in one hundred alleles in the population is d (an allele frequency of 1%), then one in every fifty persons is a heterozygous **carrier** of this allele (because each person in the population has two alleles of all autosomal genes). The probability of two carriers randomly mating is $\frac{1}{50} \times \frac{1}{50} = \frac{1}{2,500}$. The probability of one of their offspring being homozygous recessive is $\frac{1}{4}$, and therefore the probability of an affected offspring from a random mating is $\frac{1}{2,500} \times \frac{1}{4} = \frac{1}{10,000}$.

Another way of calculating this probability is to consider a random mating and the probability of a zygote receiving two d alleles, one from each parent. The probability of any parent contributing the allele to a gamete is the frequency of the allele in the population, or $\frac{1}{100}$.

Thus, the probability of a random zygote receiving two d alleles is $\frac{1}{100} \times \frac{1}{100} = \frac{1}{10,000}$. Recessive alleles in low frequency are expressed only rarely. They are maintained in the population almost exclusively in the heterozygous condition.

The key concept in the example above is *random mating*. If mating is random, and if the allele occurs in low frequency, the probability of one carrier mating with another carrier is about twice the frequency of the deleterious allele in the population. However, under some types of non-random mating, especially **consanguineous matings,** the probability of one carrier encountering another is much higher.

Descendants of an individual carrying a particular allele have a much higher probability of carrying that same allele than the population at large. Matings between close relatives greatly increase the chance of both parents having identical alleles, thereby increasing the probability of progeny carrying recessive alleles in the homozygous condition, and in turn expressing deleterious traits.

The harmful effects of consanguineous matings (inbreeding) are well known and are called **inbreeding depression.** The effect of inbreeding depression increases with the relatedness of the mated individuals. On the average, full sibling matings are far more likely to produce negatively affected progeny than random matings.

It is believed that, on average, each of us carries four or five deleterious alleles in the heterozygous condition. Inbreeding greatly increases the probability that one of these will be expressed in our descendants. In a few societies, brother-sister and father-daughter matings have been practiced, but in most societies an **incest taboo** is observed (tacit recognition of the harmful effects of inbreeding). In the United States, most states restrict the degree of consanguinity permitted in marriages. These prohibitions are generally based on religious custom, not adherence to genetic information.

Autosomal Recessive Inheritance

At least 626 traits in humans have been positively identified as the result of homozygous recessive alleles on the autosomes (table 4.1). Another 851 traits are strongly suspected to

TABLE 4.1

Some autosomal recessive genetic diseases in humans

albinism   Absence of the pigment melanin (figure 4.5).

alkaptonuria   A metabolic defect characterized by black urine due to the absence of a key enzyme in a metabolic pathway.

amyotrophic lateral sclerosis   Spasticity and muscle atrophy; Lou Gehrig's disease.

atrichia   Nearly complete absence of hair; individuals are born with hair that falls out and is not replaced.

Bloom's syndrome   Dwarfism with skin changes; characteristic rash from exposure to sun; increased susceptibility to death from cancer.

Chediak-Higashi syndrome   Partial albinism associated with death by age seven.

total color blindness   Day blindness; cones in the retina of the eye are missing. Persons affected can see better at night than in the day.

congenital heart block   Lethal heart defect leading to neonatal death.

cystic fibrosis   Exocrine gland malfunction.

deafmutism   Deaf from birth.

Ellis-Van Creveld syndrome   Dwarfism associated with polydactyly, partial harelip, and cardiac malfunction (figure 4.1d).

fructose intolerance   Inability to metabolize fruit sugar, fructose.

galactosemia   Inability to metabolize galactose, with mental retardation and blindness resulting from buildup of galactose.

hypotrichosis   Failure to replace intrauterine hair that is shed shortly before birth or after birth. No pubic or axillary hair develops at puberty.

lactose intolerance   Inability to metabolize milk sugar, lactose.

mast syndrome   Presenile dementia with onset occurring in the late teens or early twenties.

Niemann-Pick disease   Accumulation of lipids, leading to retarded physical and mental growth. Death usually occurs by age three.

phenylketonuria   Defect in phenylalanine metabolism leading to mental retardation.

Tay-Sachs disease   Onset of developmental retardation in infancy, followed by paralysis, dementia, blindness, and death by age three.

Werner syndrome   Precocious aging and senility. Death often results from coronary artery disease by age ten.

Wilson's disease   Defect in copper metabolism leading to cirrhosis of the liver.

xeroderma pigmentosum   Sun sensitivity leading to skin tumors and early death. Patients are unable to repair ultraviolet light damage to DNA (figure 11.17).

From V. A. McKusick, *Mendelian Inheritance in Man,* 8th edition. Copyright © 1988 Johns Hopkins University Press. Reprinted by permission.

be due to homozygous recessive alleles. Most, but not all, of these cause serious phenotypic conditions. They are now understood as Mendelian genetic variants because they differ qualitatively from the normal condition, just as did the traits of peas that Mendel examined. Certainly, many more or less normal conditions are also governed by Mendelian variants, but they are less clearly phenotypically defined, and therefore less easily studied. Many other traits (such as height and weight) are governed by the interaction of many alleles and the environment and thus are extremely difficult, if not impossible, to subject to normal Mendelian analysis.

Enzymes, one of the many types of proteins produced by genes, are important as catalysts in the thousands of biochemical reactions continually occurring in the body. These reactions proceed in highly controlled sequences called *metabolic pathways,* whose functions are the production of crucial chemicals such as pigments, the digestion of our foods, and the breakdown of various body waste products. In the homozygous state, many recessive alleles result in a defective or absent enzyme in a metabolic pathway, often with serious physiological consequences. The inability to produce an important enzyme, thus interrupting a metabolic pathway, is called an **inborn error of metabolism.** Many of the best known genetic diseases are inborn errors of metabolism. Alkaptonuria was the first such disease to be described in the early 1900s (table 4.1).

Examination of a pedigree often can reveal the mode of inheritance of a particular trait. A trait controlled by an autosomal gene should affect both sexes in about equal proportions. An autosomal recessive condition will be expressed in an affected child of phenotypically normal (but heterozygous) parents and can appear to skip generations (see figures 4.2, 4.4, and 4.5). Furthermore, roughly 2/3 of the normal offspring of such heterozygous parents will also carry the recessive allele. In species other than humans, the existence of traits controlled by recessive alleles can be confirmed by specific breeding experiments.

Discovery of albinism as an autosomal recessive trait in humans

Pedigree of albinism in a black family in Mississippi. Individual II-6 (blue) received his allele from his father (I-2), but both his wives were also carriers of the allele for albinism.

By 1903 breeding tests had established that albinism was an autosomal recessive trait in mice, rabbits, and guinea pigs. In that year, W. E. Castle, a geneticist at Harvard, analyzed the pedigree of a black family in Mississippi that demonstrated that albinism was almost certainly an autosomal recessive trait in humans, also (figure 4.4). Castle subsequently hypothesized that all mammals have a similar inheritance pattern for albinism.

The grandfather in this family was an albino who married a normal woman. They had three sons, all normal—recessive conditions often "disappear" for one or more generations. The first two sons had only normal children. The third son married twice, having one albino and five normal children by the first wife, and three albino and six normal children by the second. How do these data fit a Mendelian pattern?

All the children of the albino grandfather were heterozygous carriers for the albino allele. The wives of the first two sons were probably homozygous normal, resulting in all normal children for them. However, both wives of the third son were carriers. (Castle's paper notes that albinos were not uncommon in the area.) The Mendelian prediction for the mating of two heterozygotes is that $\frac{1}{4}$ of the offspring will be homozygous recessive. Because $\frac{4}{15}$ (as close to $\frac{4}{16}$ as possible) of the offspring of the third son were albino, Castle reasonably concluded that albinism is inherited as an autosomal recessive trait in humans.

Thus, only three years after its "discovery" by the scientific establishment, Mendel's work on garden peas was shown by Castle to have direct application to human health. The evolutionary relationship of all life can thus be seen to have extraordinary *practical* significance. Breeding experiments on laboratory species such as mice and guinea pigs, and also on more distantly related species such as plants, yeast, and bacteria, can help to understand conditions of medical significance in humans. This is possible only because a common evolutionary origin has resulted in all living organisms sharing mechanisms of inheritance.

Albinism

Albinism is an hereditary disorder involving an absence or reduction of melanin, the brown pigment responsible for the coloration of our skin, hair, and eyes. Albinism is one of the most common inborn errors of metabolism, having been observed in plants, insects, and all major groups of vertebrates. One of the first genetic traits identified as an autosomal recessive (box 4.1), it occurs in all major races of humans. The word *albino* is derived from the Latin word *alba,* meaning white. Due to the absence of pigment, albinos are a ghostly white in color. In some species, the eyes of albinos may appear pink from the color of *hemoglobin* (a pigment in red blood cells) visible through the unpigmented eyes.

The production of melanin requires the enzyme tyrosinase. At least three different genes are known that can interfere with the production of tyrosinase. The most common form of albinism, caused by a recessive allele at one genetic locus, affects one in thirty-three thousand

BOX 4.2

Natural selection sometimes favors the allele for albinism

Figure 4.5

An albino Hopi girl and a pedigree of a family with a high frequency of albinism among the Hopi Indians. The four carriers indicated have proven genotypes, but other normally pigmented individuals in this pedigree may also be carriers.

Field Museum of Natural History.

○ Normal female   ○ Affected female

□ Normal male   □ Affected male

○ Carrier female

□ Carrier male

The frequencies of many alleles differ greatly among human populations. Some diseases or genetic traits are restricted to certain populations or are far more characteristic of some populations than others. Throughout this book, we will note the frequency of occurrence of various genetic conditions in different human populations.

The genetic disease albinism illustrates how differing conditions in human societies may lead to very different frequencies of certain alleles. In this instance, an allele causing albinism appears to be relatively favored by natural selection in one tribe of American Indians, leading to an unusually high frequency of a normally deleterious allele.

whites in the United States and one in twenty-eight thousand blacks. Different populations and racial groups are often characterized by differing frequencies of many alleles.

Melanin serves as a protective screen against the harmful effects of sunlight in normally pigmented individuals. Lacking pigmentation, albinos are highly susceptible to sunburn and skin cancer. Albinism leaves the retina of the eye unprotected by pigment, so visual problems and blindness are also common. About 9% of the visually handicapped students in the United States are albinos.

In addition to classic (tyrosinase negative) albinism, there are several types of albinism where small amounts of tyrosinase are present. These tyrosinase positive albinisms are also inherited as autosomal recessive traits. Tyrosinase positive albinos have white hair in infancy, but with age the hair turns yellow, light brown, or red. This is the most frequent type of albinism in American blacks (one in fifteen thousand), and in some Native American tribes it occurs at truly extraordinary frequencies (box 4.2). Tyrosinase positive albinism in whites

Among most populations in North America, the common form of albinism is due to a relatively infrequent allele found in about 1 in 30,000 individuals. This is not surprising given the many harmful effects of the condition and the shortened life expectancy of albinos. However, among the Hopi Indians of Arizona, the frequency of tyrosinase positive albinism is about 1 in 200 individuals (figure 4.5), and in the Cuña Indians of San Blas, Panama, the frequency is 1 in 140. How can such a harmful allele become so abundant?

In many societies, unusual individuals are ostracized or subject to some forms of discrimination. Both the Hopi and Cuña have strong taboos against marriage between albinos. Thus, the high frequency of the disease results from matings of albinos with carriers, of carriers with other carriers, or of illicit liasons between albinos. The Cuña and Hopi are tribes of farmers, and in general the men cultivate the fields, while the women work at home. Because albinos cannot tolerate sunlight, albino men are also permitted to remain at home during the day. These circumstances apparently offer albino men increased opportunities for reproduction. Thus, cultural practices have resulted in a high frequency of a normally deleterious allele. An additional factor may be small population size, which increases the probability of consanguineous matings.

This example offers two lessons about the way in which natural selection occurs. First, alleles can be favored that may not cause increased survival, but that result in increased reproduction of the carriers. This phenomenon is called *sexual selection*. Albinos still do not survive as well as normal individuals, but in some societies they breed more frequently. Second, the selective value of an allele depends upon the environment in which it occurs. Many alleles are extremely valuable in some circumstances, but harmful in others. The way in which an allele functions depends upon the circumstances in which it occurs. In most places and societies, albinos have a reproductive disadvantage, but in a few, albinos have a reproductive advantage. The allele for albinism, however, is neither good nor bad—only different from others. Genetic diversity, the existence of many different alleles, is a sign of a genetically healthy species.

occurs at about the same frequency as tyrosinase negative albinism. The two forms are not allelic—they are controlled by separate genes.

Phenylketonuria

Phenylalanine is an amino acid necessary for human nutrition. The proper metabolism of phenylalanine requires the enzyme phenylalanine hydroxylase, whose production is genetically controlled. In the United States about one in fifteen thousand whites is born unable to produce this enzyme due to a homozygous recessive allele, causing **phenylketonuria (PKU)**. In Ireland and Scotland, PKU is more common, occurring once in every five thousand to six thousand births. Among blacks, PKU is extremely rare, affecting one in every three hundred thousand births.

One effect of PKU is the accumulation of phenylalanine in the blood and phenylpyruvic acid, a related chemical, in the urine. PKU was discovered in 1934 by a Norwegian who noticed that the urine of her two mentally retarded children had a distinctive odor which, upon analysis, turned out to be phenylpyruvic acid. The abnormal accumulation of particular chemicals in the urine and blood is characteristic of many inborn errors of metabolism.

The most serious effect of PKU is mental retardation, caused by the high levels of phenylalanine in the brain and spinal fluid in the first few weeks after birth. About 1% of all Caucasians in mental institutions suffer from mental retardation due to PKU. Infants with PKU can now be identified by a simple blood test immediately following birth. In the United States, more than forty states now require PKU screening of all newborns.

Victims of PKU, if identified at birth, can be treated with a diet carefully regulating the amount of phenylalanine. A small amount is necessary for adequate nutrition, but too much phenylalanine causes brain damage. Brain damage can be prevented if dietary regulation begins immediately after birth and is continued until about age seven. Growth of the brain is extremely rapid early in life, at which time irreversible damage can occur. However, recent studies indicate that it may be advisable for all phenylketonurics to continue a low-phenylalanine diet throughout their lives. New evidence suggests that personality and intellectual capacity may be adversely affected by the sustained high levels of serum phenylalanine typical of phenylketonurics not on special diets.

Dietary therapy for phenylketonurics is extremely effective (although any parent knows the difficulties of controlling a child's eating habits). PKU cannot be cured—an individual's genes cannot be changed—but the phenotypic consequences can be significantly ameliorated. In the late 1970s the United States government estimated that an efficient national system of screening and treatment of phenylketonurics would cost about $3.3 million annually. The result of *not* screening would be an average of 270 cases per year of preventable mental retardation, costing $189 million annually to treat. Recombinant DNA technologies (see chapter 10) make it possible to recognize more than half of all carriers of PKU. It is anticipated that soon it will be possible to identify more than 90% of all (PKU) carriers.

The effectiveness of PKU therapy has meant that many female phenylketonurics who might otherwise have been institutionalized have developed normally and begun to reproduce. However, PKU mothers who have been on diet therapy even from birth almost always produce children who are mentally retarded. The children are usually heterozygous for the PKU allele, so their condition is not due to their genotype, but to the abnormal biochemistry of their mother. The concentrations of phenylalanine in the affected mother's blood are so great (unless she is on a special diet during pregnancy) that the brain of a fetus she carries cannot develop normally. Thus, a genetically normal fetus is irreversibly damaged by its environment—the womb of its affected mother—and it exhibits the phenotype of a genetically affected child. The phenomenon, in which an environmentally caused condition mimics a genetically caused condition, is called a **phenocopy.** Women who are phenylketonurics are advised to remain on a low phenylalanine diet, starting at least three months prior to conception and throughout the entire pregnancy.

Cystic Fibrosis

The most common serious genetic disease among Caucasians in the United States is **cystic fibrosis (CF),** caused by an autosomally inherited recessive allele. Approximately one in every two thousand white births is affected by CF, an exceptionally high frequency (compared to PKU or albinism, for example), considering that affected individuals usually die before reproducing. This means that all cystic fibrosis births occur as the result of the mating of two heterozygotes; essentially no homozygotes reproduce.

The high frequency of carriers in the population (one in twenty Caucasians) may result from *overcompensation;* parents of CF children continue to reproduce until they have the number of normal children they desire. The result is that the average number of offspring in families with a CF child is about 25% higher than in normal families. Of the normal children in CF families, two-thirds are carriers and continue to transmit the harmful allele.

The gene responsible for most cases of CF was identified and sequenced in 1989. The CF gene encodes a large (about 1500 amino acids) protein that functions in cell membranes to regulate the water balance inside cells. The defective CF protein prevents chloride ions from exiting cells, resulting in excess water inside the cell. This leads to a buildup of thick dry mucous in the lungs, which characterizes the disease. CF is actually a *syndrome* (a large set of symptoms associated with one pathological condition) in which all the mucous producing glands malfunction, particularly those of the respiratory and digestive systems. The mucous produced by CF victims is extremely viscous, much like cotton candy, clogging the lungs and digestive tract. The accumulation of mucous in the lungs causes a persistent cough and difficulty in breathing. The residual mucous is a good medium for bacterial growth, making pneumonia and other infections a common and sometimes fatal complication.

The mucous also clogs the ducts of the pancreas, blocking the pathway for important digestive enzymes into the gut. Thus, even if CF children eat a balanced diet, they cannot digest their food properly and consequently suffer from malnutrition. Malnutrition can be alleviated by massive doses of vitamins and other nutrients, and such therapy has caused some increase in life span. Persons with CF now live to age twenty and some even to forty years and older.

Another effect of CF is an unusually high concentration of salt in the sweat of affected children—about two to five times normal concentrations. The disease is sometimes first detected when parents kissing their children taste the salty perspiration. The disease can be diagnosed by a *sweat test,* which measures the concentration of salt in a sample taken from a warmed patch on the child's back or forearm. Parents of CF children are heterozygotes, and their levels of salt are intermediate to levels found in affected persons and in normal persons. The different phenotypic aspects of this disease thus have different types of expression. For the ultimately fatal set of symptoms that are called CF, the allele is expressed as a recessive. For the single biochemical trait of salt concentrations in sweat, the CF allele is expressed as a codominant or incompletely dominant allele.

A prenatal diagnostic test to identify fetuses with CF is available for families where both parents are known carriers of the CF allele. However, unlike PKU, CF has no known cure. This leaves therapeutic abortion as the only means of preventing CF. Tests are also available to detect carriers of CF. Recent work has demonstrated that this test is accurate for 70–75% of all defective alleles that cause CF. Thus, at least 20–40 different alleles can cause CF and cannot be identified by the presently available test. A definitive test for CF carriers may never be realized. In 1990, gene transfer was successfully used to cure CF cells growing in culture. A virus was used to insert a normal gene into cultured CF cells, resulting in the cells having normal membrane function. Extensive research is underway to translate this *in vitro* gene therapy into a cure for patients with CF.

Sickle-Cell Anemia

The most common genetic disease among blacks in the United States is **sickle-cell anemia (SCA).** Victims of SCA possess an abnormal form of the protein *hemoglobin.* Hemoglobin occurs in the *red blood cells* (RBCs), imparts color to them, and carries oxygen to all tissues of the body. It is composed of four chains of amino acids (two alpha chains and two beta chains) and four heme groups, an iron containing portion that functions to bind and release oxygen to the tissues. Each RBC contains about 280 million molecules of hemoglobin, and each molecule contains 574 amino acids. The hemoglobin found in sickle-cell anemics differs from normal hemoglobin by one amino acid difference in each beta chain—two differences that have profound physiological effects. They cause the hemoglobin to crystallize, which in turn causes malformation of the RBCs. The RBCs assume a crescent or sickle shape, hence the name of the disease (figure 4.6).

SCA has two primary physical effects: (1) the deformed RBCs become fragile and are easily destroyed; (2) sickle-shaped cells tend to become entangled with each other, increasing blood viscosity and thus reducing blood flow. The physiological impact of these effects is felt throughout the system (figure 4.7). The high rate of cell destruction induces *anemia* (low RBC count), reflected in a decreased oxygen supply to the tissues, which causes the victim to tire easily. Anemia stresses the heart, which must work harder to pump adequate amounts of blood to the tissues, eventually leading to heart failure. The spleen is taxed trying to break

(a)

(b)

Figure 4.6

(a) Normal red blood cells;
(b) sickled red blood cells;
(c) diagrammatic
representation of molecule of
hemoglobin, showing alpha
and beta globin chains and
heme groups.

(a and b) AP/Wide World
Photos; (c) From Stuart Ira
Fox, *Human Physiology,* 3d
ed. Copyright © 1990 Wm.
C. Brown Publishers,
Dubuque, Iowa. All Rights
Reserved. Reprinted by
permission.

Beta chain

Beta chain

Alpha chain

Alpha chain

Heme group

(c)

down defective RBCs; eventually it begins to deteriorate, losing its capacity to remove bacteria from the blood. SCA victims are continually subject to infection due to failure of the spleen.

Increased blood viscosity interferes with circulation of the blood, causing further stress on the heart and localized damage to many organs. Circulatory impairment in the brain may cause strokes and paralysis. Fluids accumulate in the lungs, leading to pneumonia. Decreased circulation in the muscles leads to rheumatism. A gene that has many phenotypic effects is called **pleiotropic;** there is no better example than the allele for SCA.

SCA is usually lethal. Most victims die by age ten, but medical treatment is permitting an increasing number to survive into adulthood. As with cystic fibrosis, the incidence is extraordinarily high, considering the lethality of the allele when homozygous. Among blacks in the United States, about one birth in four hundred is a child with SCA, and one in every ten blacks is a carrier of the disease.

Although the disease is normally considered to be a recessive condition, heterozygotes show some phenotypic manifestations and are said to have **sickle-cell trait.** About 35% of the RBCs of individuals with sickle-cell trait have defective hemoglobin. However, the remaining normal cells permit the carrier to lead a relatively normal life, subject to occasional episodes of anemia and complications during times of stress. As with cystic fibrosis, this allele shows a recessive pattern of expression in terms of the disease itself, but some physiological traits show incomplete dominance.

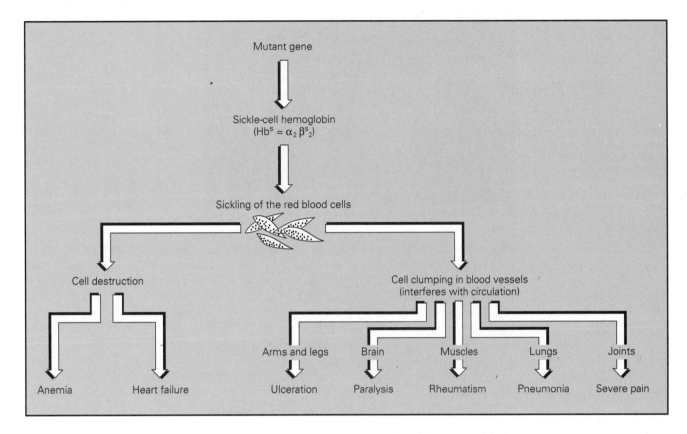

One trait of heterozygotes may explain the high frequency of the allele among blacks. Carriers of the sickle-cell allele are relatively immune to a type of malaria common in Africa. Historically, malaria has been the largest single cause of death for humans in the world, and it is still one of the largest causes of mortality in Africa. In parts of Africa where malaria was common, heterozygotes for the sickle-cell allele survived longer and produced more offspring than either homozygous genotype, thus perpetuating the sickle-cell allele in spite of its lethal effects in homozygotes.

Malaria is caused by a parasitic protozoan that lives in RBCs. The defective hemoglobin in individuals with sickle-cell trait appears to make the RBCs an unsuitable environment for the protozoan, thus conferring relative immunity on carriers. In parts of East Africa, as much as 45% of the black population may possess sickle-cell trait. In North America, immunity to malaria no longer enhances survival, and the sickle-cell allele has therefore decreased in frequency among blacks.

There is no cure for SCA. New methods of DNA analysis now provide a safe, quick, and reliable **prenatal screening** technique, which allows parents to know early in the pregnancy whether their fetus carries zero, one, or two copies of the SCA allele. The test can also screen adults who wish to know if they are carriers of the disease.

Figure 4.7

Some of the pleiotropic effects of the gene causing sickle-cell anemia. Collectively, these symptoms constitute the *syndrome* characterizing the disease.

Genetic Counseling

The ability to diagnose certain diseases as heritable in predictable ways has direct medical implications. Most people have a limited knowledge of genetics, however. In addition, genetic knowledge is expanding so rapidly that techniques of diagnosis and treatment of genetic diseases are constantly improving. In response to such rapid change, professional **genetic counseling** has arisen to make the latest information on medical genetics available to all persons.

Most commonly, a family is referred to a genetic counselor after the birth of a child afflicted with a genetic disease or when the family has a history of genetic disease. A genetic counselor helps the individual or family understand the genetic basis of specific disorders, the probable course of the disease, and available management or treatments. The counselor

Figure 4.8

Genetic counseling assists individuals and families to understand and deal with inherited illnesses. As new technologies allow improved detection and diagnosis of genetic conditions, genetic counseling is becoming an important component of medical practice.

© Will and Deni McIntyre/ Photo Researchers, Inc.

also advises clients of the risk of recurrence in specified relatives and options for dealing with the risk of recurrence. Finally, counselors help families adjust to genetic disorders in family members, suggesting courses of action appropriate to the family goals and backgrounds.

Traditionally, a counselor educates potential parents but does not make choices for them. The counselor explains medical risks and results of diagnostic procedures, as well as the probabilities for genetic affliction in future children. In all instances, final decisions rest with the couple. In many genetic counseling centers, psychologists and social workers help the geneticist and physician in the counseling process.

While advances in medical genetics offer new methods of treatment and prevention of certain diseases, they also present affected families with difficult and often highly emotion-laden choices. Perhaps the most extreme case is the prenatal diagnosis of a serious genetic defect. If the pregnancy is allowed to continue, the family must accept a severe emotional and financial burden in caring for or institutionalizing the defective child. The decision to continue the pregnancy may also impose costs on society, which must often assume the responsibility of caring for the child.

On the other hand, parents have the option of terminating the pregnancy by **therapeutic abortion.** If the condition is serious and the child will have a short, painful life, a decision to abort may seem relatively straightforward. However, many serious conditions are being treated with increasing success; children with cystic fibrosis, for example, now may live two decades or more. Thus, the decision to abort is not simple, even in very serious cases. Furthermore, because of ethical and religious reasons, some families find abortion unacceptable under any circumstance.

In cases where abortion is not a consideration, the use of genetic information may still present difficult choices. The ability to diagnose the gene for sickle-cell anemia, for example, has not been greeted with uniform enthusiasm. Some employers have suggested that SCA screening (box 4.3) be a part of the hiring procedure, but because such tests would only be conducted on blacks, these practices have been called discriminatory. Similarly, some insurance companies have required tests for SCA before issuing health and life insurance policies.

The treatment of carriers, such as individuals with sickle-cell trait, is particularly contentious because many of these individuals show few symptoms. Should they receive different treatment simply because they carry a gene that probably will cause them little harm? And, if employers and insurance companies are given access to genetic information, what are the limits? Should a possible future spouse be allowed such information?

The ethical questions raised by genetic technologies vary considerably with the specific genetic condition. As we discuss other diseases throughout the book, we will draw attention to other aspects of genetic counseling.

Autosomal Dominant Inheritance

At least 1,443 traits in humans are now known to be inherited in an autosomal dominant fashion (table 4.2), and a further 1,141 traits are suspected. Identification of autosomal dominant traits from pedigree analysis is relatively straightforward. Affected offspring must have at least one affected parent (barring new mutations), and all children of homozygotes and half the children of heterozygotes should express the trait. Dominant traits do *not* skip generations.

Achondroplasia

A well-documented autosomal dominant allele affects the growth centers of the long bones, causing a form of dwarfism called **achondroplasia.** The affected individual has a normal-sized torso, but very short limbs. Achondroplastic dwarfs average about four feet in height (figure 4.10). This condition affects less than one birth in ten thousand. Most cases are believed to result from spontaneous mutations, as most affected offspring are born to phenotypically normal parents. The homozygous condition is probably lethal. There is no treatment.

The Church of Jesus Christ of Latter-Day Saints (the Mormon church) maintains detailed genealogical records, useful in genetic studies. Achondroplasia is relatively common in several Mormon families, where achondroplastic dwarfs have routinely married and had children. In one family, seventy-six children were born of matings between normal (dd) and dwarf (Dd) individuals—a genetic testcross. The Mendelian prediction is that thirty-eight should be normal, thirty-eight dwarf. The actual result was that forty-two were normal, thirty-four dwarf. This is close enough to the expected values that the difference can be attributed to chance factors. A larger sample size would probably approximate more closely the 50% expectation of each type of child.

Nail-Patella Syndrome

Not all autosomal dominant traits have dramatic and serious effects. Carriers of the allele causing **nail-patella syndrome** have absent or malformed fingernails, toenails, and kneecaps (patellas). Some also have chronic inflammation of the kidneys. Otherwise, affected individuals exhibit no serious symptoms.

Neurofibromatosis

The course of **neurofibromatosis** (NF) caused by an autosomal dominant allele, involves the development of hundreds of non-malignant tumors over the entire body. In combination with abnormal bone growth, the victim of NF may be grossly deformed.

This disease cannot be diagnosed prenatally and cannot be cured. Surgery is sometimes used to remove masses of tumors, especially those affecting nerves. About 20% of NF victims develop tumors of the spinal cord, which can cause deafness and blindness. Plastic surgery to correct the massive deformation is traumatic, and the results are seldom totally satisfactory. The onset and progress of the disease are unpredictable. Unfortunately, this relatively common condition afflicts about one in 3,500 newborns of all ethnic groups; and over 100,000 Americans.

In the summer of 1990 the gene causing NF was mapped to chromosome 17, cloned, and found to be a member of a class of tumor-suppressor genes (anti-oncogenes, see chapter 11). This new understanding of the NF gene opens the possibility of improved diagnosis for young children and makes development of prenatal tests likely in the near future.

Earwax

Variation in the type of *cerumen* or *earwax* exemplifies a dominant Mendelian trait with innocuous phenotypic effects. Two types of cerumen occur in humans. Wet earwax, the result of a single dominant gene, W, is a sticky type of earwax found in about 85% of white Americans and 93% of black Americans. Dry earwax, genotype ww, produces a granular or scaly and non-adhesive earwax and is found in high frequency in Mongoloid populations—98% in northern Chinese, 92% in Koreans, and 92% in Japanese (see table 13.2).

Such genetic differences are common among various racial groups, but the causes are often obscure. Wet earwax is associated with axillary (armpit) odor, which in Japan is considered a pathological condition requiring medical attention. In many mammals, including

BOX 4.3

Genetic screening—sometimes a good idea

Once genetic diseases can be efficiently diagnosed, **genetic screening** programs may be an appropriate public health response. A screening program is a systematic evaluation of members of a susceptible population to identify either affected individuals or carriers of a harmful allele.

The screening program for phenylketonuria (PKU), noted earlier, seems to be an unquestioned success. Most states now require that a blood sample be taken within a day of birth for PKU testing. When identified, an affected child is immediately placed on a diet that protects it from the high levels of phenylalanine in the blood that cause severe mental retardation. Such children can lead a normal life, rather than requiring institutional confinement.

The PKU program is highly effective because it meets a number of essential criteria: (1) virtually all members of the risk group can be tested—newborns are a captive audience; (2) the program is cost-effective, in that the costs of the test and dietary treatment of the disease are enormously less than the costs of institutionalizing untreated children, not to mention the tragedy of unnecessarily wasted lives; (3) most importantly, a highly effective treatment (although not a cure) and necessary counseling are available to all affected individuals.

Unfortunately, few other screening programs for other genetic diseases meet those conditions. A relatively successful prenatal testing program for carriers of Tay-Sachs disease has been in operation since about 1970. Tay-Sachs is a particularly horrible condition, caused by a recessive autosomal gene and affecting about one in 3,600 children of Ashkenazi Jews, but having a far lower incidence in other ethnic groups. Affected children are normal for the first 4–6 months of life, but then experience severe physical and mental retardation, paralysis, blindness, and inevitable death by about age three.

No treatment exists for Tay-Sachs disease. The first goal of the screening program was to identify carriers of the Tay-Sachs allele. Couples identified as both being carriers and therefore at-risk of having affected children could be counselled about the probabilities and the effects of the disease. Furthermore, a prenatal diagnostic test of the fetus was available, offering the option of aborting affected children. In the first 10 years of its use, the number of affected children born declined by about 75%.

The Tay-Sachs screening program generally met the criteria listed above. The at-risk group could be contacted efficiently through the local community and synagogues, education and counseling were an integral part of the program, and a means of preventing unwanted births was available.

Almost concurrently with the Tay-Sachs program, a screening program for sickle-cell anemia (SCA) was introduced that was not only ineffective, but instead created serious new problems. In contrast to the Tay-Sachs program, relatively little emphasis was placed on education and counseling. Therefore, many carriers did not understand that they did not have the disease or that they would not necessarily produce affected children. In some states, the test was mandatory, but confidentiality was sometimes not maintained and some carriers subsequently could not obtain health insurance. Carriers were thus penalized rather than helped, and not surprisingly the program came under suspicion of racism because the at-risk population was black. Furthermore, no prenatal test existed, so carriers could be offered no options for preventing birth of affected children other than not to reproduce at all, compounding the suspicions of racism.

Today, due to rapid advances from DNA biotechnologies, fetuses having sickle-cell anemia can be identified (chapter 10), thus providing at-risk couples with the option of therapeutic abortion. However, due to the negative experiences of the early sickle-cell programs, this option is not frequently exercised.

A different set of problems have resulted from discovery of the cystic fibrosis (CF) gene in 1989. A national screening program for carriers was immediately envisioned for the USA. However, plans for such a program may be on hold indefinitely for several reasons. First is the magnitude of the task. Over 100 million Americans are of reproductive age, and the test alone might cost at least $50. Providing counseling both before and, for carriers especially, after the test, would be an equally enormous and expensive test, but essential to success.

Second, CF is caused by at least twenty, but potentially hundreds, of mutations at the CF locus. The most common allele (mutation) causes about 70–75% of cases, but all the others are extremely rare. A separate test is required for each allele, clearly not feasible. Testing for the common allele would identify only about half ($0.7 \times 0.7 = 0.49$) of all at-risk couples. A test that is only 50% successful is usually not considered adequate; in this instance, it would offer a false sense of security from producing affected children to half of all at-risk couples, who might subsequently be devastated to have a CF child. On the other hand, if the test is not made available, physicians might be subject to legal liability for professional negligence. Which risk should we take?

Finally, there is no cure for CF, and treatments are of only limited success. Prenatal testing can be used for at-risk couples, allowing abortion of affected children. At-risk couples who find abortion unacceptable might choose to conceive by artificial insemination, avoiding entirely the possibility of having an affected child.

Thus, it is clear that mass screening programs can sometimes be extremely effective, but that they can also compound an already difficult situation if not implemented carefully and sensitively, as shown by the SCA program. Even a major advance in knowledge, such as the extraordinary discovery of the CF gene, does not guarantee a reduction in the frequency of affected children and, ultimately, a reduction in human suffering. Identification of the CF gene is an important landmark in the struggle to find a treatment or cure for CF, but for now this new knowledge forces us to confront new and increasingly difficult choices.

TABLE 4.2

Some human traits inherited in an autosomal dominant fashion

absence of fingerprints   No dermal ridges in fingers.

achondroplasia   Dwarfism due to abnormality of ossification centers of the long bones (figure 4.10).

amyloidosis   Congestive heart failure.

brachydactyly   Short fingers and toes.

cleft chin   Bony peculiarity underlying a Y-shaped fissure in the chin.

congenital night blindness   No night vision, but normal vision in daylight.

deafness   Progressive high-tone deafness.

Ehlers-Danlos syndrome   Loose jointedness and fragile, easily bruised skin.

Huntington's disease   Choreic movements and dementia; variable age of onset (figure 1.4a,b).

juvenile glaucoma   Glaucoma developing by age eight.

Marfan's syndrome   Excessively long arms and legs with associated aneurysm of aorta. Olympic volleyball star Flo Hyman died of an aneurysm caused by this gene.

myokymia   Spontaneous muscle twitches.

myopia   Nearsightedness.

nail-patella syndrome   Dysplasia of the nails and absence of kneecaps.

neurofibromatosis   Presence of cafe-au-lait spots and numerous benign skin tumors.

osteogenesis imperfecta   Brittle bones resulting in frequent fractures, associated with loose jointedness, blue sclerae, and progressive deafness.

piebald trait   White forelock and white spotting over body.

polydactyly   Extra toes or fingers (figure 4.1a).

porphyria   Increased excretion of breakdown products of hemoglobin; increased sensitivity to sunlight.

progeria   Precocious senility (figures 4.1c and 11.18).

ptosis   Drooping eyelids (figure 4.1b).

retinoblastoma   Tumors of the retina, resulting in blindness; fatal if eye is not removed.

woolly hair   Short, tightly curled and woolly hair.

From V. A. McKusick, *Mendelian Inheritance in Man*, 8th edition. Copyright © 1988 Johns Hopkins University Press. Reprinted by permission.

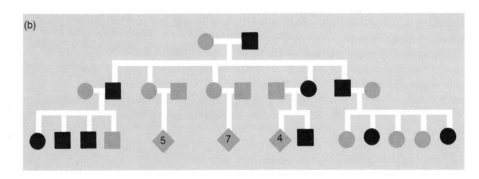

(a)

Figure 4.10

(a) An achondroplastic boy.
(b) The hypothetical pedigree of a family affected by achondroplasia over three generations.

(a) Dr. Judith Hall.

humans, scent is an important trait in the choice of a mate. Scientists have suggested that the high frequency of the allele for dry earwax among Mongoloids results from sexual selection—an historical preference for the aroma of people with dry earwax or a distaste for the aroma of people with wet earwax. This explanation suggests that the quality of earwax itself may have been of little selective importance. Rather, the allele producing dry earwax was selectively favored among Mongoloids because of one of its pleiotropic effects, absence of an offensive axillary odor.

The gene for cerumen type appears to exert a general effect on the apocrine glands, a type of sweat gland. The glands producing earwax and axillary odor are apocrine glands, as are the mammary (milk producing) glands of the breast. Yet another, more serious pleiotropic effect of the allele for dry earwax may involve breast cancer. Among Japanese women, breast cancer is about twice as common in women who have wet earwax as in women with dry earwax. Worldwide, the frequency and mortality rates for breast cancer are directly correlated with the frequency of the allele for wet earwax among all races. Although earwax type may be a relatively trivial trait, it is clear that the controlling gene probably has several important effects on phenotype and survival.

Anthropologists can often trace the origins of particular racial groups through genes that have well-defined racial differences. Native Americans are believed to have originated from Mongoloid ancestors who migrated across the Bering Straits thousands of years ago. Among full-blooded Navaho Indians of the American southwest, the frequency of the dry earwax phenotype is 70%, providing strong support for their origin from a Mongoloid stock. Among full-blooded Sioux Indians, the frequency is somewhat lower, about 47%. A possible explanation may be the fact that Sioux in the upper Mississippi Valley had contact with French trappers as early as 1700 (that is, they are not really full-blooded), while the relatively isolated Navaho have only recently come into contact with Caucasians.

Blood Groups

A person's blood type is a phenotypic character that depends upon the types of molecules—called **antigens**—that occur on the surfaces of the red blood cells (RBCs). Although the ABO **blood group** is the best known, several dozen other blood groups have been defined in humans. The existence of so many blood groups indicates that the surface of a red blood cell is indeed crowded with proteins. It is now known that there are many "public" blood groups—antigens that occur on the surface of all human RBCs, exhibiting no variation whatever. Some "private" blood groups, variants unique to only a few individuals, have also been found. Thus, the known blood groups almost certainly underestimate the immense diversity of molecules that occurs on each RBC.

The ABO Blood Group

The ABO blood group locus is characterized by three alleles, A, B, and O. O is recessive to both A and B, which are codominant to each other. A and B specify antigens on the surface of the RBC; O specifies no antigen. Although three primary alleles occur in the population as a whole, each individual possesses only two.

Discovery of the ABO group by Karl Landsteiner in 1900 depended not only upon the presence or absence of A and B antigens on the RBCs, but also upon the presence or absence of particular **antibodies** in the serum. Antibodies, a primary mechanism of defense for the body, combine with foreign proteins such as those on the surface of bacteria, neutralizing their pathological effects. Table 4.3 summarizes the combinations of antigens and antibodies of the four main ABO blood types.

Note that a person never has antibodies to his or her own antigens. For example, an individual of blood type A has antigen A, but anti-B antibodies. A person of blood type O has both anti-A and anti-B antibodies. If blood of different types is mixed, the antibodies combine with antigens of the same type (anti-A with antigen A, for example), causing a reaction called **agglutination.** Agglutination is visible to the naked eye (figure 4.11), and it was Landsteiner's observation of agglutination that resulted in discovery of the ABO blood group.

Most antibodies are produced only after a foreign antigen is introduced into the circulatory system. The antibodies of the ABO system, however, are present at birth. Landsteiner would not have made his fortunate discovery if the antibodies were not naturally occurring, and if all persons had the same ABO blood type.

Discovery of the ABO blood group was an important medical advance. During surgery or as the result of a serious wound, persons often need transfusions of blood. Prior to Landsteiner's work, transfusions had been used with variable results. Sometimes they were successful, but equally often the patient suffered a severe, sometimes fatal reaction to the transfusion.

TABLE 4.3

ABO blood types and their inheritance in humans; antigen and antibody relations, and permissible transfusions of red blood cells

Blood type		Antigen(s) on surface of red blood cells	Antibodies present in serum	Transfusion status	
Phenotype	Genotype(s)			*Can receive*	*Can donate to*
O*	OO	Neither A nor B	Anti-A, Anti-B	O	O, A, B, AB
A	AO or AA	A	Anti-B	A, O	A, AB
B	BO or BB	B	Anti-A	B, O	B, AB
AB†	AB	A, B	Neither anti-A nor anti-B	AB, A, B, O	AB

*Universal donor
†Universal recipient
There are two major type A antigens or alleles, designated A-1 and A-2. Eighty percent of persons with type A are of the A-1 genotype. Several other rare subgroups of the A antigen have also been identified. Variants of the O and B groups are also known, but are quite rare. (Often a different allelic nomenclature is used for the ABO blood groups. The gene for the blood group is indicated by I, representing "isoagglutinin," the molecule involved in the agglutination (clumping) of RBCs in blood typing. Specific alleles are represented by superscripts, I^A, I^B, I^O.)

Current medical practice matches the blood type of the donor with that of the recipient prior to the transfusion. Usually, the recipient needs the oxygen carrying RBCs of the donor, so the critical aspect of the match is to ensure that the donor's RBCs are not attacked by the antibodies in the serum of the recipient (table 4.3). Transfusion of type A blood into a type B person or type O person, for example, results in destruction of the type A cells, inducing shock in the recipient. Type B individuals are thus never given type A blood, and vice versa. Because type O persons have neither A nor B antigens, they cause no reaction in a recipient of any blood type and are considered *universal donors*. Conversely, a type AB person has neither anti-A nor anti-B antibodies and is a *universal recipient*. Donor-recipient ABO types are normally precisely matched, however, except in emergencies.

You might reasonably expect that transfusion of type O blood into a person of type B, for example, would have harmful consequences due to the presence of anti-B antibodies in the donor's serum. Usually this is not the case, however. The quantity of antibodies in the donor's serum is sufficiently diluted in the serum of the recipient to preclude much damage to the large volume of RBCs in the recipient. It is the far larger quantities of antibodies in the recipient's blood that cause injury.

Beyond its medical importance, the identification of the ABO and other blood groups has been of **forensic** (legal) importance. Because the ABO group shows strict Mendelian inheritance, blood types provide evidence in paternity suits. For example, a child of type AB cannot have a father of type O. In this case, the ABO group alone is sufficient to establish conclusively that an innocent man has been wrongfully accused. Frequently, however, ABO types provide inconclusive evidence. A type A man might be the father of a type O child, for example. The use of several blood groups can greatly increase the probability of a definitive finding of innocence. Evidence from blood groups can *never* prove paternity, although it can be highly suggestive.

Blood groups, like earwax, have been used to trace the history of particular racial groups. Many blood groups are highly polymorphic and show substantial differences among human populations. Considering again the origin of Native Americans, blood type B is extremely rare in most Native American tribes. Type B is totally absent from the Navaho, and occurs in about 4% of "full-blooded" Sioux. The occurrence of type B among Native Americans is believed to result from interbreeding with Caucasians. The combined frequency of the B and AB phenotypes among contemporary American whites is about 12%, thus supporting the hypothesis based on earwax frequencies that the Sioux have had a longer history of interbreeding with Caucasians than have the Navaho. Table 4.4 gives frequencies of ABO blood types in various human populations (see also figure 13.14).

Anti-A antibodies

Anti-B antibodies

O cells

A cells

B cells

AB cells

Figure 4.11

The reaction of the four blood phenotypes when mixed with anti-A or anti-B antibodies. Agglutination (clumping) occurs when there is a match of the antigenic type of the blood with the antibody.

Victor B. Eichler.

The MN Blood Group

Discovery of other blood groups has depended upon the use of antibodies not occurring naturally in humans. Landsteiner discovered the MN blood group by producing antibodies to human RBCs in rabbits, then using those antibodies to identify the antigens on the RBCs for which they were specific—that is, those that they caused to agglutinate. The MN blood group is controlled by two alleles, M and N, occurring at one genetic locus and expressed codominantly.

Table 4.4

Frequencies of ABO blood types in human populations

Population	Phenotype			
	O	*A*	*B*	*AB*
Armenians	.29	.50	.13	.08
Eskimos (Greenland)	.47	.45	.06	.02
Austrians	.43	.39	.12	.07
Danes	.42	.43	.10	.04
Irish	.54	.32	.11	.03
French	.42	.45	.09	.04
Russians (Moscow)	.34	.38	.21	.08
Indians (Calcutta)	.30	.26	.38	.07
Vietnamese	.42	.22	.31	.06
Chinese	.44	.27	.23	.06
Japanese (Niigata)	.30	.37	.23	.10
Nigerians (Yoruba)	.52	.21	.23	.04
Amerindians (Lima)	.86	.14	.00	.00
Amerindians (Bolivia)	.93	.05	.02	.00
U.S. whites (St. Louis)	.45	.41	.10	.04
U.S. blacks (Iowa)	.49	.27	.20	.04

See also figure 13.14.
Source: Data from A. E. Mourant, A. C. Kopec, and K. Domaniewska-Sobczak, *The Distribution of the Human Blood Groups,* 2d edition. Copyright © 1976 Oxford University Press, London.

Figure 4.12

Four phenotypes result from a cross of a female of blood type O, M with a male of blood type AB, MN. Note that the male can produce four genetically different types of sperm, while the female produces only a single type of ovum.

The MN blood group, of no medical importance, exhibits simple Mendelian inheritance. Because of codominant expression, however, two alleles at one locus produce three phenotypes (MM, MN, and NN) rather than the two phenotypes observed when one allele is dominant to the other. In the case of **codominance,** both genotypes and phenotypes from a mating of heterozygotes (MN × MN) occur in the ratio of 1:2:1 (simply a more precise expression of the 3:1 monohybrid ratio).

The MN group is inherited independently of the ABO group—that is, they are not linked. Mating of persons of blood types O,M and AB,MN, for example, produces four types of offspring in equal proportions A,M; B,M; A,MN; and B,MN (figure 4.12).

The HLA Antigens

The best studied system of cell surface antigens are those of the red blood cells, especially the ABO system. However, a complex set of antigens on the surface of most body cells, including the white blood cells, the **human leukocyte antigens (HLA),** is also of great medical importance. The individual HLA antigens are detected using antisera made to each of the known antigens. This complex system shows great Mendelian variation and governs the tissue rejection response, which will be discussed in chapter 12.

HLA antigens are the products of at least two genetic loci with major effects on the immune response and at least three loci with minor effects. All five loci are on the short arm of chromosome 6. The two major loci, designated HLA-A and HLA-B, are probably the most polymorphic loci yet identified in human populations. At least twenty-three codominant alleles encoding twenty-three unique antigens are known at the HLA-A locus, and another forty-seven alleles have been identified at the HLA-B locus.

Together, the A and B loci can generate an extraordinarily large number of unique genotypes. The A locus with twenty-three alleles generates 276 possible genotypes, while forty-seven alleles at the B locus generate 1,128 different genotypes. In combination, these two loci alone can produce 311,328 unique genotypes. At least forty-one other alleles are known at the three minor HLA loci, creating millions of unique HLA genotypes. Because not all alleles occur in equal frequency, in the population many genotypes are exceedingly rare. The single most common HLA genotype has a frequency of less than one in every hundred persons. Figure 4.13 illustrates the allelic diversity found within a family of six. Note that no two individuals share an HLA genotype.

HLA and Disease

The widespread use of HLA typing has revealed striking correlations between the incidence or severity of many diseases and certain HLA phenotypes (table 4.5). One of the best-studied associations is between the B27 antigen and a type of arthritis known as ankylosing spondylitis. This HLA antigen occurs in only 8% of the British population, but in 90% of all British patients with ankylosing spondylitis. Of sixty relatives of ankylosing spondylitis patients tested, thirty-two had the B27 antigen, further evidence of a genetic association. Similar data have been obtained in Japan. Among healthy Japanese the B27 antigen is very rare, but over 90% of affected Japanese have this antigen.

Despite the strong association, the B27 antigen cannot be considered to cause ankylosing spondylitis. Not all persons with B27 develop the disease. Perhaps the B27 antigen increases susceptibility to another causal agent, such as a virus. If such an agent could be identified, genetically susceptible persons could be vaccinated or given early treatment.

Recently HLA antigens have been associated with susceptibility to leprosy. With routine tissue typing, children found to be at risk can be given drug treatment that may prevent or diminish the severity of this disease.

Athlete's foot, a nuisance disease, is caused by a fungus. Most persons occasionally develop this condition, but it is easily treated. A small group of persons, however, suffer chronic athlete's foot. Research has indicated that these individuals share a common HLA genotype, suggesting a genetic component for susceptibility to persistent fungal infection. As with ankylosing spondylitis and leprosy, understanding of the genetic connection can mean early screening and treatment of the disease.

HLA and Forensic Medicine

HLA antigens have wide application in forensic medicine. Because the HLA system is so diverse, only a small fraction of the population shares any particular tissue type. While individual genetic uniqueness is a major barrier to transplant surgery, it also means that cell surface antigens provide each person with a "biological fingerprint" of great legal value.

As with ABO typing, HLA typing is commonly used in the estimated two hundred thousand paternity suits in the United States each year. The larger number of HLA alleles, however, greatly increases the probability of excluding from paternity males accused unfairly. In an interesting case, for example, a woman accused a man of fathering her twins. ABO, MN, and HLA types were obtained, revealing that the putative father's HLA antigens matched that of one twin, but *not* of the other (figure 4.14). Subsequently, the mother admitted involvement of a second man. His HLA type indicated that he was the probable father

TABLE 4.5

Some examples of HLA and disease associations*

Disease	Antigen	Frequency in patients (%)	Frequency in controls[†] (%)
Ankylosing spondylitis	B27	90	8
Celiac disease	B8	71	23
	A1	64	29
Insulin dependent diabetes	B8	37	22
	BW15	23	15
Multiple sclerosis	A3	36	26
	B7	34	24
	DW2	70	16
Psoriasis	B13	20	4
	B17	26	8
	B37	8	1
Ragweed hay fever (Ra5 sensitivity)	B7	50	19

Source: Data from Ryder, et al. (1974), and McDevitt and Bodmer (1974).
[†]The population at large.

Figure 4.13

Inheritance of phenotypes at the HLA-A and HLA-B loci linked on chromosome 6. The numbers designate alleles at the A locus (top) and B locus (bottom).

of the second twin. These nonidentical twins were the seventh documented case of **superfecundation,** the simultaneous birth of children fathered by different men. While rare in humans, superfecundation is common in other mammalian species. It requires that a female have intercourse with several males within a few hours of ovulating more than one egg. Superfecundation was first documented in humans in 1810 when a woman gave birth to twins, one white and the other mulatto.

In the previous case, neither ABO nor MN typing provided definitive information for establishing paternity. The mother, both twins, and both men were blood type A; the mother, one twin, and one man were MN, while the other twin and second man were type M, excluding neither male from paternity of either twin. The mother's HLA antigens were 2,44/ 11,27. Each twin had 2,44 from the mother. Additionally, one twin was 24,54 and the other 2,15. One man was 2,7/24,54, whereas the other was 2,3/2,15. Using HLA types, each man was positively excluded from fathering one child, while being strongly implicated as father of the other. The frequency of the 24,54/2,15 genotype required by a third male who could have fathered both children was estimated to be one in 140,000.

Figure 4.14

Pedigree showing ABO, MN, and HLA types that demonstrate the occurrence of *superfecundation*—a woman having twins, each with a different father.

This case also illustrates that while science can sometimes provide truth, it cannot enforce justice. After genetic analysis, one of the males accepted responsibility for *both* children, even though he was the biological father of only one of them. The mother dropped the paternity suit, and the second male was legally absolved of responsibility for the child he had fathered.

Autosomal Linkage

Simultaneous examination of two or more autosomal traits in humans usually shows that they obey Mendel's Law of Independent Assortment. Knowledge that genes controlling specific traits are linked, however, can sometimes permit detailed analysis that might not be possible otherwise, such as precise identification of the genotype of both parents.

The genes for both nail-patella syndrome and the ABO blood group are located on chromosome 9 (figure 2.10). Figure 4.15 shows two pedigrees for ABO-NP phenotypes. In figure 4.15a the mother is blood type B and normal (nn) for NP syndrome. The father has NP syndrome (Nn) and is type A. The two sons have different phenotypes: the first has NP syndrome and is type B; the second is normal and type AB. What are the parental genotypes?

The genotype of the second son must be ABnn. He must have received an An gamete from his father and a Bn gamete from his mother. The first son must also have received Bn from his mother, while receiving ON from his father. Knowing the linkage groups, we are thus certain that one of the father's chromosomes is ON, the other An. One of the mother's chromosomes is Bn, and the other is either Bn or On.

The pedigree in figure 4.15b is more complex. The father is a doubly recessive homozygote, OOnn, producing only On gametes. Because two of the children are type O and two are normal for nail-patella, the mother must be heterozygous AO, as well as Nn. Which of her alleles occur on the same chromosome?

The first daughter received On from her father and thus received An from her mother. The mother's second chromosome is therefore ON. Union of a maternal ON gamete with a paternal On gamete would produce the second daughter. A crossover between the two maternal loci could then produce an On gamete, giving rise to the son. Pedigree analysis, however, often does not give a definitive answer. In this case, the limited pedigree is also consistent with two other types of maternal gametes, AN and On. If this were the case, what children could have resulted from maternal crossover?

Figure 4.15

Pedigrees involving the linked genes, ABO and nail-patella. (Note that ABO phenotypes are given; individuals with magenta are those who have nail-patella syndrome.) (a) A pedigree where there is no indication of crossing-over; (b) a pedigree where crossing-over must have occurred.

Summary

1. Pedigree analysis is a common methodology of human genetics.
2. Genetic relatedness can be calculated between relatives and measures the proportion of shared alleles that are descended from a common ancestor. In general, relatedness decreases by a factor of one-half with each step in a relationship.
3. Consanguineous matings increase the probability of expressing recessive alleles.
4. The closer the relationship in a consanguineous mating, the more severe the effects, on average, of inbreeding depression. Incest taboos reflect societal recognition of the harmful effects of inbreeding.
5. Inborn errors of metabolism usually result from a deficiency or structural alteration of an enzyme in a metabolic or biochemical pathway.
6. Cystic fibrosis is the most common lethal genetic disease of American whites. Sickle-cell anemia is the most common lethal genetic disease of American blacks.
7. Frequencies of some polymorphic alleles differ greatly among populations. These genetic differences can be used to trace the history and origin of populations and races. For example, both cerumen type and ABO blood type have been used to show the origin of Native Americans from a Mongoloid stock, and to show the effects of Native Americans interbreeding with American whites.
8. Both red and white blood cells have many surface antigens that specify a variety of blood groups. Because these blood group phenotypes are inherited in a Mendelian fashion, they are often used to determine paternity. The more alleles that occur at a locus, the more useful it is in paternity testing. Many blood group alleles show associations with certain diseases.
9. Linkage relationships can sometimes be used to identify genotypes precisely.
10. Genetic counseling can help a family understand the genetic basis of disease and assist them in managing genetic conditions afflicting a family member. Often, genetic counselors can advise on the probability of recurrence of genetic diseases.

Key Words and Phrases

achondroplasia

agglutination

albinism

antibody

antigen

blood group

carrier

clone

codominance

coefficient of relationship

consanguineous

consanguineous mating

cystic fibrosis (CF)

dizygotic twins

forensic

genetic counseling

human leukocyte antigens (HLA)

identical by descent

inborn error of metabolism

inbreeding depression

incest taboo

index case

monozygotic twins

nail-patella syndrome (NP)

neurofibromatosis (NF)

pedigree

phenocopy

phenylketonuria (PKU)

pleiotropic

prenatal screening

sickle-cell anemia (SCA)

sickle-cell trait

superfecundation

therapeutic abortion

Questions

1. Why are humans so difficult to study genetically? What aspects of humans make them easier than some species to study genetically?
2. Could two albinos produce a normally-pigmented child? What genetic mechanism or mechanisms could produce such a result?

3. On the basis of ABO blood group evidence *alone*, should a type AB woman ever instigate a paternity suit? For the evidence to be most incriminating, what should be the ABO type of her child? If the woman were type O, would she be more or less likely to find ABO evidence in her favor than if she were type AB?

4. A man of type B is accused in a paternity suit of fathering a child, type A, by a woman of type O. Could he have been the father?

5. Suppose a woman of blood type B, M marries a man of type A, N. What possible ABO and MN types of children might they have?

6. Assume that eye color is controlled by one locus where allele B for brown eyes is dominant to allele b for blue eyes. If a female albino, genotype aaBB, marries a doubly heterozygous male, AaBb, what genotypes of children might they have? What will be the phenotypes associated with each genotype?

7. Why can the use of blood group evidence in paternity cases never *prove* that a particular male is the father of a child? How does the use of multiple blood groups increase the strength of such evidence in paternity cases?

8. If crossing-over occurred in the father in figure 4.13, what possible genotypes could be found among his children? If crossing-over occurred in both the mother and the father, how many different genotypes could be found among their offspring?

9. What is the coefficient of relationship of a person to an uncle whose identical twin is that person's father?

10. Approximately what proportion of genes are identical in two unrelated human beings? What is the coefficient of relationship (*r*) for these same two people? How does this modify your interpretation of *r*?

11. Autosomal dominant traits *never* skip generations. Nonetheless, it is possible that an autosomal dominant trait might *appear* to skip generations, i.e., the trait might be present in generation I, absent in generation II, and present again in generation III. How?

12. PKU can be treated successfully, but not cured. What is the difference between a treatment and a cure? What would a cure for PKU involve?

For More Information

Diamond, J. 1989. "Blood, Genes, and Malaria," *Natural History,* February: 8–18. (on sickle cell anemia)

Friedman, M. J., and W. Trager. 1981. "The Biochemistry of Resistance to Malaria." *Scientific American,* 244:154–66.

Guillery, R. W. 1974. "Visual Pathways in Albinos." *Scientific American,* 230:44–62.

Hartl, D. L. 1985. *Our Uncertain Heritage: Genetics and Human Diversity,* second edition. Harper and Row, Pub. Inc., New York.

Kidd, K. K. 1987. "Phenylketonuria: Population Genetics of a Disease." *Nature,* 327:282–83.

Loupe, D. E. 1989. "Breaking the Sickle Cycle." *Science News,* 136:360–62.

Luzzatto, L., and P. Goodfellow. 1989. "Sickle Cell Anemia—A Simple Disease with No Cure." *Nature,* 337:17–18.

Marx, J. L. 1985. "New Sickle Cell Test." *Science,* 230:1365 and 1350–54.

Marx, J. L. 1989. "The Cystic Fibrosis Gene is Found." *Science,* 245:923–25. (The papers describing the discovery of this gene are published in *Science,* 245:1059–80, 8 September 1989.)

McKusick, V. A. 1988. *Mendelian Inheritance in Man,* eighth edition. The Johns Hopkins University Press, Baltimore.

Quinton, P. M. 1990. "Righting the Wrong Protein." *Nature,* 347:226.

Ringe, D., and G. A. Petsko. 1990. "Cystic Fibrosis—A Transport Problem?" *Nature,* 346:312–13.

Roberts, L. 1990. "Cystic Fibrosis Corrected in Lab." *Science,* 249:1503.

Chapter

Genetics of sex

Sex Determination
Genic and Chromosomal Mechanisms of Sex Determination
The XX–XY System
A Variation: The ZZ–ZW System
The XX–XO System
Sex Determination by Ploidy
Box 5.1 Selfish Genes and Reproductive Mechanisms
Box 5.2 Sex Determination and Sea Turtle Conservation
Environmental Sex Determination
Sex Linkage
X Linkage
Red-Green Color Blindness
Hemophilia
Duchenne Muscular Dystrophy
Ichthyosis
Chicken Feathers: ZZ–ZW Linkage
Y Linkage
X Inactivation
Sex-Influenced Inheritance
Sex-Limited Inheritance
Sex Ratios

Figure 5.1

Sexual dimorphism in vertebrates: (a) Costa Rican frogs (the male is the smaller frog clasping the larger female from behind); (b) pair of wood ducks (the male is on the left); (c) mating pair of Southern Elephant Seals (*Mirounga leonina*); note differences in size as well as pronounced snout of the male, responsible for the common name of this species.

(a) Roy McDiarmid; (b) © Breck P. Kent/Animals Animals; (c) © Ben Osborne/ Animals Animals.

In many species, the most obvious phenotypic variation is associated with the sexes, **sexual dimorphism** (meaning two forms—figure 5.1). We seldom have difficulty distinguishing male and female humans, but in some species the differences between the sexes are dramatic. Male elephant seals, for example, are several times the size of females and bellow through a bulbous snout in defense of territory and harem (figure 5.1c). Males of many species possess elaborate ornamentation that appears to attract females (the peacock, for example). The sexes differ so strikingly in some fish that biologists first classified them as different species.

Sexual dimorphism is not universal among multicellular organisms, however. Many species of plants and invertebrate animals are **hermaphroditic** (from the combined names of the Greek god Hermes and goddess Aphrodite). In animals an hermaphrodite possesses both male and female gonads (testes and ovaries). The flowers of most plants, for example, have both male and female organs—each plant is both male and female. Hermaphrodites may be self-fertilizing, cross-fertilizing, or both. In some hermaphroditic annelid worms and snails, mating pairs exchange sperm, thus avoiding self-fertilization, which is the most deleterious form of consanguineous mating.

Some plants and many vertebrates have more or less clearly differentiated sexes. In some species, dimorphism is observable only in the gonads, the **primary sexual characters.** In the ginkgo tree, for example, the sexes can be distinguished only during flowering or when the female plants bear their foul-smelling fruit. Male ginkgoes are a very desirable ornamental tree, but the sexes cannot be distinguished until reproductive maturity is reached. Gardeners often must choose between chopping down a handsome shade tree or tolerating the annual nauseous stench when their promising sapling becomes a sexually mature female.

Traits other than the gonads that show sexual dimorphism are called **secondary sexual characters.** Examples in humans are the external genitalia, the internal reproductive tracts through which the gametes travel and in which the fetus develops in the female, patterns and coarseness of bodily and facial hair, patterns of fat deposition, hormonal balances, and size.

In many vertebrates, size is the only obviously dimorphic trait. Males, however, are not always the larger sex; perhaps more commonly, females are larger (figure 5.1a). In a few species such as the jacanas (a group of tropical aquatic birds), the female is not only larger but also gaudier than the male, and she maintains a harem of males whom she actively courts and defends. In many other species, the sexes are so externally similar that they can be told apart only by dissection and examination of the gonads.

Sex Determination

The genetic mechanisms associated with sex determination are extremely variable and few generalizations apply to all species. However, in those organisms with sexual reproduction and some degree of sexual dimorphism, the following principles usually apply.

1. Like all phenotypic characters, sexual phenotypes are governed by the interaction of genes and the environment. The sexual phenotype involves many morphological, physiological, and behavioral traits, indicating the probable involvement of many different genes on many different chromosomes.
2. Sex determination in many species is associated with a major dimorphism in one pair of chromosomes, the *sex chromosomes*. The sex chromosomes, however, function very differently in different species. Thus, no one mechanism explains sex determination in all species.
3. Sex determination appears to depend upon a developmental switch, that directs an individual to become either male or female. That switch may be: (1) *genic,* as in the case of the single gene (called *SRY*) located on the Y chromosome in humans and other placental mammals that initiates the developmental sequence leading to maleness; (2) *chromosomal,* as in species such as bees and wasps where the number of sets of chromosomes determines sex; or (3) *environmental,* as in many reptiles where the temperature of incubation determines sex, or as in fish where age and/or social interactions may determine sex.
4. The sex-determining genes direct the development of an individual into one of two pathways—male or female—leading to a sexual phenotype. In other words, each individual has the potential to develop into either sex. Gene mutations may cause an individual to develop into the sex opposite to that indicated by the chromosomes. Interruption of developmental pathways may cause an organism to display an ambiguous sexual phenotype.
5. A gene very similar to SRY in mammals has been identified in yeast, suggesting that the same genes may govern sex determination in many species, but that their precise mode of action differs among species.

Genic and Chromosomal Mechanisms of Sex Determination

By 1910, sexual phenotype had been shown to behave like a Mendelian character. Genes were hypothesized to occupy chromosomes (the Sutton-Boveri hypothesis), and a difference in one pair of chromosomes was shown to exist between the sexes in some species. Several persons—including Mendel himself in 1870—suggested that the 1:1 ratio of males to females at birth resembled the result of a testcross. This observation was essentially correct for humans, because the Y chromosome functions much like a dominant gene to produce maleness. However, other genetic mechanisms of sex determination have also been discovered, and much more has been learned about the Y chromosome.

Figure 5.2

(a) Chromosomes at prophase I (meiosis) of a human male. The 23 bivalents can be easily identified; the X and Y bivalent represents the end-to-end pairing of homologous regions on the X and Y chromosomes. (b) Crossing over between the human X and Y chromosome occurs only at the pseudoautosomal segment as shown.

(a) From James Thompson and Margaret Thompson, 1980. Genetics in Medicine, W. B. Saunders Company, Philadelphia. (b) Reprinted by permission from *Nature,* Volume 319, p. 258, January 23, 1986. Copyright © 1986 Macmillan Magazines Ltd.

(a)

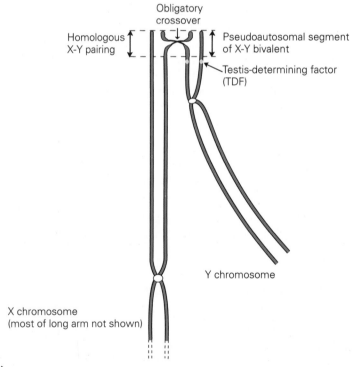

(b)

The XX-XY System

As noted in chapter 2, many species have one pair of sex chromosomes in which the homologues may differ in structure. Among most placental mammals, females possess two similar sex chromosomes, called the X chromosomes. Males, in contrast, usually possess one X chromosome and one chromosome dissimilar in morphology from all others, the Y chromosome. In humans, females are XX for chromosome pair 23, and males are XY. Because the male's sex chromosomes are different, the male is called the **heterogametic sex;** the female is the **homogametic sex.** In humans the Y chromosome is smaller than the X (figures 2.8 and 2.9).

Since 1959, the Y chromosome has been known to determine maleness in humans and other placental mammals. However, no genes could be identified on the Y chromosomes, so the precise functional mechanism of the Y chromosome remained an object of conjecture. In 1990, recombinant DNA methods finally allowed the identification of a gene in humans, mice, and other placental mammals that meets all the requirements of a sex-determining gene. Called the **SRY** gene (for sex-determining region of the Y) and located on the short arm of the Y chromosome (figure 5.2), this gene is believed to be identical with a long-postulated gene called *TDF* (for testis-determining factor). SRY (or TDF) is believed to cause the embryonic gonad to become a testis, whose hormonal products then induce the remaining components of the male phenotype. In mice, the SRY gene has been shown to be active at just the time that the gonad differentiates into a testis.

Almost certainly, many other genes located on the Y and on the autosomes control aspects of sexual differentiation, but they are normally governed primarily by the hormonal products of the gonad. Genes on the Y chromosome, for example, control spermatogenesis, but obviously can only function *after* the gonad has differentiated to become a testis.

Thus, in the **XX-XY system** the SRY gene on the Y chromosome determines maleness. This is supported by studies showing that individuals who are XX but male have the SRY gene incorporated into one of their X chromosomes. Similarly, an individual has been studied who is phenotypically female but has the short arm of the Y chromosome attached to one of her autosomes; usually the presence of the Y chromosome would make her a male, but in this instance the segment of the Y chromosome is lacking the SRY gene, thus making her female. The SRY gene, rather than the Y chromosome, is the key to maleness.

A Variation: The ZZ-ZW System

In birds and some insects, the male is the homogametic sex, and the female is heterogametic. In order not to confuse this type of sex determination with the XX-XY mechanism, the male chromosomes are labelled ZZ, the female ZW.

To be fully analogous to the XX-XY system, the **ZZ-ZW system** would require a gene on the W chromosome that induces femaleness, but such a gene has yet to be identified.

The XX-XO System

In some species, the number of *doses* of the sex chromosome (rather than the presence of a single sex-determining gene such as SRY) appears to govern sex. In the **XX-XO system,** which is common in insects, females have two sex chromosomes, just as in the XX-XY system, but the male has only one and is thus designated XO (O indicates the absence of an X). The diploid number of the male is thus one less than that of the female.

In contrast to the XX-XY system of placental mammals, it is hard to imagine a single gene unique to one sex that determines sex in an XX-XO system. The genetic, or molecular, basis for this apparently truly chromosomal sex-determining mechanism remains to be discovered.

Sex Determination by Ploidy

In some species of hymenoptera (bees and wasps), sex is determined by the number of sets of chromosomes, or **ploidy,** of the individual. Females are diploid, producing normal haploid gametes via meiosis. Most eggs are fertilized by haploid sperm from males, but a few are not. The fertilized eggs become females, while the unfertilized eggs develop into haploid males (figure 5.3). Haploid individuals, of course, cannot undergo normal meiosis, so males produce gametes via mitosis. The type of sex determination found in bees and wasps is called **haplodiploid.**

Because males do not use meiosis for gamete production, all the sperm from one individual are genetically identical to each other and to the male parent. This has the interesting consequence of increasing the genetic relatedness of a male's daughters. Remember that in this system a male produces no sons—only daughters. On average, all daughters of one mated pair of bees share 75% of their genes, rather than the normal 50% relatedness of siblings of

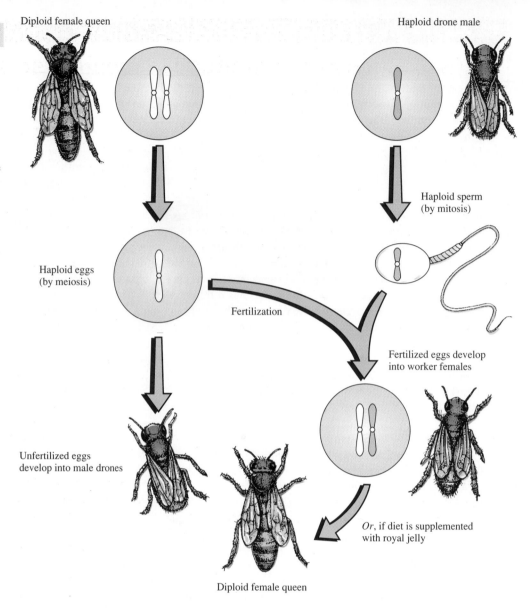

Figure 5.3

Sex determination in honeybees: a haplodiploid system. Females are diploid and males are haploid. The queen has been fed "royal jelly" and is larger than the average worker females who do not reproduce. Unfertilized eggs develop into haploid males (drones) whose only function is to provide sperm for the queen.

Diploid female queen

Haploid drone male

Haploid sperm (by mitosis)

Haploid eggs (by meiosis)

Fertilization

Fertilized eggs develop into worker females

Unfertilized eggs develop into male drones

Or, if diet is supplemented with royal jelly

Diploid female queen

most species. The daughters are identical for all 50% of the genes received from their father, plus one-half of the 50% of the genes from their mother, an additional 25%, or a total of 75% genetic relatedness.

There are thousands of species of bees and wasps, and many variations in their reproductive systems are known. One species of parasitic wasp sometimes uses a unique mechanism to achieve the haploid condition: information on a single chromosome causes the entire paternal haploid set in fertilized diploid eggs to condense and then be lost from the egg. Such offspring then develop into males (box 5.1).

In some vertebrates, unusual degrees of ploidy are associated with sex. Some species of lizards consist only of triploid (3N) females and no males. In fact, males would be superfluous, because the females have become completely *parthenogenetic*—a triploid female produces a triploid egg, which undergoes complete development without being fertilized. Haploid male bees are also the result of **parthenogenesis.**

BOX 5.1

Selfish genes and reproductive mechanisms

Figure 5.4

The parasitic wasp *Nasonia* in which the supernumerary chromosome *Psr* factor, causes fertilized (diploid) eggs to expel all paternal chromosomes, except *Psr*, resulting in all male progeny.

Samuel W. Skinner (Indiana University) and Don Alstad (University of Minnesota).

Evolutionary biologist Richard Dawkins achieved considerable public notoriety in 1976 with the publication of his book, *The Selfish Gene*. Dawkins argued that evolution can be understood most clearly from the perspective of the individual gene: evolution is like a game played by genes, where each gene selfishly "schemes" to reproduce itself as frequently as possible, even at the expense of other genes or the organism in which that gene may occur. Dawkins argues that living organisms are merely temporary vehicles to ensure efficient reproduction of genes.

Dawkins' views have not met universal acceptance. In particular, many people do not accept that human behavior is inevitably guided by selfish internal forces beyond the individual's control. To accept Dawkins' views, they claim, provides too easy an excuse for bad behavior: "I couldn't help myself—my genes made me do it!"

On the other hand, a few extraordinary genetic systems strongly support Dawkins's perspective that genes sometimes behave quite selfishly, ensuring their own reproduction at the expense of other genes and the organisms whose cells they inhabit. One such case is the parasitic wasp *Nasonia vitripennis*. This species has a haploid number of five chromosomes and haplodiploid sex determination; unfertilized eggs become males, while fertilized eggs become females.

Sex determination in *N. vitripennis*, is complicated because some, but not all, males have an additional *supernumerary* chromosome. Sometimes also called *B chromosomes*, supernumerary chromosomes are much smaller than normal chromosomes, and their functions (if any) are not well understood. In *N. vitripennis*, the B chromosome is called a *psr* factor, for paternal sex ratio, and females mated to *psr* males produce only male offspring.

This happens because, following fertilization, a gene or genes on the *psr* factor causes the entire paternal haploid set to condense and become lost from the egg. The *psr* factor is not lost, however. Thus, the egg goes from a diploid to a haploid condition, causing it to become male and to carry the *psr* factor.

This mode of transmission of the *psr* factor is entirely consistent with Dawkins' selfish gene hypothesis. Carriers of *psr* are always male (because *psr* ensures that they are haploid), and all the offspring of a carrier will be male carriers. However, transmission of *psr* ensures that no other genes of the male are transmitted, because the male's entire haploid set is eliminated. Thus, the gene secures its own reproduction, but it does so at the expense of all other genes carried by the male. In agreement with Dawkins's model, a single gene appears to be using a whole organism entirely for its own benefit (i.e., reproduction), while the organism receives no equivalent evolutionary reward.

BOX 5.2

Sex determination and sea turtle conservation

(a)

(b)

Knowledge of the reproductive systems of animal and plant species has direct application to agriculture and conservation. Environmental sex determination, for example, is a relatively unusual phenomenon in nature, especially among vertebrate animals. It was first discovered in turtles in the 1960s, and now is also known in a number of egg-laying vertebrates. Temperature-dependent sex determination is crucial information in the fight to save some species of sea turtles (figure 5.5).

Sea turtles spend virtually all their lives at sea, except for one night each year when the females laboriously haul themselves onto a beach to dig nests and lay eggs, before returning to the sea. Because turtles are so vulnerable at this time, many have been taken as food by humans at turtle nesting areas throughout the tropics, and turtle populations declined dramatically in the past century.

Conservationists began collecting eggs from sites where endangered species nested. These eggs were incubated in captivity, but many of these early efforts were of limited success. Because the eggs were carefully incubated at only one temperature, all the artificially reared young were of one sex.

Since the discovery of temperature-dependent sex determination, conservationists now incubate eggs at several temperatures or in thermally fluctuating environments that mimic the variation found at nesting sites in nature. Thus, young of both sexes are produced, and the future of these species is more secure.

Environmental Sex Determination

In some invertebrates and fish, sex is determined by the environment of the larva or juvenile. In the marine annelid worm, *Bonellia,* a larva may establish a parasitic relationship with an adult female, in which case it becomes a male. If it encounters no females, the larva develops into a female itself.

A similar phenomenon occurs in some deep-sea fish. In these species, the larva actually burrows into the side of a female and becomes little more than a parasitic testis, barely recognizable as a separate individual. These fish occur in very low numbers in the deep ocean. Such a mechanism probably maintains equal numbers of each sex and guarantees each mature individual a mate.

In other marine fish, all individuals are born as females and upon maturity become a member of one male's harem. Each male defends his harem and his territory, a coral outcrop. Females within a harem establish a dominance hierarchy. When the male dies, the highest-ranking female undergoes a sex change, becoming the male and assuming possession of the harem. Other species are known in which sex reversal occurs at a certain size or age, independent of dominance hierarchies.

TABLE 5.1

X-linked traits in humans

cleft palate   Incomplete cleft of the secondary palate.

color blindness, partial   Deutan series, insensitivity to green; accounts for roughly 75% of color blindness.

color blindness, partial   Protan series, insensitivity to red; accounts for roughly 25% of color blindness.

congenital night blindness   Night blindness associated with nearsightedness.

deafness   At least three separate X-linked forms of deafness have been described for congenital and progressive deafness.

Duchenne muscular dystrophy   Onset of muscle wasting by age six, chair ridden by age twelve, and death usually by age twenty. A heart defect is associated with this X-linked form of muscular dystrophy (figure 5.9a).

ectodermal dysplasia   Absence of teeth and sweat glands, accompanied by scanty body hair and early baldness (figure 5.15).

glucose-6–phosphate dehydrogenase deficiency   Reduced activity of a key enzyme of glucose metabolism. Usually this condition is not serious, but when persons of this genotype ingest certain drugs, their red cells lyse, resulting in anemia. Removal of the offending substances brings complete recovery.

hemophilia A, classical hemophilia   Defect in antihemophilic factor.

hemophilia B, Christmas disease   Defect in Christmas factor, plasma thromboplastic component.

ichthyosis   Rough, scaly skin; present from birth (figure 5.9c).

Lesch-Nyhan syndrome   Enzyme deficiency resulting in mental retardation, palsy, and self-mutilating behavior, involving compulsive biting of fingers and lips (figure 5.9b).

ocular albinism   Absence of pigment in the eye accompanied with rolling of the eyeball.

retinitis pigmentosa   Degeneration of the retina of the eye.

testicular feminization syndrome   XY males with female external genitalia, breast development, blind vagina, and abdominal or inguinal testes. — no testosterone receptors.

vitamin D resistant rickets   A dominant X-linked mutation producing a defect in vitamin D metabolism. This disease is not cured by high doses of vitamin D (figure 5.9d).

Xg(A) blood group antigen   A blood group antigen inherited as a dominant X-linked trait. This antigen is useful as a genetic marker for linkage studies and for determining where nondisjunction occurred in cases of X chromosome aneuploidy.

From V. A. McKusick, *Mendelian Inheritance in Man,* 8th edition. Copyright © 1988 Johns Hopkins University Press. Reprinted by permission.

Physical, rather than biological, factors may also influence sex determination. For many species of reptiles, eggs incubated at relatively high temperatures become one sex; at low temperatures, the other sex is produced (box 5.2). This type of sex determination may have arisen several times in reptiles, as high temperatures produce males in some species, but females in others.

Sex Linkage

Genes on the sex chromosomes are **sex linked;** a gene that is on only the X chromosome is **X linked;** one found only on the Y chromosome is **Y linked** or **holandric** (meaning entirely male). Homologous genes that occur on both the X and Y chromosomes are *partially X linked* or **pseudoautosomal.** They should exhibit the same pattern of inheritance as autosomal genes. Pseudoautosomal segments of the X and Y chromosomes exhibit some pairing at meiosis (figure 5.2).

X Linkage

During meiosis in human males, the X and Y chromosomes do not line up side by side in normal synapsis. Instead, they pair end to end. Because of dissimilarity in gross morphology as well as the absence of normal pairing, the X and Y chromosomes probably share few genes.

At least 140 genes have been positively assigned to the X chromosome (table 5.1), and 171 more are suspected of being X linked. Because so few genes are known on the Y chromosome, the terms "X linked" and "sex linked" are often used synonymously.

X-linked traits have a unique mode of inheritance because females have two doses of an X-linked gene, while males have only one. Males are thus **hemizygous** for X-linked traits. Dominant X-linked genes are always expressed in both sexes, just as with autosomal traits.

Figure 5.6

The crisscross pattern of inheritance of the sex chromosomes. The two X chromosomes of the female and the one X chromosome of the male are distinguished by their color; they can carry the same or different alleles of all the X-linked genes. The mother contributes one of her contributes one of her chromosomes to her sons, who will express any sex-linked genes on the X because they are hemizygous. The daughters receive an X from each parent and will not express their father's X-linked genes, unless they are dominant X-linked genes or unless they also receive the same X-linked recessive from their mother. The crisscross refers to the mother-to-son, and the father-to-daughter, transmission pattern.

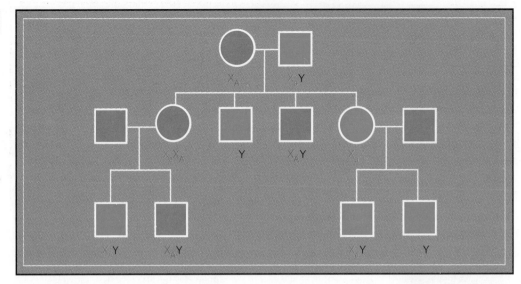

Because females have two X chromosomes, the frequency of dominant X-linked traits is always higher in females than in males. Such traits are expressed in both homozygous and heterozygous females.

The opposite is true for recessive alleles. Because they are hemizygous, males always express recessive X-linked alleles. Females, however, express recessive alleles only when they are homozygous. Thus the frequency of X-linked recessive traits is always lower in females than males.

Most X-linked genes have been discovered because they have recessive alleles with deleterious effects appearing in males. There are probably many common recessive traits whose effects are innocuous and thus escape the attention of geneticists.

Males transmit their X chromosome to every daughter, their Y chromosome to every son. Recessive X-linked traits thus show a pattern of inheritance in which the phenotype is usually expressed only in males of alternate generations. A male bearer transmits the recessive allele to a daughter, who does not express the allele because it occurs in the heterozygous condition. However, each of her male offspring has a 50% chance of receiving that allele and expressing the phenotype. The trait should thus appear in $\frac{1}{4}$ of her offspring ($\frac{1}{2}$ of her offspring are expected to be male, and $\frac{1}{2}$ of her sons receive the recessive allele: $\frac{1}{2} \times \frac{1}{2} = \frac{1}{4}$). The heterozygous female is a *carrier* of the allele. X-linked inheritance is often said to show a *crisscross* pattern of inheritance because of its pattern in pedigrees: the allele is transmitted male to female, female to male, but the trait is expressed only in males in alternate generations (figure 5.6).

Red-Green Color Blindness

Many persons cannot see certain colors. The most common such defect is an inability to distinguish red from green. This condition, *partial* **color blindness,** affects more than five million American males. Two X-linked loci are now known to cause partial color blindness. One locus causes *deutan-type color blindness* (about 75% of the cases), while the other causes *protan-type color blindness.* Each gene affects different pigments in the retina, but both produce confused color perception. A far rarer condition is total color blindness, inherited as an autosomal recessive, where persons have no color perception at all.

Partial color blindness was recognized as an hereditary condition in the eighteenth century, but the mode of inheritance was not demonstrated until 1911. The recognition that genes are located on chromosomes soon led to association of partial color blindness and the X chromosome, both of which show the crisscross pattern of inheritance. Color blindness and hemophilia, both X linked, were the first genes in humans assigned to a specific chromosome. In 1986 the genes corresponding to the red, green, and blue visual pigments were isolated and sequenced.

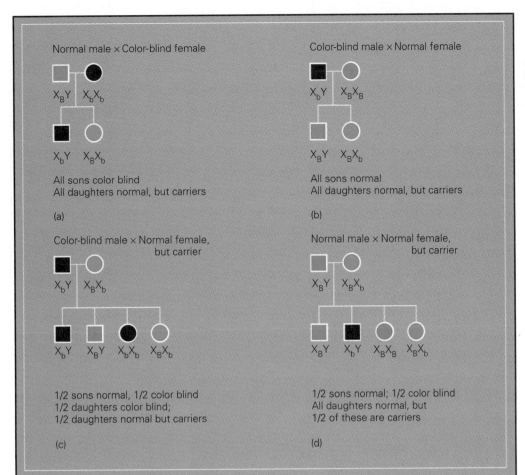

Figure 5.7

Four possible matings involving inheritance of the sex-linked recessive trait, partial color blindness: (a) normal male × color-blind female; (b) color-blind male × normal female; (c) color-blind male × normal female who is a carrier; (d) normal male × normal female who is a carrier.

Normal male × Color-blind female

X_BY X_bX_b

X_bY X_BX_b

All sons color blind
All daughters normal, but carriers

(a)

Color-blind male × Normal female

X_bY X_BX_B

X_BY X_BX_b

All sons normal
All daughters normal, but carriers

(b)

**Color-blind male × Normal female,
but carrier**

X_bY X_BX_b

X_bY X_BY X_bX_b X_BX_b

1/2 sons normal, 1/2 color blind
1/2 daughters color blind;
1/2 daughters normal but carriers

(c)

**Normal male × Normal female,
but carrier**

X_BY X_BX_b

X_BY X_bY X_BX_B X_BX_b

1/2 sons normal; 1/2 color blind
All daughters normal, but
1/2 of these are carriers

(d)

Patterns of sex-linked inheritance do *not* conform to Mendel's Law of Parental Equivalence. Figure 5.7 shows the results of reciprocal matings of affected and unaffected parents. The mating of a normal male with a color-blind (recessive homozygous) female produces normal but heterozygous daughters. All sons have the disease (figure 5.7a). The reciprocal mating demonstrates crisscross inheritance. A color-blind male (hemizygous) mated with a normal (homozygous) female produces no affected offspring; the sons receive a normal X from their mother, while all the daughters are carriers (figure 5.7b). A color-blind male mated with a carrier female is analogous to a testcross: 50% of the offspring of both sexes are normal, 50% are color blind (figure 5.7c). Another possibility, a normal male mated with a carrier female, produces all normal female offspring (half of whom are carriers), but 50% of male offspring are affected (figure 5.7d). The mating of two color-blind individuals results in all color-blind offspring, if they have the same type of color blindness.

Hemophilia

Hemophilia, also known as "bleeders' disease," is a condition characterized by the inability to form blood clots. An affected individual is in continual danger of bleeding to death from the smallest cut or from internal hemorrhaging after even a minor bruise. Rheumatism is a common complication due to bleeding within the joints.

At least three different genetic loci control different forms of hemophilia. The rarest is controlled by an autosomal recessive gene, but the other two result from recessive alleles at two X-linked loci. One of these, *Christmas disease,* or *hemophilia B,* comprises about 20% of all hemophilia and is caused by an abnormality or deficiency of plasma thromboplastin component.

Figure 5.8

The Royal Hemophilia. Partial pedigree of the descendants of Queen Victoria showing carriers (green) and hemophiliacs (black), as well as normals (blue) and possible carriers (magenta). The present royal family of Great Britain is free of the hemophiliac allele. The photograph is of Czar Nicholas II, the Czarina Alexandra, their four daughters, and their son Alexis, who had hemophilia.

Historical Pictures Service, Chicago.

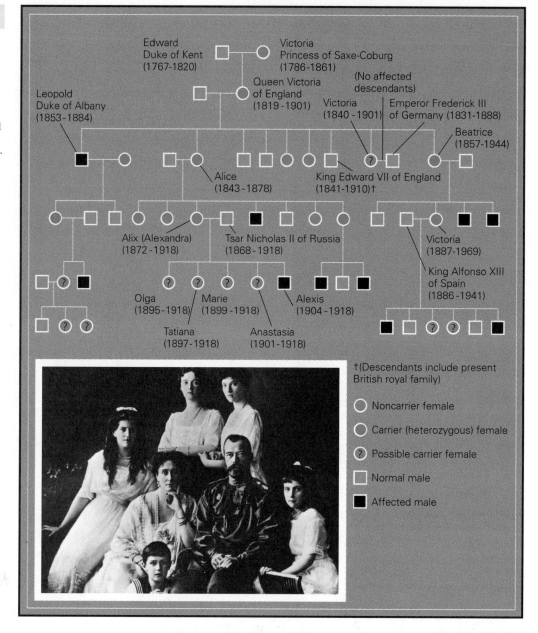

Edward Duke of Kent (1767-1820)

Victoria Princess of Saxe-Coburg (1786-1861)

Queen Victoria of England (1819-1901)

(No affected descendants)

Victoria (1840-1901)

Emperor Frederick III of Germany (1831-1888)

Leopold Duke of Albany (1853-1884)

Beatrice (1857-1944)

Alice (1843-1878)

King Edward VII of England (1841-1910)†

Alix (Alexandra) (1872-1918)

Tsar Nicholas II of Russia (1868-1918)

Victoria (1887-1969)

King Alfonso XIII of Spain (1886-1941)

Olga (1895-1918)

Marie (1899-1918)

Alexis (1904-1918)

Tatiana (1897-1918)

Anastasia (1901-1918)

†(Descendants include present British royal family)

○ Noncarrier female

◔ Carrier (heterozygous) female

? Possible carrier female

□ Normal male

■ Affected male

Hemophilia A, classical hemophilia, is caused by an abnormality or deficiency in antihemophilic factor, AHF. Until recently, hemophilia A was untreatable; only about 25% of affected males reached age twenty-five. Treatment with AHF now results in a more normal life span, but the cost is high (six thousand dollars to ten thousand dollars per year).

The frequency of hemophiliacs is about one in ten thousand males, but much lower in females, perhaps one in 100 million or less. A female hemophiliac can result only from the mating of a heterozygous female with an affected male. Such a mating is highly unlikely because, until recently, few male hemophiliacs survived long enough to reproduce. Females were also believed to die at the onset of menstruation, but some female hemophiliacs have now survived to maturity.

Hemophilia may have been understood as an inherited disease longer than any other genetic condition. The Talmud, the Hebrew book of law, noted that when excessive bleeding occurred during circumcision of two male infants of one mother, future male offspring were exempt from this procedure. When sons of three sisters exhibited bleeding, sons of other sisters were also exempt. However, sons of brothers were not exempt, implying an understanding of the crisscross pattern of inheritance.

(a)

(b)

Figure 5.9

Examples of conditions caused by sex-linked genes. (a) Young boy exhibiting muscle deterioration typical of Duchenne muscular dystrophy; (b) a boy with Lesch-Nyhan syndrome, characterized by compulsive self-destructive behavior; (c) a boy afflicted with the skin disorder ichthyosis; (d) a mother who is heterozygous for the dominant vitamin D-resistant rickets and her children.

(a) Stanley Willard; (b) with permission of W. L. Nyhan and N. A. Sakati. *Diagnostic Recognition of Genetic Disease:* Lea & Febiger, 1987, Philadelphia; (c) Lester V. Bergman and Associates, Inc.; (d) E. W. Lovrien.

(c)

(d)

   Hemophilia A has been called the "Royal disease" because it affected males in the royal families of Europe. Queen Victoria, a carrier of the hemophilia allele, had nine children (figure 5.8). Her eighth child, Leopold, was a hemophiliac who died at age thirty-one. Her other three sons were unaffected, so we know they did not receive the allele. One daughter had no children; her status as a carrier cannot be assessed. Two daughters had children, none of whom were hemophiliac, indicating that the mothers probably were not carriers. Two other daughters were carriers giving birth to hemophiliac sons.

   The possible historical influence of hemophilia is tantalizing. Victoria's third child was Princess Alice, whose daughter Alix married Czar Nicholas II of Russia. The Czarina, Alix, had four daughters before giving birth to the long-awaited son, Alexis, heir to the Russian throne. Unfortunately, Alexis had the hemophilia allele, a legacy from his great-grandmother, Queen Victoria. Distressed over their son's condition, the Czar and Czarina turned to the monk Rasputin, a charlatan who claimed he could help Alexis. While affairs of state deteriorated, culminating in the Russian revolution, the Czar was preoccupied with the health of his son and manipulated by Rasputin.

Figure 5.10

A pedigree demonstrating the occurrence of an X-linked recessive trait, ichthyosis, in females, as a result of consanguineous mating.

Among Victoria's descendants, eight of twenty-five males in four generations were hemophiliac. Queen Victoria almost certainly received the gene for hemophilia A as the result of a mutation on the X chromosome she received from her father, Edward, Duke of Kent. He was fifty-two years old at the time of Victoria's birth, and, the frequency of such mutations may be increased in the germ cells of older males.

Hemophilia is a disease for which direct genetic treatments or cures are particularly desirable. Because traditional treatments involve transfusions of blood or blood products, hemophiliacs are at risk of acquiring such serious diseases as hepatitis and acquired immune deficiency syndrome (AIDS) as a result of treatment. In a few tragic instances, male hemophiliacs have developed AIDS as a result of transfusions, subsequently transmitting the AIDS virus to their wives, who have then transmitted it to one of their children. In most countries, donated blood is now routinely screened for hepatitis and AIDS, but these tests are never 100% effective, so a risk always exists.

 Duchenne Muscular Dystrophy

Most forms of muscular dystrophy are controlled by autosomal genes. **Duchenne muscular dystrophy (DMD)**, however, is controlled by an X-linked recessive allele and thus primarily affects young males (figure 5.8a). Muscular deterioration begins between the ages of three and five years; by the time affected individuals reach their teens, they are confined to wheelchairs; and they die in their early twenties from atrophy of the respiratory muscles. Few affected males reproduce, so the condition is transmitted mainly by female carriers. Duchenne muscular dystrophy occurs in roughly one in every 4,000 newborn males; perhaps one-third of these are due to new mutations. This is several times more frequent than the occurrence of hemophilia; other X-linked disorders are far rarer.

In 1986 the defective gene that causes Duchenne muscular dystrophy was identified. This led to the identification in early 1988 of a protein (dystrophin) whose absence causes DMD. While this finding does not constitute a cure for DMD, it offers new research avenues that may eventually allow treatment. As with hemophilia, a longer-term hope may be direct modification of the defective DNA.

Ichthyosis

Consanguineous matings can greatly increase the frequency of expression of X-linked traits in females. In consanguineous pedigrees containing X-linked recessive alleles, females have a high probability of carrying the X-linked allele, as they can receive the allele from either parent. In turn, matings of carrier females and affected males (figure 5.7c) produce daughters and sons with an equal likelihood (50%) of being affected.

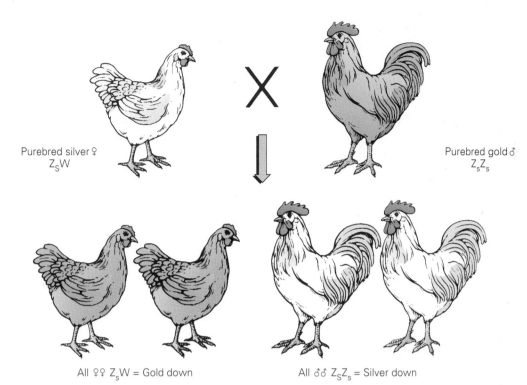

Figure 5.11

Feather color in chickens, a sex-linked trait used by poultry breeders to select breeding chickens. Down feather color is controlled by a pair of alternate alleles, S, producing a silver color down, and s, producing a gold color. Gold-feathered males are bred with silver-colored females. All male offspring will have silver down and all female offspring will have gold down. Thus breeders can select the females for raising, rather than raising all chicks.

Purebred silver ♀
Z_SW

Purebred gold ♂
Z_sZ_s

All ♀♀ Z_SW = Gold down

All ♂♂ Z_SZ_s = Silver down

Ichthyosis is a disorder characterized by extreme dryness, roughness, and scaliness of the skin (*ichthy* meaning fish-like). One form, usually present at birth, is caused by an X-linked recessive allele (figure 5.9c). Figure 5.10 shows a pedigree of a family in which consanguineous mating resulted in expression of this X-linked trait in females.

The allele for ichthyosis first appeared in male I-1. His first four daughters all transmitted the allele, without expressing it themselves. Male III-11 mated with his first cousin, once removed, IV-3, who must have been a carrier, having received the ichthyosis allele through two generations of females. One of their three sons and two of their three daughters exhibited ichthyosis, a highly unlikely result without consanguineous mating.

Chicken Feathers: ZZ-ZW Linkage

Sex-linked patterns of inheritance in birds are identical to those in mammals, except that the female, not the male, is the hemizygous sex. A dominant Z-linked allele, designated S, has been discovered in chickens. Bearers of the S allele have silver-colored down at hatching. The alternate allele, s, produces a gold down feather. This locus has been put to practical use by many chicken farmers.

For egg production, male chicks are of no value. In fact, males cost the farmer money, but the sexes cannot be distinguished until they reach sexual maturity. The S allele can be used to distinguish the sexes immediately after hatching, thus permitting farmers to dispatch males and raise only females (figure 5.11). If a hemizygous (SW) silver female is mated to a gold (ss) male, crisscross inheritance produces an obvious phenotypic distinction between the sexes. Female chicks are all gold (sW), while males are silver (Ss). Males can then be sorted at birth, saving the farmer the expense of raising them to maturity.

Y Linkage

Y linked, or *holandric,* genes should have the simplest possible pattern of inheritance. Because only males have a Y chromosome, Y-linked traits should appear only in males. Because they are hemizygous, every Y-linked gene should be expressed. Each male contributes his one and only Y chromosome to every male offspring, so Y-linked traits should show strict father-to-son inheritance.

The only confirmed Y-linked gene is the SRY gene that governs differentiation of the gonad to a testis in males. This gene had long been postulated to be on the Y chromosome,

Figure 5.12

Barr bodies are darkly staining structures seen in the nuclei of nondividing cells of normal females. They are inactivated X chromosomes. Turner females (XO) do not have Barr bodies, while Klinefelter males (XXY) have a single Barr body. XXX females have two Barr bodies. (a) Normal male, no Barr body; (b) normal female, one Barr body; (c) XXX female, two Barr bodies; (d) XXXX female, three Barr bodies; (e) Klinefelter male, one Barr body.

(a and b) Wilson and Foster, *Williams Textbook of Endocrinology,* 7th ed. © W. B. Saunders 1985; (c–e) National Jewish Hospital and Research Center.

(a)   (b)

(c)   (d)

(e)

as it is difficult to imagine that the trait of maleness could not be Y-linked. Genes affecting spermatogenesis, stature, and other aspects of maleness are also likely to be on the Y chromosome.

X Inactivation

It is sometimes possible to distinguish the phenotypes of individuals who are homozygous or heterozygous for autosomal dominant alleles. Consider a gene producing compound A. Individuals who are AA normally produce twice as much of this compound as individuals who are Aa. Thus, it is reasonable to ask if homozygous normal females produce twice as much gene product for X-linked genes as hemizygous normal males. Alternatively, are females homozygous for a deleterious X-linked allele affected twice as severely as affected males? The answer appears to be no: X-linked traits are usually phenotypically similar in males and females, due to a phenomenon known as **dosage compensation.**

Early in embryonic development in female mammals, one of the two X chromosomes is turned off or inactivated. In spite of possessing twice as many X chromosomes as a male, a female is functionally hemizygous at X-linked loci, just like males.

Giemsa staining of somatic cells reveals **X inactivation** as a distinct sexual dimorphism. In cells of females, one darkly staining region is found at the edge of the nucleus (figure 5.12). This region, named a **Barr body** (after its discoverer, M. L. Barr), is the inactivated

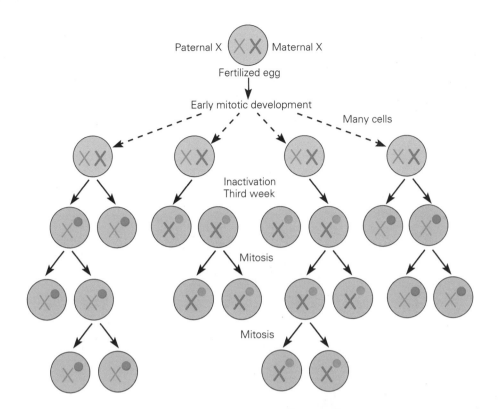

Paternal X **X X** Maternal X

Fertilized egg

Early mitotic development

Many cells

Inactivation
Third week

Mitosis

Mitosis

Figure 5.13

Barr body formation in
human females. During the
third week of embryonic
development, one of the two
X chromosomes in each cell
becomes inactivated to form a
Barr body. All descendants of
these cells have the same
chromosome inactivated, so
females are functional
mosaics for their maternally-
derived and paternally-
derived X chromosomes. In
the germinal tissues Barr
bodies are not seen. Both X
chromosomes may need to be
functional in these cells.

heterozygous female

Figure 5.14

Tortoiseshell cat. The patches
of rust and black fur result
from random inactivation of
X chromosomes bearing
alternate alleles for the two
colors. The amount of white
fur is determined by alleles at
a different locus.

© William Ferguson.

X chromosome. One X chromosome is essential for normal development in both sexes, but
if an individual (of either sex) has more than one, all but one are inactivated and visible in
stained somatic cells as a Barr body. Thus, somatic cell nuclei of normal males have no Barr
bodies, and those of normal females have one.

The hypothesis that all but one X chromosome are inactivated in each cell was proposed
by geneticist Mary Lyon in 1961 and is known as the **Lyon hypothesis.** Crucial evidence
supporting this hypothesis was provided by sexually aneuploid individuals. **Aneuploidy**
(meaning not true number) refers to the possession of an abnormal number of chromosomes.
Aneuploid individuals have the normal diploid number, *plus or minus* one or more chromosomes. Females lacking one X chromosome exhibit **Turner's syndrome,** designated 45, XO
(for 45 chromosomes, with one X missing). Males with an extra X have **Klinefelter's syndrome,** designated 47, XXY. Cells from 45, XO females have no Barr bodies, while those
from 47, XXX females have two; 47, XXY males have one Barr body, 48, XXXY males
have two (figure 5.12).

exclude

Testing for the number of Barr bodies is simple and inexpensive compared to preparation
of a complete karyotype. Examination of Barr bodies can be done easily to screen for sex-
chromosome abnormalities. The mandatory "sex tests" that have been required of Olympic
athletes include a count of Barr bodies in epithelial cells scraped from the inside of the cheek.
Males disguised as females have no Barr bodies.

Figure 5.15

Mosaic phenotypes in three generations of females, all heterozygous for the X-linked gene for ectodermal dysplasia. Stippling indicates areas in which sweat glands are missing. The affected areas differ in each woman because the controlling gene is inactivated randomly, even in the identical twins in generation III.

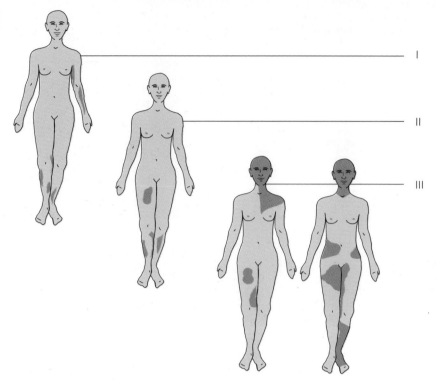

The inactivation of X chromosomes during development apparently occurs at random. Early in development, the maternally-derived X is inactivated in some cells, while the paternally-derived X is inactivated in others. Thereafter, descendants of a particular cell have the same X inactivated (figure 5.13). If a female is heterozygous for an X-linked gene, she is **mosaic** for that trait. One of her X chromosomes is active in roughly half of her cells, while the second X is active in her other cells.

Tortoiseshell cats exhibit mosaicism due to dosage compensation. Several loci control coat color in cats, but only one X-linked locus is involved in producing tortoiseshell individuals. Two alleles occur at that locus, R and R′. In hemizygous males, R produces rust coat color, R′ black. In heterozygous females, X inactivation produces clones of R-bearing rust fur intermixed with R′-bearing black fur—the tortoiseshell cat (figure 5.14). Thus, almost all tortoiseshell cats are females. The only male tortoiseshell cats result from sex-chromosome aneuploidy. XXY males also undergo X inactivation, so an occasional tortoiseshell male occurs.

When a female is heterozygous for a deleterious X-linked allele, the effects of the cell lines bearing the normal allele may compensate for the harmful effects of the cell lines bearing the deleterious allele. In females heterozygous for partial color blindness, for example, some cell clones in the retina are in fact color blind, but the presence of other normal clones results in normal color vision.

X inactivation in humans can sometimes be seen in females heterozygous for certain X-linked traits. In 1875 Darwin described *ectodermal dysplasia,* a condition now known to be X linked that involves the lack of some teeth and sweat glands. Heterozygous females show a mosaic of areas of the jaw with and without teeth, and patches of skin with and without sweat glands. In figure 5.15, the females in generation III are identical twins; they have developed from a single fertilized egg and are therefore genetically identical. However, due to the *random* inactivation of different X chromosomes during development, they show considerable difference in the location of patches of skin lacking sweat glands. X-chromosome inactivation is one example of a developmental process that can produce *phenotypic* differences in *genotypically* identical individuals. Thus, even clones may differ significantly.

Figure 5.16

Rocky Mountain sheep showing sexual dimorphism in horn development. A sex-limited gene on the autosomes causes the large horns in males. Females also carry this gene, but do not express it.

Science VU/Visuals Unlimited.

Sex-Influenced Inheritance

Two types of inheritance exhibit sexually related effects, although neither is sex linked. **Sex-influenced** genes show differing patterns of expression in each sex—usually, the trait behaves as a dominant in one sex and a recessive in the other. Sex-influenced genes occur only on autosomes.

The best documented example of sex-influenced inheritance involves *pattern baldness*. Individuals expressing pattern baldness begin to lose their hair relatively early in life, often in their twenties. Affected individuals are not totally bald: a distinct rim of hair surrounds the head in patterns varying from person to person. This trait is controlled by an autosomal gene. The allele for pattern baldness is dominant in males, and heterozygous males therefore express pattern baldness. In females, however, the gene is recessive.

Sex-Limited Inheritance

Some autosomal genes are expressed only in one sex or the other and are thus **sex limited.** The most obvious examples are the external genitalia and other secondary sex characteristics. Only females, for example, possess fully developed breasts, and only males develop full beards. Nonetheless, the genes controlling the development of these characters occur in both sexes, although the traits are only expressed in one. In chapter 6, we describe how all individuals possess the potential for developing the phenotype of either sex, indicating clearly that all of us possess genes for traits of the opposite sex that are normally never expressed. The horns of male sheep and goats are a non-human example of sex-limited traits (figure 5.16).

Sex Ratios

Because sex is determined by the Y chromosome, and because males produce X- and Y-bearing gametes in approximately equal numbers, Mendel's Law of Segregation predicts that the sexes should occur in equal proportions, or as commonly expressed, in a male:female ratio of 1:1. In most species most of the time, the number of males and females is about equal, but that is not always the case. In humans, for example, different **sex ratios** occur for different age groups. The **primary sex ratio** is the sex ratio at conception. It is extremely difficult to measure, but it certainly favors males. Estimates as high as 1.60:1.00 have been made. The **secondary sex ratio** (at birth) in white Americans is about 1.06:1.00, again favoring males.

Tertiary sex ratios may be measured at any time after birth. In humans, the sexes are about equally abundant around age twenty, but females increasingly outnumber males at later ages. The mean life expectancy of white females in the United States is about ten years longer than for males, which is reflected in the unbalanced sex ratio at later ages.

Several factors seem to affect the sex ratio. For unknown reasons, males of many species have a higher death rate than females at all ages. In humans, it has long been postulated that male hemizygosity for the X chromosome could be responsible. Any deleterious X-linked genes are expressed in males, but may be hidden in females, if recessive. That hypothesis has not been proven, however, and there is some evidence against it. Remember that in all female cells except oocytes, one of the X chromosomes is inactivated (the Barr body), so that deleterious genes on the active X should be expressed, just as in males. Further, in many species without dimorphic sex chromosomes, males still have a higher death rate.

Natural selection apparently favors equality in the sex ratio at the time of sexual maturity. The ratio is 1:1 in humans at about age twenty, for example. Thus, the proportion of males at birth is increased sufficiently that the higher male death rate produces a balanced sex ratio at about puberty.

Given the apparently equal production of X- and Y-bearing gametes by meiosis, how can the sexes occur in other than equal proportions? As usual, both environmental and genetic factors can alter the balance. As described earlier, in sea worms and deep-sea fish, juveniles develop into a female if they cannot find a female upon whom they can become parasitic males. Such a mechanism virtually guarantees one male per sexually mature female. As we noted earlier, the temperature at which eggs of some turtles are incubated alters the proportion of the sexes produced.

Several genetic factors can also alter sex ratios. Any sex-linked recessive lethal gene causes the death of hemizygous males, thus unbalancing the ratio in favor of females. Additionally, single genes are known in many species that cause unequal production of gametes with respect to different characteristics.

Finally, physical characteristics of X- and Y-bearing sperm may make a difference. Y sperm are lighter and thus may swim faster. If Y sperm regularly reach the egg first, the probability of male offspring should be increased. Dairy cattle breeders have attempted to use the difference in weight to separate the two types of sperm by centrifugation of bull semen. The heavier X-bearing sperm are then used for artificial insemination to increase the probability of conceiving milk-producing heifers. This technique has met with varied success.

Summary

1. In many species the sexes exhibit marked sexual dimorphism in both primary and secondary sexual characteristics.
2. Hermaphrodites possess sexual characteristics of both sexes.
3. Mechanisms of sex determination probably involve the interaction of environmental factors with many genes located on many chromosomes.
4. In many species a major chromosomal dimorphism is associated with the two sexes. Three common chromosomal mechanisms of sex determination are the XX-XY, XX-XO, and ZZ-ZW systems.
5. Placental mammals have an XX-XY chromosomal mechanism of sex determination. In males, a single Y-linked gene called SRY probably causes the embryonic gonad to differentiate to a testis. In turn, hormones produced by the testis lead to development of other male secondary sexual characteristics.
6. In bees, wasps, and some vertebrates, the sexes exhibit different degrees of ploidy.
7. For some species, environmental factors exert primary control over the sex of an individual.
8. In species with an XX-XY mechanism, genes on the sex chromosomes may be X linked, Y linked, or partially sex linked.

9. Recessive X-linked traits show a crisscross pattern of inheritance. Female heterozygotes are carriers who pass the trait to 50% of their male offspring. Recessive X-linked traits are expressed far less commonly in females than males.

10. The only traits that seem likely to be Y linked are some that directly affect certain sexual dimorphisms.

11. In female mammals, dosage compensation occurs by the inactivation of one of the X chromosomes in all somatic cells. This is called the Lyon hypothesis, and its effects are seen clearly in female mosaics.

12. Sex-limited and sex-influenced traits are the result of genes on the autosomes. The expression of these genes is sexually dimorphic.

13. For most species, the sex ratio is about 1:1 most of the time, but many factors are known that can greatly alter this ratio.

Key Words and Phrases

acquired immune deficiency syndrome (AIDS)

aneuploidy

Barr body

color blindness

dosage compensation

Duchenne muscular dystrophy (DMD)

haplodiploid

hemizygous

hemophilia

hermaphrodite

heterogametic sex

holandric

homogametic sex

Lyon hypothesis

mosaic

parthenogenesis

ploidy

primary sexual characters

pseudoautosomal

secondary sexual characters

sex-influenced inheritance

sex-limited inheritance

sex linked

sex ratios: primary, secondary, tertiary

sexual dimorphism

SRY gene

X inactivation

X linked

XX-XO systems

XX-XY systems

Y linked

ZZ-ZW systems

Questions

1. A woman has parents with normal color vision, but her brother is color blind. What is her genotype?

2. A woman with normal color vision, whose father was color blind, marries a man with normal color vision. What proportion of her children will be color blind? If her husband had been color blind, what proportion of her children would be color blind?

3. A man who has protan-type color blindness marries a woman who has deutan-type color blindness. What phenotype or phenotypes will their offspring have?

4. A non-bald man marries a non-bald woman whose mother was bald. What is the probability that their first child will be bald?

5. Woolly hair is inherited as an autosomal dominant trait. If a woolly-haired man marries a woman with normal hair, what proportion of their sons will have woolly hair?

6. The diploid number in cats is 38. How many chromosomes would you expect in a female tortoiseshell cat? In a male tortoiseshell cat?

7. It is possible for an individual to be a mosaic of two or more cell lines. In an individual whose karyotype is mosaic 45, XO/47, XXY, how many Barr bodies would you predict to find in the somatic cells?

8. The size and shape of female breasts are sex-limited traits. A small-breasted mother produced a daughter with large breasts. To what do you attribute this?

For More Information

Beam, K. G. 1988. "Duchenne Muscular Dystrophy: Localizing the Gene Product." *Nature,* 333:798.

Burgoyne, P. S. 1986. "Mammalian X and Y Crossover." *Nature,* 319:258–59.

Diamond, J. M. 1986. "Variation in Human Testis Size." *Nature,* 320:488–89.

Gubbay, J., *et. al.* 1990. "A Gene Mapping to the Sex-Determining Region of the Mouse Y Chromosome is a Member of a Novel Family of Embryonically Expressed Genes." *Nature,* 346:245–50.

Hrdy, S. B. 1988. "Daughterless Sons." *Natural History,* 97:63–83.

McKusick, V. A. 1965. "The Royal Hemophilia." *Scientific American,* 213:88–95.

McKusick, V. A. 1988. *Mendelian Inheritance in Man,* eighth edition. The Johns Hopkins University Press, Baltimore.

McLaren, A. 1990. "What Makes a Man a Man?" *Nature,* 346:216–17.

Miller, J. A. 1986. "The Genes Behind Vision's Palette." *Science News,* 129:246.

Mittwoch, U. 1963. "Sex Differences in Cells." *Scientific American,* 209:54–70.

Molton, J. D. 1986. "Understanding Color Vision." *Nature,* 321:12–13.

Morse, G. 1984. "Why Is Sex?" *Science News,* 126:154–57.

Nur, U., J. H. Werren, D. G. Eickbush, W. D. Burke, and T. H. Eickbush. 1988. "A 'Selfish' B Chromosome that Enhances its Transmission by Eliminating the Paternal Genome." *Science,* 240:512–14.

Page, D. C., E. M. C. Fisher, B. McGillivray, and L. G. Brown. 1990. "Additional Deletion in Sex-Determining Region of Human Y Chromosome Resolves Paradox of X, t(Y;22) Female." *Nature,* 346:279–81.

Rushton, W. A. 1975. "Visual Pigments and Color Blindness." *Scientific American,* 232:64–74.

Sinclair, A. H., et al. 1990. "A Gene from the Human Sex-Determining Region Encodes a Protein with Homology to a Conserved DNA-Binding Motif." *Nature,* 346:240–44.

Slater, C. R. 1987. "Muscular Dystrophy: The Missing Link in DMD." *Nature,* 330:693–94.

Weiss, R. 1988. "Muscular Dystrophy Protein Identified." *Science News,* 133:4.

© David Scharf/Peter Arnold.

Human sexuality: genotype and phenotype

Figure 6.1

A human egg at the time of conception: (a) with the polar body visible; (b) covered with sperm, only one of which will achieve fertilization; (c) after the sperm has penetrated the egg, but prior to uniting with the egg nucleus (both sperm and egg nuclei are visible).

(a) Dr. Landrum Shettles; (b) © David Scharf/Peter Arnold; (c) Z. Dickman, T. H. Cleve, W. A. Bonney, and R. W. Noyes, "Anatomical Records," 152 (1965).

Human sexual function and identity is a complex mixture of genetic, morphological, physiological, and psychological traits. Understanding human sexuality thus requires knowledge of the contribution and relationships of these varied factors. In this chapter, we discuss aspects of human sexuality that demonstrate the complexity of the genetic and biological processes that control the sexual phenotype.

Human Reproductive Anatomy

The primary sexual structures are the *gonads,* testes in males and ovaries in females. Gonads of both sexes serve two functions: (1) production of haploid gametes and (2) production of hormones that control a variety of physical and physiological traits. Testes and ovaries share a common origin from undifferentiated gonadal tissue that appears in the abdomen of the developing fetus about one month after conception. By the end of the first three months of gestation, the embryo has begun to exhibit clear differentiation in secondary as well as primary sex structures.

Female Anatomy

In an adult female, the **ovaries** are paired oval structures about four to five centimeters long. One is located on each side of the upper part of the pelvic cavity (figure 6.2), held in place by suspensory ligaments. Internally, each ovary contains about four hundred thousand *ovarian follicles,* which were formed during fetal development. Each follicle consists of one **primary oocyte** (meaning first egg cell), surrounded by supportive cells.

Following *menarche,* the onset of menstruation and puberty, one or two primary oocytes begin development per month, continuing meiosis and **oogenesis** to produce one egg or **ovum** per oocyte. A female thus ovulates only about four hundred to five hundred eggs throughout her lifetime, one or two per month for thirty-five to forty years. The remaining oocytes never develop, and ovulation ceases at *menopause.* The ovaries produce the sex hormones, **estrogen** and **progesterone,** throughout a female's lifetime.

The remainder of the female reproductive system is adapted for receiving the male penis and sperm, for the passage of sperm and egg, and for the development and birth of a fetus, should fertilization occur. Lying over each ovary is the fringed entrance to the *fallopian tube* or *oviduct.* At ovulation, an ovum is released by one of the ovaries and enters the fallopian tube. If sperm are present, fertilization normally occurs in the fallopian tube. If sperm do not reach the egg within about twenty-four hours after ovulation, the egg dies.

The ovum (figure 6.1), whether fertilized or not, travels down the fallopian tube and enters the *uterus.* The uterus is a muscular, hollow organ located low in the pelvic cavity. The fallopian tubes enter on each side near the broad upper end. If the egg has been fertilized, it implants in the highly vascularized lining of the uterus, the *endometrium.* If the egg has not been fertilized, the endometrium is sloughed off about once every twenty-eight days as part of the menstrual flow.

(a)

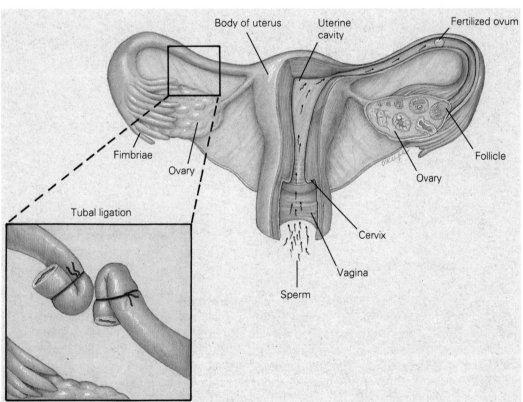

(b)

Figure 6.2

Human female reproductive anatomy: (a) sagittal section showing reproductive organs; (b) frontal view indicating path of sperm passage and usual site of fertilization. Insert: A contraceptive tubal ligation prevents sperm from reaching the egg.

(a) From John W. Hole, Jr., *Human Anatomy and Physiology,* 5th edition. Copyright ©1990 Wm. C. Brown Publishers, Dubuque, Iowa. All Rights Reserved. Reprinted by permission.

Implantation occurs about seven to eight days after fertilization. The uterine wall grows to surround the developing zygote, and, as the zygote also grows, tissues of both zygote and mother form a complex, intertwined structure called the **placenta.** The placenta provides a structure for the passage of nutrients from the mother to the fetus and of waste products

Urinary bladder

Symphysis pubis

Vas deferens

Urethra

Penis

Glans penis

Prepuce

Rectum

Ampulla

Seminal vesicle

Ejaculatory duct

Prostate gland

Bulbourethral gland

Anus

Epididymis

Testis

Scrotum

Figure 6.3

Human male reproductive system in sagittal section. Insert: Contraceptive vasectomy prevents sperm transport.

From John W. Hole, Jr., *Human Anatomy and Physiology,* 5th edition. Copyright © 1990 Wm. C. Brown Publishers, Dubuque, Iowa. All Rights Reserved. Reprinted by permission.

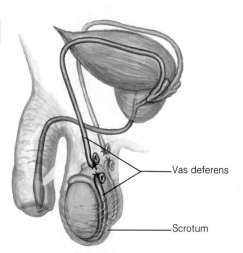

Vas deferens

Scrotum

from the fetus to the mother. The placenta also serves as a filter, isolating the fetus from the larger molecules in the mother's circulatory system. Nonetheless, the fetus is not perfectly protected. Toxins such as alcohol and nicotine readily cross the placenta, and some of the mother's own proteins and antibodies may also cross the placenta, damaging the fetus as in the case of Rh incompatibility (see chapter 12).

The lower, constricted portion of the uterus, the *cervix,* opens into the *vagina,* a tubular passageway from the uterus to the surface of the body. The vagina serves not only as a birth canal, but also as a passageway for menstrual flow and, in the opposite direction, for sperm.

Externally, the opening of the vagina is surrounded by two sets of folds of skin and adipose tissue, the *labia majora* (the external lips) and the *labia minora* (the internal lips). Where the labia minora join anteriorly is the small, cylindrical *clitoris.* The clitoris is homologous to the male penis in structure and function. It is erectile tissue, densely innervated and functioning in tactile stimulation during intercourse. Located within the folds of the labia minora are two sets of glands, which produce mucus for lubrication during intercourse.

The testes

The testes illustrate the delicacy and precision of the reproductive process. Male gametogenesis requires a temperature of about 35°C (95°F), several degrees lower than core body temperature. Therefore, the testes are suspended from the body in a pouch, allowing sufficient cooling for successful gametogenesis. The walls of the scrotum contain the *dartos muscle,* which contracts when the temperature is cold, warming the testes, and relaxes when the temperature is hot, cooling them. Failure of the testes to descend during gestation is called *cryptorchidism,* and it results in sterility due to lack of gametogenesis. Cryptorchidism can be treated easily by hormone administration or surgery, if necessary.

However, the location of the delicate testes in an exposed sac leaves them subject to injury. Because males are capable of continuous reproductive activity following puberty, the testes remain vulnerable throughout life.

In humans the *inguinal canal* normally closes permanently following descent of the testes. However, it may open as a result of trauma, causing an *inguinal hernia.* Some humans, however, actively attempt to keep the canal open throughout life. In Japan, the parents of future sumo wrestlers massage their sons' testes from birth, passing the testes into and out of the abdomen. This maintains an open canal, and when the youths begin wrestling, the scrotum is tightly bound, forcing the testes into the abdomen and protecting them from injury.

Unlike humans, most mammals reproduce only seasonally, often just once a year. In males of many species, the testes are retracted into the abdominal cavity during most of the year, descending into the scrotum only for the few weeks of annual reproductive activity.

Male Anatomy

The male reproductive system (figure 6.3) does not have to fulfill as many functions as the female system. The male system is designed primarily for the production of the male gametes (the sperm) in the testes and the transport of sperm to the female reproductive tract.

Like the ovaries, the **testes** originate high in the pelvic cavity, but during fetal development they descend into the *scrotum* (box 6.1). Internally, each testis is divided into a series of *lobules.* Within each lobule are long, convoluted *seminiferous tubules.* The walls of the seminiferous tubules are the site of **spermatogenesis,** and the immense surface area on the interior of the seminiferous tubules allows production of millions and millions of sperm (figure 6.4). Interstitial cells between the seminiferous tubules secrete **testosterone,** the male sex hormone.

Developing sperm cells are released into the lumen, or hollow interior, of the seminiferous tubules, where they travel to the *epididymis* on the surface of the testis. The epididymis is a long, coiled tube in which sperm are stored as they complete maturation.

Sufficient sexual stimulation results in *ejaculation,* the expulsion of mature sperm and accessory fluids. The sperm are forced through the system by the rhythmic contraction of muscles in the walls of the epididymis and other portions of the male reproductive tract. The sperm themselves have long tails for "swimming" that begins only after deposition in the female reproductive tract.

As the epididymis contracts, sperm are forced into the *vas deferens,* which passes through the *inguinal canal* and up into the lower pelvic cavity. Each of the vasa deferentia actually loops over the pubic bone, one passing on each side of the urinary bladder. They then enter the *prostate gland,* within which they join the *urethra.* The urethra, the tube that drains the urinary bladder, also passes directly through the center of the prostate. Beyond the prostate, the products of the urinary and reproductive systems share a common passageway, which is still called the urethra. In females, the urethra is always separate from the reproductive tract, although it opens within the space between the labia minora.

Human Sexuality: Genotype and Phenotype

Figure 6.4

(a) Cross section through a testis. (b) Details of a single seminiferous tubule.

(a) From Kent M. Van De Graaff, *Human Anatomy,* 3d edition. Copyright © 1992 Wm. C. Brown Publishers, Dubuque, Iowa. All Rights Reserved. Reprinted by permission; (b) © Gary Retherford/Visuals Unlimited.

(a) Seminiferous tubules

(b)

The urethra passes through the length of the *penis*. The shaft of the penis consists of three parallel columns that, during sexual excitement, become engorged (filled with blood), resulting in *erection*. The head or *glans* is enlarged and richly innervated. The opening of the urethra is through the glans.

The ejaculate, called **semen,** consists not only of sperm cells, but also the products of several *accessory glands*. The *seminal vesicles* open into the vasa deferentia just as they enter into the prostate. They secrete a viscous alkaline fluid that comprises over half the volume of the semen and provide nutrients for the sperm. The prostate gland also secretes an alkaline fluid into the semen. The environment within the epididymis is acidic, primarily due to the waste products of the developing sperm cells, and the sperm are relatively inactive as long as they remain in acidic conditions. The alkaline secretions of the seminal vesicle and prostate thus activate the sperm at the time of ejaculation.

The vagina is an acidic environment that sperm alone could not tolerate, but the alkaline seminal fluid produces an environment in the female reproductive tract more favorable for the sperm. The ducts of the two bulbourethral or Cowper's glands enter the urethra just below the prostate. These glands secrete a small amount of alkaline fluid that lubricates the glans of the penis prior to sexual intercourse.

Each ejaculate contains about two to six milliliters of semen and about 50–100 million sperm per ml of semen. Below a concentration of 20 million sperm per ml, a male is considered *clinically sterile*. This means that the odds of fertilization occurring are greatly reduced at low concentrations, i.e., below 20 million per ml. Only one sperm actually penetrates the membrane of the ovum and fertilizes it, but apparently millions of sperm are required before fertilization becomes likely. It is indeed extraordinary that an ejaculate might contain 100 million sperm, but the probability of fertilization remains low.

Medical attention has recently focused on a significant decline in the average **sperm count** of American males. The average sperm count decreased from about 90–100 million per ml in 1929 to 60–70 million per ml in 1979. About 23% of all American males have a sperm count less than 20 million per ml, the level of clinical sterility.

The causes of this extraordinary decline are unknown. Present hypotheses suggest that several chemicals widely used in our society may be responsible: the pesticide DDT, the industrial chemicals called PCBs, and dioxin, a highly toxic contaminant of many herbicides. All of these chemicals are now found in seminal fluid, and dioxin has been implicated in causing a variety of birth defects.

Gametogenesis

The process of gamete formation involves not only meiosis but also the maturation of the products of meiotic division. There are significant differences between spermatogenesis and oogenesis, so we discuss them separately.

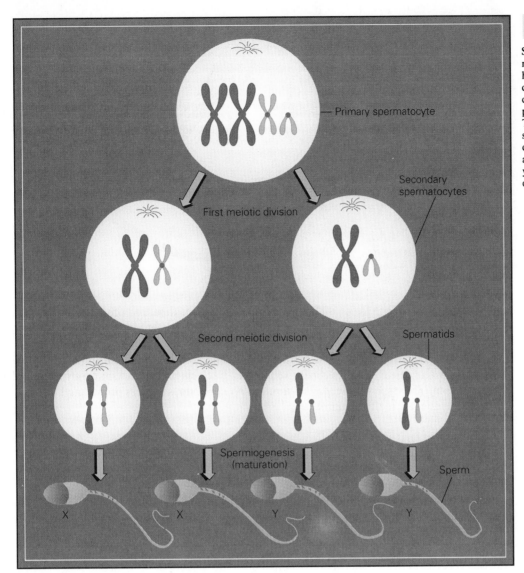

Primary spermatocyte

Secondary spermatocytes

First meiotic division

Second meiotic division

Spermatids

Spermiogenesis (maturation)

Sperm

X   X   Y   Y

Figure 6.5

Spermatogenesis in a human male. Primary spermatocytes have the normal diploid complement of forty-six chromosomes, twenty-two pairs of autosomes plus XY. These are represented by a single maroon pair of chromosomes for the autosomes and a second yellow pair of sex chromosomes.

Spermatogenesis

Male gametogenesis begins with *spermatogonia,* precursor cells just inside the wall of the seminiferous tubules. Each diploid spermatogonium divides mitotically to produce one daughter cell spermatogonium that remains on the periphery of the tubule as a precursor cell and one daughter cell that is pushed toward the interior of the tubule. As the diploid daughter cell moves inward, it undergoes a developmental change to become a **primary spermatocyte,** in which meiosis begins (figure 6.5).

In the primary spermatocyte, replication of the chromosomes occurs prior to meiosis I. Meiosis I produces two daughter or **secondary spermatocytes** that are haploid with respect to the number of centromeres, but retain chromosomes in the duplicated condition. The secondary spermatocytes then undergo meiosis II, producing a total of four haploid **spermatids** of two kinds, X bearing and Y bearing. The spermatids are then released into the lumen of the seminiferous tubules, where they move to the epididymis for completion of **spermiogenesis,** becoming mature **spermatozoa** (literally, "seed animals") (figure 6.6).

Mature spermatozoa are highly specialized cells, each consisting simply of head, body, and tail, but almost no cytoplasm. The head contains the chromosomes that the father contributes to the zygote if the sperm achieves successful fertilization (the odds are slight). The body contains rows of mitochondria that provide energy for the long journey the sperm will attempt to the upper reaches of the oviduct—a swim of 180 millimeters (seven inches) for

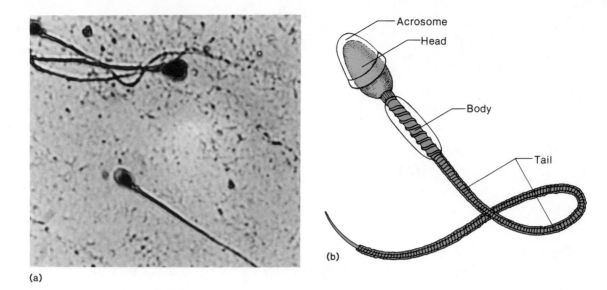

Figure 6.6

Morphology of human sperm:
(a) photo of sperm;
(b) diagrammatic
interpretation of sperm.

(a) Kirk Kreutzig/Photo
Graphics. (b) From John W.
Hole, Jr., *Human Anatomy
and Physiology,* 5th edition.
Copyright © 1990 Wm. C.
Brown Publishers, Dubuque,
Iowa. All Rights Reserved.
Reprinted by permission.

an organism only .06 mm long (.0023 inches), or almost three thousand times its body length, equivalent to a swim of three miles for an adult human. The tail, a long, whiplike structure, propels the sperm.

The rate of spermatogenesis is extraordinary. The production of mature spermatozoa from a spermatogonium requires about two months. Each ejaculate contains several hundred million sperm. Thus, spermatogenesis is a relatively continuous process producing billions of sperm throughout a male's lifetime.

Oogenesis

Development of mature ova also involves meiosis, but the timing and results differ considerably from spermatogenesis. Recall that at birth all the potential eggs a female will ever possess are already present in ovarian follicles. Each follicle consists of numerous supportive cells surrounding one oogonium, which may become an ovum.

Prior to birth, each oogonium develops into a **primary oocyte,** which replicates its chromosomes and begins meiosis I. However, development stops near the end of prophase I, and the oocytes remain in this condition from birth until the onset of puberty at about age twelve. Then, one or two oocytes per month become active and resume meiosis. In contrast to the millions of sperm that mature daily in a male, a female produces only four hundred to five hundred mature ova throughout her entire reproductive life span. The remaining million or more oocytes simply degenerate.

When a follicle resumes development, meiosis I continues from the stage of prophase I where it ceased, and the first meiotic division is completed (figure 6.7). However, in contrast to spermatogenesis, the two daughter products are not of equal size. Both contain a duplicated haploid set of chromosomes, but the division of cytoplasmic contents is highly unequal. One daughter cell, the **secondary oocyte,** receives virtually all the cytoplasm from the primary oocyte, while the other, the *first* **polar body,** receives essentially none.

The follicle matures at the end of the first meiotic division, and the secondary oocyte, polar body, and some surrounding cells are released by the ovary, or *ovulated.* The secondary oocyte moves into and down one of the fallopian tubes. If it encounters a sufficiently large number of sperm, fertilization occurs (figure 6.1). Although only a single sperm penetrates the ovum, the presence of many sperm within twenty-four hours after ovulation is required for penetration to occur.

Meiosis II begins only *after* fertilization. The polar body may or may not divide. The secondary oocyte undergoes equatorial division, and again the cytoplasm is divided unequally. One daughter cell, the *second polar body,* receives essentially no cytoplasm, while the other, the ovum, retains all the cytoplasm. The ovum, of course, has already been fertilized, and it continues development into a zygote, embryo, and fetus. The two or three remaining polar bodies degenerate.

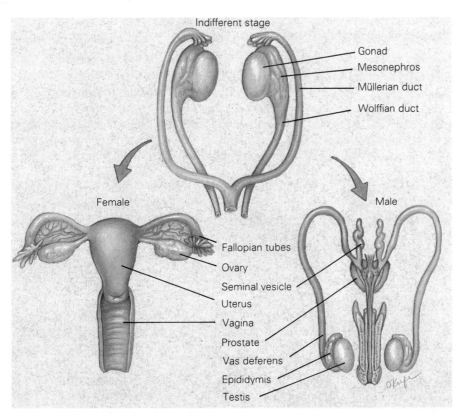

Figure 6.7

Oogenesis in a human female. Primary oocytes have the normal diploid complement of forty-six chromosomes, twenty-two pairs of autosomes plus XX. These are represented by a single pair of shaded autosomes marked with the A locus (2 alleles–A, a) and a second pair of unshaded X chromosomes.

Figure 6.8

Differentiation of internal reproductive organs in human males and females. Every fetus passes through an indifferent stage of reproductive anatomy. As the fetus subsequently becomes either male or female, some structures develop further while others regress entirely.

The results of meiosis are thus very different for males and females. Male meiosis produces four haploid and equal (but not genetically identical) gametes. Female meiosis produces only one mature ovum, plus two or three polar bodies of no further function. Female meiosis is not completed unless fertilization occurs.

Development of the Sexual Phenotype

For at least the first month of gestation, human embryos exhibit no observable sexual dimorphism. At this stage, the embryonic reproductive system has three components: (1) undifferentiated gonads; (2) two systems of genital ducts, *Müllerian ducts* and *Wolffian ducts;* and (3) externally, a set of genital folds (figures 6.8 and 6.9).

Control of the Male Phenotype

Differentiation begins earliest in the male embryo. At about one month of age, the **SRY gene** (sex-determining region of the Y, or testis determining factor, TDF—chapter 5) apparently becomes active, "instructing" the gonads to differentiate to testes.

The next step of phenotypic development depends upon the products of the testes. The testes secrete two controlling hormones, Müllerian inhibiting substance and **testosterone.** Müllerian inhibiting substance causes the regression of the Müllerian duct system. Testosterone induces *virilization* or differentiation of embryonic tissue to masculine structures. In the embryo, this results in alteration of the Wolffian duct to produce the epididymides, vasa deferentia, and seminal vesicles. Primordial cells differentiate to produce the prostate gland and Cowper's glands. Genital folds elongate and form the penis, while swellings surrounding the folds enlarge to form the two halves of the scrotum (figure 6.9). Descent of the testes into the scrotum is usually complete by birth.

Testosterone performs a variety of important functions throughout the life of a male. In addition to its function in sexual differentiation, testosterone is continually required for maintenance of secondary male structures. Testosterone regulates bone growth; muscle development; fat deposition; sperm production; sexual behavior; patterns of facial, axillary, body, and pubic hair; and enlargement of the thyroid cartilage that results in deepening of the voice.

Control of the Female Phenotype

Female sexual differentiation begins somewhat later than in the male, becoming apparent by about the second month of development. In the absence of the SRY gene, the primitive gonads develop into ovaries. In the absence of testosterone and Müllerian inhibiting substance, the Wolffian ducts degenerate. The Müllerian ducts develop into the fallopian tubes, uterus, and part of the vagina (figure 6.8). The genital tubercle, which gave rise to the male glans, develops into the clitoris. The genital folds, which became the shaft of the penis, become the labia minora; and the genital swellings, precursors to the scrotum, become the labia majora (figure 6.9). The primordial cells that became the prostate develop into Skene's glands; and the primordia of the male Cowper's glands become Bartholin's glands.

Thus, a broad set of homologies exists between male and female sexual characteristics. Without the action of the SRY gene and testosterone, the fetus develops a female phenotype by default. With them, it develops a male phenotype. The male phenotype is thus an *inducible* condition, the female *noninducible.* Individuals of both karyotypic sexes possess the capacity to develop the phenotype of either sex.

Overview of Sexual Development

Development can be viewed as a sequence of diverging pathways (figure 6.10). As development proceeds, genes are turned on and off in a precise sequence. A gene active early in life produces a developmental stage that leads to activation of other genes and tissues, whose effects may then predominate. Such a system is called a **developmental cascade.**

In the case of the sexual phenotype, the presence of the SRY gene commits an individual to a male developmental pathway by causing the gonad to become a testis. Testosterone produced by the testis then causes masculinization of various tissues, especially the external genitalia and other secondary sexual characteristics. In the absence of the SRY gene, ovaries are produced and levels of testosterone are low, so the female phenotype results.

The sexual phenotype is extremely complex, of course, so development must depend on the interactions of many genes and their products. For example, the hypothalamic portion of the brain of both males and females produces a hormone called GnRH (gonadotropin-releasing hormone). GnRH production begins at puberty, probably as a result of a genetic signal. In turn, GnRH induces production of other hormones called gonadotropins (also in the brain), which first induce puberty and then govern the reproductive cycle throughout life.

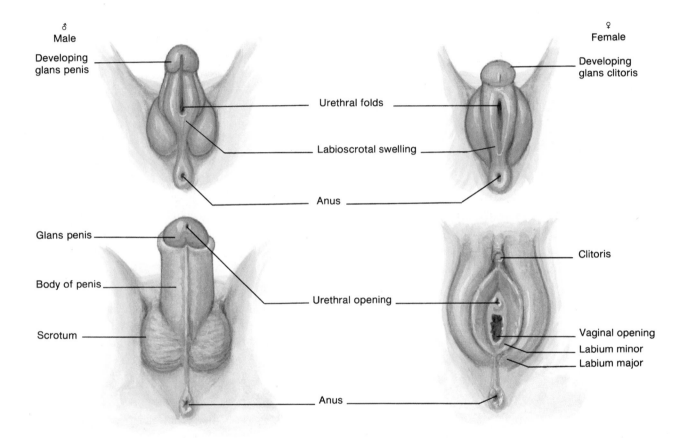

Another gene important to the sexual phenotype occurs on the long arm of the Y chromosome and produces the **H-Y antigen.** The H-Y antigen is a protein that is found on cells of males, and its presence is required for normal spermatogenesis. The H label refers to its discovery as a determinant of tissue, or *histological,* incompatibility between the sexes; the Y label refers to its location on the Y chromosome.

Disturbance of any of the genetic or hormonal components of the developmental cascade can alter the sexual phenotype. For example, women athletes and ballerinas who undertake intensive physical conditioning programs often stop menstruating, apparently because of decreased GnRH production. The juvenile phenotype of young women gymnasts may result from delayed puberty due to strenuous training. When the training schedule is reduced, menstruation usually returns to normal, or puberty begins.

Figure 6.9

Differentiation of external reproductive organs in human males and females.

From Kent M. Van De Graaff, *Human Anatomy.* 2d edition. Copyright © 1988 Wm. C. Brown Publishers, Dubuque, Iowa. All Rights Reserved. Reprinted by permission.

Figure 6.10

Development of the sexual phenotype. An individual follows one of many possible developmental pathways, depending upon genotype, environment, and their interactions. Both sexes encompass a wide range of "normal" phenotypes, but some individuals follow pathways leading to sexually intermediate phenotypes. T = Testosterone produced or present; SRY = possessing the SRY gene; gold circles represent individuals with XX karyotypes; beige circles represent XY karyotypes.

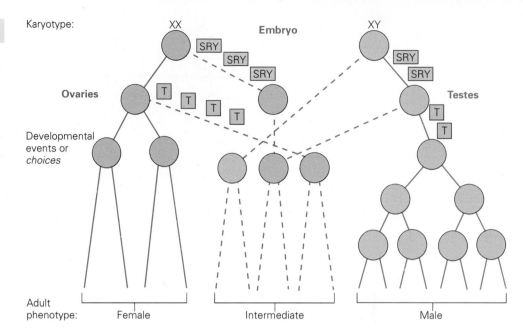

The sexual phenotype can also be changed by administration of synthetic hormones, particularly a type called **anabolic steroids.** Anabolic steroids are variant types of testosterone, and over 3,000 are now known. They are designed to increase muscle mass, but with relatively small masculinizing effects.

Anabolic steroids were first developed in Germany in the 1930s as part of Hitler's programs to create Nordic supermen. Beginning in the 1950s, however, athletes of both sexes began using them in the hope of improved performance. Although prohibited in most sports, their use is surprisingly widespread: 5% of college athletes in America and between 7% and 8% of professional football players test positive for illegal steroid use.

The unwanted consequences of sustained use of anabolic steroids can be severe, particularly when taken in large doses. Women may develop deep voices and grow facial hair. Men may grow breasts and the testes may shrink, because high levels of artificial testosterone induce low levels of production of natural testosterone. The testosterone level of the Canadian sprinter Ben Johnson, who lost his Olympic gold medal for illegal steroid use, was only 15% of normal. In both sexes, sterility, liver cancer, and other serious diseases are among the risks of anabolic steroid use. The effects of steroid use may not be reversible when their use stops. Ironically, it is not certain that steroid use actually improves athletic performance.

Anabolic steroids may have opposite effects on the sexes, masculinizing for women and feminizing for men. This demonstrates that effects of a gene or its products (which anabolic steroids mimic) clearly depend upon interactions with other genes and the developmental history of the individual. Anabolic steroids also illustrate one way in which normal developmental pathways can go askew, producing individuals of intermediate sexual phenotype. In the next section, we describe several genetic conditions where the normal developmental pathway cannot be followed, resulting in alterations of the sexual phenotype.

Genetic Sexual Disorders

Many genetically based variations on normal sexual development have been discovered. We discuss some of these because they illuminate the mechanism of sex determination and the definition of sex.

Single Gene Disorders

Although the chromosomal basis of sex determination is well known, single genes are sometimes able to cause massive alteration of the sex phenotype.

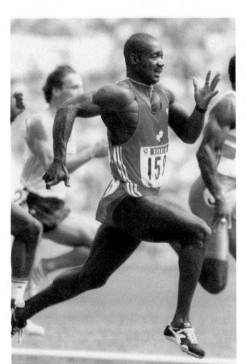

Figure 6.11

In 1988, Ben Johnson forfeited his Olympic gold medal when tests showed he had been taking illegal steroids. Steroids have been shown to increase muscle mass when their consumption is accompanied by appropriate exercise. Because of known harmful side effects associated with steroid intake and long term use has unknown (and potentially devastating) effects, athletes are prohibited from using steroids.

Bettmann Archives.

Testicular Feminization

Bearers of this rare syndrome possess a normal male karyotype (designated *46, XY* for 46 chromosomes and the normal XY male sex chromosomes). However, they possess internal testes and no other internal reproductive organs. Externally, they are phenotypically well-developed females—so well developed, in fact, that they are often characterized as voluptuous. Their psychological orientation is clearly female. They usually come to medical attention because they fail to begin menstruation as teenagers.

The cause of this condition appears to be a recessive mutation at an X-linked locus designated Tfm. Persons with an XY karyotype and the Tfm mutation develop testes, due to the presence of the SRY gene. Hormone production by the testes is normal, and Müllerian inhibiting substance results in degeneration of the Müllerian ducts, as expected.

As a result, all tissues respond as if testosterone were not present, differentiating into female secondary structures. The Wolffian ducts therefore regress, leaving no internal reproductive organs at all. The **testicular feminization** syndrome illustrates the extraordinary effect only one locus can have on the sexual phenotype.

Guevodoces

True **hermaphrodites** possess both types of primary sex organs, testes and ovaries. Some humans possess both ovarian and testicular tissue, but both types are almost never functional. **Pseudohermaphrodites** have only one type of gonad or none at all, but they show ambiguous genitalia, i.e., indeterminate secondary characteristics or secondary characteristics of both sexes. In the village of Salinas in the Dominican Republic, twenty-four pseudohermaphrodites have been detected showing a 46,XY karyotype. These individuals, called *guevodoces,* are born with ambiguous external genitalia—the scrotum appears to be labia, a blind vaginal pouch is present, and the penis resembles a clitoris. Until recently, these children were raised as girls.

At puberty, however, they show a distinct virilization of the external secondary sex structures. The voice deepens, the muscular development is masculine, and the apparent clitoris enlarges to become a penis. (Guevodoces means, literally, "penis at twelve.") The individuals become fully functional as males, their psychological orientation is masculine, and they are fertile.

This condition results from an autosomal recessive allele regulating the use of testosterone. Testosterone acts directly on the Wolffian ducts, but before causing virilization of the external genitalia it must be biochemically altered to a related chemical, *dihydrotestosterone*. A male homozygous for the recessive allele controlling the enzyme that catalyzes testosterone to dihydrotestosterone does not show virilization of the external genitalia. Apparently, the effect of testosterone alone is sufficient to induce virilization of the genitalia at puberty. Again, as with testicular feminization, one gene exerts massive influence over the sex phenotype.

The high frequency of guevodoces in this village in the Dominican Republic is the result, as might be expected, of consanguineous matings. Salinas is an isolated village of 4,300 persons, yet this rare condition has appeared in twenty-four males, twenty-three of whom trace their ancestry through twelve families back to one woman.

Chimeras and Mosaics

Rarely, an individual is found who possesses both ovarian and testicular tissue and is thus a true hermaphrodite. Usually, these individuals are identified at birth because of ambiguous genitalia. Cytological and histological examination usually reveals that their tissues consist of two different cell types. These persons have two distinct karyotypes, one per cell line. Such an occurrence can be explained if the individual resulted from the fusion, early during development, of two or more different zygotes. Such an individual is called a **chimera.** Alternatively, an hermaphrodite can arise as a result of early mitotic nondisjunction (see the following section) of an XY or an XXY embryo, producing a **mosaic** of cell lines, XO/XY, XX/XXY, and so on.

Most chimeras are discovered because the fusing zygotes are of different sexes. The karyotype is designated chi46,XX/46,XY. Chimeras resulting from fusion of zygotes of the same sex are discovered far more rarely because they do not exhibit obvious defects, but they are probably equally common.

Chimeras and mosaics may occur in a variety of ways. In some instances, a polar body is fertilized by a sperm at the same time the ovum is being fertilized by a different sperm. If one of the sperm is X bearing and one Y bearing, the resulting zygotes are of opposite sexes. Subsequent fusion of those zygotes would give rise to one individual bearing two different genetic types of cells. While chi46,XX/46,XY is most commonly identified, other chimeras discovered include: chi45,XO/46,XY; chi46,XX/47,XXY; and chi45,XO/46,XY/47,XYY.

Chromosomal Disorders

Several syndromes have been discovered that result from **aneuploidy** of the sex chromosomes. Aneuploidy results from an error during cell division called **nondisjunction** (figure 6.12). As the chromosomes segregate during anaphase, one pair of homologues does not separate and divide equally. Instead, one daughter cell receives both members of an homologous pair, while the other daughter receives none. Nondisjunction usually involves only one pair of chromosomes at a time.

Nondisjunction can occur during either meiotic division. If it occurs during the first division, it is called *primary nondisjunction* (figure 6.12b1). None of the gametes produced by primary nondisjunction are normal: two possess an extra member of the affected pair of homologues, two have none of that pair. *Secondary nondisjunction* (figure 6.12b2) occurs during the second meiotic division, producing two normal gametes, one gamete with one extra chromosome, and one gamete with a missing chromosome.

Most nondisjunctions affecting autosomes are lethal, especially those involving loss of an autosome. However, a few aneuploid karyotypes are viable, and those affecting autosomes are discussed in chapter 11. A number of aneuploidies of the sex chromosomes also are viable.

Turner's Syndrome—XO

About one in every 2,000 female births is deficient in one X chromosome, symbolized 45,XO (figure 6.13). Phenotypically, the individual is female, but the ovaries are poorly developed, the individual is short and has a webbed neck, and the secondary sexual characteristics are incompletely developed.

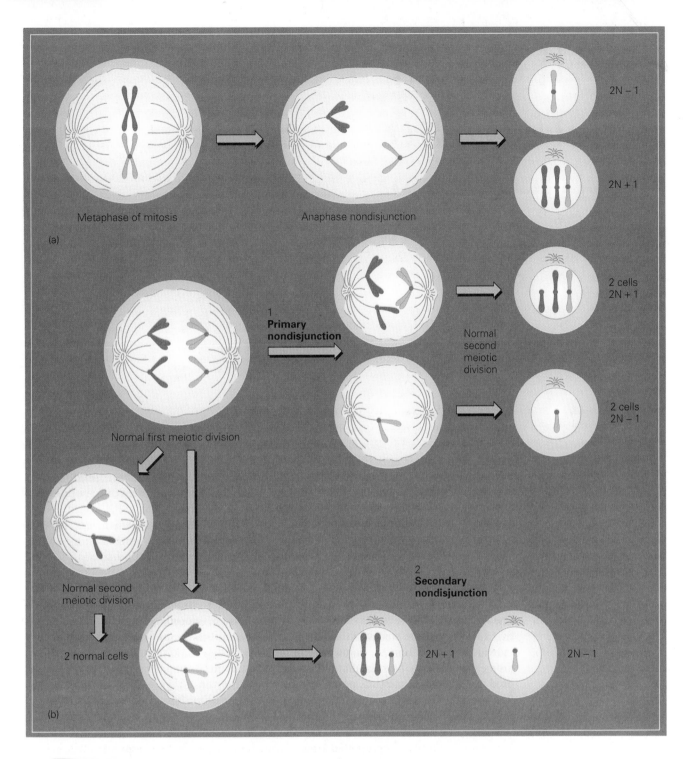

(a)

Metaphase of mitosis

Anaphase nondisjunction

2N – 1

2N + 1

(b)

Normal first meiotic division

1
Primary nondisjunction

Normal second meiotic division

2 cells 2N + 1

2 cells 2N – 1

Normal second meiotic division

2 normal cells

2
Secondary nondisjunction

2N + 1

2N – 1

Turner's females have no Barr bodies. The inactivation of one X chromosome in normal female cells might lead to the expectation that loss of one would have limited effect. In fact, X inactivation may explain why *monosomy* (meaning one chromosome) for the X is the only monosomy that is viable at all in humans. However, X chromosomes are not inactivated in female germinal tissue; both Xs are required for female gametogenesis. Therefore, XO females are sterile.

Turner's females usually arise from meiotic nondisjunction during gametogenesis (figure 6.14). They can also arise from mitotic nondisjunction (figure 6.12a) during early embryonic development. In this case, they are a mosaic of XX and XO tissues. Because Turner's females, like males, are hemizygous for the X chromosome, they show an increased frequency of expression of X-linked traits.

Figure 6.12

Nondisjunction during cell division leading to chromosomal aneuploidy: (a) mitotic nondisjunction occurs with centromere misdivision at anaphase; (b) meiotic nondisjunction may occur at either anaphase I or anaphase II, with differing results.

(b)

Klinefelter's Syndrome—XXY

Approximately one of every five hundred males born has an extra X chromosome, a 47,XXY karyotype (figure 6.15). Because the Y is male determining, these individuals show a male phenotype. However, they tend to have longer limbs than normal, and some of the secondary sex characteristics are feminized. They usually have some degree of mental retardation.

Klinefelter's males are also usually sterile. One Barr body is present in somatic cells of XXY males (remember, there is always one less Barr body than there are X chromosomes). The presence of two Xs is associated with sterility in XXY males. Nonetheless, the male phenotype of Klinefelter's males is convincing evidence that the Y chromosome—and not the dosage of X chromosomes—is the controlling factor for maleness in humans.

XYY Males

It is estimated that one male in a thousand has an extra Y chromosome and is thus 47,XYY. Phenotypically, XYY males are fertile and normally developed, with the single exception that they tend to be somewhat taller on average than the normal male population. These individuals result from the same type of meiotic nondisjunction that causes Turner's syndrome (figure 6.14). Nondisjunction of the Y chromosome during the second meiotic division produces one O gamete (i.e., no sex chromosome) and one YY gamete. Fertilization by the first gamete causes Turner's syndrome, XO, while fertilization by the second gamete produces an XYY male.

Attention has focused on XYY males in recent years because of the discovery that they are disproportionately represented among white inmates of mental/penal institutions. XYY males are twenty times more likely than normal XY males to be incarcerated in such an institution; about 2% of all inmates appear to be XYY.

This fact led to the hypothesis that XYY males are more aggressive than XY males and thus tend to criminal behavior. While XYY males may have an increased probability of incarceration, the causes of this phenomenon have yet to be determined (box 6.2). No proof has been found that an extra Y chromosome leads to criminal behavior. Until more definitive data are found, it is best to remember that the vast majority of XYY males lead apparently normal lives.

Fragile X Syndrome

One recognized abnormality of the X chromosome produces phenotypic effects very different from those associated with sex chromosome aneuploidies. Called the **fragile X syndrome,** this

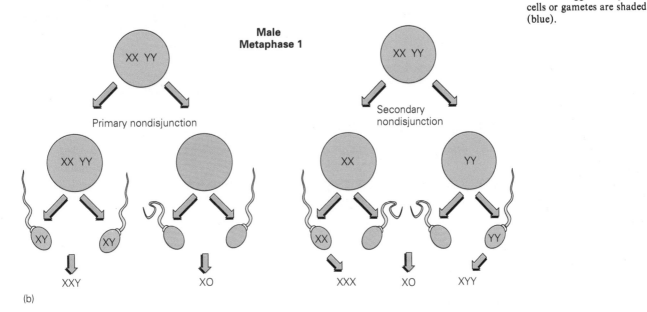

Figure 6.14

Examples of how meiotic nondisjunction can give rise to persons with aneuploidies of the sex chromosomes. (a) primary or secondary nondisjunction in females produces eggs with two X chromosomes or no X chromosomes leading to XXX, XXY, XO, or YO fertilized eggs; YO is lethal; (b) primary nondisjunction in males leads to sperm with no sex chromosomes, or with both an X and Y chromosome giving XXY and XO fertilized eggs. Secondary nondisjunction produces sperm with no sex chromosomes, with XX and with YY; these sperm produce XXX, XYY, and XO fertilized eggs. Aneuploid cells or gametes are shaded (blue).

condition is diagnosed by a constriction near the tip of the long arm of the X chromosome (site Xq27.3—figure 6.15) in mitotic white blood cells grown in culture. The chromatin apparently does not condense, producing a weak point in the chromosome that is prone to breakage. Direct DNA assays can also now test for this condition.

About one person in 1,000–1,500 possesses a fragile X chromosome. Phenotypic effects include a variety of growth abnormalities, including protruding ears, large head, long face, high birth weight, flat feet, and, in males, large testes. The behavioral symptoms are more serious, including hyperactivity, hand biting, learning disabilities, autism, and mental retardation. About 10% of autistic males have the fragile X chromosome, and the syndrome is the second most common chromosomal cause (after Down's syndrome—chapter 11) of inherited mental retardation in the United States.

Fragile X syndrome is particularly puzzling because of the variability in expression of symptoms. Because it is an X-linked condition, all males should be affected. This is not the case, however; about 20% of males with the chromosome show no symptoms. These males are carriers for the fragile X chromosome and transmit it to their children. Strangely, daughters of fragile X carriers produce some children with the chromosome and expressing the symptoms. This suggests that the genes causing the disease were turned off in the carrier, but then somehow activated in the daughter, resulting in affected children.

Human Sexuality: Genotype and Phenotype   **139**

Figure 6.15

Karyotype and individual
with Klinefelter's syndrome.

(a) A. Redding and K.
Hirschhorn, "A Guide to
Human Chromosome
Defects," D. Bergsma (ed.)
New York: The National
Foundation–March of Dimes,
BD:OAS IV (4) 1968;
(b) F. A. Davis Co. and Dr.
R. H. Kampmeier.

(a)

(b)

Females are less frequently and severely affected than males, possibly as a result of X inactivation. About one-third of females with the fragile X are mentally retarded, while others show mild disturbances in social or learning behaviors, or no symptoms at all.

The screening, counselling, and ethical problems associated with fragile X syndrome are similar to those for XYY males (box 6.2). Identification of the fragile X in a newborn might result in that child being stigmatized or treated in ways that actually induce unwanted behaviors, even if no symptoms would otherwise have been expressed. On the other hand, the effectiveness of behavioral therapies and other treatments is probably greatest if started early in life. Furthermore, many parents would be relieved to know that their child's unusual appearance and behavior had an identified physical cause and were not their own fault. Autism, for example, is an extremely difficult and frustrating condition for a parent to manage, yet it has sometimes been claimed to be caused by harsh or emotionally distant parents.

Other Sex Chromosome Aneuploidies

Females have been discovered who are triplo-X (XXX), tetra-X (XXXX), and penta-X(XXXXX). Triplo-X females are relatively normal phenotypically, while tetra-X and penta-X individuals exhibit mental retardation. Triplo-X and tetra-X females are often fertile. All females with extra X chromosomes show extra Barr bodies—two Barr bodies for XXX individuals, three for XXXX, and four for XXXXX.

Males showing extreme symptoms of Klinefelter's syndrome have been discovered who are 48,XXXY or 49,XXXXY. They are almost always mentally retarded. Also known are 48,XXYY and 49,XXXYY males. They tend to be tall, more aggressive than normal, and mentally deficient. Like females with extra X chromosomes, their somatic cells have one less Barr body than the number of X chromosomes.

Gender

Sex or *gender* is a complex phenomenon. Numerous variations have been observed in both karyotypic and phenotypic measures of sex. In some instances, an individual cannot be unambiguously classified as either male or female. The causes of these anomalies are complex and may include genetic, hormonal, and environmental factors. Testicular feminization, for example, involves a clearly karyotypic male exhibiting a clearly female external phenotype. Which sex is the individual? The answer is less relevant than an understanding of the factors that may cause this condition. There is no legal definition of sex.

BOX 6.2

A long-term study of XYY males

In 1968 a long-term study of male infants with karyotypic abnormalities was begun at the Boston Hospital for Women. The goal was to identify XXY (Klinefelter's) and XYY males, then to study their development and behavior. Klinefelter's males present no ethical problem. They are often mentally retarded, prone to institutionalization, and therefore clearly require special care.

The case of XYY males, however, provides a genuine dilemma. About one male in a thousand is XYY, and most lead perfectly normal lives. The primary effect of the condition is that many XYY males are unusually tall. Concern has been raised that XYY males may be prone to aggressive or antisocial behavior, but no conclusive evidence yet supports that contention. The main evidence indicating such a possibility is that the risk of an XYY male being incarcerated in a hospital for the criminally insane is twenty times greater than the risk for an XY male.

The Boston study was designed, in part, to determine if XYY males were in fact prone to antisocial behavior. The program was terminated in 1975, however, due to public opposition. Many people believed that identifying a child as XYY would create a self-fulfilling prophecy: Because the child would be labeled from birth as potentially antisocial, he would be treated differently—and the different treatment, *not* the extra Y chromosome, would produce aberrant behavior. As with the IQ controversy, separating nature from nurture is extraordinarily difficult.

The ethical dilemma is obvious. If an XYY male is not genetically prone to aggressive behavior, then identification serves no positive purpose. Instead, the stigma of being labeled genetically different could promote undesirable behavior. Even parents would be unlikely to avoid treating an XYY child differently. Normal childhood rowdiness would always be under suspicion as overaggressive behavior in an XYY son. Parents might then punish the XYY child with undue severity, eventually producing the undesirable behavior.

On the other hand, if XYY children are indeed subject to producing antisocial behavior, their identification at birth could permit treatment. For example, XYY males might experience greater than normal reading and learning disabilities, frustration over which could in turn generate overtly aggressive behavior. Knowledge that a child is XYY could enable parents and teachers to provide special education—including much patience and understanding—that could ultimately lead to normal development. In this case, failure to identify XYY males at birth could actually *deprive* them of a normal life.

Other possibilities also exist. The increased height of XYY males may elicit unusual reactions to them. The effect of differences in treatment due to height, combined with different treatment due to, for example, learning disabilities, might also produce antisocial behavior. While the opponents of the Boston study may be correct that identification of XYYs could create a self-fulfilling prophecy of aggressive behavior, the only way that the truth will ever be determined is by long-term studies such as the Boston group was attempting.

The determination of gender is further complicated by consideration of psychological factors. An individual's perception of his or her sexual identity is referred to as *gender identity*. In most cases, gender identity agrees with karyotypic and phenotypic sex, but it need not. Some karyotypic and phenotypic males *feel* as if they are females, and vice versa. Obviously, such internal conflict can cause serious emotional and psychological problems.

A person whose psychological identity is at odds with his or her phenotypic identity is called a *transsexual*. While karyotypic sex cannot be altered, phenotypic sex can. Since the mid-1950s transsexual surgery has been performed to alter phenotypic sex to correspond to psychological identity. Normally, six or more months of hormone treatments are given prior to surgery. Administration of large doses of testosterone, for example, results in reduction in size of breasts, redistribution of fat to enlarge shoulders and reduce hip size, growth of

Figure 6.16

(a) Young boy with fragile X syndrome. Note the characteristically protruding ears. (b) An electron micrograph of a fragile X chromosome. The arrow points to the fragile site near the tip of the long arm.

D. M. Barnes. *Science*, 243:171–2, 1989.

(a) (b)

facial and body hair, and deepening of the voice. Doses of estrogen have an opposite effect, and even prior to surgery, hormone treatment can cause major changes in secondary sex structures.

Transsexual surgery is less complicated for the male to female transition. The testes and the penis are removed, and the tissue from the penis and the scrotum are used to create an artificial vagina. In female to male surgery, the breasts, ovaries, and uterus are removed. A scrotum and, sometimes, a penis are constructed. The extent to which such surgery can affect psychological adjustment and happiness is a matter of ongoing debate. A stigma is still attached to individuals who have undergone such surgery.

Yet another complication arises with respect to sexual preference or *object choice*. Most persons develop love and sexual relationships with persons of the opposite sex and are thus *heterosexual*. The object choice of an *homosexual* is a person of the same sex; while *bisexuals* may choose a person of either sex. The causes of sexual preferences, like gender identity, are poorly understood. As with transsexuals, society only partially accepts individuals who exhibit non-heterosexual behavior.

Accurate data are extremely difficult to collect, but recent studies indicate that a minimum of 20% of American males have at least one homosexual experience in their lives. Does one such experience permanently identify a person as homosexual? Or warrant lifelong stigmatization? Such questions are of considerable social significance because of the historical discrimination accorded homosexuals, and because understanding patterns of sexual behavior is essential for stopping the spread of sexually transmitted diseases such as AIDS.

Knowledge that a zygote of either XX or XY karyotype may differentiate either to male or to female phenotypes has helped science redefine the meaning of maleness and femaleness. Testicular feminization demonstrates that an individual can be psychologically healthy with a sexual orientation different from karyotypic sex. Guevodoces successfully change psychological orientation midway through life—they are normal little girls before virilization, normal adult males afterward. Such information persuasively argues the importance of developmental factors in producing sexually stereotyped behaviors. The definition of gender will continue to increase in complexity as the many factors governing sexual identity and differentiation continue to be elucidated.

Summary

1. The human embryo possesses primordial structures that may differentiate into the sexual structures of either sex. Both the internal and external reproductive structures of males and females are homologous.

2. The SRY gene on the Y chromosome appears to govern the process of sexual differentiation. It induces differentiation of embryonic gonadal tissue to testes. Females do not have the SRY gene, and in its absence the gonads differentiate to ovaries.

3. If testes are formed, they produce testosterone, which induces formation of male secondary sexual structures. In the absence of testosterone, female secondary characteristics develop.

4. Spermatogenesis results in the formation of four spermatozoa from each primary spermatocyte. Two of the four carry a Y chromosome, two carry an X chromosome.

5. Oogenesis results in the formation of only one mature ovum from each primary oocyte, plus two or three polar bodies.

6. A male produces billions of sperm throughout his reproductive life span. A female produces only a few hundred mature ova.

7. Expression of the sexual phenotype may be altered as the result of mutations involving a single gene on either an autosome or a sex chromosome. Syndromes having major phenotypic effects result from aneuploidy of the sex chromosomes.

8. X-chromosome aneuploidies are always reflected in the number of Barr bodies in somatic cells.

9. Chimerism is the result of fusion of two or more zygotes and may cause true hermaphroditism.

10. Nondisjunction of chromosomes occurs during anaphase of cell division and results in aneuploid gametes and mosaic individuals.

11. Gender is a phenomenon involving the complex interaction of genotype and phenotype.

Key Words and Phrases

anabolic steroids	ovary	sperm count
aneuploidy	ovum	spermatid
chimera	placenta	spermatogenesis
developmental cascade	polar body	spermatozoan
estrogen	primary oocyte	spermiogenesis
fragile X syndrome	primary spermatocyte	testicular feminization
hermaphrodite	progesterone	testis
H-Y antigen	pseudohermaphrodite	testis determining factor (TDF)
Klinefelter's syndrome	secondary oocyte	testosterone
mosaic	secondary spermatocyte	Turner's syndrome
nondisjunction, primary and secondary	semen	
oogenesis		

Questions

1. Why is a woman able to conceive only one day of each month?
2. What would you infer about the potential fertility of an adult Japanese sumo wrestler?
3. What are the functions of the accessory glands to the reproductive tract of each sex?
4. A normally pigmented female whose mother was an albino is a carrier for Lesch-Nyhan syndrome. She is married to an albino male. Considering these two traits, how many different types of eggs can she produce? What is the probability of the first child being an albino with Lesch-Nyhan syndrome?
5. A color-blind woman marries a normally visioned man, and they have a color-blind daughter. How do you explain this? How might this same couple produce a child with normal vision?
6. What is the difference between a Barr body and a polar body?
7. Persons with either Klinefelter's or Turner's syndrome are usually sterile. However, because of X inactivation, one might expect that phenotypically these individuals would otherwise be normal. What does the presence of a variety of phenotypic abnormalities in these individuals suggest about either the extent of X inactivation (i.e., is it complete in somatic cells?) or the presence of genes on the X chromosome (i.e., do genes on the sex chromosomes govern "non-sexual" traits?)?
8. Imagine that you have a family history of a serious genetic disease. Prenatal screening reveals that the fetus that you (or your spouse) are carrying is normal for the disease, but is XYY or fragile X. Would you consider the XYY or fragile X sufficiently serious to warrant abortion? Why, or why not? Suppose you raise the child to age ten (or fifteen or twenty), and the child exhibits none of the symptoms often associated with XYY or fragile X. Do you then tell the child of his or her condition? Why, or why not? Should other members of the family be told? Do they have a right to know? Why or why not?

For More Information

Barnes, D. M. 1989. " 'Fragile X' Syndrome and its Puzzling Genetics." *Science*, 243:171–72.

Beckwith, J. 1975. "The Harvard XYY Study." *Science*, 187:298.

Chandra, H. S. 1985. "Is Human X Chromosome Inactivation a Sex-Determining Device?" *Proc. Nat'l. Acad. Sci.*, 82:6947–49.

Davies, K. 1991. "Breaking the Fragile X." *Nature*, 351:439–40.

Diamond, J. M. 1986. "Variation in Human Testis Size." *Nature*, 320:488–89.

Epel, D. 1977. "The Program of Fertilization." *Scientific American*, 237:128–40.

Fay, R. E., C. F. Turner, A. D. Klassen, and J. H. Gagnon. 1989. "Prevalence and Patterns of Same-Gender Sexual Contact Among Men." *Science*, 243:338–48.

German, J. 1970. "Studying Human Chromosomes Today." *American Scientist*, 58:182–201.

Hodgkin, J. 1988. "Everything You Always Wanted to Know About Sex." *Nature*, 321:300–301.

Hoffman, M. 1991. "Unraveling the Genetics of Fragile X Syndrome." *Science*, 252:1070.

Miller, J. A. 1986. "X Chromosomes: Too Few and Too Many." *Science News*, 129:358.

Naftolin, F. 1981. "Understanding the Bases of Sex Differences." *Science*, 213:1263–64.

Ohno, S. 1980. "Two Major Regulatory Genes for Mammalian Sex Determination and Differentiation." *Genetica*, 52/53:267–73.

Roberts, L. 1988. "Zeroing in on the Sex Switch." *Science*, 239:21–23.

Segal, S. J. 1974. "The Physiology of Human Reproduction." *Scientific American*, 231:52–62.

Reproduction: technologies and choices

Increasing knowledge of human reproductive biology has led to the development of a wide array of medical techniques to manipulate, control, and—some would say—intrude upon conception, gestation, and fetal health (figure 7.1). In this chapter, we consider some of these techniques and their genetic and ethical relationships.

Birth Technologies

Techniques for enhancing fertility are an important aspect of contemporary medical practice. Some of these procedures have received considerable attention in the public media, often being incorrectly labeled as a form of genetic engineering. They are not genetic engineering because the genomes of neither parent nor child are affected by such procedures. We discuss them here because reproduction is so closely linked to other aspects of genetic health.

Artificial Insemination

Artificial insemination (AI) refers to the introduction of human sperm into the female vagina and cervical canal by means other than normal sexual intercourse. The first recorded use of AI in humans was in 1790. AI is most often used by couples when the husband is sterile, but the wife is fertile. In the past the doctor would arrange for a suitable anonymous sperm donor, often a medical student. The donor male collects the semen by masturbation, and the semen is then introduced into the woman's reproductive tract by the doctor.

The probability of successful fertilization is increased by careful monitoring of the woman's ovulatory cycle, so the sperm can be introduced at the most opportune time. Generally, healthy women conceive after one to four AI attempts. In some cases of AI the sperm donor is the husband, who may have a low sperm count. Combining semen from several ejaculations can increase the sperm count and increase the probability of fertilization. This is possible because of the technique of *sperm banking*.

In the United States, AI is used on about 172,000 women annually. The result is about 65,000 babies, or 1% of all births. About 35,000 of these use sperm donated by the husband or partner of the mother, while 30,000 use donor sperm.

Given the large scale on which AI is used, it is surprising that governmental regulation is relatively limited. One survey of private physicians administering AI showed that only 44% routinely tested sperm donors for the HIV virus believed to cause AIDS; 30% tested for syphilis, gonorrhea, or other sexually transmitted diseases; and 48% screened for common genetic diseases such as sickle-cell anemia or Tay-Sachs disease. Women in Australia and Canada are now known to have acquired an HIV infection as a result of AI.

Sperm Banks

Human sperm can be maintained indefinitely at ultracold temperatures in **sperm banks,** just as can the sperm of other animals. Sperm banks were established commercially in the United States in 1970. A fresh sample of semen is usually distributed among several ampules and then frozen in liquid nitrogen ($-196°$ C).

Most commonly, semen for artificial insemination has been donated anonymously by medical students. More recently, men undergoing sterilization by vasectomy have placed samples in sperm banks as a form of fertility insurance, in the event they later wish to reproduce.

In the 1960s, geneticist H. J. Muller advocated that AI not rely upon randomly donated and chosen sperm. Instead, he urged that sperm should be collected from potentially high-quality donors and not be used until after the donor's death, so that the donor could be evaluated properly. The crux of such a program would be the criteria used to evaluate donors; understandably, agreement on qualities of donors would be difficult to achieve in a diverse democratic society. Muller contended, however, that children conceived by AI should be given every possible opportunity for a high-quality life.

In 1980 a California businessman established an exclusive sperm bank. All donors were required to have exceptional IQs, and Nobel prizewinners were particularly solicited. Several Nobel laureates have contributed thus far, all anonymously except William Shockley, a co-winner of the 1956 Nobel Prize for Physics. Shockley was in his seventies at the time of his deposit, a factor that might run counter to Muller's high-quality criteria, as mutations in the germ cells of both males and females accumulate with age (see chapter 11).

Sperm from this exclusive bank are made available only to selected intelligent, healthy females under thirty-five years of age who are married to sterile males or are single. The sperm bank was advertised in the bulletin of MENSA, an organization of members whose IQs are among the top 2% of the nation. In 1982 the first child born to a woman using the exclusive sperm bank was announced. Many more have been born since, and many appear to be very bright children. However, most have been raised in privileged homes and in conditions highly conducive to rapid development and learning. Thus, even if these children prove to be more intelligent than average, which remains to be seen, environment is as likely to be the cause as genes.

Although it is a private citizen's right to open a selective sperm bank, such an act implies both that there is a large genetic component to human intelligence and that extremely high intelligence is desirable. The first implication is a matter of intense debate (see chapter 9), and the second is a matter of personal values.

Of far greater significance to public health is the security that sperm banks can afford those using AI. Initially, freezing of sperm was feared to cause mutations that might lead to birth defects and genetic diseases. Extensive experience with frozen semen of humans and domesticated animals such as sheep, cows, and pigs now indicates that freezing is extremely safe.

In marked contrast to private physicians who use fresh sperm for AI, virtually all sperm banks test for HIV infections of the donor, and the large majority test for genetic and transmissible diseases. The U.S. Food and Drug Administration now recommends that semen used for AI be stored for at least six months prior to use and that the donor be given a blood test for HIV infection after donating the semen sample. Fresh semen should be used only when the donor and recipient each have a long-term mutually monogamous relationship. Thus, far from being a risk, freezing has instead come to be an important part of safety procedures.

Surrogate Motherhood

When a wife is sterile, her husband's semen can be used to fertilize the ovum of a **surrogate mother** by artificial insemination. Following birth, the child is given to the couple to raise as their own.

A variation on this procedure, called **gestational surrogacy,** is sometimes used for women whose ovaries are functional but who cannot carry a fetus, for example, because they have had the uterus removed. One or several eggs are removed from the woman, subjected to *in vitro fertilization* (see next section), and the fertilized egg(s) then implanted in another woman.

In both types of surrogacy, the surrogate mother is usually paid a large fee for her services and for surrendering all legal claims to the child; ten thousand dollars is a common amount. All her medical expenses are usually paid by the couple receiving the child.

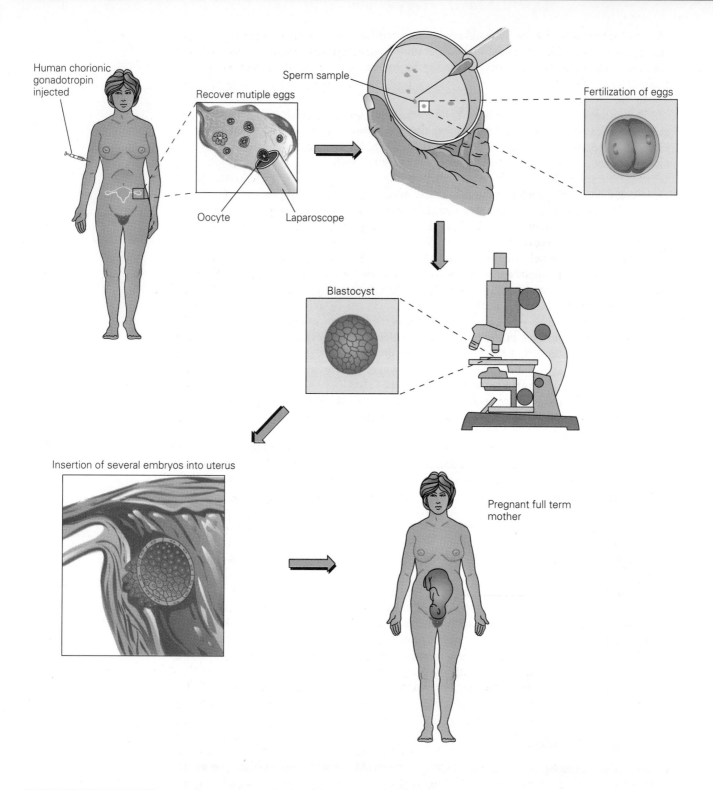

Human chorionic gonadotropin injected

Recover mutiple eggs

Oocyte

Laparoscope

Sperm sample

Fertilization of eggs

Blastocyst

Insertion of several embryos into uterus

Pregnant full term mother

Figure 7.2

Diagrammatic interpretation of *in vitro* fertilization.

Surrogate motherhood is now practiced openly in many countries. It has probably long been practiced secretly, as well. Excluding cases associated with *in vitro* fertilization, about 100 babies are born in the United States each year by surrogate mothers.

Nonetheless, surrogate motherhood does not have wide public acceptance. Many people object to the financial aspects, believing it to be like buying a baby. Some people are offended by the class differences associated with surrogacy arrangements; commonly, a relatively wealthy couple pays a poorer woman to incubate their child. On the other hand, normal adoption often involves large amounts of money, and the children often come from lower class families. Some hospitals participate in surrogacy arrangements only when the surrogate mother receives no fee. One case in Britain received lurid publicity because the contracting parties decided to avoid the medical expenses of artificial insemination and used traditional methods of conception. The surrogate mother revealed details of the arrangement in an article that she wrote for a daily newspaper, for which she received almost as much money as for being a surrogate.

A specialty of law has developed to deal with the many problems created by surrogate maternity. Some surrogate mothers, for example, have changed their minds following birth and wished to keep the child. If the child is deformed or genetically abnormal (Turner's syndrome, for example), has the surrogate mother fulfilled her part of the contract, or must the child be normal? What assurances can be made that the AI sperm actually fertilized the egg, and not the sperm of the husband of the surrogate mother? Contracts negotiated between the couple and the surrogate mother attempt to specify unambiguously the rights of all parties, but it will likely be decades before most legal ramifications of surrogate motherhood are fully resolved.

A case that received much publicity in the United States illustrates how complex and soul-wrenching surrogate motherhood can become. A couple who had no children due to the wife's infertility contracted with a married woman to be the surrogate mother of the husband's child. Following artificial insemination, the contracted surrogate became pregnant and gave birth to a child with a serious defect, *hydrocephalus*, an accumulation of fluid in the center of the brain that causes an enlarged skull and compressed brain. The contracting couple then refused to accept the child, not on the grounds that the child was defective, but that the husband was not the father of the child—in other words, that conception had not occurred as a result of the artificial insemination. Subsequent paternity analysis showed that the contracting husband was indeed not the father of the child; the results supported the probability that the surrogate's husband had fathered the child. The surrogate mother then accepted responsibility for the child.

Most people find such conflicts appalling and feel particular sympathy for the innocent child at the center of the debate. Although this particular case was resolved by a successful paternity test, such will not always be true. Furthermore, this case does not set a legal precedent for the rejection of a child *because* it is defective, as that was not the issue.

In vitro Fertilization

Some women are sterile because their oviducts are blocked. Although they ovulate normally, the blocked passageway precludes access of sperm to the ovum. The technique of **in vitro fertilization** (IVF—figure 7.2) can often provide an effective cure for sterility for such women.

In theory, the technique is simple. A physician collects freshly ovulated ova from the woman. These eggs are then fertilized in a dish in the laboratory—*in vitro* (in glass)—using the husband's sperm or that of a donor if the husband is sterile. About a week after fertilization, one or several viable embryos are placed in the mother's uterus. If implantation occurs, a normal pregnancy usually follows.

In vitro fertilization was developed in the 1950s for use in animal husbandry (box 7.1). Doctors in Britain and Australia pioneered the use of these techniques on humans, and since the first successful birth in 1978, thousands of babies worldwide (including some twins and triplets) have been born as a result of IVF. About 600 IVF children are born each year in the United States. IVF children appear to be normally healthy. The technology is advancing rapidly; some embryos have now been stored in ultracold embryo banks for as long as six months prior to successful implantation. Being frozen as an embryo appears to have no harmful effects on the child.

The popular media apply the provocative term "test-tube babies" to these children, but it is a misnomer. Gestation does not occur in a test tube. Other than the brief period of fertilization and early development, usually less than a week, gestation occurs totally within the uterus of the mother, just as in other pregnancies.

Although IVF is a technique of great importance to animal husbandry, its application to human fertility is fraught with both practical and ethical problems.

BOX 7.1

Birth technologies have revolutionized animal husbandry

Figure 7.3

In vitro fertilization as practiced on cattle.

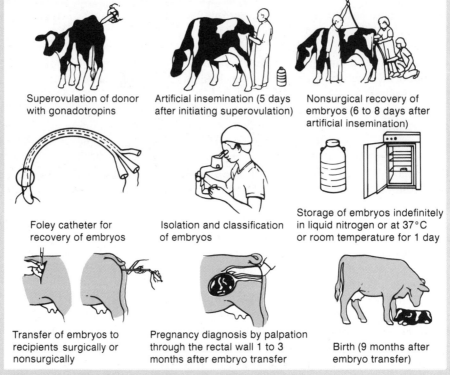

Superovulation of donor with gonadotropins

Artificial insemination (5 days after initiating superovulation)

Nonsurgical recovery of embryos (6 to 8 days after artificial insemination)

Foley catheter for recovery of embryos

Isolation and classification of embryos

Storage of embryos indefinitely in liquid nitrogen or at 37°C or room temperature for 1 day

Transfer of embryos to recipients surgically or nonsurgically

Pregnancy diagnosis by palpation through the rectal wall 1 to 3 months after embryo transfer

Birth (9 months after embryo transfer)

The spread of useful new traits such as increased fertility or the propagation of new breeds of domestic livestock has traditionally been a slow process. The normal reproductive rates of most animal species allow new characteristics to be spread through existing herds only over many generations, usually involving five, ten, or more years. Beginning in the 1950s, the techniques of **superovulation** and **embryo transfer** were developed to increase the reproductive rates of valuable livestock (figure 7.3). Since the 1970s, these techniques have been used to respond to market factors and to improve the genetic quality of many herds. They have also served as a laboratory for development of IVF techniques for humans.

The techniques used on animals are very similar to those used for humans; the primary difference is that fertilization is not external. Injections of gonadotropic hormones cause a cow to superovulate, i.e., to ovulate many eggs rather than just one. Shortly after ovulation, the cow is inseminated, either naturally or artificially, with the sperm of a high-quality bull. About a week after insemination, the fertilized embryos are collected from the oviducts prior to implantation in the uterus and stored in liquid nitrogen ($-196°$ C).

Later, the frozen embryos are thawed and implanted in other cows, the *surrogate mothers*. Surrogate mothers are less valuable as breeding stock than the genetic mother. They may, for example, produce less meat or milk than females used as the genetic mothers, but they are physiologically well suited to the rigors of pregnancy and birth. About 50% of all embryos implanted in surrogates produce viable offspring. Recalling that more than half of all human embryos abort spontaneously in the first trimester, the success of this technique is remarkably good. All ten calves in figure 7.4 are the progeny of a single superovulation, and as many as twenty calves have been produced from a single superovulated cow.

Since its commercial application in the 1970s, this technology has produced tens of thousands of calves worldwide. Superovulation and embryo transfer are also widely used in the husbandry of horses, swine, goats, sheep, and rabbits. Frozen embryos of these species can

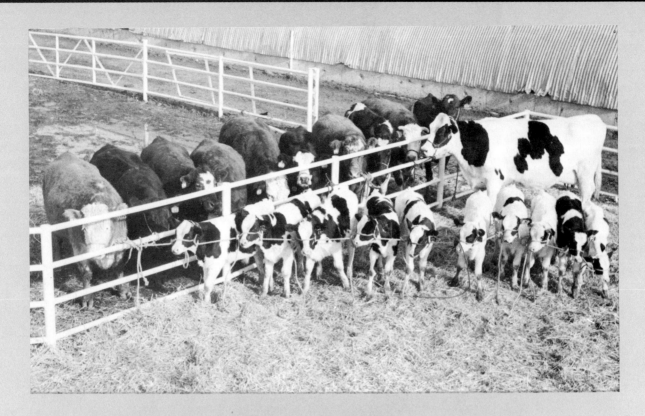

be temporarily implanted in a rabbit and transported worldwide with relative ease. An entire herd of cattle can be readily introduced from one country to another in just a few small rabbits, without worry of transmitting disease and at relatively low cost.

Future uses of this technology appear likely to lead to truly extraordinary reproductive rates. For example, scientists have successfully removed single cells from a sixteen-cell sheep embryo. Three cells were then injected into separate unfertilized sheep eggs, producing three normal sheep. These were all *clones,* genetically identical copies, of the original embryo. Combined with existing technologies, this type of cloning could allow the cow in figure 7.4 to produce one hundred or more offspring per reproductive cycle, rather than ten. In other words, the most desirable animals may soon be able to parent an entire herd in each reproductive cycle. These techniques may also offer new hope for rapid propagation of endangered species (figure 7.5).

Figure 7.4

A "genetic" mother and her ten offspring produced in one breeding season by superovulation and embryo transfer. Some of the ten surrogate mothers are shown on the left.

George Seidel.

Figure 7.5

Mary Alice, the world's first tiger produced by *in vitro* fertilization. Born in April, 1990, Mary Alice is a rare Siberian tiger with only 200 of her subspecies remaining in the wild. She represents a costly investment (over $2 million dollars) in an attempt to use IVF to preserve endangered species. She is currently on display at the National Zoo in Washington, D.C.

AP/Wide World Photos.

BOX 7.2

"Excess" fetuses and scientific research

Figure 7.6

A fourteen-day-old human embryo. Only the tissues in the middle (the primitive streak; arrow) of the embryo will form a human being. The remaining embryonic tissues form membranes and the placenta that support the developing fetus.

R. O'Rahilly and Müller, "Developmental Stages in Human Embryos," *Science,* 246:1554–5, 1991.

An unexpected result of *in vitro* fertilization has been the production of large numbers of fetuses that will not be used by the couples who donated the sperm and eggs. These fetuses have been either allowed to die without being implanted, or they have been ultrafrozen for possible future use. Or, they may have been used for scientific research.

Unpleasant though the notion of experimentation on fetuses is, research using living fetuses or tissues from dead fetuses has been conducted for many decades, with significant medical benefits for millions of people. The polio vaccine was developed in the 1940s and 1950s using tissues from spontaneously aborted fetuses. Vaccine research still requires the use of fetal tissues. Some research is only possible using fetuses. Drugs administered to a pregnant woman, for example, may have adverse effects on her fetus, but these drugs can only be identified conclusively through tests on "excess" fetuses. More such research might have prevented the severe deformities affecting 3,000 children born in the 1960s whose mothers were given the drug thalidomide. Procedures for prenatal diagnosis of genetic diseases can only be developed through experiments involving fetuses.

Fetal cells differ significantly from cells taken from adults. They grow well in culture, still have the potential to differentiate into a variety of cell types, and do not induce immune responses when transplanted into other people. These characteristics make them ideal for one of the most exciting avenues of medical experimentation now in progress, **fetal-cell surgery.** This technique involves implanting fetal cells into defective tissues and organs of persons suffering from conditions such as diabetes, Parkinson's disease, and the blood disease thalassemia. If all goes well, normal fetal cells may then assume the functions of the diseased tissues, producing insulin, dopamine, or normal hemoglobin, and thus cure the disease. Literally millions of people stand to benefit if fetal-cell surgery proves successful. Dr. Robert Edwards, who developed IVF techniques for use on humans in collaboration with Dr. Patrick Steptoe, justified his work by the many advances in medical treatments likely to result from the ability to control and manipulate early human embryos.

The first fetuses used in research were obtained from spontaneous abortions. But the supply was not reliable, and many of the fetuses aborted because they were in some way defective and thus often not suitable for research purposes. In recent years, fetuses from therapeutic abortions have been the primary source for medical research.

Thus, the objections to use of living fetuses or fetal tissue in research derive largely from arguments against abortion. Opponents of abortion argue that human life begins at conception, and abortion is therefore murder. In this view, fetal research amounts to complicity in murder; research is likened to medical torture of the type Nazi scientists employed. Some opponents argue that medical research should be conducted only on persons giving informed consent, which a fetus cannot do. People who do not necessarily oppose abortion sometimes object to fetal research using a "slippery slope" argument, i.e., that fetal research may not be inherently bad, but that it might lead to worse practices, such as research on advanced fetuses or babies, or the creation of embryos solely for research purposes. Some people have argued that no special knowledge can be gained by fetal research, a view that most scientists would dispute.

Proponents of fetal research almost always adopt a utilitarian or pragmatic position. Research on fetuses offers the possibility to reduce so much human suffering that it would be immoral not to pursue it. Fetal-cell surgery, they note, is analogous to normal organ transplantation, which is relatively non-controversial.

In the United States, all fifty states and the federal government permit tissues taken from dead fetuses to be used for research. Guidelines are taken from the recommendations of the 1978 National Commission for the Protection of Human Subjects of Biomedical and Behavioral Research. They allow use of aborted fetuses for medical research if (1) they are less than twenty weeks of gestational age, (2) the research does not alter their life spans, (3) the knowledge is important and cannot be gained by other means, (4) suitable preliminary experiments have been conducted on other animals, and (5) the research has been approved by appropriate review boards.

Nonetheless, divided opinion in the United States is reflected in regulations prohibiting funding of research involving fetal tissue. Because most medical research is extremely expensive, limiting funding is almost as effective as prohibiting the research. In 1988, a federal review panel recommended that research on fetal tissue be eligible for federal funding as long as it complied with the above guidelines and two other key safeguards: (1) tissues from fetuses could not be sold (e.g., to wealthy persons needing fetal-cell surgery); and (2) a woman's decision to have an abortion must be entirely independent of the decision to use the tissue for research or transplantation. The latter requirement would prevent, for example, a woman aborting a fetus to provide tissues for transplant to another of her children, or even for herself. However, the recommendations of the review panel have not been implemented, and federal funds are still prohibited from supporting fetal research.

Internationally, laws are equally divided. In 1990, the British government followed the recommendations of the Warnock Commission and decided to allow research on fetuses up to the age of fourteen days, with appropriate regulations. France is adopting similar laws. Denmark, on the other hand, prohibits any research on fetuses, and countries such as Ireland and Spain have similar positions. Because of the ethical horrors of medical research under the Nazis, leading German research institutions have established their own prohibitions on fetal research.

The disparities in law and practice among countries are unlikely to be resolved easily or soon. One European diplomat summarized the complexities of the present situation by observing that "the same research might lead to a Nobel Prize in some [countries], and to prison in others."

TABLE 7.1

The incidence of different categories of genetic diseases

Type of Condition	Frequency Per Thousand
Chromosome abnormalities	6.9
Defects caused by single genes of major effect	
Autosomal dominant	1.9–2.6
Autosomal recessive	2.2–2.5
Sex linked	0.8–2.0
Subtotal	11.8–14.0
Congenital malformations*	(19.0–22.0)
Complex disorders†	(7.0–10.0)
Total genetic contribution (approximately)	(38.0–46.0)

From *Report of the Task Group on WHO Human Genetic Programme Health for All by the Year 2000: The Contribution of Human Genetics,* WHO Report HMG/TG/81.3.1981. Reprinted by permission.
*One-half of malformations such as spina bifida, congenital heart disease, and cleft lip with or without cleft palate are assumed to be genetic in origin.
†One-third of disorders such as schizophrenia and diabetes mellitus are estimated to be genetic.

Figure 7.7

Amniocentesis. (a) Ultra-sound analysis is used to visualize the fetus and determine its position, to allow safe sampling of amniotic fluid. Ultra-sound analysis can also be used to monitor the growth of the fetus during pregnancy. (b) A needle and syringe are used to withdraw 10–30 ml of amniotic fluid, which is then centrifuged to separate fetal cells and fluid. The cells are cultured for karyotypic, biochemical, and DNA analysis; some biochemical tests can be made directly on the amniotic fluid. Alternatively, a catheter is inserted through the mother's cervix and fetal cells are sampled directly. Karyotypes can be made, biochemical tests run, and/or DNA from these fetal cells can be cut with restriction enzymes for RFLP analysis or amplified by PCR for direct gene sequencing.

(a) © Will McIntyre/Photo Researchers.

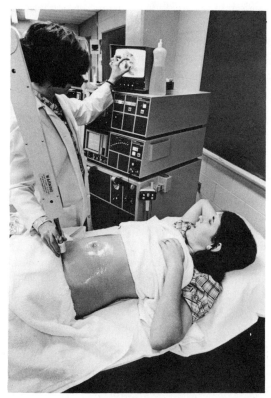

(a)

Practical Problems of IVF

IVF has only limited application to curing infertility. The success rate is low. Only 17% of couples attempting IVF get pregnant, and only 10% bring home a baby. The treatment involves risks, and it is expensive. Early fears of the manipulation or freezing of embryos causing birth defects have not been realized.

IVF is suitable only for treating women with blocked oviducts, about 2% of all cases of infertility in women. Although IVF does allow some of these women to have children, it does not cure the blocked oviduct. Thus, IVF must be used for each subsequent pregnancy attempt. Furthermore, fertilization is seldom achieved on the first attempt. Six to twelve monthly attempts are often required before pregnancy occurs. Each attempt involves hormone treatment (of unknown risk) and the risk of minor surgery. Also, spontaneous abortions are as common as with a normal pregnancy.

Amniocentesis

Amniotic fluid (contains some fetal cells)

Chorionic villus (entirely fetal cells)

Chorionic villus sampling (by suction)

Centrifugation

Supernatant fluid

Cell pellet

Supernatant fluid

Cell culture

DNA tests

A — DNA isolated

B — DNA cut and/or amplified by PCR

RFLPs

Fingerprints

Biochemical tests

Chromosomes

karyotype

(b)

Financial costs for IVF are high. Each attempt may cost ten thousand dollars, and women requiring multiple attempts may spend fifty thousand to one hundred thousand dollars, with less than a 10% chance of success.

Ethical Problems of IVF

The use of IVF has been questioned in two ways. First, ethically the high cost and low success rate of the technique virtually guarantee that only the wealthy will have access to IVF, unless public funds are used to make it widely available. In the United States, IVF is undertaken only at private clinics because the technique is considered to involve experimentation on fetuses, for which federal funding is prohibited (box 7.2). IVF has thus been criticized as another example of disparity in health care between the wealthy and poor.

More broadly, IVF is not cost-effective in treating infertility. It is a modestly effective treatment suitable for only a tiny fraction of infertile women. Opponents of IVF argue that infertility could be *prevented* (as opposed to treated) by research and education programs designed to reduce the effect of, or cure, the primary causes of infertility, such as sexually transmitted diseases. Such programs would be of particular benefit to the poor who could never afford IVF. Some opponents also argue that the primary beneficiaries of IVF are not infertile women, but rather medical professionals, who receive large financial rewards and public adulation, despite the low applicability and success rate of IVF.

A second ethical constraint on IVF is the problem of "excess" embryos. Women undergoing IVF receive human chorionic gonadotropin, a hormone that controls ovulation (so the ova can be collected at the proper time) and induces superovulation. This allows many eggs to be fertilized and multiple embryos to be introduced to the uterus. Extra embryos are also frozen for later use.

The fate of these frozen embryos or other embryos not implanted is the source of controversy. What should be done with them, particularly after the woman either becomes pregnant or stops IVF treatment? Those who believe that all human life is sacred and begins at conception view disposal of these embryos as a form of abortion or, in their words, murder. Other people take a more utilitarian view, believing that termination of the embryos is appropriate and that under appropriate controls the embryos may be used for medical experimentation. Views on both sides are very strongly held and expressed (box 7.2).

Controversy over the use of IVF embryos can be expected to continue for many decades. The unpredictability of the course of this controversy is shown by a case in Australia in the mid-1980s involving a wealthy couple undergoing IVF treatment. The couple had yet to achieve a successful implantation when they were both killed in an accident, leaving some embryos frozen in the IVF laboratory. A suit was filed on behalf of the embryos as the rightful and only heirs to the wealth of the couple, presumably to require that they be implanted in a surrogate mother. A court ruled against the suit, however, and the embryos were allowed to die.

Prenatal Diagnosis

Diseases of genetic origin are a major human health problem. Some 3–5% of all humans, perhaps 250 million persons, are seriously affected by genetic diseases (table 7.1). In the United States, about one hundred fifty thousand children are born each year with genetic diseases.

In this section we discuss briefly the primary techniques now available for sample tissues *in utero* for genetic analysis. These tissues can then be subjected to **prenatal screening** for chromosome, biochemical, or DNA effects. Over 400 conditions can be diagnosed prenatally. Diagnostic techniques are developing so rapidly that by the turn of the century it may be possible to prenatally diagnose most common genetic diseases and also to identify carriers of deleterious recessive alleles.

Amniocentesis

Developed in the 1960s as a test for maternal-fetal Rh incompatibility, **amniocentesis** is a procedure that samples the amniotic fluid surrounding the fetus. A needle is inserted through the mother's body wall and uterus, and about ten to twenty milliliters of fluid are withdrawn (figure 7.7). Amniotic fluid contains fetal secretions and living epidermal cells from the digestive and respiratory tracts. The living cells are grown in laboratory culture and then subjected to both karyotypic analysis and biochemical tests. Over one hundred single gene defects can be identified from tissues collected by amniocentesis. The risk of complications due to amniocentesis causing loss of a pregnancy is about 0.5% in the United States.

Timing of amniocentesis is crucial. Before the fifteenth week of pregnancy there is insufficient amniotic fluid to permit safe extraction for culturing. It takes two to three weeks to grow cells and culture chromosomes for karyotype analysis and an additional four to six weeks to grow enough cells for successful biochemical tests. Therapeutic abortion is legal in most states only through week twenty-eight. If results of the first amniocentesis are inconclusive, little time is available for a second testing. Furthermore, danger to the mother during abortion increases with duration of the pregnancy. According to the U.S. National Academy of Science Institute of Medicine, the risk of maternal death due to legal abortion in the first trimester is roughly 1.8 per one hundred thousand. Abortions following amniocentesis are performed in the second trimester when the risk of death has risen to 12.2 per one hundred thousand, still slightly lower than the risk during a noncaesarian live birth of 12.5 per one hundred thousand.

The most common conditions detected by amniocentesis are Down's syndrome (chapter 11; about 5,000 cases annually, identified by the presence of an extra chromosome 21) and neural tube closure defects (about 6,000–8,000 cases per year, detected by the presence of elevated levels of alphafetoprotein in the amniotic fluid). The latter condition results in anencephaly (absence of the brain and spinal cord) and spina bifida (an open spine, due to failure of normal closure of the spinal cord during embryonic development).

Chorion Villus Sampling

The technique of **chorion villus sampling** (CVS) allows more rapid diagnosis of genetic conditions than amniocentesis. As early as the eighth week of pregnancy, a plastic tube can be inserted through the vagina and cervix, the entrance to the uterus. A tiny piece of the chorionic membrane of the fetus is snipped off painlessly and extracted.

Because the chorionic tissue is rich in cells, karyotypes can be prepared and biochemical tests conducted within twenty-four hours, rather than waiting for cells to be cultured. Abortions of seriously affected fetuses can thus be performed with increased safety in the first trimester, if necessary. The risk of the CVS procedure itself is slightly higher than for amniocentesis.

Fetoscopy

Fetoscopy is a technique introduced in the early 1980s that allows a physician to visualize a fetal or umbilical blood vessel *in utero*. A needle can then be inserted into the vessel to obtain a sample of fetal blood that can be examined for genetic conditions such as hemophilia. Additionally, intravenously administered therapies rely upon fetoscopy. For example, vitamin deficiencies, which can cause mental retardation, can now be treated prenatally; in addition, blood transfusions can be administered to fetuses.

Any technique allowing visualization of the fetus—fetoscopy or ultrasound, for example—can also be used to diagnose morphological defects prenatally. Sometimes, as in the case of blocked ureters, prenatal surgery may allow the fetus to be saved.

Use of Prenatal Diagnosis

Who should have prenatal diagnosis? At first glance, widespread screening of all pregnant mothers for the dozens of genetic conditions that can now be detected prenatally might seem a good idea. Prevention of all genetic diseases would be an obvious benefit to society, but three factors presently preclude the application of these techniques to all pregnant women: risk, cost, and use of the information.

Risk

Amniocentesis and chorion villus sampling are associated with a number of risks. Amniocentesis may lead to leakage of amniotic fluid. In addition, the needle may cause injury to the fetus, and lung and limb problems sometimes result. Both amniocentesis and chorion villus sampling can lead to infections; the spontaneous abortion rate following either procedure is less than 1%, much lower than the incidence rate for serious genetic anomalies. No such risks are known to be associated with ultrasound observation of the fetus, but some scientists caution that the technique has not been used long enough to discount the possibility of long-term risks.

Cost

Prenatal sampling of tissues is usually an out-patient procedure requiring approximately one hour in a clinic. Total costs for ultrasound, amniocentesis or chorion villus sampling, karyotype analysis, and biochemical and DNA tests are usually at least $1,000. A single biochemical test may cost several hundred dollars.

Use of Information

Although many genetic diseases can be diagnosed (and the number is increasing rapidly), many cannot be treated. The most common use of diagnostic techniques is to detect serious genetic conditions such as Tay-Sachs disease (table 4.2) or major chromosomal anomalies (chapter 11) and to provide information upon which a decision for therapeutic abortion can be based.

However, routine prenatal diagnosis would also discover many genetic conditions that have only minor effects or whose expressions are variable or uncertain. For example, karyotypic analysis would regularly reveal XYY males, but as discussed in box 6.2, XYY males present a dilemma. The condition cannot be treated, and XYY males usually lead normal lives, although there is some possibility that they may be prone to aggressive behavior. In addition, the *knowledge* that an individual is XYY may itself induce the feared aggressive behavior.

Thus, a spectrum of genetic conditions exists, some causing such serious problems that abortion seems the reasonable choice, but many others in which diagnosis itself may compound or even create problems. This consideration, alone and in conjunction with risk and cost factors, argues strongly that wide-scale, indiscriminate prenatal screening is inappropriate. Future reductions in risk and cost combined with the ability to treat most genetic conditions might warrant a different view, but for today's society such screening cannot be justified.

Criteria for Prenatal Diagnosis

In general, and with a certain amount of latitude on the part of the physician, prenatal diagnosis is recommended and permitted for cases in which all of the following criteria are satisfied:

1. *The fetus is known to be at special risk for certain genetic conditions.* If a family has a history of a serious genetic disease such as hemophilia or sickle-cell anemia, prenatal diagnosis is recommended. This is particularly true if a mother has already given birth to an affected child, and thus the parents are known to be carrying deleterious genes. Other special risk groups might include women taking certain mutagenic or teratogenic drugs such as thalidomide, women contracting rubella during pregnancy, or women over thirty-five, who have an increasing probability of giving birth to children with Down's syndrome.

2. *The genetic condition under screening is serious.* Because the most common result of a positive finding following prenatal screening is therapeutic abortion, screening for nonserious conditions is pointless. Screening for blood type, as an extreme example, is of no merit.

   While such a requirement (that the genetic condition under screening be serious) might not need explicit statement, at least one example suggests otherwise. In India, abortions are strictly limited and are legal only when the life or health of the mother is threatened. Studies of abortions in India revealed that over 99% of abortions— 7,997 of 8,000 cases, in one instance—involved female fetuses. Clearly, prenatal diagnosis of sex is practiced on a large scale, and the results are used illegally to pursue an ancient social prejudice.

   Other societies also have biases against female children. In China, where government policy imposes penalties for families with more than one child, female infanticide is sometimes claimed to be a common practice. In Western societies, more prospective parents state preferences for male babies than for females. Although physicians in Western societies will not conduct prenatal diagnosis to determine the sex of the fetus alone, diagnosis of sex is used for serious X-linked diseases. If the parents are known carriers or are at risk of being carriers for hemophilia or Lesch-Nyhan syndrome (table 5.1), for example, a karyotypic sex test is used; male fetuses can be aborted, and females can be permitted to continue gestation.

In vitro fertilization offers even more opportunity for sex testing when required. It is possible to remove one or two cells from the embryo at the six or eight cell stage, before it has been implanted in the mother. DNA in those one or two cells can then be amplified sufficiently to test for the presence of Y-specific segments of DNA that would identify the embryo as male. Embryos lacking those sequences are female and therefore unlikely to express any X-linked diseases. In two instances, this procedure has been used to produce normal pregnancies by implanting female embryos in women who had previously given birth to boys affected by serious X-linked conditions.

3. *There is no treatment for the disease.* If the condition is treatable, therapeutic abortion cannot be justified, and the parents gain little from knowing in advance of their child's illness. Such knowledge may cause needless worry during pregnancy. Screening for PKU, for example, would not be permitted.

4. *The parents must be willing to terminate the pregnancy following a positive diagnosis.* Again, diagnosis for its own sake has limited virtue and may cause needless psychological trauma. If the parents are not willing to abort an affected fetus, the risk of the diagnostic procedure is added to the risk of genetic disease with little or no benefit.

In some cases, willingness to terminate the pregnancy is not a requirement for prenatal diagnosis. Some physicians believe that knowledge that an unborn child is affected can help those parents unwilling to abort, by allowing them to make both emotional and physical preparations prior to birth, which may aid them in caring for their seriously affected child.

Bioethics and Therapeutic Abortion

In this chapter and in chapter 4, we have noted that **therapeutic abortion** is one response to knowledge that a fetus is at risk of or suffering from a serious genetic disease. For many persons the decision to abort a fetus affected by a serious genetic condition such as Lesch-Nyhan syndrome is straightforward. These people support such a decision because it has many consequences that they see as beneficial to all involved—the fetus, the parents, the family, and society at large. The fetus is spared a short, painful life. The family is spared the overwhelming burden of disruption, emotional anguish, and medical expenses. Society is spared the high financial cost of caring for the child if the family cannot do so.

Others, however, find abortion unacceptable under any circumstance. They often support their views on the grounds that abortion is murder and that murder is never acceptable.

The conflicting views expressed over the suitability of abortion are referred to as a philosophical **dilemma,** a choice between alternate courses of action that each can be fully justified by its proponents. A dilemma poses an apparently insolvable predicament. The study of human conduct, especially in terms of distinguishing good and evil, lies within the philosophical field of **ethics.** Difficult choices between conflicting behaviors are said to be **moral decisions.**

Advances in biological knowledge have led to the development of a subdiscipline of ethics called **bioethics.** Bioethics deals with moral decisions arising from choices now available to humans because of new biological technologies and information. Many of these choices relate to human health and medicine, such as the decision to abort or to allow gestation to continue. (Other bioethical decisions arise from our treatment of the environment—for example, what limits should be set on exploitation of nature?)

Utilitarian or Absolutist?

Bioethical analysis often cannot determine what is right or wrong. Instead, conflicting choices presented by new biological technologies are analyzed within an ethical framework. Furthermore, the rapid development of new technologies is outracing our ability to conduct bioethical analysis and achieve a societal consensus on the use of new techniques. Thus, although bioethics can help us to identify and understand issues, the final choice on particular issues must be made by the individual based on his or her personal values. Scientific knowledge alone is never sufficient to provide answers to the most difficult ethical questions.

The two most common types of justification of bioethical choices are **utilitarian** and **absolutist.** A utilitarian decision is made by examining the probable consequences of a decision and attempting to maximize the benefits to all persons affected by the decision. In the case of therapeutic abortion, a decision to abort might be based on a desire to spare the fetus, family, and society needless pain and suffering. Refusing to abort a seriously affected fetus might be expected to cause pain to all involved.

On the other hand, a utilitarian view could be used to oppose abortion by employing a *slippery slope* argument. A slippery slope argument often claims that while the immediate effects of a decision may be harmful or unpleasant, the long-term consequences justify such a choice. For example, a slippery slope (and utilitarian) argument against abortion is that abortion conflicts with the Hippocratic Oath, which requires physicians to save lives. In the long-term, allowing abortion might undermine the entire basis of medical practice. A very different utilitarian argument might claim that abortion should be acceptable only if supported by public funds, making it as available to the poor as to the wealthy, unlike the present situation in the United States. Otherwise, the inequity could lead to widespread disrespect for all laws, or to medically unsupervised abortions with a high risk to the mother.

Absolutist ethical arguments are based on the belief that some moral laws are always true and must never be violated. In the case of abortion, the moral law is "murder is evil." Abortions are seen to constitute murder and thus should be opposed, regardless of any practical consequences. This is the most common basis for opposing abortion.

An absolutist position is often associated with a religious viewpoint, but need not be. Absolute rules may be moral values that are self-evident to any thinking civilized person. Such a moral absolute, for example, is the rule "suffering is wrong." Acceptance of this rule could lead to an absolutist justification for abortion, if abortion is seen to reduce suffering. The Hippocratic Oath also requires physicians to reduce suffering.

Societal Standards

It is difficult to comprehend and balance the many aspects of bioethical choices, but the issues are pressing and decisions must be made. Most people will confront bioethical choices during their lives. Abortion, for example, is widely used in Western societies for both health and social reasons, and most families are likely to have a member confronted with this choice at some time. About 1.5 million abortions occur annually in the United States. Worldwide, the highest abortion rate is found in the Soviet Union, mainly because other types of contraception are in short supply. It is common for a Soviet woman to have ten or more abortions in her lifetime. The abortion rate is also high in Japan, where the contraceptive pill is not legal.

Fortunately, society gives some guidance with respect to accepted standards of conduct. These guidelines may take the form of generally accepted religious codes or they may take the form of laws, legal limitations on behavior. Laws, of course, work only so long as they embody widely accepted social and ethical standards, and prohibition of both abortion and alcohol are examples of laws that were widely ignored because they violated this precept.

The U.S. law on abortion derives from the Supreme Court decision of *Roe v. Wade* of 1973. This ruling states that decisions on abortion in the first trimester of pregnancy are the sole right of the pregnant woman in consultation with her physician. Abortions later in the pregnancy are legal if they threaten the woman's health, and in the third trimester the state may intervene to protect the life of the fetus. Subsequent decisions have specifically prohibited the father of the child from interfering with the woman's right to decide, and parents of young unmarried mothers also have no right to interfere.

The role of the law is central in social consensus on ethical matters, but the law is usually conservative (in the general rather than political sense) in its approach. That is, the law responds to changing values only *after* the change. The law usually reflects social consensus at the time a specific law originated, often decades ago, and is subsequently only changed when a new consensus is achieved.

The rapid advance in technology emphasizes the conservative nature of the legal process and the need for an increasingly informed citizenry. Legal codes can only change and stabilize when a consensus is reached. For this to occur, the average citizen must be knowledgeable about new technologies and their ethical implications.

The present diversity of views regarding abortion makes it highly unlikely that a social consensus will be achieved in the near future. In fact, a number of subsidiary issues to *Roe v. Wade* remain to be resolved, for example, the appropriateness of public funding to support abortion. Additionally, many groups object to *Roe v. Wade* and continue to seek constitutional changes to prohibit abortion. And, as medical technology permits survival of fetuses earlier and earlier in pregnancy, the precise rights of fetuses and the determination of when a fetus is to be legally declared a human being are likely to be sources of increasingly fierce debate.

Summary

1. Artificial insemination, introduction of sperm into the uterus and cervical canal by other than natural means, is widely used when a male is sterile or has a low sperm count, but the female is fertile.

2. Surrogate motherhood is sometimes used when a woman is infertile. Her husband's semen is used to artificially inseminate the surrogate, who surrenders her rights to the child upon birth. However, the numerous legal and ethical complications of this technique have limited its acceptance.

3. *In vitro* fertilization is a treatment for female infertility due to blocked oviducts. While increasingly widely used, the low success rate, high cost, and ethical problems related to use of excess embryos pose unresolved problems.

4. Prenatal diagnosis refers to a set of technologies that allow the genetic health of the fetus to be examined prior to birth. Diseases that can be assayed in chromosomal, biochemical, or DNA tests can be diagnosed from amniotic fluid via amniocentesis or from samples of the chorionic villi.

5. Prenatal diagnosis is costly and risky and is not used on all pregnant women. It is applied most often when the fetus is known to be at risk for treatable serious genetic conditions and the parents are willing to terminate the pregnancy following a positive diagnosis.

6. Therapeutic abortion exemplifies the type of bioethical dilemma that many people now confront. Compelling arguments both for and against the suitability of therapeutic abortion can be advanced from many different viewpoints.

7. The two most common bases for bioethical decisions are utilitarian and absolutist arguments. Bioethical decisions must be based on our personal value systems.

Key Words and Phrases

amniocentesis	fetal-cell surgery	sperm bank
artificial insemination	fetoscopy	superovulation
bioethics	*in vitro* fertilization	surrogate mother
chorion villus sampling	prenatal diagnosis	therapeutic abortion
embryo transfer	prenatal screening	

Questions

1. Would you recommend that every pregnant woman undergo amniocentesis? Why or why not? For which groups of pregnant women would amniocentesis be most strongly recommended?

2. Considering *in vitro* fertilization: What happens to a fertilized ovum in the first week following conception in a normal pregnancy, i.e., where is the embryo and what is it doing?

3. How does H. J. Muller's concept of an incubator female differ from surrogate motherhood as it is now practiced?

4. What risks do you think ultrasound analysis might pose to fetuses?
5. Outline both utilitarian and absolutist arguments for and against allowing medical research on fetuses. Construct as many arguments of each type, and on each side of the question, as you can.
6. Justify our statement that "scientific knowledge alone is never sufficient to provide answers to the most difficult ethical questions." Can you give examples?

For More Information

Beardsley, T. 1985. "Problems of Prenatal Testing." *Nature,* 314:211.

Broad, W. J. 1980. "A Bank for Nobel Sperm." *Science,* 207:1326–27.

Brody, B. 1975. *Abortion and the Sanctity of Human Life: A Philosophical View.* MIT Press, Cambridge, Massachusetts.

Byrne, G. 1988. "Artificial Insemination Report Calls For Regulation," *Science,* 241:895.

Djerassi, C. 1989. "The Bitter Pill." *Science,* 245:356–61.

Feinberg, J. 1980. "Abortion." In *Matters of Life and Death,* ed. T. Regan. Random House, New York.

Handyside, A. H., E. H. Kontogianni, K. Hardy, and R. M. L. Winston. 1990. "Pregnancies From Biopsied Human Preimplantation Embryos Sexed By Y-Specific DNA Amplification." *Nature,* 344:768–70.

Hansen, J. T., and J. R. Sladek, Jr. 1989. "Fetal Research." *Science,* 246:775–79.

Harron, F., J. Burnside, and T. Beauchamp. 1983. *Health and Human Values: A Guide to Making Your Own Decisions.* Yale University Press, New Haven.

Kuhse, H., and P. Singer. 1985. *Should the Baby Live? The Problem of Handicapped Infants.* Oxford University Press, Oxford.

McLaren, A. 1989. "IVF: Regulation or Prohibition?" *Nature,* 342:469–70.

Miller, J. A. 1986. "New Molecular Analysis for Genetic Disorder." *Science News,* 129:84.

Pierce, B. A. 1990. *The Family Genetic Sourcebook.* John Wiley & Sons, New York.

Rao, R. 1986. "Move to Stop Sex-Test Abortion." *Nature,* 324:202.

Reich, W. T., ed. 1978. *The Encyclopedia of Bioethics.* The Free Press, New York.

Rensberger, B. 1981. "Tinkering with Life." *Science 81,* November:45.

Saltus, R. 1986. "Biotech Firms Compete in Genetic Diagnosis." *Science,* 234:1318–20.

Seligman, J., J. Gordon, J. Foote, L. Drew, P. King, and S. E. Katz. 1987. "The Grueling Baby Chase." *Newsweek,* Nov. 30:78–82.

Shurkin, J. 1983. "Operating on the Unborn." *Science 83,* May:71–74.

Weaver, D. D. 1989. *Catalog of Prenatally Diagnosed Conditions.* Johns Hopkins University Press, Baltimore.

Genetic Information and Expression

© Biology Media/Photo Researchers.

Informational macromolecules

Figure 8.1

(a)

(b)

(a) An electron micrograph of a cell of *E. coli* that has been treated with mild detergent to disrupt the cell wall and release its DNA (magnification = 20,000×). The small circle of DNA at 1:00 (arrow) is a plasmid that the cell contained, demonstrating the contrast in size of genomic and plasmid DNA. (b) An electron micrograph of a frog virus that has been treated with mild detergent, causing it to release its DNA molecule (magnification = 6,233×).

(a) Dr. Jack Griffith; (b) © Dr. Gopal Murtl/Photo Researchers.

Like most good science, Mendel's work answered some questions, but also raised new ones. Through careful study of phenotypic characters, Mendel described a genetic mechanism and demonstrated the existence of discrete genes underlying phenotypic characters. He also demonstrated that those genes passed unchanged between succeeding generations. Mendel's work, however, shed no light on the physical and chemical nature of the gene. For decades after Mendel's work, many geneticists concentrated on establishing the chemical identity of the genetic material, as well as describing the chemical structure that could explain the Mendelian behavior of genes.

The Search for the Genetic Material

We now know that the hereditary material for all species is **nucleic acid.** In most species it occurs in the form of deoxyribonucleic acid, DNA; for a few viruses, it occurs as ribonucleic acid, RNA. The primary functions of nucleic acid relate to the hereditary process: the storage and transmission of information; and the expression of hereditary information as proteins. Less than a century ago, however, **protein** rather than nucleic acid was the prime candidate for the hereditary molecule.

The Structure of Protein

Proteins were a good guess for the genetic substance. About 66% of the weight of the human body is water; 47% of the remainder (16% of total weight) is protein. Proteins fulfill a variety of crucial functions for the living organism. Of particular importance, one category of proteins—called **enzymes**—serves to *catalyse* the thousands of necessary biochemical reactions in the body. Catalysts accelerate chemical reactions, permitting them to occur in a fraction of a second.

Some proteins function to provide physical support, both for individual cells and for the entire organism. Keratin is a structural protein that is the primary component of skin and hair; tubulin forms part of the "skeletal" or supportive system of an individual cell. Muscle tissue consists of protein specialized for movement. *Antibodies* are proteins that protect the body from bacteria and other foreign substances. Hemoglobin is a protein specialized for the transport of gases in the blood. Some *hormones* are proteins that provide communication among some tissues of the body. A brief introduction to the chemical structure of proteins can help illustrate how proteins fulfill so many different functions and why they once seemed the most likely chemical to bear hereditary information.

Proteins consist of long sequences or chains of molecules known as **amino acids** (figure 8.2). Twenty amino acids are commonly used in the proteins of most species. Humans are able to synthesize twelve of these twenty amino acids, while the remaining eight *essential* amino acids must be supplied from the diet.

Two amino acids are bound together by a **peptide bond** to form a molecule known as a dipeptide. A chain of many amino acids is known as a **polypeptide.** Proteins may consist of

Figure 8.2

The primary structure of proteins: (a) amino acids, the components of polypeptides; (b) the formation of the peptide bond between two amino acids; (c) a polypeptide chain of amino acids held together by peptide bonds; (d) the protein molecule hemoglobin, which transports oxygen to all body tissues. The primary structure is the sequence of the amino acids in the alpha and beta chains. The secondary, tertiary, and quaternary structures reflect the coiling and folding of the chains and their association into a tetrameric molecule with the oxygen-carrying heme groups.

(d) From Stuart Ira Fox, *Human Physiology*, 3d ed. Copyright © 1990 Wm. C. Brown Publishers, Dubuque, Iowa. All Rights Reserved. Reprinted by permission.

only one polypeptide chain, several polypeptide chains, or polypeptide chains combined with other molecules. Hemoglobin, for example, is a protein consisting of four polypeptide chains, each bound to an iron-containing molecule (figure 8.2).

The sequence of amino acids in a polypeptide chain is called its **primary structure.** Most proteins consist of a single polypeptide chain one hundred to five hundred amino acids long, but some are considerably longer. Because a primary characteristic of an hereditary molecule must be the capacity to encode large amounts of information, it seemed possible that the sequence of amino acids in a long protein could easily fulfill this requirement. (As an analogy, the English language—which is quite complex and carries enormous amounts of information—is constructed from an alphabet of only twenty-six letters.)

TABLE 8.1

DNA content of selected plant and animal somatic and gametic cells (DNA measured in picograms = 10^{-12} gram; one-trillionth of a gram).

Organism	2N Cells	Gamete
Fern	326.4	163.2
Fungus	3.0	1.5
Onion	36.7	18.4
Pine tree	138.7	69.4
Human	6.4	3.2
Cow	6.6	3.3
Rat	6.0	3.0
Chicken	2.6	1.3
Frog	15.0	7.5
Salamander	336.0	168.0
Snake	4.2	2.1
Fruit fly	0.4	0.2
Diatom	0.08	0.04

The number of different possible sequences of amino acids in a protein is 20^n, where n is the number of amino acids in the protein. For a polypeptide only four amino acids long, for example, 160,000 different messages are possible; ten amino acids can produce over 10 trillion unique messages. For a protein one hundred amino acids long, the number of messages is unimaginably large. Biochemists of the nineteenth century knew that such a large molecule could possess sufficient variety of structure to fulfill the many functions of proteins, as well as encode the immense amount of hereditary information needed to direct the biochemistry of life.

The complexity of protein structure does not end with its primary structure. Commonly, the proteins in a polypeptide chain coil into a spiral or *helix* (from the Latin word for ivy), which is the *secondary structure* of a protein (figure 8.2). In turn, the helix is twisted and folded precisely, the *tertiary structure,* into a complex, three-dimensional conformation that is crucial to the protein's function. Finally, many proteins consist of linked chains of two or more polypeptides; the arrangement of multiple polypeptides provides such proteins with a *quaternary structure.*

Nuclein

In 1869—only a few years after the publication of Mendel's work, but decades before the recognition of its importance—a Swiss biochemist, Johann Friedrich Miescher, studied the nuclei of pus cells. Miescher isolated from the pus cells a phosphorus rich, nonprotein material that he called *nuclein,* named for its site of origin. (The suffix *-in* usually identifies a chemical as protein.) Nuclein proved to be a nucleic acid, **DNA.**

Work by Miescher and other scientists subsequently demonstrated that the nuclei of cells of all species contain nuclein, and that nuclein was an important constituent of chromosomes. Quantitative analysis of chromosomes showed that they consisted of about 60% protein and 40% nuclein (DNA)—quite reasonably suggesting to scientists at the time that protein was probably the genetic material. Further evidence supporting this view was the fact that proteins are composed of about twenty types of amino acids, while DNA has only six types of molecular components. In other words, the hereditary alphabet seemed to be substantially larger in proteins. Thus, it seemed likely that the apparently more complex molecules (proteins) should be the carriers of the genetic message.

Amounts of DNA and Protein

Description of the meiotic behavior of chromosomes supported their role in heredity. The diploid number was constant in the somatic cells of all members of a species, but was halved in the germinal cells, returning to the diploid condition only after fertilization.

BOX 8.1

DNA is the genetic material—evidence from bacterial transformation

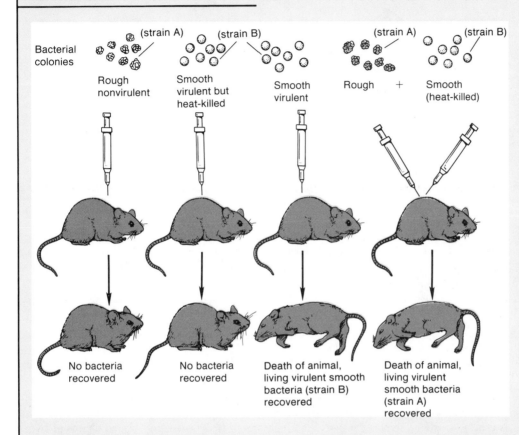

Figure 8.3

Griffith's experiments with the bacterium *Diplococcus pneumoniae* and mice.

F. Griffith studied two pneumonia-causing strains of the bacterium *Diplococcus*. When grown *in vitro* (meaning *in glass*) in a laboratory culture flask or on an agar plate, these virulent strains are smooth in appearance. Mutants of both strains, however, appear rough when grown in culture because they lack the normal bacterial capsule, nor do these mutant (rough) strains cause pneumonia (figure 8.3).

Smooth bacteria, which cause pneumonia in mice, can be recovered from mice who have died from pneumonia. When rough or smooth bacteria are heated, a process that kills most strains of bacteria, and injected into mice, all the mice survive, and no bacteria can be recovered from the mice. When rough bacteria of strain A were injected into mice simultaneously with heat-killed smooth bacteria of strain B, the mice died of pneumonia, and smooth bacteria of strain A were recovered. Thus, inside the mice, the live (rough) strain A bacteria were astonishingly transformed to smooth (virulent) bacteria. We now know this transformation occurred because the live strain A bacteria incorporated DNA coding for the production of the smooth capsule from the dead strain B bacteria. The rough strain A bacteria were transformed to smooth, pneumonia-causing bacteria, resulting in death of the mice and bacteria capable of producing a bacterial capsule.

Careful measurements of the amounts of DNA and protein in somatic and sperm cells again indicated the likely hereditary role of DNA, but not protein (table 8.1). Within a species, the amount of protein in a sperm cell varied considerably. Furthermore, the amounts of protein in sperm and somatic cells bore no simple relationship to each other. On the other hand, the amount of DNA in sperm was relatively constant. Additionally, the amount of DNA in sperm cells was exactly one-half that in somatic cells.

BOX 8.2

DNA is the genetic material—evidence from viruses

Figure 8.4

The Hershey–Chase "blender experiments" that confirmed that DNA is the hereditary material.

Virus grown in culture with S^{35} as only source of S

S^{35} label only

All viruses labeled with radioactive label

Virus grown in culture with P^{32} as only source of P

P^{32} label only

"Hot" viruses infect fresh sample of unlabeled *E. coli*

10 minutes after infection placed in blender to interrupt infective process

Contents of blender separated into two fractions—one with the radioactive label (either S^{35} or P^{32}) and one fraction with no radioactivity. Each fraction put in petri dish with fresh media and scored for growth of *E. coli* and liberation of new virus.

S^{35}

No label

P^{32}

No label

Labelled protein
No growth

DNA
Growth followed by lysis

Labelled DNA
Growth followed by lysis

Protein
No growth

Transformation

An English microbiologist, F. Griffith, began the sequence of experiments that eventually led to the establishment of DNA as the genetic material. Griffith showed that bacteria of one strain could incorporate genes from a second strain. This phenomenon caused the first strain of bacteria to display phenotypic traits of the second, a process called **transformation** (box 8.1). Transformation is now a widely practiced procedure forming the foundation for the field of *biotechnology,* but in Griffith's time, transformation was regarded as absolutely astonishing.

Subsequent elegant work in 1943 by three American biologists, O. T. Avery, C. M. MacLeod, and M. McCarty, separated the contents of bacterial cells of one strain into four basic biochemical components: DNA, proteins, lipids, and carbohydrates. Each of the four components was mixed with a separate bacterial culture of a different strain. Only the DNA was capable of transforming the bacterial culture; none of the other biochemical components could do so.

Analogous experiments with viruses in the early 1950s by A. D. Hershey and M. Chase also gave results pointing exclusively to DNA as the genetic material (see box 8.2). Viruses are perfect experimental organisms for such a study, as they are composed of only two com-

In 1952, A. D. Hershey and M. Chase presented even more convincing evidence that DNA was indeed the genetic material. Hershey and Chase took advantage of the simple morphology and life history of a virus to compare the roles of DNA and protein in viral reproduction (see chapter 12; figure 12.9).

Both DNA and protein contain the elements most characteristic of living organisms—carbon, hydrogen, oxygen, and nitrogen. They differ in one major respect, however: sulfur occurs only in protein and not DNA, while phosphorus occurs only in DNA and not protein. Hershey and Chase began their experiments by growing cultures of the virus T-2 (figure 8.6a) in suspensions of *E. coli* rich in one of two radioactive isotopes, sulfur-35 (S^{35}) or phosphorus-32 (P^{32}). All of the T-2 derived from these cultures thus had either the DNA labeled with P^{32} or the protein coats labeled with S^{35}.

Next, an unlabeled culture of *E. coli* was infected with S^{35} labeled T-2, but growth was permitted to proceed for only about ten minutes, half the lytic cycle of T-2. Growth was interrupted by placing the mixture in a blender, which separated the empty protein coat (the *viral ghost*), from the surface of the bacterial cells. The mixture was then separated into two fractions, one containing radioactively labeled sulfur and one with no label. The two fractions were placed in separate cultures and permitted to resume growth. The culture containing the S^{35} fraction showed no growth and produced no T-2. Examination showed that it contained only the viral ghosts (figure 8.4).

On the other hand, if *E. coli* were infected with P^{32}-labeled T-2 and the experiment repeated, the phosphorus-bearing fraction contained *E. coli* and the DNA of the T-2 that had been injected into the bacteria. Hence the fraction continued to grow, ceasing only when the T-2 completed their lytic cycle, killing all the bacteria. Thus, phosphorus-containing DNA had been transmitted between generations of bacteriophages. These results excluded protein as the genetic material.

ponents, DNA surrounded by a protein coat. When the DNA and protein were labeled with unique radioactive isotopes, only the labeled DNA was shown to be capable of infecting bacteria and producing more viruses. The viral protein alone was unable to direct the production of more viruses.

The Chemical Structure of DNA

By the early 1950s the accumulating evidence favored nucleic acids as the hereditary material. However, proof of their role depended upon determination of a structure for DNA sufficient to permit it to fulfill the difficult functions of an hereditary molecule. The search for the structure of DNA is one of the best known and most fascinating detective stories in scientific history (box 8.3). It involved some of the most famous scientists of the time, but the race was won by two relatively unknown scientists—James Watson and Francis Crick (figure 1.1c). The discovery of the structure of DNA won them (and biophysicist Maurice Wilkins) the 1962 Nobel Prize for Physiology or Medicine and a permanent place in the history of science.

BOX 8.3

The race to discover the structure of DNA

Other key participants in the DNA race. (a) Linus Pauling, who earned two Nobel Prizes—one for his work on the nature of the chemical bond and also a Nobel Peace Prize, and (b) Rosalind Franklin, whose X-ray pictures of DNA provided Watson and Crick with crucial information leading to their model for DNA.

(a and b) Cold Springs Harbor Laboratory.

(a)   (b)

By the early 1950s scientists in several laboratories were intensely involved in a *de facto* race to determine the chemical structure of the hereditary material. Although evidence strongly supported DNA, knowledge of the precise structure of the molecule was necessary to understand how it could fulfill the exacting requirements of inheritance.

Some aspects of the components and structure of DNA were well known. *Guanine* had been isolated from crystallized bird feces (guano) in 1844, and *thymine* had been purified from calf thymus, also in the nineteenth century. Erwin Chargaff, a biochemist at Columbia University, had demonstrated that in DNA the amount of adenine always equalled the amount of thymine, and the amount of cytosine always equalled the amount of guanine—relationships now known as *Chargaff's rules*.

Such famous scientists as Linus Pauling actively pursued the structure of DNA, but two young and relatively unknown scientists made the discovery. James Watson had received his Ph.D. from the University of Indiana at the age of twenty-three after being refused admission to graduate school at Harvard University and the California Institute of Technology. In 1951 he arrived at Cambridge University, where he began his now-famous collaboration with Francis Crick.

Crick had received his university degree in physics in 1938, but was working towards a Ph.D. in biology when he met Watson. A key piece of the puzzle was supplied by X-ray crystallographic studies of DNA by Rosalind Franklin and Maurice Wilkins. Using the X-ray studies in conjunction with previously published work, Watson and Crick deduced the probable structure of DNA, publishing their model in a short paper in 1953. Watson, Crick, and Wilkins were awarded the 1962 Nobel Prize for Physiology or Medicine for their historic discovery.

Although the X-ray crystallography that provided crucial evidence was actually done by Rosalind Franklin, she was not awarded the Nobel Prize because of several technical restrictions. No more than three persons can share one prize, and they must all be living. Franklin died of cancer in 1958. Nonetheless, the role of Rosalind Franklin in solving the structure of DNA has become a matter of historical debate. She had an uneasy professional relationship with Wilkins, who apparently showed her work to Watson and Crick without her permission. There may thus be some irony in Wilkins' receipt of the Nobel Prize. Very different accounts of Franklin's role are given by Anne Sayre (Franklin's biographer) and by James Watson.

The hereditary molecule could not be a simple one; it must fulfill at least four functions. The hereditary molecule must be able (1) to contain all the needed information to direct the development and life of an organism; (2) to transmit that information to all parts of the body, translating the information into a usable form in all the tissues; (3) to transmit that information to its descendants with high accuracy; and (4) to be mutable, in spite of the normally precise transmission of information—that is, it must permit a small number of changes to occur routinely, creating genetic variability in a population.

In 1953, Watson and Crick published a short paper in the journal *Nature* describing a model for the structure of DNA, now known as the **Watson-Crick model.** At the conclusion of this paper, they appended a sentence understating the importance of their hypothesis: "It has not escaped our notice that the [structure] we have postulated immediately suggests a possible copying mechanism for the genetic material." Indeed, the Watson-Crick model of DNA, verified and elaborated by many subsequent studies, is the foundation for modern molecular genetics. Within a decade after the Watson-Crick hypothesis, DNA was shown to fulfill all requirements of the hereditary molecule.

The Watson-Crick model describes DNA as an extremely long, two-stranded molecule twisted like a rope into the shape of a **double helix** (figure 8.6). The basic building block of DNA is a molecule called a *nucleotide*. Each nucleotide has a short backbone of a small sugar molecule linked to a phosphate group. The five-carbon sugar is known as **deoxyribose.**

Within a DNA molecule, the deoxyribose and phosphate groups show no variation. Long chains of thousands of alternating sugar and phosphate groups from thousands of nucleotides are linked together, forming two parallel backbones. The two backbones are joined by the third component of each nucleotide, a **nitrogenous base.** One of four different nitrogenous bases is attached to each of the sugar molecules in every nucleotide: *adenine* (A), *cytosine* (C), *guanine* (G), or *thymine* (T). These four bases comprise the only letters in the genetic alphabet, and the major variation between the DNA of different individuals is in the sequence of bases along the length of the molecule.

On one side, each nitrogenous base is attached to a sugar molecule in the long chain of alternating sugars and phosphates. On the other side, each base attaches to a nitrogenous base extending from a sugar on the parallel strand. Thus, the two strands are bound at each sugar by two nitrogenous bases, bridging the gap between the parallel chains much like rungs of a ladder. The entire molecule is twisted into a two-stranded or double helix, just as if a long ladder had been twisted about its longitudinal axis.

The nitrogenous bases of one strand do not join randomly with those of the parallel strand. Rather, only two pairings are possible, adenine with thymine, and cytosine with guanine, A-T and C-G. Thus, one strand is *complementary* to the other (**complementary base pairing),** explaining Chargaff's rules (see box 8.3). The sequence of nucleotide bases on one strand reflects the sequence of bases on the other.

The sequence of bases and their complementary pairing on opposing strands are the keys to the function of DNA. The sequence of bases encodes hereditary information, while specific pairing provides the mechanism for accurate replication and transmission of this information between cells of one individual and between members of consecutive generations.

It was immediately apparent that DNA possessed the capacity to meet the first requirement of an hereditary molecule—that it be capable of encoding an immense amount of information. The number of unique messages that can be contained by a DNA molecule is 4^n, where n is the number of nucleotide pairs in the DNA. A DNA molecule only 4 base pairs long can encode 256 unique messages; 10 pairs can encode 1,048,576 messages; 100 pairs can encode over 10^{60} unique messages. The human genome contains over three billion nucleotide pairs. Thus, even with an alphabet of only four letters, the information-bearing capacity of DNA is far more than sufficient to encode all the information for life.

The Watson-Crick model resolved the search for the hereditary material. Each chromosome is now known to consist of immensely long chains of DNA surrounded by protein. While the protein probably serves an important function in the expression of the genetic message, it is DNA that bears the message. The Watson-Crick model also moved the science of genetics into the realm of molecular biology. Geneticists no longer studied only the phenotypic expression of genes whose location and nature were unknown. Instead, geneticists began to study the genetic message itself.

(a)

(b)

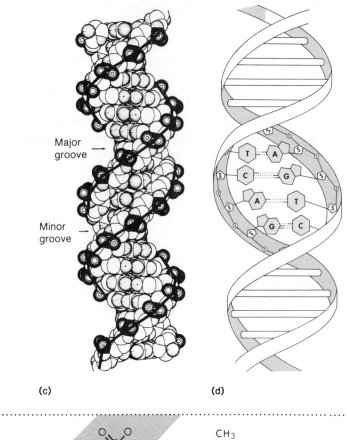

Major groove

Minor groove

(c)   (d)

Figure 8.6

An electron micrograph of the phage T-2, (a) intact and (b) with its DNA liberated from the viral head by osmotic shock; (c) Watson-Crick model of DNA; (d) a schematic representation of the spiralling ladder of the DNA molecule held together by base pairing between adenine(A) and thymine(T), cytosine(C) and guanine(G); (e) an enlarged section indicating how the sugar and phosphate molecules form the ladder backbones of the DNA.

Phosphate

Sugar

Thymine

Phosphate

Sugar

Cytosine

Phosphate

Sugar

Adenine

Phosphate

Sugar

Guanine

(e)

Figure 8.7

A generalized structure for the informational macromolecule RNA has a single-stranded sugar-phosphate backbone and four nitrogen bases. Note the sugar, ribose (arrow), differs from the deoxyribose of DNA, and the base uracil is found in place of the base thymine that occurs in DNA. In portions of RNA where base pairing can occur, adenine pairs with uracil and cytosine pairs with guanine.

RNA

RNA (ribonucleic acid) resembles DNA in structure and also plays a major role in the hereditary mechanism. Three structural differences distinguish RNA from DNA (figure 8.7). First, RNA contains **ribose,** a five-carbon sugar with one more oxygen atom than *deoxyribose.* Second, RNA is usually single rather than double stranded, although it is also helical in shape. Third, the nucleotide base thymine does not occur in RNA. Rather, it is replaced by a similar base, *uracil* (U).

In some viruses, RNA rather than DNA serves as the hereditary molecule. Because of its analogous structure, it obviously has the same information-carrying capacity as DNA. However, all organisms have three forms of RNA, all of which are crucial to genetic function. All are transcribed directly from nuclear DNA, and they function as crucial intermediaries in the process of protein formation. They will be considered in detail.

Figure 8.8

DNA and Chromosomes

Every cell in the body contains an identical set of chromosomes. In most species, each chromosome consists of one very long molecule of DNA. In eukaryotes, the DNA is surrounded and enclosed by several types of nuclear proteins, especially histones. Chromosomes containing only one copy of the DNA molecule are *unineme*. Some tissues of some species have *multineme* chromosomes (chromosomes consisting of many identical copies of DNA arranged in parallel), but little is known about the use of such an arrangement.

The DNA in a chromosome is remarkably long. If unravelled, the DNA from a normal-sized chromosome would stretch several inches, many times the diameter of the cell itself. Thus, DNA can be packaged into a cell only as a result of extreme coiling. One of the functions of histones is probably to organize the dense coiling of DNA.

Both meiosis and mitosis produce daughter cells that contain genetic information derived from the parent cell, and both involve one episode of DNA duplication. Recall that replication of DNA occurs during the S or synthesis phase of the cell cycle—not at the time of cell division. Thus, at the beginning of cell division each chromosome consists of two identical double helices, joined at a single centromere.

Throughout most of the life of the cell, however, chromosomes are far more diffuse than during the brief period of cell division. It is probable that the absence of coiling during interphase is crucial to the function of DNA during the vegetative portion of the cell's life.

Function of DNA

The Watson-Crick model of DNA is a cornerstone of modern biology. In the few decades since the discovery of the chemical structure of the gene, molecular genetics has become one of the most intensely exciting and rapidly advancing fields of modern science. Since 1953, a large body of knowledge has accumulated that verifies the double-helical nature of DNA and describes in considerable detail how this structure permits DNA to fulfill necessary hereditary functions.

The Central Dogma

In 1958 Francis Crick described the modern concept of the relationship between genotype and phenotype as the **central dogma** (figure 8.8). Expression of the hereditary information contained in DNA requires several steps. The genetic message is first copied precisely to form **messenger RNA**—the process of **transcription.** In eukaryotes, messenger RNA then moves from the nucleus to the cytoplasm, where its encoded information is read precisely by both **ribosomal RNA** and **transfer RNA** to assemble proteins—the process of **translation.** The proteins themselves, particularly in the form of enzymes, mediate the subsequent processes of phenotypic development.

Geneticists of the late nineteenth and early twentieth centuries were correct that proteins were intimately involved with heredity, but they were not correct in believing that proteins actually were the hereditary molecule. The sequence of amino acids in a protein is itself determined by the sequence of nucleotide bases in DNA. Like DNA, protein is an **informational macromolecule;** but it is not the controlling molecule. Genetic information flows from DNA to proteins, and not vice versa.

The central dogma also describes the flow of genetic information between generations. In the cell nuclei DNA replicates itself during interphase. The precise duplication and division of chromosomes during mitosis ensure that each cell contains a set of genetic instructions identical to those in the newly fertilized zygote. Meiotic processes govern the recombination and independent assortment of the genetic material in germinal cells, with each gamete receiving a complete haploid set of chromosomes.

Figure 8.9

Semi-conservative replication in DNA. Complementary base pairing ensures the two new DNA molecules are identical to the parent molecule.

From Sylvia S. Mader, *Inquiry Into Life,* 2d edition. Copyright © 1979 Wm. C. Brown Publishers, Dubuque, Iowa. All Rights Reserved. Reprinted by permission of the author.

Replication

Complementary base pairing was recognized by Watson and Crick as a mechanism that permits precise replication of the DNA molecule. As a result of complementary pairing, each strand of a DNA molecule can direct the synthesis of a complete new molecule of DNA.

During replication, each strand of DNA serves as a *template* for the formation of a new parallel and complementary strand. As in most biochemical processes, the replication of DNA is controlled by a series of enzymes with precise functions. One type of enzyme breaks the bonds between complementary nucleotide pairs—in effect, the halves of the DNA molecule separate much like the opening of a zipper (figure 8.9). As the two strands unwind, nucleotide bases are added to each, complementary to the nucleotides of the original strands. Each single strand thus provides the pattern for the formation of a new complete double helix, each identical to the original DNA molecule.

Each daughter DNA molecule contains one strand derived from the parent and one strand assembled from free nucleotides. Since each of us arose from a single fertilized egg, it is possible (but highly unlikely) that one strand of DNA from each of our parents is still in some cell of our body, even after numerous mitotic divisions. This type of replication, in which each daughter double helix receives one complete parental strand of DNA, is called **semi-conservative replication.**

The Genetic Code

The variable sequence of nucleotides in DNA immediately suggested that nucleotides comprise the letters in a genetic alphabet. Therefore a dictionary was necessary to understand this new language. Many geneticists immediately focused on decoding the genetic message.

TABLE 8.2

Genetic dictionary of RNA base codons and the amino acids they encode. In turn, RNA codons are complementary to DNA codons.

First base	Second base				Third base
	U	**C**	**A**	**G**	
U	UUU } Phenylalanine UUC } UUA } Leucine UUG }	UCU } UCC } Serine UCA } UCG }	UAU } Tyrosine UAC } UAA† } Stop UAG† }	UGU } Cysteine UGC } UGA† Stop UGG Tryptophan	U C A G
C	CUU } CUC } Leucine CUA } CUG }	CCU } CCC } Proline CCA } CCG }	CAU } Histidine CAC } CAA } Glutamine CAG }	CGU } CGC } Arginine CGA } CGG }	U C A G
A	AUU } AUC } Isoleucine AUA } AUG* Methionine	ACU } ACC } Threonine ACA } ACG }	AAU } Asparagine AAC } AAA } Lysine AAG }	AGU } Serine AGC } AGA } Arginine AGG }	U C A G
G	GUU } GUC } Valine GUA } GUG }	GCU } GCC } Alanine GCA } GCG }	GAU } Aspartic Acid GAC } GAA } Glutamic Acid GAG }	GGU } GGC } Glycine GGA } GGG }	U C A G

*Initiation codons. The methionine codon AUG is the most common starting point for translation of a genetic message.
†Terminator codon.

DNA and proteins have analogous structures. Both are comprised of long chains of molecular subunits—nucleotides in the case of DNA, and amino acids in the case of proteins. The simplest model is that a one-to-one correspondence exists between the two sequences; that is, that they are **colinear.** In 1958, Crick proposed colinearity as an hypothesis to explain the structure of proteins: the sequence of nucleotides in DNA directly specifies the sequence of amino acids in a protein. Subsequent work has verified this hypothesis.

Because the DNA alphabet consists of only four different bases (letters), each base alone, a one-letter word, is obviously not sufficient to specify uniquely twenty different amino acids. Assuming a common word length for all amino acids and a four-letter alphabet, the number of amino acids that could be specified by a word of n letters is 4^n. Thus, a dictionary of two-letter words would also be inadequate: $4^2 = 16$. However, the use of three letters to identify an amino acid provides $4^3 = 64$ possibilities, far more than enough words to encode uniquely all amino acids.

Working on the hypothesis that the dictionary contained three-letter words, M. W. Nirenberg broke the **genetic code** in the early 1960s. Nirenberg added all of the protein synthesizing apparatus of *E. coli* to a test tube, as well as all twenty amino acids. He then synthesized an artificial genetic message (messenger RNA) consisting only of uracil: U-U-U-U-. . . . He subsequently examined the mixture for the presence of polypeptides and found only one type, a polypeptide consisting of phenylalanines and no other amino acids. He concluded that the messenger RNA sequence UUU encodes the message or word for phenylalanine. (The RNA sequence UUU mirrors the DNA sequence AAA. Messenger RNA is discussed later in this chapter.) Similar experiments showed that the RNA message AAA encoded the amino acid lysine; CCC encoded proline; and GGG encoded glycine.

Further detailed work soon elaborated the entire genetic code (table 8.2). The code consists, as hypothesized, of three-letter words called **codons** or *triplets*. There are many more codons than amino acids, and most amino acids are specified by more than one codon. Leucine and arginine are each specified by six different codons. The code is said to be **degenerate** because of this redundancy of message. The biological functions, if any, of degeneracy in the code are a matter of debate. Only methionine and tryptophan are specified by a single codon.

TABLE 8.3

Types of mutations

Original DNA		CGATCGCAA
Messenger RNA		GCUAGCGUU
Codes for		ala/ser/val/

Frameshift mutation	DNA	* CGGATCGCAA
	mRNA	GCCUAGCGUU
Now codes for		ala/STOP
Substitution mutation	DNA	* AGATCGCAA
	mRNA	UCUAGCGUU
Now codes for		ser/ser/val
Same sense mutation	DNA	* CGGTCGCAA
	mRNA	GCCAGCGUU
Still codes for		ala/ser/val/

* = Mutation

Four codons serve as *punctuation* for the genetic message. AUG always specifies the amino acid methionine, but it is also known as the *initiator codon,* because most messages specifying a protein begin with AUG. Three triplets are **nonsense** or **terminator codons** because they specify no amino acids at all. *Nonsense* is probably a misnomer, however, because these codons (UAA, UAG, and UGA) all terminate the reading of the genetic message. They signify the end of a sequence of codons specifying a particular polypeptide.

The genetic message is thus encoded in a sequence of codons along the length of a DNA molecule. The sequence of amino acids in a polypeptide is specified one at a time by the sequence of codons in DNA. Production of a particular protein or polypeptide stops when a terminator codon is encountered.

The genetic code has been shown to be **nonoverlapping;** that is, for a single gene the sequence of bases is always read in discrete groups of three per codon. Each base participates in only one codon (some rare exceptions are discussed at the end of this chapter). The organization of nucleotide bases into sequential triplets is analogous to the frames of a motion picture. The frames must be read precisely one at a time, or the message is blurred. It also suggests one possible mechanism of mutation, **frameshift.** If at some point in the nucleotide sequence a base is either added or deleted, the entire reading frame is shifted, altering the subsequent message (table 8.3).

Yet another type of mutation is **substitution** of one nucleotide base for another (table 8.3). The consequence of base substitution can be far less drastic than frameshift mutation. A simple change in one nucleotide base alters at most one amino acid in a protein. Because of the degeneracy of the code, many substitutions are **same sense;** that is, they do not change the amino acid specified by the affected codon (table 8.3).

A final and most important characteristic of the genetic code is that it is **universal.** With only minor exceptions (such as the DNA of mitochondria of some species, including humans) the code is identical among all species. UUU encodes phenylalanine in humans, snails, cotton, and *E. coli.*

The universality of the code has been shown by experiments in which mRNA from rabbits encoding the structure of hemoglobin was combined with the protein synthesizing apparatus of *E. coli.* The result: rabbit hemoglobin. More dramatically, the gene encoding the

(a)

(b)

Figure 8.10

Example of the universality of the genetic code. This tobacco plant has had the gene for firefly luciferase incorporated into its genome by genetic engineering techniques. (a) The plant is seen in ordinary light, and (b) it has been "watered" with a solution demonstrating the plant expressing the firefly luciferase gene.

Keith V. Wood, UCSD.

firefly enzyme luciferase (responsible for the firefly's glow) has been inserted into tobacco plant DNA. When the tobacco plant was "watered" with a solution of the right compounds, the plant glowed, demonstrating that the firefly gene was expressed by the plant (figure 8.10).

The universality of the code is important in at least two ways. As we have already noted, humans are a difficult species to examine genetically. Because all organisms speak the same genetic language, experimental geneticists have been able to select from the thousands of available species those that have qualities favoring laboratory analysis. Most knowledge of the molecular mechanism of inheritance has been derived from a few well-studied laboratory species: viruses, bacteria, fungi, *Drosophila*, and mice. Because all species share a common genetic apparatus, studying the genes of *Drosophila* tells us much about the genes of humans that could not be learned if scientists studied only humans.

Additionally, the universality of the code is some of the strongest evidence for a common evolutionary origin of all living species. Such an intricate mechanism, using the same code for all species, argues strongly that the genetic apparatus originated in the progenitors of modern life very early in the history of the earth. Although several billion years have passed for the development and modification of the genetic mechanism, the remarkable similarities shared by all species are clearly the result of a common evolutionary origin.

The Structures and Functions of RNA

Although DNA is the hereditary material for most species, RNA is fundamental to the function of DNA. DNA is a master file of information, but processing that information to a usable form, making proteins, requires RNA as an intermediary. The structural similarity of RNA and DNA allows DNA to serve as a template for RNA synthesis, just as it does for its own replication. An understanding of protein synthesis requires that we first examine the structure and function of the major types of RNA.

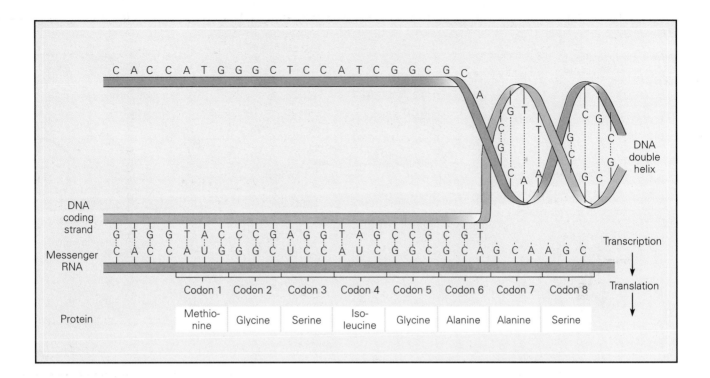

Messenger RNA

The DNA template resides in the nucleus of eukaryotic cells, but proteins are synthesized in the cytoplasm. DNA does not directly participate in protein formation. Instead, a copy of the information encoding the sequence of amino acids for a protein is *transcribed* from nuclear DNA in the form of a complementary molecule, **messenger RNA (mRNA).**

Synthesis of mRNA occurs primarily during the G1 and G2 phases of the cell cycle in similar fashion to the replication of DNA itself. (During the S phase, DNA is busy replicating itself; during cell division, the chromosomes are too coiled for transcription to occur.) The double helix is believed to unwind at the site of mRNA synthesis (figure 8.11). However, in contrast to DNA replication, mRNA is transcribed from only one of the two strands of DNA, which is often called the **sense strand.** Remember also that uracil (U) rather than thymine (T) pairs with adenine on the sense strand in transcription of RNA. This process is appropriately called *transcription* because it is analogous to a scribe copying a message. The information transcribed into mRNA is a precise, complementary copy of the hereditary message of DNA.

Proteins vary considerably in length. Consequently, mRNAs also vary in length. Most are five hundred to one thousand nucleotides long. However, initial transcription of mRNA in eukaryotic cells usually involves the production of a precursor called **HnRNA, heterogeneous nuclear RNA.** Found only in the nucleus, HnRNA is many times longer than mRNA. For most eukaryotic genes the final formation of mRNA involves removal of segments of nucleotides scattered along the length of the HnRNA molecule.

The existence of HnRNA indicates that many segments of DNA are transcribed, but, because they are excised from HnRNA, they are not later translated into proteins. Thus, a segment of eukaryotic DNA encoding a polypeptide contains two types of nucleotide sequences: **exons,** portions of the template for protein synthesis, and **introns,** which are never translated into amino acid sequences (figure 8.12). Introns vary in length from one hundred base pairs to over one thousand base pairs. The number of introns also varies from one or two in some genes to sixteen in a chicken egg white protein. In some genes, the total length of the introns may even exceed that of the exons. DNA sequences controlling production of most polypeptides consist of alternating sequences of introns and exons.

Figure 8.11

Transcription. A portion of the DNA molecule unwinds, exposing the base pairs for use as a template for the formation of mRNA or HnRNA from the *sense strand* only. Following transcription, the DNA assumes its normal conformation.

Figure 8.12

Difference in formation of
mRNA in prokaryotic and
eukaryotic cells.

The existence of non-translated DNA sequences was first recognized in 1977. Introns are extraordinary because they suggest a new level of complexity in the function of DNA. Little is known regarding the function of introns, but present hypotheses argue that introns may regulate the complex functioning of the DNA of eukaryotes.

Bacterial mRNAs seem to have a very short lifetime, usually only a few minutes. Apparently, the various types of mRNA are transcribed as they are required by the metabolic processes of the organism. In eukaryotes, in contrast, some mRNAs are known to exist for relatively long periods. It is possible that HnRNA serves as a relatively stable form in which mRNA can be stored until needed.

The initial sequence of all mRNAs, the *leader sequence,* provides a binding site for the **ribosomes,** the actual site of protein synthesis. Following the leader is the *initiation codon* AUG, one of four words in the genetic code used as punctuation for the genetic message. Consequently, polypeptides always begin with the amino acid methionine, although this amino acid is often removed upon completion of polypeptide synthesis.

All mRNAs terminate with one of the three nonsense codons, UAA, UAG, or UGA, which stop protein synthesis. Many eukaryotic mRNAs also have long sequences of as many as two hundred adenines at their terminal ends, called *poly-A sections.* The function of poly-A sections is not known, although it is possible that they may help stabilize some of the longer lived eukaryotic mRNAs.

Ribosomal RNA

Messenger RNA carries the information specifying the sequence of amino acids in a polypeptide, but actual synthesis of a protein requires two additional types of RNA. About 80% of the RNA in a cell is **ribosomal RNA (rRNA),** which is transcribed directly from nuclear DNA and moves to the cytoplasm after formation. In the cytoplasm rRNA associates with proteins to form one of the major types of intracellular organelles, the *ribosomes.* In prokaryotes, ribosomes occur scattered freely throughout the cytoplasm, but in eukaryotes many ribosomes are found along the complex network of the *endoplasmic reticulum* (figures 2.3 and 2.5). The proportions vary somewhat between eukaryotes and prokaryotes, but in general ribosomes are about half rRNA and half protein.

Ribosomes are the site of protein synthesis. Each eukaryotic ribosome is composed of two subunits of different sizes. During protein assembly, a ribosome attaches to an mRNA molecule, then moves along it, reading the codons as it goes and attaching amino acids to the growing protein in the proper sequence—the process of **translation.**

Figure 8.13

A molecule of transfer RNA. Although the "cloverleaf" conformation is characteristic, its three-dimensional structure is actually more complex. Dotted lines indicate hydrogen bonding between complementary bases. The anticodon is the portion of the molecule that interacts with the mRNA. The amino acid is asparagine.

From Stuart Ira Fox, *Human Physiology,* 3d ed. Copyright © 1990 Wm. C. Brown Publishers, Dubuque, Iowa. All Rights Reserved. Reprinted by permission.

In the early 1980s, the study of ribosomal RNA led to an entirely unexpected finding: RNA can function like an enzyme to catalyze chemical reactions. Previously, only proteins had been known to have catalytic functions. Excision of introns from precursors (HnRNA) to ribosomal RNA is achieved by the RNA itself and not, as had been expected, by enzymes. These catalytic RNAs are now called **ribozymes,** and they have been identified in every species that has been examined thus far. They are apparently the main agent for excision of introns and splicing of loose ends for all three types of RNA. Ribozymes are also abundant in mitochondria, which produce their own RNA independently of nuclear RNA.

The existence of ribozymes also answers a key question about the origin of life. Biologists had speculated that the first living organisms might have had proteins for hereditary material; without proteins as catalysts, how could nucleic acids function, even to reproduce themselves? On the other hand, no living organisms are known to use proteins as the hereditary material. The capacity of RNA to catalyze parts of its own synthesis supports the view that RNA was the essential ingredient in the origin of life 4 billion years ago, both storing and processing hereditary information for the first living organisms.

Transfer RNA

Amino acids are brought to the ribosomes for assembly into proteins by the smallest type of RNA, **transfer RNA (tRNA).** Like all RNAs, tRNA is single stranded; but some portions of the molecule appear double stranded, because it doubles back on itself. The resulting structure is similar to a cloverleaf with one small loop and three large loops, but it is certainly far more complex than that (figure 8.13). The tRNAs are about eighty nucleotides long, and like the other RNAs they are synthesized in the nucleus, directly from DNA.

The two free ends of a tRNA lie adjacent to each other. One end attaches to a particular amino acid, which the tRNA then carries, or transfers, to the ribosome for incorporation into a growing polypeptide.

The loop of the cloverleaf opposite the free ends has a triplet **anticodon,** complementary in sequence to a specific mRNA codon and identical in sequence to the template DNA codon, except that all thymines in DNA are replaced by uracils in tRNA (figure 8.13). A tRNA with a particular anticodon can attach to only one type of amino acid, and the anticodon sequence of the tRNA for a particular amino acid is complementary to the mRNA codon specifying that amino acid. Thus, when a tRNA bearing an amino acid moves to a ribosome, the match between the mRNA codon and tRNA anticodon ensures that the amino acid is inserted in the proper order to manufacture a particular protein. In other words, the ribosome facilitates insertion of an amino acid into a developing protein only when the tRNA anticodon matches the mRNA codon specifying the next amino acid.

Over fifty types of tRNAs have now been discovered, each specifying only one amino acid. Most of the twenty major amino acids can be transported by more than one type of tRNA.

Figure 8.14

Protein assembly: (a) as a series of ribosomes move along an mRNA molecule, each generates an increasingly long polypeptide chain; (b) a closer look at an individual ribosome shows how the chain grows.

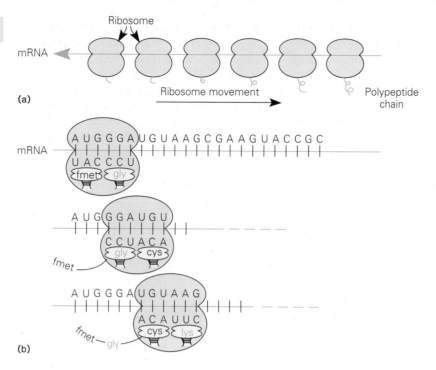

(a)

(b)

Figure 8.15

Translation. In this electron micrograph of cytoplasm from an insect larva, a strand of messenger RNA begins at the lower right and forms a long loop ending at the upper right. Dark round ribosomes are attached along its length, and increasingly long polypeptide chains are clearly visible hanging from the ribosomes as they move toward the end of the mRNA. The entire structure of mRNA, ribosomes, and growing polypeptide chains is called a polyribosome.

E. V. Kiseleva, *Nature,* Febs Letts, 251–253, 1989.

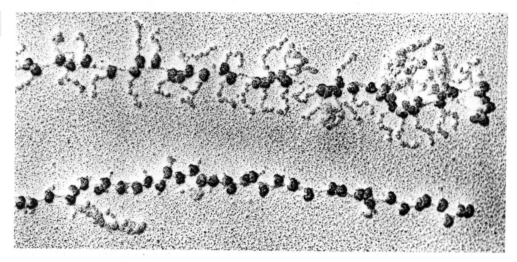

Protein Synthesis

Let us now review protein synthesis in its entirety (figure 8.14). Assembly of proteins requires three types of RNA, each performing its own specialized job. The sequence of nucleotide bases in nuclear DNA encoding the *colinear* sequence of amino acids in a protein is *transcribed* to *messenger RNA,* which moves to the cytoplasm. In the cytoplasm, several ribosomes attach to the mRNA, forming a structure known as a **polyribosome** (figure 8.15). A ribosome binds to the leader sequence of the mRNA, and beginning at the initiator codon (AUG) *translation* of the mRNA nucleotide sequence into a sequence of amino acids (a polypeptide) begins.

A tRNA with the appropriate anticodon and amino acid attaches to each triplet codon on the mRNA as the ribosome moves along its length. Each tRNA carries a specific amino acid, which is joined by a *peptide bond* to the growing polypeptide chain. The tRNA is then released back into the cytoplasm, where it may attach to and transport another amino acid. When a ribosome encounters one of the three terminator codons, it releases the polypeptide chain into the cytoplasm. Translation is then completed. No known molecular apparatus is associated with formation of the three-dimensional structures of a protein. The sequence of amino acids, the primary structure alone, determines the secondary and tertiary structures of the protein.

Proteins and messenger RNAs are long molecules. Many, perhaps most, mRNAs survive only a few minutes, indicating a rapid rate of transcription and translation. The speed of both activities varies according to a number of factors, but can be truly exceptional. Ten or more amino acids may be added to a polypeptide chain in one second. Translation of the beta chain of hemoglobin, consisting of 146 amino acids, occurs in about ten seconds. Even with a lifetime of only two minutes, a single mRNA molecule could serve as the template for 30 or more proteins that are each 300 amino acids in length. The transcription of mRNA occurs at a rate of about 28 nucleotides per second in *E. coli*.

A Unified View of the Gene

We have seen three models of the hereditary mechanism: Mendelian, chromosomal, and molecular. Each describes differing aspects of the genetic process. Together, they offer a comprehensive basis for understanding the mechanism of inheritance.

Mendelian principles postulate the existence of genes, discrete particles transmitted between generations that control phenotypic traits. These laws were deduced directly from observations of the phenotype itself. Mendel knew nothing of the actual physical nature of genes.

Analysis of meiotic behavior of chromosomes revealed that chromosomes were the likely location of the genes. Meiosis is a cellular process that explains Mendel's Laws of Segregation and Independent Assortment. Knowledge that chromosomes were the physical location of the genetic material did not, however, explain the molecular mechanism of inheritance.

Over a period of several decades in the twentieth century, molecular geneticists first identified DNA as the hereditary material and then described its chemical structure. The central dogma summarizes contemporary understanding of control of the phenotype by the molecular apparatus of inheritance. As the meiotic behavior of chromosomes suggests, the genetic material, DNA, resides in the chromosomes. The information encoded in DNA is transcribed into RNA, which in turn directs assembly of proteins in the cytoplasm.

The contemporary view of a gene incorporates aspects of all three lines of genetic investigation. Chromosomal and molecular processes are mechanisms explaining the hereditary principles discovered by Mendel. Proteins provide the crucial link between Mendelian genes and the phenotype. Proteins are a primary phenotypic product of the genes, which in turn direct further aspects of phenotypic expression.

This principle is illustrated clearly when mutations in DNA cause formation of a defective protein, as in the case of sickle-cell anemia. Recall that normal hemoglobin consists of two alpha polypeptide chains and two beta polypeptide chains, plus iron-containing heme groups (figure 8.2d). Each alpha chain is 141 amino acids long; each beta chain is 146 amino acids long. Sickle-cell anemia results from a single nucleotide substitution in the DNA encoding the beta chain. The sixth codon in the beta chain is altered so that, during synthesis of the chain, valine is added rather than the normal glutamic acid (figure 8.16). Homozygotes for this substitution have the sickle-cell syndrome of more than a dozen pathologies (figure 4.7).

The One Gene—One Polypeptide Hypothesis

The fundamental importance of proteins has led to the **one gene–one polypeptide hypothesis** of genetic expression. This hypothesis states that each gene codes for the production of one polypeptide chain, which is often an entire protein molecule. This polypeptide, or the protein of which it is a constituent, in turn participates in a series of biochemical reactions that are ultimately expressed as some aspect of the phenotype. For example, in one form of albinism

Figure 8.16

The sickle-cell anemia mutation is an example of the colinearity of the gene and the protein. A single base substitution in the DNA causes a codon change that results in the wrong amino acid being incorporated into the beta chain of the hemoglobin molecule.

the abnormal DNA sequence leads to the production of a defective form of the enzyme tyrosinase. Tyrosinase normally catalyzes formation of the pigment melanin (figure 8.17). Numerous other genetic diseases, the *inborn errors of metabolism,* are now also understood to result from mutations affecting the production of a key protein. Phenylketonurics, for example, do not have the enzyme necessary to metabolize the amino acid phenylalanine, which accumulates in the nervous tissues and results in mental retardation. The immediate cause of such diseases is a defective protein, but the ultimate defect lies in the sequence of the DNA itself.

Present knowledge of the structure and function of DNA supports many aspects of the one gene–one polypeptide hypothesis: a gene is a sequence of nucleotide bases on a chromosome that determines a colinear sequence of amino acids in a protein. It is clear, however, that some genes have functions other than producing proteins. One of those functions is controlling the function of genes that do produce proteins, the subject of the next chapter.

The Central Dogma Revisited

Since the central dogma was first stated, research has shown clearly that DNA meets all the requirements of the hereditary material stated earlier in this chapter:

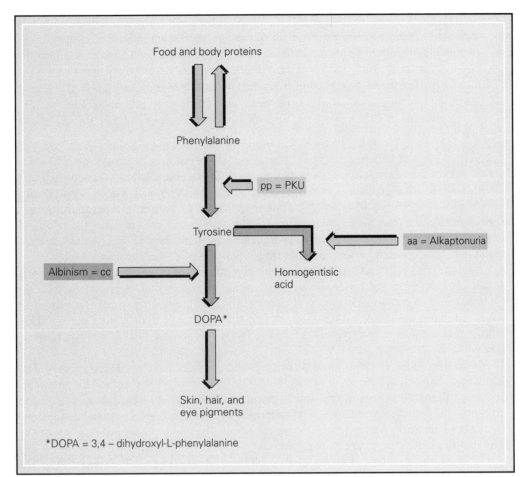

Figure 8.17

Metabolic pathways for some compounds involved in genetic inborn errors of metabolism. Absence of the enzyme converting phenylalanine to tyrosine results in phenylketonuria (PKU). Absence of the enzyme converting tyrosine to DOPA results in classical albinism. Absence of the enzyme converting tyrosine to homogentisic acid results in alkaptonuria.

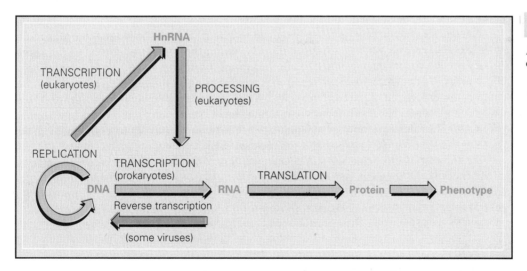

Figure 8.18

A contemporary model of the central dogma.

1. The sequence of bases in DNA has an immense capacity for encoding information, certainly more than sufficient information to direct the development and life of an organism.

2. A complete set of hereditary information is available in every cell of the body as the result of mitosis. The information is translated into a usable form through the production of necessary proteins, mediated by the action of RNA.

3. As a result of the replication of DNA and the subsequent reduction division of the chromosomes during meiosis, a diploid, sexually reproducing individual transmits a complete haploid set of chromosomes to each offspring. For both mitotic and meiotic division, complementary base pairing of DNA provides an extremely accurate mechanism for the transmission of hereditary information.

4. Addition, loss, or change of nucleotide bases in a DNA molecule provides a mechanism for mutation of a gene, ensuring long-term renewal of genetic variability in a population.

Verification and understanding of DNAs role in the hereditary process is one of the great achievements of modern science. The research that allowed verification of the central dogma has also led to some surprises that require modifications of the model. For example, the first step in protein synthesis involves activation of the gene to be transcribed, the process of *regulation.* As we have seen, production of mRNA in eukaryotes involves the additional *processing* of HnRNA. Protein production in eukaryotes, therefore, requires four basic steps—regulation, transcription, processing, and translation—rather than two (figure 8.18).

Research has also established some clear exceptions to the simple rules of the central dogma. For example, some viruses use RNA rather than DNA as their genetic material. Completion of their parasitic life cycle requires these viruses to construct DNA from the RNA template; the DNA is then inserted into the genome of their host organism. Thus, information can flow in reverse—from RNA to DNA—in a process called **reverse transcription.**

Other variations are even more surprising. Some viruses have **overlapping genes.** That is, a single segment of DNA may encode two or three different proteins; protein production begins at different locations in the same segment of DNA, and different frames are read for each protein. Some genes, known as **palindromes,** encode the same message whether read forward or backward. And, a few instances have been discovered where both strands, rather than just the sense strand, are transcribed.

DNA is also far more dynamic than envisioned in the original Watson-Crick model. Mutations are not uncommon, but DNA repair mechanisms exist to restore the original sequence. Most surprisingly, some genes do not occupy fixed locations in the genome, but can move between locations within the genome or even between species. The biology of **mobile genes** will be discussed briefly in the next chapter.

Finally, the genetic code is often described as an alphabet. Codons are the words of the genetic language. But words have meaning only within the context of the entire language. A book is not a collection of random words, but words assembled according to complex rules of grammar. Contemporary genetic research is now focusing on the genetical grammar. What controls the expression of particular words? What makes a sentence or a paragraph? How do complex phenotypes result from the interactions of dozens of proteins and biochemical reactions? The rules governing gene regulation are only just beginning to be understood, but they are the key to understanding the genetic language.

For the rest of this book, we will develop aspects of the relationship between genotype and phenotype. Although increasing knowledge shows that expression of genetic information contained in DNA is far more complex than the original Watson-Crick model implied, the central dogma remains the primary framework for understanding this relationship.

Summary

1. Proteins fulfill many important functions for the living organism. In particular, proteins that are enzymes catalyze the biochemistry of life.
2. Proteins consist of amino acids joined together by peptide bonds into polypeptides. Either one polypeptide chain or several chains bound together may form a protein. The structure of a protein is described not only by its sequence of amino acids, but also by the twisting and folding of the chains.
3. Proteins were once thought to be the hereditary molecules because they are long molecules of variable amino acid sequence and therefore capable of encoding immense amounts of information.

4. The nucleic acid DNA was discovered in 1869 and recognized as a major constituent of chromosomes, although its role as the genetic substance was not immediately recognized.

5. A series of important experiments eventually demonstrated the hereditary role of DNA. Griffith, Avery, MacLeod, and McCarty showed that DNA was the transforming principle of bacteria. Hershey and Chase showed that nucleic acid was the hereditary material of viruses.

6. Watson and Crick described the structure of DNA in 1953. They proposed that DNA is a long molecule in the shape of a double helix, and that the two sugar-phosphate backbones of the double helix are bound together by bonds formed between paired nitrogen bases. Adenine always binds to thymine; cytosine always binds to guanine. Information is encoded in the sequence of bases. Accurate replication results from specific complementary base pairing.

7. RNA is similar in structure to DNA. It is constructed from a sugar with one more oxygen atom than the sugar in DNA; it is single stranded; and it uses uracil rather than thymine. RNA is the hereditary material in some viruses, and three forms of RNA are required for protein synthesis in all organisms.

8. The central dogma describes the flow of hereditary information. Between generations, the accurate replication of nuclear DNA results in precise transmission of genetic instructions. Within the cells of an individual, nuclear DNA is transcribed to messenger RNA, which in turn is translated into proteins. Both mRNA and proteins may be considered phenotypic characters that exert a primary influence over other phenotypic characteristics.

9. Complementary base pairing ensures accurate replication of DNA. Each parent strand of DNA serves as a template for the production of a new double-stranded DNA molecule. Replication is semi-conservative, because each parent strand forms one-half of a daughter molecule.

10. The genetic code is contained in the sequence of nucleotide bases in DNA. It is a triplet code, consisting of three-letter words, or codons, each specifying either a particular amino acid in a polypeptide chain or punctuation of the genetic message. Because some amino acids are encoded by more than one codon, the code is degenerate. Essentially the same code is used by all species on earth: It is universal.

11. A gene is a sequence of nucleotides encoding information for the production of a colinear sequence of amino acids in a polypeptide. In most species, genes are nonoverlapping. Some viruses, however, have overlapping genes.

12. Frameshift mutations result from the addition or deletion of a nucleotide base in a DNA sequence, which alters the three-letter reading frame of subsequent codons. Mutations involving the substitution of one nucleotide base for another will affect only the amino acid produced by that codon. A substitutional mutation may have major phenotypic effects, as in the case of sickle-cell anemia. Alternatively, the degeneracy of the code results in many same sense mutations, which have no apparent phenotypic effect.

13. Messenger RNA (mRNA) is transcribed only from the sense strand of DNA. In eukaryotes, heterogeneous nuclear RNA (HnRNA) is first transcribed, from which segments are excised to form mRNA. Segments of eukaryotic DNA that are transcribed but not translated are called introns; segments that are translated are called exons.

14. Ribosomes, formed of ribosomal RNA (rRNA) and proteins, are the sites of translation. Translation often involves the attachment of many ribosomes to one mRNA, forming a polyribosome.

15. More than fifty types of transfer RNAs (tRNAs) are known. Each carries a specific amino acid to the ribosome for attachment to the growing polypeptide chain. Insertion of an amino acid into the correct location in a polypeptide depends upon the match between the mRNA codon and the anticodon on a tRNA.

16. The modern view of a gene and the hereditary mechanism combines discoveries from three models of genetic study: Mendelian, chromosomal, and molecular. The one gene–one polypeptide hypothesis relates the molecular definition of a gene to its immediate phenotypic product (the protein) that in turn mediates further phenotypic development.

Key Words and Phrases

amino acid

anticodon

central dogma

codon

colinear

complementary base pairing

degenerate

deoxyribose

DNA

double helix

enzyme

exon

frameshift

genetic code

heterogeneous nuclear RNA (HnRNA)

informational macromolecules

intron

messenger RNA

mobile genes

nitrogenous base

nonoverlapping genes

nonsense codon

nucleic acid

one gene–one polypeptide hypothesis

overlapping genes

palindrome

peptide bond

polypeptide

polyribosomes

primary structure

protein

reverse transcription

ribosomal RNA

ribosome

ribozymes

RNA

same sense mutation

semi-conservative replication

sense strand

substitution

terminator codon

transcription

transfer RNA

transformation

translation

universal code

Watson-Crick model

Questions

1. What is the minimum number of nucleotide bases needed to code for the 135 amino acids that comprise the enzyme cytochrome c of some species?
2. If a segment of DNA has the sequence ACACGATTCCCGACG, what is the sequence of bases in the complementary strand? What would be the sequence of a molecule of mRNA transcribed from the first strand? What polypeptide or polypeptides would be encoded by this sequence?
3. What would be the anticodons on the tRNA molecules that would service the message transcribed in the previous question?
4. Genetic engineering techniques now permit scientists to transfer genes from one species to another—often distantly related species. What will the universality of the genetic code mean, in practical terms, to the success of such procedures?
5. Why is it unlikely that a segment of DNA from an individual's father or mother still exists in some cell of the individual's body?
6. Compare and contrast the roles of mRNA, rRNA, and tRNA in protein synthesis.
7. If normal human cells have 46 chromosomes and roughly 6.4 picograms of DNA (table 8.1), could a male produce sperm containing 6.4 picograms of DNA? How?
8. RNA is a single-stranded molecule. How could RNA form double-stranded regions like DNA? What would hold these regions together?
9. Specify the quantity of DNA that you would find in a single human cell at the following stages of cell division: a) prophase of mitosis, b) late anaphase of mitosis, c) prophase of meiosis I, d) metaphase of meiosis II, e) secondary oocyte.
10. The number of unique messages that can be encoded by a segment of DNA is 4^n, where n is the number of bases in the DNA segment. What does the number four signify?

For More Information

Chambon, P. 1981. "Split Genes." *Scientific American,* 244:60–71.

Choi, Y. D., P. J. Grabowski, P. A. Sharp, and G. Dreyfuss. 1986. "Heterogeneous Nuclear Ribonucleoproteins: Role in RNA Splicing." *Science,* 231:1534–39.

Cornish-Bowden, A. 1985. "Are Introns Structural Elements or Evolutionary Debris?" *Nature,* 313:434–35.

Crick, F. 1966. "The Genetic Code: III." *Scientific American,* 215:55–64.

Crick, F. 1988. *What Mad Pursuit. A Personal View of Scientific Discovery.* Basic Books, New York.

Gilbert, W. 1985. "Genes-in-Pieces Revisited." *Science,* 228:823.

Kornberg, R. D., and A. Klug. 1981. "The Nucleosome." *Scientific American,* 244:52–80.

Lake, J. A. 1981. "The Ribosome." *Scientific American,* 245:84–97.

Lewin, R. 1985. "Reverse Transcriptase in Introns." *Science,* 229:1083.

Ptashne, M. 1986. "Gene Regulation by Proteins Acting Nearby and at a Distance." *Nature,* 322:697–701.

Racker, E. 1987. "Structure, Function, and Assembly of Membrane Proteins." *Science,* 235:959–61.

Rich, A., and S. H. Kim. 1978. "The Three-Dimensional Structure of Transfer RNA." *Scientific American,* 238:52–62.

Rosenfeld, I., E. Ziff, and B. Van Loon. 1983. *DNA for Beginners.* Writers and Readers Publishing, Inc., W. W. Norton & Co., New York.

Sayre, A. 1975. *Rosalind Franklin and DNA.* W. W. Norton & Co., New York.

Watson, J. D. 1968. *The Double Helix.* Atheneum, New York.

Yanofsky, C. 1967. "Gene Structure and Protein Structure." *Scientific American,* 216:80–96.

© George Lefevre and
Academic Press

Control of gene expression

Fruit fly

Mouse embryo

Figure 9.1

Homeobox genes regulate the development of animals as diverse as the fruit fly and the mouse. These genes have a common evolutionary origin, and their location along the chromosome corresponds with the head to tail body axis of each animal. The correct regulation and expression of these genes determine the shape of the body, as cells are programmed to become specific tissues and organs.

From DeRobertis, et al., *Scientific American*, Volume 263, July 1990. All rights reserved.

Each cell in an organism possesses a complete set of chromosomes containing all the information necessary for the metabolic and physiological functions of that organism. Not all of that information is used all the time. Most cells share a small, common set of *background* genes that function more or less continuously to meet basic needs—genes controlling carbohydrate metabolism, for example, necessary to provide a regular energy supply for the cell.

Other genes are turned on for only a short period of one's lifetime. Genes controlling the umbilical cord and the fetal portion of the placenta function only during the first nine months of life. Embryonic and fetal hemoglobin genes (see figure 9.3) function exclusively or primarily during gestation. The SRY gene may function only once, early in fetal development, when it instructs the embryonic gonad to differentiate to a testis.

Other genes function at irregular intervals, often in response to environmental stimulation. A woman may express the genes controlling the maternal portion of the placenta several times over the course of a few decades. Body-building exercises induce a genetically controlled developmental response in affected muscles. The onset of many genetic diseases occurs only at predictable times of life.

The branch of genetics dealing with the control of gene expression is known as *regulatory genetics*. Although the importance of the regulation of gene function has long been recognized, the techniques for studying regulatory processes have developed slowly over the past several decades. The Watson-Crick model of DNA, for example, immediately suggests mechanisms by which DNA can both contain and transmit immense amounts of genetic information. It does not, however, directly suggest factors controlling the expression of that information.

Knowledge of regulatory mechanisms has developed slowly, through painstaking studies of the structure of specific segments of DNA. Geneticists now commonly distinguish two categories of genes. **Structural genes** encode the amino-acid sequences for polypeptides destined to become normal proteins, i.e., catalysts, hormones, and supportive, protective, and transport molecules. **Regulatory genes,** on the other hand, function to control the expression of either structural or other regulatory genes. Regulatory genes may produce a polypeptide product or an RNA molecule with a regulatory function.

Secondary and tertiary structures of DNA may also be involved in control of gene expression, and doubtless most phenotypic traits are the result of interactions of many structural loci. Although it is clear that gene regulation is a fundamental aspect of life, the many forms of regulation and their complex interactions are just beginning to be understood.

The rewards of mastering gene regulation will be enormous. Recombinant DNA techniques, for example, depend not just on being able to move genes between species, but to turn them on and off at will. Similarly, medical practice may dramatically change if genes controlling development can be manipulated and controlled: wound-healing could be made quicker, more effective, even in old age (figure 9.1); cancer, a failure of gene control, could be cured; perhaps even the aging process can be reversed or modulated.

Gene Regulation in Prokaryotes

Some of the best-defined regulatory systems are in the bacterium *E. coli*. Recall that prokaryotes have only one chromosome, a single ring of DNA, with no associated proteins. In 1961 French scientists François Jacob and Jacques Monod proposed the **operon** hypothesis to explain gene regulation in *E. coli*. The operon hypothesis, for which Jacob and Monod received the 1965 Nobel Prize for Physiology or Medicine, was the first complete model of regulation of gene function in any organism. It provides the basis for understanding regulation in all prokaryotes. With the advent of genetic engineering techniques in the late 1970s, the operon model has also been put to practical use, providing a technique for controlling the artificial production of synthetic gene products (see chapter 10).

The Lac Operon

E. coli can derive energy from the metabolic decomposition (digestion) of the milk sugar lactose. Three enzymes produced by *E. coli* are involved in the digestion of lactose, but the enzymes are produced only when lactose is present in the growth medium. Jacob and Monod proposed that the production of these three enzymes is controlled by a structural and functional unit of the DNA of *E. coli* called an *operon* (figure 9.2).

The operon consists of a sequence of contiguous genes in one segment of the bacterial DNA. Three structural genes, designated Z, Y, and A, encode the amino-acid sequence of the three enzymes involved in the breakdown of lactose. Immediately adjacent to gene Z are two regulatory genes—gene O, the **operator,** and gene P, the **promoter.** The entire sequence of promoter, operator, and three structural genes comprises the lac operon, so called because of its role in lactose metabolism.

For transcription of Z, Y, and A to occur, the enzyme **RNA polymerase** (the enzyme that catalyzes the formation or *polymerization* of mRNA from a DNA template) attaches to the promoter region of the operon. The RNA polymerase then moves along the operon, passing the operator. Upon reaching gene Z, it begins transcription of a single, long mRNA

Figure 9.2

The lac operon. (a) In the absence of the inducer substance (lactose), the active repressor protein binds to the DNA at the operator region, preventing mRNA transcription. No enzymes are produced. (b) When the inducer, lactose, is present, it binds to the repressor protein, removing it from the DNA. Now mRNA can be transcribed from the entire operon, and the enzymes used in lactose metabolism are synthesized. When all of the lactose is metabolized, the repressor protein again binds to the DNA, shutting off mRNA production, leading to a cessation of enzyme production.

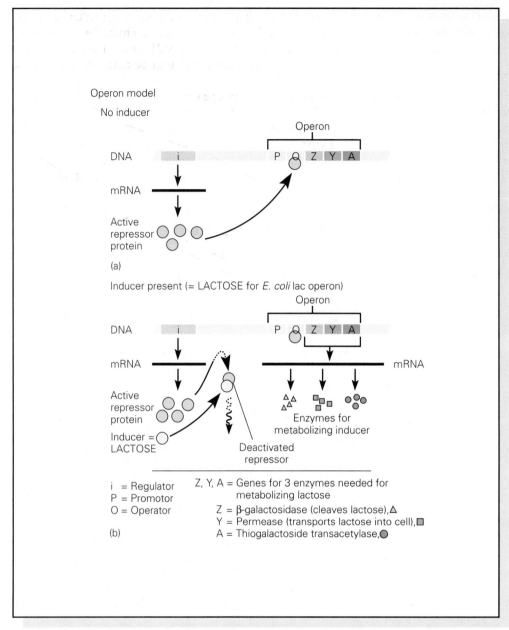

Operon model

No inducer

Operon

DNA   i   P O Z Y A

mRNA

Active repressor protein

(a)

Inducer present (= LACTOSE for *E. coli* lac operon)

Operon

DNA   i   P O Z Y A

mRNA   mRNA

Active repressor protein

Inducer = LACTOSE

Deactivated repressor

Enzymes for metabolizing inducer

i = Regulator
P = Promotor
O = Operator

Z, Y, A = Genes for 3 enzymes needed for metabolizing lactose

Z = β-galactosidase (cleaves lactose), △
Y = Permease (transports lactose into cell), ■
A = Thiogalactoside transacetylase, ●

(b)

molecule, which encodes the amino-acid sequence of the three enzymes required for lactose metabolism. Subsequent translation of this single mRNA produces the three necessary enzymes, which are not attached to each other, though they were produced from a single mRNA.

Transcription of the Z, Y, and A genes does not occur continuously. Transcription is controlled by a *regulator gene,* designated i, located at some distance from the operon and not considered to be a part of the operon itself. The regulator gene, like structural genes, encodes an mRNA that is translated into a polypeptide chain. When four polypeptide chains are produced by the regulator, they bind together loosely, forming a single active protein called the **repressor.** The repressor protein binds to the operator site of the operon, physically interfering with the movement of the RNA polymerase and inhibiting transcription. *Transcription inhibition,* then, is the major mode of regulation of the lac operon.

Most of the time, a repressor protein is bound to the operator gene, and transcription does not occur. This is presumably an adaptive trait of *E. coli.* After all, lactose is not continuously present in its surroundings, so continual synthesis of lactose metabolizing enzymes would be inefficient. However, when lactose is present, enzyme synthesis begins. How does this occur?

Lactose in the growth medium of *E. coli* crosses the cell membrane, entering the cytoplasm. There, it combines with the repressor protein, causing a physical deformation of the repressor that releases it from the operator site (figure 9.2). Once the operator site is free, movement of RNA polymerase on the promoter is no longer blocked, and transcription of Z, Y, and A commences.

Transcription continues only as long as the cytoplasm has sufficient lactose to bind repressor proteins, leaving the operator free. Note that lac operon activity (transcription and subsequent translation) reduces the amount of the controlling molecule. In other words, the presence of lactose *induces* transcription, which then leads to production of lactose metabolizing enzymes. In turn, these enzymes break down cytoplasmic lactose, freeing the repressor proteins that then bind again with the operator and turn off transcription, at least until more lactose enters the cell.

Regulatory Characteristics of the Lac System

Many regulatory mechanisms, particularly in eukaryotes, differ substantially from the lac operon, but the elegance and simplicity of the lac operon illustrate principles of regulatory design that may be common in many organisms:

1. The lac operon is an **inducible** system. Production of enzymes occurs only when necessary and is induced by the presence of the substrate upon which those enzymes are active, in this case, lactose. Such a system is efficient for metabolic pathways that are used only periodically.

2. Although the lac operon has many components, a single regulatory system controls all genes induced in one metabolic pathway. Thus, for example, the lac operon produces only one mRNA, encoding all three enzymes needed for lactose metabolism. Under normal circumstances, either all three enzymes involved in lactose metabolism are present, or none are present.

3. In the lac operon, control occurs at the transcriptional level. In general, however, gene function may be regulated at any point between the gene and its ultimate product. Gene regulation occurs at the level of transcription, translation, or post-translational modification of protein structure. Control of gene function also includes the control of replication of DNA, as well as protein synthesis.

4. The lac operon exemplifies **negative feedback control.** A feedback system is one in which a product of the system exerts control over its future level of function. In a negative feedback system, high levels of the product result in a decrease in future production levels. Low levels of the product cause increased future levels of production. Just as a thermostat maintains temperature at a set level, so the lac operon maintains a relatively constant amount of lactose-metabolizing enzymes in the cytoplasm. Negative feedback is probably a crucial component of many systems of genetic regulation.

5. Regulation of the lac operon involves both the primary structure (nucleotide sequence) and the secondary and tertiary structures (physical conformation) of the DNA molecule. Transcription is inhibited by the repressor molecule physically altering the shape of the DNA molecule. We have previously stressed the overriding importance of the primary structure of DNA, but higher level structures also play a role in gene function.

Gene Regulation in Eukaryotes

Control of gene function in eukaryotes is presently an area of intense scientific and medical research. The structure of DNA is now reasonably well known, but precise knowledge of its function is developing only slowly.

There is, however, no question that gene regulation is the essence of life. Nothing demonstrates the astonishing precision and complexity of gene regulation in eukaryotes more dramatically than the orderly **development** of an individual organism from the fertilized egg

Figure 9.3

Timing of production of different polypeptide chains of hemoglobin during human development. The vertical axis shows the proportion of hemoglobin molecules containing the various chains.

Figure from *Our Uncertain Heritage: Genetics and Human Diversity* by Daniel Hartl. Copyright © 1977 by Harper & Row Publishers, Inc. Reprinted by permission of HarperCollins Publishers Inc.

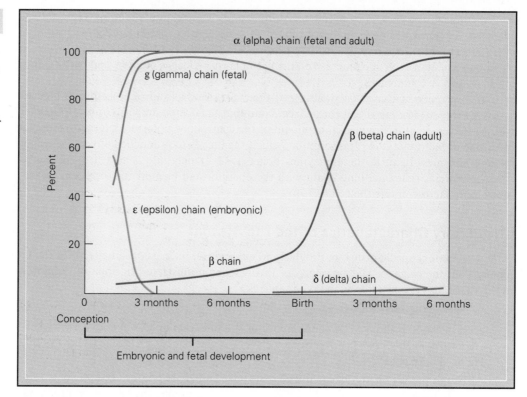

to maturity to inevitable death. The basis of developmental events is differential gene activity—the regular turning on and off of genes. One expression of the loss of normal regulatory mechanisms is cancer, uncontrolled growth of cells.

A primary goal of developmental genetics is to understand the specific mechanisms controlling developmental sequences. The ability to turn off harmful genes or turn on beneficial genes has applications to medicine, agriculture, and business.

Because the essence of gene control is the precise turning on and off of genes, we begin this section with an example of regulation at the biochemical level—the sequential production of various types of hemoglobin during early human development. Then, we consider other mechanisms of regulation and allelic interactions.

Development and Regulation at the Biochemical Level: Hemoglobin

Hemoglobin is a physiologically important molecule whose genetics is relatively well known. It combines reversibly with oxygen, which it transports from the lungs to all tissues of the body. The complete hemoglobin molecule consists of four polypeptide chains, each bound to a heme (iron containing) group (figure 8.2d). At least seven different genes in humans encode subunits or chains of hemoglobin, but not all are equally active throughout life. The differential activity of the various hemoglobin genes exemplifies gene regulation at the biochemical level (figure 9.3).

Most types of hemoglobin contain two alpha chains (α_2). The gene controlling the alpha chain is turned on early in embryonic life and maintains a high level of activity until death. A gene controlling the epsilon (ϵ) chain also becomes functional in the yolk sac, an embryonic structure, very early during gestation, but its function ceases about three months into development. Embryonic hemoglobin consists of two α chains and two ϵ chains ($\alpha_2\epsilon_2$).

At about three months of gestation, the gamma gene (γ) becomes active in the fetal liver, maintaining a high level of activity until birth, after which it diminishes in function until about six months of age. Fetal hemoglobin ($\alpha_2\gamma_2$)—designated Hb F—and embryonic hemoglobin both have a very high affinity for oxygen, a characteristic advantageous for a developing fetus who must be able to pull oxygen across the placenta from the red blood cells of the mother.

Also during gestation, the gene for the beta (β) hemoglobin chain becomes active, achieving a high level of function postnatally. About 90% of all adult hemoglobin is designated Hb A ($\alpha_2\beta_2$). A small portion of adult hemoglobin is type Hb A2, consisting of two alpha chains and two delta (δ) chains. The δ gene becomes active at birth and maintains a low level of activity throughout life. Both the beta and delta genes are expressed in red blood cells, delta expressed in young red blood cells and beta in older red blood cells.

Other types of hemoglobin, containing other chains, have been identified. Additionally, many genetic variants of hemoglobin, often involving only one amino acid substitution, have been discovered. Over 250 different forms of Hb A alone are known. Some of these types are associated with pathological conditions—sickle-cell anemia, for example.

Thalassemia refers to a group of genetic diseases involving reduction or absence of one type of hemoglobin subunit. Consequently, certain types of hemoglobin are reduced or absent in affected individuals. Beta thalassemia is caused by a gene that inhibits synthesis of the β chain. In homozygotes, the β chain is not produced at all. These individuals are totally deficient in Hb A, causing a serious form of anemia that is usually lethal by the teens. Heterozygotes produce some β chain and thus are less severely affected. Increased amounts of Hb A2 and persistent Hb F compensate somewhat for reduced amounts of Hb A in the heterozygote.

Thalassemia reflects a change in the amount of protein produced, rather than a change in the structure of the protein. Studies have shown that there is a mutation in the gene regulating the synthesis of the β chain of Hb A. The mutation changes one short nucleotide sequence in the regulator gene causing it to resemble a base sequence that terminates transcription. When the transcriptional apparatus reaches this portion of the regulator, it ceases transcription and releases the forming mRNA molecule before the structural gene has even been transcribed. Because the genetic basis of diseases involving defective or deficient hemoglobin is well understood, these diseases are hoped to be among the first to be successfully treated by genetic engineering.

Mechanisms of Eukaryotic Regulation

Although operons have not been identified in eukaryotes, certain aspects of operon structure and function—the regulatory gene and the regulator molecule, for example—are integral components of eukaryotic gene regulation. However, the greater size and complexity of eukaryotic cells probably require more complex types of regulation than are found in prokaryotes.

Control points regulating gene function affect virtually every process governing the transition from genotype to phenotype. In this section we describe some of the better known regulatory mechanisms.

Organization of the Nucleus
The physical relationships of genes and chromosomes probably play an important role in gene function. Meiosis demonstrates that chromosomes can identify each other and that intranuclear organization is not random: homologues pair and remain in close proximity until fission.

Assessing physical relationships of chromosomes is extremely difficult, because experimental techniques often severely disrupt cell and nuclear structure. Nonetheless, it appears likely that functionally related genes on different chromosomes may be activated (i.e., initiate transcription) when their chromosomes move to adjacent positions, or when the chromosomes actually attach to one another.

Modifications of DNA Structure
The physical structure of eukaryotic chromosomes changes predictably during the cell cycle. During mitotic division, chromosomes are tightly coiled and condensed. During the G1, S, and G2 stages of the cell cycle, the chromosomes are relatively diffuse. Replication of chromosomal DNA occurs only during the S phase, while transcription occurs during the G1 and G2 phases. Neither replication nor transcription occurs during mitotic division itself.

BOX 9.1

Evolutionary relationships of human globin genes

Figure 9.4

Chromosomal location of genes of the alpha and beta hemoglobin families and the phylogenetic relationships of these genes based on gene sequence data.

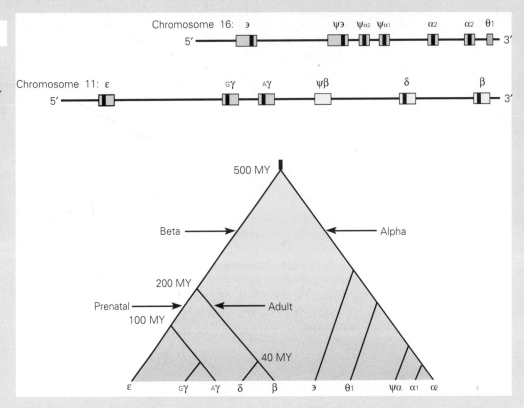

Thus, transcription and replication occur only in those phases of the cell cycle when DNA is uncoiled. This suggests that the physical conformation of DNA, particularly the extent of coiling, plays a major role in gene expression. Enzymes that direct transcription and replication may have access to DNA only when it is unwound.

Chromosome puffs. Studies of *Drosophila* have provided some of the best information regarding gene regulation in eukaryotes. One of the great advantages of *Drosophila* and some other insects for chromosomal studies is that in some tissues, particularly the salivary glands, the chromosomes are *polytene*. Recall that in most cells, the chromosomes are unineme, or consist of only one strand of DNA plus associated nuclear proteins. **Polytene chromosomes,** on the other hand, are spectacularly large and easily observed microscopically (figure 9.5). Their large size is the result of the side-by-side arrangement of as many as one thousand identical strands of DNA.

Drosophila salivary chromosomes are not uniform in appearance, but exhibit about five thousand darkly staining *bands.* These bands correspond to *chromomeres,* bead-like bulges along the length of all chromosomes (figure 9.5). The lightly staining segments between bands are called *interbands.* Each *Drosophila* chromomere appears to contain several dozen structural genes.

During development particular chromomeres expand or *puff* in a predictable sequence (figure 9.6a). The puffs result from uncoiling of the chromomeres within each puff. Uncoiling is apparently necessary for transcription of RNA, and **chromosome puffs** have relatively high concentrations of RNA.

The ability to clone and sequence genes (chapter 10) has permitted us to understand the evolutionary relationships of the major globin genes expressed throughout our lives. Two gene families have been described based on the residual similarities in their primary structure (amino acid sequences), an alpha and a beta family. The alpha gene family is located on chromosome 16 and the beta family on chromosome 11. Most of the functional globin genes sequenced are composed of both exons (coding regions) and introns (see chapter 8). In figure 9.4 the intervening sequences (introns) that are spliced out of the messenger RNA prior to transcription by the ribosomes, are indicated in blue. A relatively high proportion of most of these genes appear not to be expressed. Incomplete and defective gene duplicates, or pseudogenes, are recognized by their sequence similarity to transcribed genes and are not translated into protein.

Based on comparative gene sequencing from numerous species whose divergence times are known from fossil information, it was determined that a gene duplication event gave rise to the alpha and beta gene families roughly 500 million years ago, shortly after the divergence of jawless fish (Agnatha) from the ancestor of all other vertebrates. All of the globins that are formed by combination of an alpha and a beta family gene have different oxygen-binding capabilities, differ in other physiological properties, and are expressed at different times in development.

In the alpha family, the embryonic zeta chain is most divergent (oldest) and the two alpha chains have almost identical DNA sequences suggesting a very recent divergence. In the beta family there was an early separation of the genes coding for prenatal and adult globin chains and the most recent duplication and divergence occurred after the human lineage diverged from that leading to the New World Monkeys about 35 million years ago.

Gene duplication and divergence resulting in new function of the duplicate gene, or refinement of a physiological function (such as better oxygen binding ability) have been discovered in other gene families throughout evolutionary history because of our recent technological advances permitting sequencing of genes including non-expressed pseudogenes. Some experimental gene therapy has already been proposed whereby chemicals can be used to turn fetal globin genes back on for individuals suffering from sickle cell anemia, to provide a functional hemoglobin in place of a defective one.

Lampbrush chromosomes. Oocytes often exhibit high levels of metabolic activity, including transcription. A special type of chromosome has been discovered in the oocytes of salamanders (figure 9.6b). **Lampbrush chromosomes** have a brush-like appearance due to the presence of many loops of DNA extending out from the main axis of the chromosome. Like chromosome puffs, lampbrush chromosomes are sites of transcription, and high concentrations of RNA are associated with them. Again, physical uncoiling of chromosomes is associated with high levels of transcription.

Chromosome inactivation. One of the most spectacular examples of conformational changes affecting gene regulation is **X inactivation** (chapter 5). Very early in development in cells of all placental and most marsupial female mammals, one of the two X chromosomes is turned off by **heterochromatinization,** an essentially permanent condensation of the chromosome. In its heterochromatic state, the chromosome stains heavily during interphase and is visible as a *Barr body* near the margin of the nucleus.

X inactivation is a major regulatory mechanism that equalizes the expression of genes carried on the X chromosome in the two sexes of mammals, hence its common designation as *dosage compensation*. The extreme condensation of the inactivated chromosome seems to be the mechanism by which transcription is turned off, a finding complementary to the uncoiling and high levels of transcription in chromosome puffs and lampbrush chromosomes.

In most female placental mammals, inactivation of one X chromosome appears to occur at random. In some cells, the paternally derived X is inactivated, while the maternally derived X is inactivated in others. In mules (hybrids between a donkey and a horse) preference

Puffing

(a)

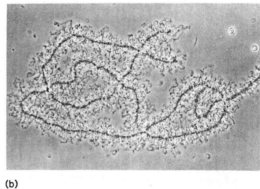

(b)

has been demonstrated for inactivation of the donkey X chromosome. The importance of
non-random inactivation of X chromosomes, as well as the mechanisms controlling it, has
yet to be understood.

A final twist to the X chromosome story lies in the sex cells. In the oogonia of sexually
mature female mammals, both X chromosomes are metabolically active, participating fully
in meiotic processes. However, X inactivation occurs in *all* cells of the embryo during the
first few days of development. Later the inactive X is reactivated in those cells that differentiate to become ovaries. This phenomenon is of considerable importance, because many
developmental changes are normally irreversible. Discovery of the mechanism of X chromosome reactivation could eventually lead to control and manipulation of other developmental processes.

Mobile Genes.   Changes in the sequence of nucleotide bases in DNA or in the structure
of chromosomes is called **mutation,** a topic that will be examined in detail in chapter 11.
Some of the strongest evidence for the role of the physical relationships of DNA regulating
gene activity lies in the serious phenotypic consequences that often accompany gene rearrangements, even when no DNA is gained or lost.

I

Transposon Ds · Gene for color → Purple

II

Transposon · Mutant gene → Colorless

III

Transposon Ac · Ds

Ac · C locus

(a)

(b)

Stable mutant · Spm transposon · Ac transposon

(c)

(d)

Figure 9.7

Mobile genes, or transposons, and their effects on color in maize. (a) Two transposable genes, Ds and Ac can influence the color of maize kernels. When Ds "jumps" into the pigment gene, the latter stops producing pigment. When Ac is also present, it can cause Ds to "jump out" of the pigment gene during development of some cells, producing pigmented and colorless sectors on the kernel. (b) Ears of corn with and without transposable gene mutations. (c) Nobel laureate, Barbara McClintock at work in 1947, and (d) after being awarded the Nobel Prize in 1983.

(b) Mrs. F. Goro, executive of the estate of Fritz Goro (photographer), *Scientific American,* 1984. (c) AP/ Wide World Photos; (d) Brian F. Alpert/ Keystone.

In this section, we note one type of gene whose activities are equivalent to repeated mutation, but that are a component of the genome in every species in which they have been sought. Called **transposable elements,** or **transposons,** or **mobile genes,** these genes can have major effects by regulating gene expression (figure 9.7).

Transposable elements were discovered in the late 1940s by Barbara McClintock (figure 9.7c,d), as a result of her painstaking studies of color in maize plants. The idea that genes could move to different sites among the chromosomes was so radical that her ideas were not accepted until similar genes were found two decades later in bacteria and subsequently in every species examined. Transposons can move to virtually any site on any chromosome, but most commonly they move small distances to a nearby site on the same chromosome. Insertion of a transposon into an active structural gene terminates transcription, and its subsequent removal has the opposite effect, allowing transcription to be re-initiated. Such on-off activity has the effect of producing maize kernels that are mosaic (figure 9.7b).

McClintock was awarded a Nobel prize in 1983 for discovering such a widespread, and entirely unexpected, genetic mechanism. Transposons comprise only a very small fraction of the genome, but one type appears to be involved in causing some cancers (chapter 11), so the practical significance of McClintock's discovery is large indeed.

Regulatory Molecules

Just as coiling and other conformational changes can govern DNA activity, regulatory chemicals that attach to, or interact with, DNA can also control transcription and replication. Eukaryotic DNA is usually bound to histones, proteins that may play a significant role in

205

regulation. In this section, we discuss three types of molecules that are known to have major regulatory effects.

Hormones.   **Hormones** are molecules that are a primary means of intercellular communication in most eukaryotes. *Hormone* is derived from a Greek word meaning "to set in motion," an apt description of its effect in regulating transcriptional activity. In mammals and probably most eukaryotes, hormones are produced by many tissues and organs throughout the body, then released into the systemic circulation, where they are transported to many different sites of activity. Some hormones, for example, interact with histone and non-histone proteins to initiate transcription in target cells.

Among the hormones that affect transcriptional activity are the steroids, secreted by the adrenal glands, testes, ovaries, and the placenta. Sex hormones such as estrogen and testosterone are steroid hormones. In chapter 6, we discuss the strong influence that testosterone, in particular, exerts over sexual development. Testosterone is produced in the testes, but circulates throughout the entire body. Its effects are not limited to reproductive anatomy, but include aspects of the anatomy and development of many systems. Skin, muscles, and skeleton are particularly affected by testosterone, as are many aspects of physiology and behavior. Clearly, testosterone is a molecular signal controlling gene expression throughout the body.

Other steroids have equally large effects. Most tissues respond to estrogens, the female hormones, as well as to testosterone; both sexes produce male and female hormones, although in different proportions. The balance of male and female steroids is delicate, and many traits respond throughout one's lifetime to alterations in the balance. Sexual function changes regularly, for example, in response to tiny changes in hormonal level. The masculinizing effects of testosterone and the feminizing effects of estrogens demonstrate a fundamental genetic role of these hormones: differentiation of tissues resulting from initiation of transcription in response to a chemical signal.

Pheromones.   Rather than being released into the internal environment like hormones, molecules called **pheromones** are released to the external environment by many species. There, pheromones may play a role in communication between members of the same or different species. Pheromones may chemically resemble hormones. The active ingredient in catnip, for example, is similar to a sex hormone of cats. Truffles, mushrooms that are a highly prized luxury food, produce a chemical that resembles a sex hormone of pigs; in turn, pigs are used to sniff out the mushroom growing underground on the roots of trees.

The various bodily odors of humans probably originated as powerful pheromones that identified the sex of the individual and, perhaps, served to attract mates. Most perfumes, with which we try to conceal our own natural odors, are derived from the pungent pheromones contained in the anal glands of certain wild species of cats and civets; we thus substitute their pheromones for ours, but to achieve the same purposes. Another human physiological phenomenon is likely to be due to the regulatory effects of pheromones: Women living together exhibit *menstrual synchrony;* that is, their menstrual cycles begin at about the same time. Presumably, menstruating women produce odors (pheromones) that can induce menstruation in other women.

Pheromones are of particular importance in insect control programs. Sexually mature females of many insect species release pheromones, which attract males. Genetic engineering techniques now allow mass production of synthetic pheromones that can be used in several ways to limit insect reproduction: Traps can be baited with pheromones to attract males; or pheromones can simply be sprayed over wide areas, overwhelming and confusing the sensory systems of males who consequently cannot find mates (see chapter 10).

Methylation.   The attachment of methyl groups (CH_3) to nitrogenous bases in DNA, especially to cytosine, may be a key method of regulating gene activity. **Methylation** of genes may block their transcription, while removal of the methyl groups allows gene activity.

Methylation of DNA may in fact represent a type of inheritance that is superimposed upon Mendelian mechanisms. Present hypotheses suggest that patterns of methylation are transmitted via the gametes from parent to offspring, ensuring that expression of genes in

(a)

(b)

Figure 9.8

Development requires both paternal and maternal genomes. (a) The mouse embryo is the result of normal fertilization of an egg by a sperm while (b) developed in the same mother but is derived from an egg with two maternal genomes and no paternal genome.

J. L. Marx, "A Parent's Sex May Affect Gene Expression," photo Sheila Barton and M. Azim Surahi, *Science*, 239:352–353, 1988.

the offspring is similar to that in the parents. Methylation is thus a mechanism for marking, or **imprinting,** a particular gene and thus governing its expression in the next generation.

Amounts of methylation in the maternally-derived genome often exceed that of the paternally-derived genome. Methylation may thus explain some unusual patterns of gene expression that appear to depend upon the origin of a gene from the mother or father. As noted in chapter 6, for example, about 20% of males with the fragile X chromosome are carriers and do not express any symptoms. This might result from imprinting by methylation of the X chromosome in one parent. Later, the imprinting pattern is changed in daughters of carriers, whose male children then express the fragile X condition.

Another surprising finding is that mammalian embryos manipulated to have either two male or two female nuclear parents do not develop to full-term (figure 9.8). If expression of genes were independent of gametic origin, such an embryo might be expected to develop normally. In fact, in species such as frogs, the paternal genome appears to be entirely unnecessary; a needle prick can stimulate an unfertilized egg to begin normal embryonic development.

In mammals, however, the paternal genome appears to be necessary for placental development, while the maternal genome is required for development of the embryo itself. In humans, pregnancies occasionally result from eggs that incorporate two paternal genomes but lose the maternal genome. Known as *hydatidiform moles,* these embryos usually die early in gestation; the placenta is overdeveloped, while the embryo itself is extremely underdeveloped. The DNA content of eggs and sperm is similar except for the presence or absence of X and Y chromosomes, so the influence of parental genomes may result from differing patterns of paternal and maternal methylation of individual chromosomal genes.

Variations in Amounts of DNA

Gene function may be controlled by alterations in the amount of DNA in the genome. Unneeded genes may simply be eliminated. Genes whose product is needed in large quantities may be replicated and come to comprise a relatively large proportion of the genome.

Elimination of DNA. In a variety of taxa, gene regulation involves elimination of entire chromosomes from the genome. Recall that mitochondria and chloroplasts possess their own DNA, separate from nuclear DNA. In protozoa, organellar DNA inherited from one of the parents is selectively eliminated from the genome. When cells of different species are hybridized in the lab, the chromosomes of one or the other parent are slowly eliminated throughout the life of the cell.

In some marsupials, one X chromosome is eliminated rather than simply inactivated. However, the most extraordinary example of chromosome elimination occurs in some coccid insects (mealy bugs and their relatives) in which the entire paternal genome is destroyed.

The resulting individuals are haploids. This system also provides the sex-determining mechanism for these insects: the haploids are male, while diploids—in which no chromosome elimination occurred—are female.

The most interesting point regarding chromosome elimination is its selectivity, which implies that cellular mechanisms exist for precise recognition of specific chromosomes. Chromosome elimination is perhaps the ultimate mechanism of gene regulation. Genes that are destroyed or eliminated can never be expressed.

Increased amounts of DNA.   Relatively large amounts of particular RNAs may occasionally be needed during the life of a cell. One way of meeting such a need is to have many copies of the required gene. In the oocyte of many amphibian species, many copies of genes that produce rRNA are present, and the genes are said to be **amplified.** The increased synthesis of rRNA during oogenesis provides for post-fertilization protein synthesis of the zygote.

In fact, the number of copies of particular base sequences varies immensely within the eukaryotic genome. This variation provides clues to the relative organization of the genome into structural and regulatory portions. *Non-repetitive* or *unique sequences* occur only once in each haploid complement. Structural genes encoding proteins probably occur primarily as unique sequences. *Middle repetitive sequences* have ten to one hundred thousand copies of a particular sequence. Middle repetitive sequences encode rRNA, tRNA, and histones, and may also have regulatory functions. The amplified rRNA genes described in amphibian oocytes are an example of middle repetitive sequences. *Highly repetitive sequences* have more than one hundred thousand copies of a particular sequence. Highly repetitive sequences are often associated with heterochromatic regions such as centromeres. They also occur in satellite DNA, where they may play a role in recognition of homologues during meiosis.

The proportion of unique, middle, and highly repetitive DNA in the genome shows considerable variation among species, but in general repetitive DNAs comprise more than half of the entire genome. Unique sequences (i.e., structural genes) occupy 10% or less of the nuclear DNA. This suggests that repetitive sequences play a major role in gene regulation.

Interactions of Alleles and Loci

Thus far, we have examined gene regulation as a molecular and biochemical phenomenon. The most obvious expressions of gene regulation, however, are those affecting gross phenotypic traits, particularly those of medical importance.

In this section, we examine gene regulation at the level of the Mendelian gene and its associated phenotypic traits. Phenotypic variation reflects not only the expression of differing alleles of a gene, but also the way that variation in one gene controls the expression of another. Most major phenotypic traits are the result of interactions of many Mendelian genes. We use an example of a medical condition—PKU—to relate variation in a single Mendelian gene to its many phenotypic expressions. This in turn allows understanding of the key Mendelian concepts, dominance and recessiveness.

Dominance and Recessiveness

Dominance and **recessiveness** have thus far been treated primarily as qualitative characteristics of alleles. In fact, the extent to which a particular gene is expressed to the exclusion of another gene often shows quantitative variability. The phenotype of heterozygotes may occur at any location along the continuum of phenotypic variation. If the heterozygote exhibits a phenotypic value identical to that of one of the homozygotes, the controlling allele shows **complete dominance.** If the phenotypic value is intermediate to the two homozygous types, the allele shows **partial** or **incomplete dominance.** Note that *intermediate* can mean any value between the two homozygous values, not just half way between them.

Many, perhaps most, genes exhibit *pleiotropic* or multiple effects (figure 4.7). One cause of pleiotropy is that each allele produces one type of polypeptide, which may subsequently participate in many biochemical pathways, affecting many different traits. One gene affecting different phenotypic traits may show different patterns of dominance relationships with respect to each of the different traits.

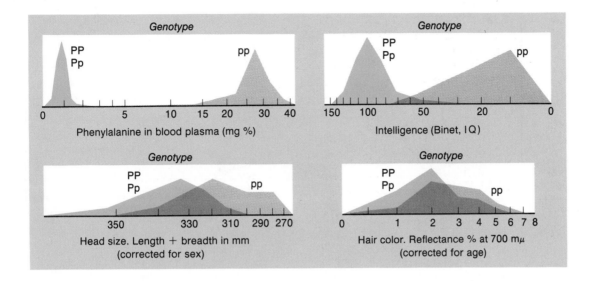

The amount of polypeptide produced by the alleles at a locus also explains some aspects of dominance relationships. At the level of protein production, some alleles may produce an inactive or defective form of the protein. Heterozygotes thus may have half as much of a functional protein as do normal homozygotes; that is, they exhibit partial dominance for production of that protein. In some cases, aspects of the phenotype dependent upon those proteins will not be significantly affected by reduced amounts of the protein, as long as *some* protein is present. In such cases, heterozygotes exhibit complete dominance for the traits produced by the protein. In other cases, reduced levels of certain proteins quantitatively alter phenotypic traits; in other words, partial biochemical dominance causes partial dominance for associated traits. Phenylketonuria illustrates both possibilities.

An Example: PKU

Phenylketonuria (PKU) is caused by an allele that, when homozygous, results in the inability to produce the enzyme phenylalanine hydroxylase, needed to metabolize the amino acid phenylalanine. Heterozygotes produce half as much enzyme as normal homozygotes, so when considering enzyme production, the normal allele shows partial dominance to the PKU allele.

Only small amounts of enzymes, however, are required for normal phenylalanine metabolism. Remember that enzymes are *not* used up—they are used again and again to catalyze reactions. Even half the normal quantity of phenylalanine hydroxylase is usually sufficient to maintain normal phenylalanine metabolism, so despite reduced levels of phenylalanine hydroxylase, a heterozygote usually has the same level of plasma phenylalanine as a homozygote. The normal allele is dominant to the PKU allele in terms of plasma phenylalanine levels (figure 9.9).

Phenylalanine levels, in turn, affect many other traits, including intelligence, head size, and hair color. Because heterozygotes have plasma phenylalanine levels similar to those of normal homozygotes, the phenotype for the associated traits usually shows the same range for normal and heterozygous individuals (figure 9.9). On the other hand, heterozygotes differ from normal homozygotes in at least one respect (figure 9.10). Phenylalanine tolerance tests show that heterozygotes metabolize phenylalanine more slowly than homozygotes. Heterozygotes can usually be identified on the basis of elevated blood levels of phenylalanine immediately after *loading* (ingesting high levels of phenylalanine). The high levels are only transient, but they permit clinical detection of carriers.

Norm of Reaction

The data for PKU illustrate an important principle of phenotypic variation. Note that the genetic types that may occur in a population are discrete; that is, an individual may possess zero, one, or two PKU alleles. Phenotypic values, however, may be more or less continuous with a relatively broad range. The range of phenotypic responses for a particular gene over

Figure 9.9

Dominance for some of the phenotypes associated with variation at the locus controlling phenylketonuria. In these cases, the normal homozygous genotype (PP) and the heterozygous genotype (Pp) show the same range of variation; thus for these traits, the P allele is dominant to p. The results for individuals homozygous pp are for phenylketonurics who were not placed on dietary therapy.

Figure 9.10

Response curves to phenylalanine loading for two genotypes. At time zero, individuals ingest large quantities (0.1 g/kg body weight) of phenylalanine. The subsequent change in blood levels of phenylalanine is measured hourly. Heterozygotes for the PKU allele take significantly longer on average to clear high levels of phenylalanine from the blood than do normal individuals.

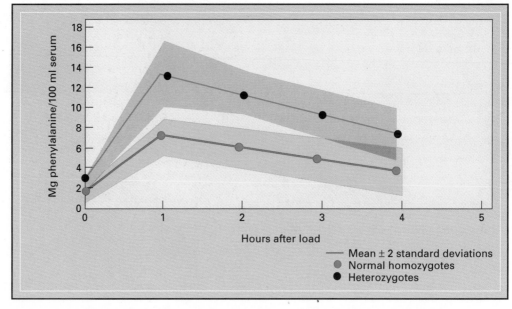

all environments is called the **norm of reaction.** The norm of reaction comprises all possible phenotypic values that may be assumed by a particular genotype.

The range of variation expressed by a genotype depends upon both the external and the internal, or genetic, environment within which the genotype functions. Even assuming an identical external environment (and two environments almost always differ in some respect), if two individuals have identical alleles at one genetic locus, the modifying influence of other genes will produce variable phenotypic responses. Presumably, the interaction of alleles at a locus with both external and internal environmental factors transforms qualitative genetic variation into quantitative phenotypic variation.

Figure 9.9 illustrates the broad range of phenotypes realized by individuals identical at one genetic locus, but presumably differing for many *background* loci. Untreated phenylketonurics have IQs in the range of zero to seventy-five, while their hair color spans the spectrum of reflectance from 0%-70%. In the first case, IQ distinguishes phenylketonurics from the normal population with considerable accuracy; in the latter case, hair color separates them relatively poorly. In both cases, some overlap in phenotype occurs between normal and affected individuals, even for a relatively serious condition.

The phenotype may be altered dramatically by the external environment, as the norm of reaction of the IQ of phenylketonurics illustrates. The data in figure 9.4 refer only to untreated individuals. Dietary therapy of phenylketonurics results in a range of IQs similar to that for unaffected individuals. Thus, the norm of reaction for IQ for PKU individuals encompasses essentially the same range of IQ values shown by non-PKU individuals.

Until dietary therapy was recognized as an effective preventive measure for mental retardation, the norm of reaction of IQ for phenylketonurics would have been considered to be only seventy-five and below. However, dietary therapy provides the appropriate environment in which the norm of reaction can be greatly extended to include the range of normal IQs. This exemplifies one result of contemporary medicine: Understanding the genetic and biochemical bases of diseases often permits affected individuals to be treated (provided a suitable environment) so that they can lead more or less normal lives.

However, because an individual's genes cannot yet be changed, environmental modification has some limitations. For example, women who are homozygous for the PKU allele, but of normal intelligence because of dietary therapy, can marry and reproduce, but they still are unable to provide a normal uterine environment for their children. As described in chapter 4, offspring of such women often are retarded, even though they may only be heterozygous for the PKU allele. The ability to change one phenotypic trait such as intelligence does not necessarily mean a change in all traits affected by one gene.

Expressivity

Yet another term associated with phenotypic variability is **expressivity.** Expressivity measures the extent to which an individual demonstrates various phenotypic effects of a gene.

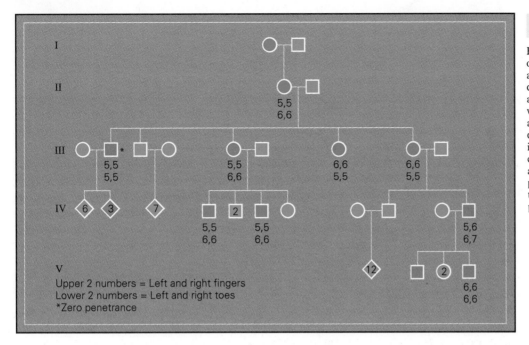

Figure 9.11

Pedigree of five generations of a family with polydactyly, a trait controlled by a dominant allele that is not always penetrant and that has variable expressivity. Where available, the numbers of digits are shown for affected individuals (magenta boxes or circles). Individual III-2 almost certainly exhibits no penetrance, as his mother and three of his children show polydactyly.

It may refer to one or several characters. For example, expressivity may be high for a phenylketonuric who is severely retarded and very blond, but low for one whose diet is low in phenylalanine. Conversely, most heterozygotes would have low expressivity of the deleterious allele, except where they are subjected to phenylalanine loading.

For many genetically based diseases, the degree of expressivity is extremely variable. Diabetes, for example, may require minimal or major clinical intervention, depending upon the extent of phenotypic manifestations (the symptoms of the disease).

Penetrance

Some alleles have such extreme variation in expression that they may not manifest their presence at all, in which case they are not **penetrant.** With respect to a single individual, a gene is penetrant if it has any detectable phenotypic expression. For an individual, a gene is either penetrant or not. Quantitative variations in the extent of phenotypic effect are discussed in terms of expressivity.

A gene may be common in a family, for example, but not be penetrant in all who have it. Lack of penetrance is often observed when a trait skips generations. Figure 9.11 shows the pedigree of a family with the allele for polydactyly over five generations. Polydactyly (meaning many digits) occurs when an individual has more than five digits on a hand or foot and results from a single dominant gene. If the gene showed complete penetrance, all persons having that gene would be expected to exhibit some degree of polydactyly.

The gene for polydactyly controls the number of bony rays formed in the hand and foot buds of the developing embryo. If the gene is penetrant, it still may exhibit variable expressivity. Individuals III-8 and IV-25 show how variable the expression of this gene can be. III-8 had six fingers on each hand and a normal number of toes. Her son, IV-25, had five fingers on the left hand, six on the right, six toes on the left foot, and seven on the right.

Note individual III-2. Because his mother and three of his children had the gene for polydactyly, it is reasonable to assume that he also had it, but it was not penetrant. Penetrance may depend upon the environmental circumstances at the time the gene is normally expressed.

Delayed Onset

Many developmental traits and some genetic diseases are not expressed until relatively late in life. Perhaps the best-known disease with **delayed onset** is **Huntington's disease.** The disease is controlled by a single dominant gene, but onset occurs most commonly between the ages of twenty-five and forty-five (figure 9.12). Huntington's disease affects the central nervous system, and the symptoms include loss of motor control and mental deterioration.

Figure 9.12

Age-specific expression of the dominant gene for Huntington's disease. Of 204 affected individuals, about 90% expressed the disease by age fifty; only two were first diagnosed over the age of sixty-five.

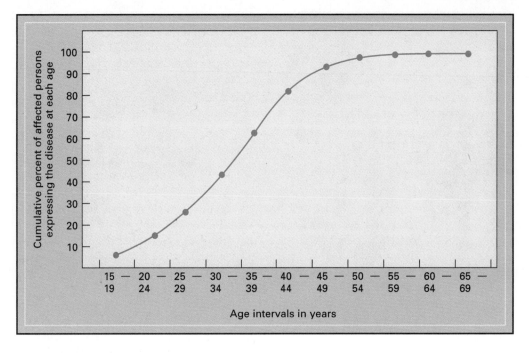

Serious diseases with delayed onset are particularly insidious. The disease may not develop until after normal ages of reproduction, so potential carriers must face the choice of having children without knowing if they may transmit the gene for the disease or even if they will live long enough to raise their children to adulthood.

Direct DNA tests for the Huntington's gene have been available since 1986. This has many benefits, allowing persons at risk (i.e., with a parent who has the disease) to be tested to see if they carry the gene and are thus likely to develop the disease later in life. Carriers of the gene can use prenatal diagnosis following amniocentesis or chorion villus biopsy to ascertain if a fetus also carries the gene.

However, DNA testing is not a cure; it simply identifies the presence or absence of the gene and still not with 100% certainty as there are subtle differences among several alleles that cause Huntington's disease. Even then, onset of the disease may occur late enough that the carrier can live a relatively normal life, or that the person dies from other causes. At best, results of DNA testing can allow informed choices to be made, but the choices differ among individuals. Most persons with a family history of Huntington's would probably choose to abort an affected fetus; after watching a parent or grandparent slowly and horribly decline into dementia and inevitable death, the decision to abort is generally straightforward. On the other hand, a positive test for their fetus would carry the bad news that the parent is affected also, a fact they may not wish to confront.

Confirming the presence of the Huntington's gene prior to onset can allow some people to organize their lives in view of a high probability of an early incapacity and death. But those responses can be very different. Some people might wish to travel or lead a life of adventure, knowing they had only limited time; they might also decide not to marry and risk transmitting the gene or leaving a bereaved spouse. Others might decide exactly the opposite: to have a family as soon as possible, knowing that prenatal diagnosis can now ensure that the gene is not transmitted.

Finally, some people prefer not to have the carrier test because they could not bear the consequences of a positive result, even though they might live a long life prior to onset. People in the early stages of Huntington's disease have a relatively high rate of suicide, presumably reflecting their despair and unwillingness to endure the consequences of the disease. A person with the gene might be prone to interpreting normal illnesses as the first sign of onset. For some people, a negative test may have unhappy consequences, inducing a burden of guilt for being free of a condition that a parent or sibling has suffered.

Codominance

When the alleles in a heterozygote produce different gene products, as opposed to a reduced amount of the normal product, the pattern of expression is **codominant.** The phenotypic results may be normal or pathological. The following examples illustrate both possibilities.

Blood groups. High levels of variation exist for the major blood groups, and in many cases the mode of expression of the gene product is codominance. Genes for blood groups encode complex molecules on the surface of red blood cells, and the presence or absence of particular antigens is not a pathological condition. Individuals who are type AB or type MN express both alleles at the controlling loci. The heterozygote also has a unique phenotype (positive agglutination reactions to each of two types of antibodies) so that both the biochemical and phenotypic expressions are codominant.

At the ABO locus, dominant as well as codominant expression occurs. The O allele encodes no antigenically active blood group molecule. Individuals who are genotype AO are phenotype or blood group A, since they produce only the A antigen, the same as homozygous AA individuals. The cell surface of AO individuals has sufficient type A antigen to produce agglutination with anti-A antibodies. Both genotypes AA and AO show indistinguishable agglutination reactions, so they exhibit phenotypic dominance of A over O. In a similar fashion BB and BO genotypes are indistinguishable phenotypically.

Sickle-cell anemia. The allele for sickle-cell anemia causes a change in one amino acid of the β chain. Normal individuals produce only normal hemoglobin (Hb A) while victims of SCA produce only Hb S, with two abnormal β chains. Heterozygotes (persons with the sickle-cell trait) produce both Hb A and Hb S. Production of the two types, however, is not equal. About 25%-40% are Hb S, the remainder are Hb A.

Because a heterozygote exhibits both Hb S and Hb A, expression at the biochemical level may be considered codominant. However, in terms of the disease, the sickling allele is recessive. Only homozygotes have sickle-cell anemia, and under normal circumstances heterozygotes seldom express any symptoms.

Epistasis

When several loci affect the same trait, alleles at one locus may mask the effect of alleles at other loci. Such interactions are termed **epistatic.**

Recall from chapter 3 that a monohybrid cross of heterozygotes, where one allele is dominant to the other, yields a 3:1 phenotypic ratio. When more than one locus affects the same trait, many different ratios and more than two phenotypes may occur, depending upon the interactions of the loci.

As an example, consider congenital deafness in humans. Evidence suggests that alleles at one locus affect the development of the auditory nerve, and alleles at a second locus affect the normal development of part of the inner ear, the cochlea. Homozygosity of a recessive allele at either of these two loci causes deafness. If we represent the normal allele at these loci by D and F, then any of the following genotypes produces deafness: DDff, Ddff, ddFF, ddFf, or ddff. Thus, the presence of the recessive homozygous type at either locus is sufficient to override the effect of one or two dominant alleles at the other locus. In a dihybrid cross of double heterozygotes, $\frac{9}{16}$ of the offspring will be expected to be normal, $\frac{7}{16}$ deaf (figure 9.13a)—not the 3:1 expected with only one locus or the 9:3:3:1 expected with two non-epistatic loci. Note that a 9:7 ratio is simply a modification of a 9:3:3:1 ratio, in which the epistatic interaction combines the three least-frequent phenotypic categories into only one (3+3+1=7).

Note also that control of a character by more than one locus provides a mechanism by which phenotypic traits do not *breed true.* Mating of two deaf persons whose deafness is a function of different loci produces heterozygous offspring of normal phenotype (figure 9.13b).

Another example illustrates how loci can interact to produce a variety of phenotypes. Coat color in rabbits, as in many other species, is controlled by several loci. Consider the following interaction of two loci, each with two alleles. Allele C at one locus is dominant to

Figure 9.13

Two epistatic loci control deafness in humans: (a) mating of two doubly heterozygous individuals produces children in the ratio 9 normal: 7 deaf; (b) two deaf parents can produce all normal offspring.

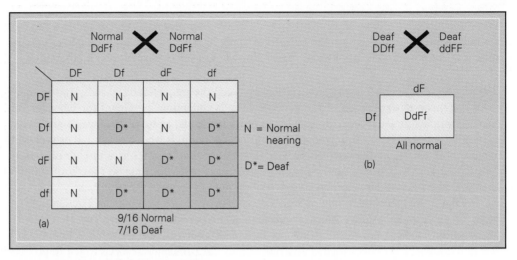

Figure 9.14

Epistatic interactions in the inheritance of coat color in rabbits, assuming only two genetic loci: CC or Cc = color; cc = albino; **BB** or **Bb** = black; bb = brown.

c. C __ individuals produce pigment, while cc individuals produce no pigment at all and are therefore albino. At the second locus, allele B for black is dominant to b for brown. However, neither B nor b can be expressed if the individual is genotype cc. Thus, the expression of any color is dependent first upon the possession of pigment. Otherwise, the individual is albino. In this instance, a dihybrid cross of heterozygotes yields the ratio of 9 black:3 brown:4 albino (figure 9.14). Imagine the ratios obtained from epistatic interactions of three, four, or more loci.

The types of interactions that may occur, even considering only two loci, are complex indeed. Only carefully planned breeding tests can sort out the precise patterns of inheritance that involve more than one locus. As the number of loci increase, the complexity also increases. Albinism in humans, for example, can be caused by any one of at least three loci. The allele for albinism is recessive at each locus, and the recessive homozygous condition at any one of the loci interrupts the synthesis of melanin. One of the three loci is X linked, however, so different ratios are obtained for each sex.

Quantitative Inheritance

The effect of many genes contributing to a single trait is often additive, analogous to partial dominance when considering the interactions of two alleles at one locus. The phenotypic result, **quantitative inheritance** (introduced in chapter 3), is distinguished from epistasis by

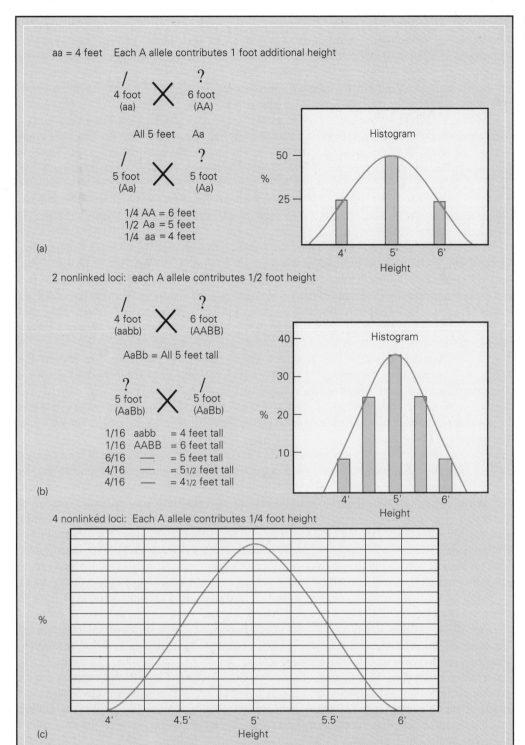

Figure 9.15

Examples of how polygenic inheritance controls quantitative traits: (a) variation due to a single locus with two alleles; (b) the results of two nonlinked loci, each with two alleles; (c) four nonlinked loci, again with two alleles each.

the mode of interactions of alleles. Epistasis is characterized by discontinuous patterns of variation due to the dominance of one locus to others. Quantitative inheritance involves the interaction of two or more loci that, individually, exhibit discrete Mendelian patterns of inheritance. However, the pattern of variation shown by these loci is quantitative rather than qualitative. Figure 9.15 illustrates how this might happen.

Consider that in a study population height is controlled, first, by one locus with two alleles (A and a). Individuals of genotype aa are four-feet tall. Each A allele in the genotype contributes one extra foot in height. Thus, Aa individuals are five-feet tall; AA individuals are six-feet tall. Allele A shows partial dominance to a. Figure 9.15a illustrates the frequency distribution of offspring of a mating of Aa × Aa. The frequencies of each genotype and phenotype are derived from a Punnett-square analysis, but displayed in a bar graph to illustrate the relative abundance of each phenotypic class. Note that only three phenotypic categories are observed, and that the intermediate height (five feet) is the most frequent. The mean height of the individuals in this population is also five feet.

Figure 9.15b shows the pattern of variation obtained if height were controlled by two loci rather than one. Again, the height of a completely homozygous recessive individual (aabb) is four feet. Each dominant allele contributes an additional one-half foot in height. The pattern is similar to that in figure 9.15a. The range of heights is still four to six feet, the mean height is five feet, and extreme heights are comparatively uncommon. The most obvious effect of adding one locus is that two additional phenotypic categories, $4\frac{1}{2}$ feet tall and $5\frac{1}{2}$ feet tall, now occur. These two phenotypes partially fill in the gaps that separated the clearly distinct phenotypes in figure 9.15a.

The addition of more loci continues the trend of changes shown between the first two figures. While the range and mean heights remain the same, the number of possible phenotypes steadily increases (figure 9.15c). The result is that clearly discrete or qualitative differences in height become quantitative; one height grades imperceptibly into the next. The dotted line in the three figures connects the peaks of the various frequency classes. Note that as the number of loci increases, the shape of the dotted curve comes to approximate a bell-shaped curve.

A bell-shaped frequency distribution is well known to mathematicians as a **normal distribution.** The frequency distribution of many traits characterizing natural populations shows an approximately normal distribution. When a trait such as height is normally distributed, most members of the population are close to some central value (the *mean* height, five feet in this case). Relatively few individuals show extreme deviation from the mean value—hence, the bell shape of the curve.

Most complex traits of living organisms are governed by the interactions of many genes. Length, weight, color, metabolic rate, and many behaviors, for example, all tend to show a normal distribution of variation.

While such examples illustrate a basic pattern of quantitative inheritance, remember that nature is seldom as simple as the models used by scientists to describe nature. Certainly, almost all continuously varying characters are affected by factors other than those in our simple model. For example, some loci would have more than two alleles, perhaps many more, with each allele potentially making very different contributions. Some loci would likely show epistatic interactions, some of which would produce novel phenotypes beyond the range of normal quantitative variation, such as albinism or dwarfism. The contributions of each of the interacting loci are unlikely to be equal. **Modifying loci,** for example, enhance or suppress the expression of other loci. Thus, a trait may be controlled by a series of primary loci interacting quantitatively and epistatically, plus a number of modifying loci that selectively alter further the expression of the primary loci.

Finally, remember the influence of environmental factors. The norm of reaction of any genetic combination is defined by the environments to which it is exposed. Given the multitude of possible interactions of many environmental factors, many loci, and many alleles, it is hardly surprising that the most obvious variations among natural populations are quantitative.

The Phenotype: Consequence of a Genotype/ Environment Interaction

As we have seen throughout this chapter, the phenotype is the final result of a complex process of genetic regulation that begins at fertilization and ends with death. In other words, the phenotype of an individual is always changing. Genes are continually being turned on and off; alleles at a locus are interacting; alleles at different loci are interacting. In countless ways, alleles and loci are influencing and modifying the expression of other alleles and loci, and in turn being influenced and controlled themselves.

Remember, too, that some loci are never expressed. None of us exhibits the full range of phenotypes of which we are capable. Instead, the genome functions under the direct influence of the external environment. Genes respond to both internal and external stimuli. The environment provides the context within which genes act, and full understanding of the development of phenotype requires knowledge of the influence of environment on specific gene function.

Unfortunately, the complexity of the interactions of environment and genome renders separation of their independent effects virtually impossible. Attempts to extricate their influences have resulted in a longstanding conflict within genetics, psychology, and many of the social sciences called the "**nature-nurture**" debate. Virtually none of the combatants in this debate openly claim that only genes (often called "nature" or "biology") or environment ("nurture" or "culture") entirely control phenotypic manifestations, but individuals differ greatly in the weight they place on the two types of factors in governing phenotype.

The biologist Stephen J. Gould has written humanely and perceptively on "the stupefying sterility of nature-nurture arguments." Gould observes that knowledge will advance only when we recognize the "inextricable interpenetration of genes and the environment" instead of trying to quantify precisely how much nature or nurture independently contributes to a phenotypic trait. Both are required to produce any phenotype, and any attempts to place priority on one alone are inadequate.

Sadly, the nature-nurture debate is not purely academic. In this century, the atrocities of Nazi Germany arose from a mistaken belief that genes of "inferior peoples" threatened those of a ruling group. Other western countries imposed quotas limiting migration of these same "inferior peoples" trying to escape Germany. For much of this century, Britain limited access of large segments of its population to higher education because of fraudulent data claiming that intelligence is largely hereditary. In Singapore in the 1980s, the prime minister argued for incentives for well-educated women to increase their reproductive rate; this, he erroneously argued, would counter declining genetic quality of his population because uneducated women tended to have more children. In the United States, public policy on numerous social, educational, and medical issues has been decided on the basis of public perceptions of the relative roles of genes and the environment.

No single controversy better illustrates the complexity of phenotypic control and its social importance than that surrounding IQ. We conclude this chapter with an extended discussion of the IQ controversy to illustrate the inextricability of genetic and environmental factors in phenotypic development and some of the related social controversies.

The IQ Controversy

During the past century, assessing intelligence and attempting to understand the causes of differences between individuals or various groups have developed into an important field of study. Psychologists, statisticians, and geneticists have all made important contributions to this knowledge. Millions of dollars have been spent to develop suitable intelligence tests and implement them on a wide scale. While intelligence testing has become highly refined, the debate over the causes of intelligence remains unresolved.

Several factors make this problem especially vexing. First, mental capacities clearly differ between persons, but defining, much less measuring, meaningful and objective measures of mental capacity has proven extremely difficult. In the last century, brain size (specifically, volume of the brain case) was used as an index of intelligence. In this century, the most common measure of intelligence has been the Stanford-Binet Intelligence Scale, a series of standardized questions that most of us have taken to determine **intelligence quotient (IQ).** Many tests other than the Stanford-Binet have been developed to assess intelligence, but all suffer from a set of common problems. In discussing intelligence and mental capacity, we define IQ strictly as the score a person obtains on a standardized test such as the Stanford-Binet. As such, IQ is assumed not to be equivalent to intelligence, although the two may be related.

Perhaps the greatest limitation of all intelligence tests lies in defining intelligence. Most scientists agree that intelligence has something to do with the ability to understand abstract concepts, to solve problems, to express one's self verbally and quantitatively. But those capacities are themselves imprecise and almost certainly embody a variety of mental qualities. A person gifted in mathematics may have limited capacities with languages and vice versa. Intelligence is a multi-faceted phenotypic trait that is poorly represented by a single numerical scale.

Another difficulty in measuring intelligence is that IQ, unlike eye color or blood type, has a wide norm of reaction. Extreme variations in IQ have been shown to occur as a result of environmental factors, particularly during the first decade of life. Just as a person's size as an adult depends upon physical activity and diet during the growth years, so IQ is dependent upon environmental conditions during the psychologically formative years: Cultural opportunities, education, diet, and even birth order and number of siblings affect IQ as an adult.

If IQ tests have such severe limitations, why does their use continue? Because, among other reasons, they seem to provide a relatively objective measure of *something*. IQ as defined by the Stanford-Binet and other tests may provide little information about intelligence, but IQ is a good predictor of professional success in our society. The higher one's score on an IQ test, the more likely that person is to succeed in school, and the more likely that person will obtain a professional, well-paid job. Successes in school and on the job, of course, clearly depend upon many factors other than intelligence—creativity, social finesse, and persistence, for example. Nonetheless, IQ as measured on a test is a useful index to the probability of success in particular endeavors, despite the fact that IQ attempts to measure an intrinsically unmeasurable and even unidentifiable quantity.

An important factor confounding the study of intelligence is the persistence of an undercurrent of racism. The goal of some persons involved in the study of intelligence has been to demonstrate the superiority of some races, nationalities, or sexes. No scientist doubts that *differences* exist between racial groups. Various races are characterized by differences in skin color, hair color, eye color, size, proportion, blood group, and countless other phenotypic traits (figure 9.16), so it is reasonable to assume that differences in psychological characteristics may also exist. Such differences reflect only the extraordinary variety found throughout nature, the result of evolutionary processes affecting all living organisms.

The recognition of variety in nature is dangerous when such differences are assigned qualitative values and one trait is claimed to be of higher worth than another. Philosophers call such bad logic the *"naturalistic fallacy"* (the "is-ought problem"), meaning that a fact of nature is used wrongly to establish moral principle. Quite often, an "is-ought" argument is used for racist or sexist purposes. For example, because a woman has a uterus and ovaries (biological fact), some people claim that her destined and appropriate role is to stay home and have babies (moral judgment). Or, because men are on average larger than women (fact), they ought to have the dominant role in society (judgment). Neither statement is logically true, but reflects only the prejudice of those who believe it. Thus, despite the unequivocal fact that some people are more intelligent than others, no relative individual worth can be assigned on that basis.

Some racist applications of presumed intelligence assessments now seem ludicrous. Following Darwin's statement of the theory of natural selection in the late nineteenth century, for example, the public's imagination was captured by the search for the *missing link* between humans and their primate ancestors. For some years, no such link was found, which led some scientists to search in unusual and self-serving directions. The most famous scientists of the time were white and male, just as Western civilization was at the time dominated by white males. Some scientists claimed that the search for the missing link had overlooked the obvious: they claimed that missing links were to be found among the various races of humans still living.

As proof of this notion, evidence on brain size was offered. Ranked on a scale of relative brain size, white males were claimed to have the largest brains. White females and blacks of both sexes exemplified intermediate sizes; chimpanzees, were used as the small end of the spectrum. Thus, missing links were found by searching groups of society other than white males. The comparisons were designed, consciously or otherwise, to indicate a regular progression from chimpanzees through blacks and women to reach a peak—clear superiority—in white males!

Figure 9.16

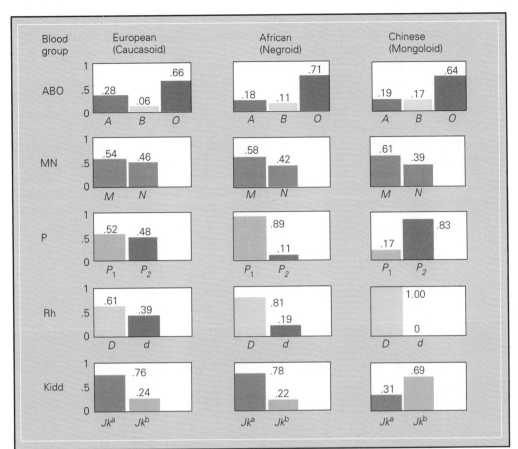

How should these data be interpreted? First, you should be aware that data collection was highly biased. Brain volumes for white males were taken from soldiers (young men in their twenties) killed in battle. Brain volumes for blacks were taken mostly from men and women dying at old ages in public hospitals. Brain volume for all persons decreases throughout life, as calcium is deposited within the brain case. Many degenerative diseases also cause a decrease in volume of the brain case. A comparison of young and healthy whites with old and sick blacks is obviously biased, whether measuring speed in the hundred meter dash or brain size.

A second point concerning these data is that they assume that brain size in fact is a measure of intelligence. That clearly is not the case. Within the range of normal human brain sizes, there is no correlation of intelligence or IQ and brain size. In fact, the greatest determinant of brain size is overall body size. Like arm length and waist size, brain volume in humans is larger in large persons than in small persons. Thus, pygmies have smaller brain volume, but not lower intelligence, than most other humans, and women on average have lower brain volume than males.

The assessment of relative brain sizes states more about the culture from which it originated than about any meaningful differences in intelligence between various human groups. Such an example seems both silly and pernicious from our viewpoint at the end of the twentieth century, but it clearly illustrates the energy that has been expended attempting to demonstrate intergroup differences in intelligence. It also demonstrates the mystique that humans impart to intelligence. Perhaps because high intelligence seems to distinguish humans from all other animals, intelligence is one trait persistently used to try to demonstrate purported racial superiority.

The absurd search for the missing link among groups low in the social hierarchy is one of the least directly harmful aspects of the IQ controversy. In the past century, intelligence assessments have been used for purposes ranging from pernicious to vicious. Presumed low intelligence was a major criterion for excluding immigrants to the United States in the early twentieth century. Often, the immigrants could not speak English; naturally their answers

Figure 9.17

IQ frequency distributions. (Top) Theoretical normal curve, with mean of one hundred; because of rounding errors, total is slightly more than 100%. (Middle) Actual distribution of 2,904 native-born white children (from California, Nevada, Colorado, Texas, Kansas, Minnesota, Indiana, Kentucky, Virginia, New York, and Vermont) taking the 1937 Stanford–Binet. (Bottom) Actual distribution of 1,800 black schoolchildren (from Tennessee, Alabama, Georgia, South Carolina, and Florida) taking the 1960 Stanford–Binet.

Source: Terman and Morrow, *Measuring Intelligence.* Copyright © 1937 Houghton Mifflin, Boston, MA.

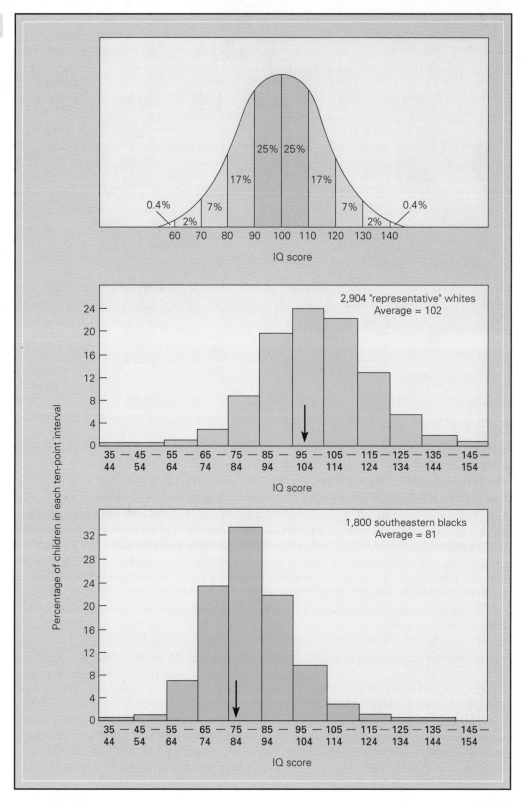

to questions asked in English were often incorrect. Decades later, Hitler and the national socialists (Nazis) claimed that mental superiority was a trait that distinguished Aryans from other races. In fact, Aryan is not even a racial group, but a prehistoric language spoken by peoples of several ethnic groups.

In the United States, differences in intelligence and the causes of those differences are common—although perhaps implicit—areas of contention in determination of amounts of public funding directed at particular disadvantaged groups. Those who believe that differences in IQ result primarily from differences in environment generally favor the expenditure of public aid for programs that increase the opportunities available to the disadvantaged. Project Head Start originated in the 1960s as such an effort.

Another school of thought claims that IQ differences have relatively little to do with environmental differences, but instead reflect genetic differences. If true, then improvements in the environment, they argue, can do little good in increasing IQs and, therefore, public money should not be wasted in the attempt.

Differences in IQ among Racial Groups in the United States

The theoretical distribution of scores on the Stanford-Binet is shown in figure 9.17. Like many quantitative traits, the distribution is normal. The mean value (the expected average intelligence) is 100, but variation is substantial. About 50% of the population scores between 90 and 110 and about 95% scores between 70 and 130.

The theoretical distribution of IQs is based on the white population of the United States. However, notable if slight deviations from the expected pattern occur. More persons seem to score below 70 or above 130 than expected, and various subdivisions of the white population yield significant differences in IQ scores. Differences have been found between nationalities, ethnic groups, religions, regions of the country, professions, and many other categories.

The most controversial differences are those that have been measured between racial groups. The mean score for whites on the Stanford-Binet is 100; for blacks, about 85; for orientals, about 107. The controversy does not surround the quality of the data. The differences between races are repeatable when different tests are given. Furthermore, the IQ differences are clearly correlated with various measures of achievement. Average family income, for example, is highest among orientals in the United States, lowest among blacks, and intermediate among whites. The controversy, rather, surrounds the *causes* of the differences detected. Do differences in IQ scores reflect genetic and therefore potentially unchangeable differences? Or do they reflect serious environmental and therefore changeable differences to which the races have been exposed? No simple answer exists to such questions, but it is worthwhile to review a small sample of the vast but not yet definitive data collected on this subject.

Evidence for the Role of Genes in Intelligence

Specific genes that contribute to normal levels of intelligence have not been identified. Therefore, data supporting the role of genes in IQ have been derived from indirect, often mathematical, analyses of genetic variation among different groups.

Among the most compelling data is the increasing similarity in IQ of persons of increasingly high genetic relatedness (figure 9.18). The average difference in IQ between genetically unrelated persons chosen at random is about 18 points. Therefore, knowledge of

Genetic and nongenetic relationships studied		Genetic correlation	Range of correlations	Studies included
Unrelated persons	Reared apart	0.00		4
	Reared together	0.00		5
Fosterparent—child		0.00		3
Parent—child		0.50		12
Siblings	Reared apart	0.50		2
	Reared together	0.50		35
Twins — Two-egg	Opposite sex	0.50		9
	Like sex	0.50		11
Twins — One-egg	Reared apart	1.00		4
	Reared together	1.00		14

A comparison of twin pairs, or sib pairs, for three quantitative traits*

	One-egg Twins*		Two-egg Twins*	Sibs*
	Raised Together *50 pairs*	Raised Apart *19 pairs*	*50 pairs*	*50 pairs*
Height	1.7 cm	1.8 cm	4.4 cm	4.5 cm
Weight	4.1 lb	9.9 lb	10.0 lb	10.4 lb
Stanford-Binet IQ	5.9	8.2	9.9	9.8

*The figures in the four columns are simply the average differences within the pairs (rather than the statistical variances)

Figure 9.18

Evidence for a genetic component to IQ from analyses of IQ test results in persons of varying genetic relatedness: (a) data are summarized from fifty-two independent studies, and mean correlation coefficients are indicated by vertical lines; (b) comparisons of height and weight in contrast to IQ from twin studies.

(a) From L. Erlenmeyer-Kimling, "Genetics and Intelligence: A Review," *Science,* Vol. 142:1477–1479, 13 December 1962. Copyright © 1961 by the American Association for the Advancement of Science. Reprinted by permission. (b) Source: Data from Newman, et al., *Twin: A Study of Heredity and Environment.* Copyright © 1937 University of Chicago Press, Chicago, IL.

the IQ of one person provides little basis for predicting the IQ of any other person chosen at random—a correlation of 0. However, as the genetic relatedness of pairs of individuals increases, the average IQ difference decreases and the correlation increases. Those persons most closely related are monozygotic twins, who are genetically identical. The correlation of IQ between monozygotic twins is almost .90; average IQs of monozygotic twins differ by an average of only about 3–4 points. (Experimental error is 2 points; maximum possible correlation is 1.0.)

These data strongly indicate a genetic influence on intelligence, but influence of the environment is also clear. The correlation of IQ values among persons reared together for a particular degree of genetic relatedness is always higher than for persons reared apart, even for monozygotic twins. Increasing the environmental similarity always increases the similarity in IQ, just as does increasing genetic relatedness.

A second line of support comes from studies of **heritability.** Heritability (h^2) provides a measure of the proportion of variation observed in a population attributable to genetic rather than environmental variation. Heritability is an easily misunderstood concept. It refers to the contribution of genes to *differences* between individuals, rather than to the actual value of the trait itself in one individual. Heritability is a value describing one aspect of the variation within one population; it is *not* the proportion of a particular IQ ascribable to one causative factor or another.

Recent analyses indicate that heritability of IQ lies between about 30% and 70%; 50% is the present best estimate. This means that about 50% (or 9 points) of the average 18 point difference between unrelated individuals is explained by genetic factors, and the other 9 points are explained by environmental factors. Heritability does *not* refer to total IQ; $h^2 = 0.50$ does *not* mean that 50 of the normal 100 points are determined by genes.

Furthermore, heritabilities change. Heritabilities are always calculated on a specific population at a specific time in a specific environment. Measured at different ages or in different environments, the same population is likely to express a different heritability.

The concept of heritability has been widely misunderstood, and not always for good reasons. For most of this century, heritability of IQ has been thought to be about 80%, rather than 50%. This estimate came from the twin studies of the influential British psychologist, Sir Cyril Burt. Burt's "data," published over many decades, are almost certainly fraudulent; his raw data have never been found, the subjects of many of his studies have never been found, and even his supposed professional collaborators apparently never existed! Nonetheless, Burt's fraudulent results formed the basis for educational "streaming" of millions of British school chldren. On the basis of IQ tests at age eleven, British children were directed either to higher education and professional training (about 20%!) or to a blue-collar vocation (about 80%). Millions of people were thus denied reasonable control of their destinies on the basis of fraudulent data.

Additionally, many people mistakenly interpreted $h^2 = 0.80$ to mean that 80% of intelligence was determined by genes. As noted above, h^2 refers only to the *average* difference in IQ between randomly chosen individuals, and not to 80 (or 50) of the points of total IQ. In other words, heritability is commonly misinterpreted to give excessive weight to genetic factors, and Burt's fraudulent data simply compounded the misinterpretation.

A final misinterpretation also led to undue importance being attached to genetic factors underlying IQ. Many people believe that genetically-caused traits or phenotypes cannot be changed. Therefore, $h^2 = 0.80$ must mean that intelligence is almost entirely due to genetics and that education or environmental improvement can do little to improve their intellectual capacities. Such a view—that our destiny depends entirely on our genes—is called **genetic determinism.**

And, genetic determinism is wrong, as shown by many genetic traits of extremely high heritability. PKU and some forms of diabetes and near-sightedness have heritabilities approaching 100%, but simple environmental interventions (low phenylalanine diets, insulin injections, and glasses) allow people with genes for these conditions to live normal lives. Genetic determinism has long been used as a mindless justification for stigmatizing certain groups as genetically inferior, or for excusing aggression or other undesirable behaviors.

Throughout much of this century, some geneticists and psychologists have argued that genes play no meaningful role in intelligence. That view is now clearly discredited in two ways. First, the present estimate of heritability of IQ, 50%, is surprisingly high for a complex genetic trait. Other types of psychological and personality tests have also shown significant correlations with relatedness (e.g., schizophrenia) and high heritabilities (e.g., 50% for neuroticism and extraversion). Second, many single gene or chromosomal defects clearly cause or are directly related to abnormal (by any standard) behavior and reduced intelligence: Down's syndrome, fragile X syndrome, PKU, Huntington's disease, Turner's and Klinefelter's syndromes, and many others.

Thus, genetics plays a central role in human intelligence. But understanding the development and expression of intelligence requires an examination of equally significant environmental factors.

Evidence for the Role of Environment in Intelligence

A long list of environmental factors are associated with differences in IQ between various groups. A brief review of some of these factors illustrates the certainty of an important environmental component to intelligence.

Culture

Differences in IQ have been detected among populations representing different races, religions, nationalities, and regions of the country. Genetic differences also occur between these same groups, so it is almost impossible to determine whether IQ differences should be assigned to genes or to environment. However, the potential role of culture is often ignored.

TABLE 9.1

IQs of Israeli children of different genetic backgrounds raised in a home environment or in a communal environment

Ancestry	Home*	Kibbutz
European	105	115
Middle Eastern	85	115

*Home environment refers also to the region in which the particular ethnic group occurred. Children of European ancestry were raised in Europe; children of Middle Eastern ancestry were raised in the Middle East.

For example, some of the first studies showing black–white differences in IQ compared blacks from the southeastern United States with whites from the northeastern United States. The blacks scored lower than the whites. Whites from the southeast, however, also scored lower than whites from the northeast. Differences in IQ among blacks and whites in these studies almost certainly reflect regional cultural and educational differences.

Another relevant line of evidence derives from children raised in a culture different from that of their genetic heritage. In Israeli kibbutzim, or farm communes, children are raised collectively rather than by their parents in a traditional home environment. Comparative IQ studies show that the communal environment has a major effect on IQ (table 9.1).

These data illustrate several points. If only the data from the first column were known, many persons might conclude that European Jews were genetically more intelligent than Middle Eastern Jews. A mean difference of 20 points is highly significant. The second column, however, provides perspective on the data of the first. Children raised in kibbutzim have a relatively uniform environment, whatever the racial background. The identical IQ scores for both groups indicate the primary importance of environmental factors. Sharing a common language, education, and social environment produced similar scores on IQ tests.

It is even more interesting that the IQ scores of both groups increased dramatically in the kibbutz environment. The communal environment may have enriched the learning opportunities of both groups, reflected in the higher IQ scores.

Age and Early Education

IQ is normally not measurable until about age two. From then until the early teens, IQ tends to increase regularly. The amount of increase is generally related to educational and cultural opportunities available to the child during this period.

Project Head Start was begun during the Johnson administration as a part of the Great Society innovations. Its object was to provide children from disadvantaged segments of society with a culturally and educationally enriched environment early in life. The program also included at least one nutritious meal per day for participating children.

The effectiveness of Head Start is still being assessed, but some studies indicate favorable results. Children from disadvantaged homes that participated in Head Start had significantly elevated IQ scores when compared to children not in the program. However, the long-term effects on IQ are uncertain, as a one-year Head Start program cannot be expected to entirely override later environmental influences. Some data indicate that Head Start children have less need for remedial courses and an improved chance of graduating from high school than non-Head Start children.

Project Head Start may have other beneficial effects, such as improved social behavior. Head Start children have fewer legal problems than non-Head Start children. Children exposed to an enriched social environment early in life may be more receptive to educational opportunities. Education is a cooperative endeavor, an interaction between the student and the environment, as is test taking. Increased facility in social interactions may play a major role in increasing IQ, which depends not only on intelligence and knowledge but equally upon the ability to *display* intelligence in an examination. Altered social behavior could also explain the data from the kibbutzim in table 9.1.

A different interpretation of the success experienced by participants in Project Head Start is that the school officials' knowledge that children had participated in a preschool program kept these children out of the special education classrooms upon entering the public school system. It is well documented that students in special education programs seldom return to the mainstream of public education. They have less expected of them and eventually come to have poor self-images and, consequently, expect less of themselves. In part, the perceived success of Project Head Start may be attributed to the extra attention the program focused on the participants and to their improved self-images.

Test Design and Majority Effect

When IQ tests are constructed, they are standardized against the test populations to which they are directed. A series of potential test questions is administered to an experimental group. Only those questions are selected for inclusion in the IQ test that (1) are correctly answered by the majority of experimental subjects and that (2) do not distinguish between various subgroups being tested. For the first IQ tests, questions were excluded that gave different results for males and females, and for different adult age groups. Thus, *by design* these tests did not discriminate on the basis of age or sex.

Tests were not standardized on the basis of socioeconomic status or race. It is hardly surprising, then, that significant differences between races have been detected. Tests now have been designed that attempt to minimize the bias against particular races. Such attempts have been partly successful, but the effort has been criticized on the basis that IQ scores are most useful in predicting success of individuals in the school system and in professions. Because American society is still primarily white in orientation, there is some value in retaining a test with predictive power with regard to success in white society. Recognizing this difficulty, one black leader has noted that the use of IQ tests, even if biased toward white culture, is more fair than the use of no tests at all because "at least [blacks are] being excluded by an objective standard."

Considering the sexist bias of society, it is somewhat surprising that IQ tests were standardized to reduce differences between sexes. The comparative mental capacities of males and females is a topic of serious scientific debate. Females commonly score higher on IQ tests than males until puberty, after which scores decline in comparison to males. This decline has been blamed on social interactions, as the roles of the sexes become more clearly defined at about the time of puberty. Young females, for example, may be reluctant to score higher than males whom they wish to date.

Some scientists claim that the sexes have significant genetic differences in mental capacity. In particular, females have been claimed to be less proficient in mathematical skills than males. Substantial evidence has been raised on both sides of this controversy. No resolution appears in sight.

A final effect relates to majority influence. One experimental study compared the performance of genetically different strains of rats in learning to navigate a maze, a standard measure of intelligence in rats. Six inbred strains were used to establish six different study populations. Each population consisted of a *majority* of a single strain of rat, plus individuals of different *minority* strains. The object of the study was to determine the predictability of performance of minority groups of rats in terms of their abilities to learn to navigate a maze. In each case, the rats belonging to the majority always learned the maze most rapidly. If genes controlled intelligence, one of the strains would have been expected to learn fastest, whether it was in the majority or minority of the population. Instead, a single environmental factor—which strain was most abundant—seemed to govern the measure of intelligence.

The extent to which data from nonhuman species relate to humans is always debatable. Nevertheless, a majority effect may explain certain aspects of IQ comparisons. Immigrant populations, for example, almost always score lower on IQ tests than the resident population. In only a generation or two, most immigrant populations are integrated into American society and show no differences on IQ tests; perhaps they have become one of the majority. Perhaps, also, they have acquired the English language and shared the opportunities of the American educational system. Blacks, however, have not become part of the majority, even after ten or more generations as Americans. Color differences have maintained them as an identifiable and therefore distinct minority group.

Nutrition

The influence of diet upon phenotype is unarguable. Many pathological conditions (rickets, scurvy, diabetes) can be cured or ameliorated by dietary modification. The average height of Americans has increased by over half a foot in the last one hundred years, almost certainly due to improvements in diet rather than to genetic changes. The diet of Americans is so good, in fact, that as a nation we may now suffer more from the effects of overabundance of food and drink than from malnutrition; heart disease, some cancers, obesity, and diabetes have become major American health problems.

Similarly, diet may significantly effect IQ in many ways. During the first few years of development adequate dietary protein is essential for the proper growth and maturation of nervous tissue. Protein-deficient children do not produce sufficient nervous tissue for full mental development. Certain biochemical defects can be treated dietarily. PKU, for example, causes serious mental retardation if untreated; restriction of phenylalanine intake produces normal intelligence.

Recent studies of retarded children indicate that biochemical deficiencies that can be treated dietarily may be one cause of retardation. Groups of mentally retarded children, including victims of Down's syndrome (chapter 11), when given large supplements of vitamins and minerals, have shown increases in IQ of 10 points or more. In the most spectacular example, a child of age seven was still in diapers, had not learned to speak, and had an IQ estimated to be 25–30. After supplements of vitamins and minerals, the child showed major improvements. By age nine he was leading a far more normal life. He could speak, read and write, do arithmetic, play basketball, and ride a bicycle; his IQ was estimated to be about 90.

A related aspect of biochemical defects is their influence on social behavior. Manic-depressive psychosis is now routinely treated by administration of lithium salts. A protein-deficient diet in monkeys induces various forms of antisocial behavior. Some scientists believe that hyperactivity in children may result, in part, from excess dietary sugar. Antisocial behavior may or may not directly influence IQ. It may, however, affect several factors clearly important to IQ, particularly learning. Where behaviors result in inattention at school, or early departure from the educational system due to criminal or psychologically aberrant actions, the IQ almost certainly suffers. Because testing involves cooperation from the test subject, even the determination of IQ of severely antisocial or retarded persons is frequently impossible.

Society's Response

The nature-nurture debate is over. As recently as the 1970s, people on both sides continued to claim primary importance of genes or environmental factors in determining IQ. The purpose—and the fallacy—was often to make **deterministic** arguments. Genetic determinists were opposed by **cultural determinists,** who claimed that genes have no meaningful influence on behavior or intelligence. But determinism, of whatever form, is a simplistic, invalid, and futile argument. People are not automatons driven inevitably by either their genes or the environment. The precise contributions of nature and nurture—even if we could identify them—to a particular phenotypic trait are irrelevant to social actions, once we accept that both matter.

As Stephen J. Gould observed, our proper concern in learning about behaviors such as IQ ought to be human potential. Factors contributing to human behaviors matter, because understanding the many causes of behavior and their interrelationships can allow an enlightened society to provide environments that help all people to achieve their full potential. In the past, the nature-nurture debate has often been a way of classifying people in order to limit, rather than expand, their opportunities.

The genetic basis of IQ and other human behaviors will continue to expand in the coming decades. If used properly, this knowledge will allow medical treatment for specific genes that have harmful effects on intelligence and personality. But stigmatizing carriers of such genes has no place in an enlightened society. Differences in IQ between individuals or races reflect only that: differences. Differences in IQ should be treated like any other phenotypic differences—a reflection of the diversity of life, not a measure of its moral worth.

Changing the genetic composition of human populations is neither practical nor desirable. Environmental improvement, on the other hand, is not only humane, but provides tangible results to society through improved social behavior and personal development of its citizens. In either moral or practical terms, an affluent society has little justification for denying its citizenry environmental opportunities for the fullest expression of their psychological capacities.

Summary

1. Regulatory genetics is the study of mechanisms governing the expression of genes.
2. The operon is the basic structural and functional unit unique to prokaryotic DNA. An operon consists of an operator, a promoter, and several structural genes. Transcription of the structural genes into a single message depends upon the interactions of the operator and promoter with a distant regulator gene.
3. Probably due to the greater size and complexity of eukaryotic cells and chromosomes, eukaryotic gene regulation is far more diverse than prokaryotic regulation. Eukaryotic chromosomes differ from prokaryotic chromosomes in being multiple rather than single, consisting of about 60% histones and other nuclear proteins, and probably in possessing a high level of spatial organization within the nucleus, where they are segregated from the remainder of the cell.
4. Physical changes in eukaryotic chromosomes are important to gene function. Uncoiling is associated with transcription, as evidenced by chromosome puffs and lampbrush chromosomes.
5. Hormones and patterns of methylation control gene function by regulating transcriptional activity.
6. Dominance, recessiveness, and codominance are characteristics of the expression of alleles at a locus. They are meaningful only in reference to particular phenotypic traits. A pleiotropic allele, for example, may be partially dominant at the biochemical level, but recessive at a visible phenotypic level.
7. The norm of reaction is the range of phenotypic responses that a genotype expresses over many different environments.
8. The degree of expression of a gene is called expressivity. Penetrance is an all-or-none measure of phenotypic detectability of a particular allele.
9. Epistasis is a type of interaction between loci in which a gene at one locus controls and may mask the expression of the genes at another locus or loci. Epistasis usually results in departures from Mendelian ratios in hybridization experiments.
10. Most complex traits are governed by alleles at many loci and exhibit continuous or quantitative inheritance. Many quantitatively varying traits exhibit a bell-shaped or normal distribution.
11. Some quantitatively varying traits, such as IQ, are strongly influenced by environmental factors. Data demonstrate that both genetic and environmental factors affect a particular trait, but precisely defining the role of each factor is extraordinarily difficult.
12. Genetic and phenotypic diversity are a fundamental characteristic of all living organisms. No scientific, biological, or moral basis exists for assigning relative values to different individuals on the basis of genotypic differences.

Key Words and Phrases

amplified

chromosome puff

codominance

complete dominance

delayed onset

development

dominant

epistasis

expressivity

genetic determinism

hemoglobin

heritability

heterochromatinization

hormone

Huntington's disease

imprinting

inducible

intelligence quotient (IQ)

lampbrush chromosome

methylation

mobile genes

modifying loci

mutation

nature-nurture debate

negative feedback control

normal distribution

norm of reaction

operator gene

operon

partial or incomplete dominance

penetrance

pheromone

polytene chromosome

promoter gene

quantitative inheritance

recessive

regulatory gene

repressor

RNA polymerase

structural gene

thalassemia

transposable elements

transposons

X inactivation

Questions

1. Why are the physical relationships of chromosomes difficult to study?
2. Of what evolutionary significance is the fact that the lac operon in *E. coli* is inducible?
3. In the absence of genetic engineering technology, why do you suppose that we cannot test a developing fetus for the presence of sickle-cell anemia, when we know that both parents have sickle-cell trait and, therefore, the developing child has a one in four chance of having SCA? (Assume that there is no difficulty in obtaining a sample of fetal blood and tissues.)
4. How were heterozygotes for the PKU allele identified for the study shown in figure 9.6? Why were PKU homozygotes not shown in this figure? What would their phenylalanine response curve look like?
5. In seed corn, two different purebreeding lines of plants with white seeds produce an F1 all with purple seeds. Inbreeding of the F1 progeny produces plants with purple seeds ($\frac{9}{16}$) or white seeds ($\frac{7}{16}$). What are the likely genotypes of the parental lines, the F1 and the F2?
6. Albinism is controlled by the homozygous recessive condition of a gene or genes involved in pigmentation. What are possible genotypes of two inbred and purebreeding albino rats who, upon mating, produce a litter with no albino offspring?
7. Would you expect to find higher heritability for a particular trait in a population that is relatively homozygous for the genes involved, or in a population that is relatively heterozygous for those genes?
8. Some of the strongest arguments for a genetic basis for many behavioral traits come from the study of monozygotic twins raised in different households. What are the assumptions of such arguments? How valid are those assumptions?
9. What is your opinion of the value and validity of IQ tests? On what do you base your opinions? What kind of evidence would make you alter your opinions?
10. Figure 9.11 is a pedigree of polydactyly. What is the genotype of III-3? How do you know? How can you be sure that III-2 is really heterozygous for the allele for polydactyly and that her children are not just the result of mutation?

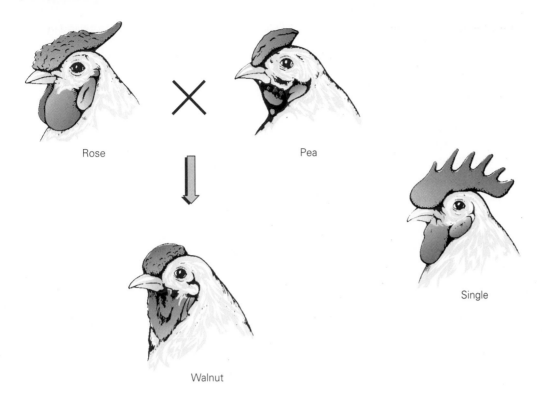

Rose

Pea

Single

Walnut

11. "Stiff little finger" is a condition where the little finger is permanently bent due to defects in the attachments of some of the muscles controlling the joint. This condition is caused by a dominant gene. Some affected individuals have only the right little finger stiff, others have only the left little finger stiff, and some have both bent and stiff. What is the most likely explanation for the observed variation?

12. Hydatidiform moles are embryos formed from two paternally-derived genomes and lacking a maternally-derived genome. They normally abort spontaneously by the second month of development. How can such a pregnancy be diagnosed after the abortion? Might a normal fetus have developed if one sperm had carried an X chromosome and the other had carried a Y chromosome?

13. Describe molecular regulatory mechanisms that might underlie such phenomena as non-penetrance or delayed onset of a gene?

14. Early studies of inheritance of combs in chickens showed that if a purebreeding Wyandotte chicken (rose comb) is crossed with a purebreeding Brahma chicken (pea comb), all offspring have a different type of comb, designated walnut (see the following figure). If two of these F1s are crossed, about $\frac{9}{16}$ have a walnut comb, $\frac{3}{16}$ a pea comb, $\frac{3}{16}$ a rose comb, and $\frac{1}{16}$ yet a different type of comb termed single. What is the most likely mode of inheritance?

For More Information

Barlow, H. B. 1983. "Intelligence, Guesswork, Language." *Nature,* 304:207–9.

Benbow, C. P., and J. C. Stanley. 1983. "Sex Differences in Mathematical Reasoning Ability; More Facts." *Science,* 222:1029–31.

Berg, D. E., and M. M. Howe. 1989. *Mobile DNA.* American Society for Microbiology, Washington, D.C.

Bouchard, T. J., Jr., and M. McGue. 1981. "Familial Studies of Intelligence: A Review." *Science,* 212:1055–59.

Bower, B. 1987. "IQ's Generation Gap." *Science News,* 132:108–9.

Britten, R. J., and D. E. Kohne. 1970. "Repeated Segments of DNA." *Scientific American,* 222:15–23.

Bryant, P. J., S. V. Bryant, and V. French. 1977. "Biological Regeneration and Pattern Formation." *Scientific American,* 237:66–82.

Cedar, H. 1988. "DNA Methylation and Gene Activity." *Cell,* 53:3–4.

Davidson, E. H. 1965. "Hormones and Genes." *Scientific American,* 212:36–45.

De Robertis, E. M., and J. B. Gurdon. 1979. "Gene Transplantation and the Analysis of Development." *Scientific American,* 241:60–80.

Fedoroff, N. V. 1984. "Transposable Genetic Elements in Maize." *Scientific American,* 250:84–99.

Gould, S. J. 1988. *An Urchin in the Storm.* W. W. Norton & Co.: New York. This is a series of book reviews. Read especially chapter 8, which conclusively discredits the work of the hereditarian psychologist Arthur Jensen; chapter 9, which summarizes the central issues in the "nature/nurture" debate; and chapter 10, which discusses mobile genes and the remarkable contributions of Barbara McClintock.

Grady, D. 1987. "The Ticking of a Time Bomb in the Genes." *Discover,* (June):26–39.

Holliday, R. 1989. "A Different Kind of Inheritance." *Scientific American,* 260:60–73.

Kolata, G. 1984. "Closing in on a Killer Gene." *Discover,* (March):83–87.

Kolata, G. 1987. "Early Signs of School Age IQ." *Science,* 236:774–75.

Lewin, R. 1981. "How Conversational Are Genes?" *Science,* 212:313–15.

Marx, J. L. 1988. "A Parent's Sex May Affect Gene Expression." *Science,* 239:352–53.

Plomin, R. 1990. "The Role of Inheritance in Behavior." *Science,* 248:183–88.

Ristow, W. 1978. "IQ Tests on Trial." *New Scientist,* 80:337–39.

Roychoudhury, A. K., and M. Nei. 1988. *Human Polymorphic Genes: World Distribution.* Oxford University press, New York.

Sapienza, C. 1990. "Parental Imprinting of Genes." *Scientific American,* 263:52–60.

Weiss, R. 1989. "A Genetic Gender Gap." *Science News,* 135:312–15.

Weiss, R. 1990. "Uneven Inheritance." *Science News,* 138:8–11.

Wessler, S. R. 1988. "Phenotypic Diversity Mediated by the Maize Transposable Elements *Ac* and *Spm*." *Science,* 242:399–405.

Applied Genetics

Virgil Sweeden, cover
of Science (10
November 1989)
© 1989 by the AAAS.

Genetic engineering: a miracle made reality

When Watson and Crick identified the structure of DNA in 1953, many thought it unlikely that the sequence of nucleotide bases for any gene in any organism could ever be known. How could a single gene be isolated from the millions of base pairs in a single cell? And how could the base sequences of an isolated gene be read like a book?

In less time than one scientific career, biologists have learned to do just these things, and more. Single genes can be identified, their sequences read (figure 10.1a), and the gene itself moved between organisms for an astonishing variety of practical purposes. And James Watson is now leading an ambitious program to learn the sequence of the 3 billion base pairs in the human genome, a project expected to be complete by the year 2005. The revolution in genetic knowledge and technologies has become so well established, producing such spectacular results despite its brief history, that it is easy to forget how miraculous this achievement is and how limitless its potential seems.

A new set of terms has quickly entered the language to describe these new scientific and technological methods. **Genetic engineering** and **gene splicing** are phrases commonly used to describe any type of manipulation of DNA for any purpose. Both imply the ability to cut segments of DNA from one or more organisms and recombine these segments to create novel genetic forms. A more precise and less emotive term for these methods is **recombinant DNA technologies.**

Another long-used word has acquired new life and added meaning: **biotechnology.** Biotechnology is the industrial use of biological systems and processes. Biotechnology is as old as human civilization; fermentation, for example, is a biological process used to manufacture such products as wine, beer, alcohol, cheese, vinegar, and bread. Other types of biotechnology allow the production of soap and medicines and the treatment of waste products.

To many people, the term biotechnology has become synonymous with genetic engineering because recombinant DNA technologies offer so much promise for industrial use. Recombinant DNA methods have been used commercially and already affect the lives of most of us. Fortunes have been made many times from various aspects of recombinant DNA work, and in the future recombinant DNA methods will be a routine and pervasive part of human life. The rapid expansion of these techniques is difficult to comprehend. Japan, for example, had no commercial application of recombinant DNA techniques in 1975, but by the year 2000 Japan plans to generate 10% of its enormous gross domestic product directly from recombinant DNA methods.

In this chapter we introduce briefly some aspects of these methodologies, then consider some of the applications of these techniques to everyday human life. Additionally, we note some of the hazards and ethical concerns raised by this rapid technological advancement.

Recombinant DNA Technology

Three obvious technical problems confront anyone wanting to manipulate and recombine DNA. The first problem is *concentration.* A single cell is visible only under a microscope, and most cells contain only two sets of chromosomes, which comprise a small portion of its mass. How can molecules occurring in such tiny quantities be studied? The answer proved to be a process called **molecular cloning,** in which DNA from any organism is fragmented, incorporated into a host bacterium, and then amplified to concentrations enormously larger than in the original cell.

The second problem is *identification* and *isolation* of a gene or DNA segment of interest. This requirement was solved by development of genetic **probes** that sort through the fragments in a molecular clone and isolate specific fragments.

The third problem is *analysis* of the isolated fragment. A probe isolates a segment of DNA, but does not identify its contents precisely. Geneticists have developed a variety of techniques for assessing the contents of a gene. Some provide a quick but limited assay for variation, while the most precise and detailed technique provides the nucleotide base sequence for the gene.

These techniques have been used worldwide in industrial, government, and university laboratories since the 1970s for direct DNA analysis. The applications seem endless. In the 1980s, a technique called the **polymerase chain reaction (PCR)** has allowed significant improvements in all three stages of DNA analysis.

Figure 10.1

How DNA sequences are read. Each vertical lane in both figures shows fragments of DNA from different individuals. The fragments have been separated in a supporting gel using electrophoresis. (a) A "sequencing" gel for three individuals (labelled 18, 19, and 20); the sequence of individual number 18, reading from the bottom up would be GGTCTATCCT-CATTTTCGGCGAGT. (b) RFLP patterns of polymorphic human DNA fragments from six unrelated British Caucasians (1–6) and twelve members from a large British Asian-Indian pedigree (7–18). These bands are compared to size standards indicated in kilobases (kb).

(a) Dr. Jack Griffith; (b) Dr. A. Jeffreys, *Nature* 314:67–73.

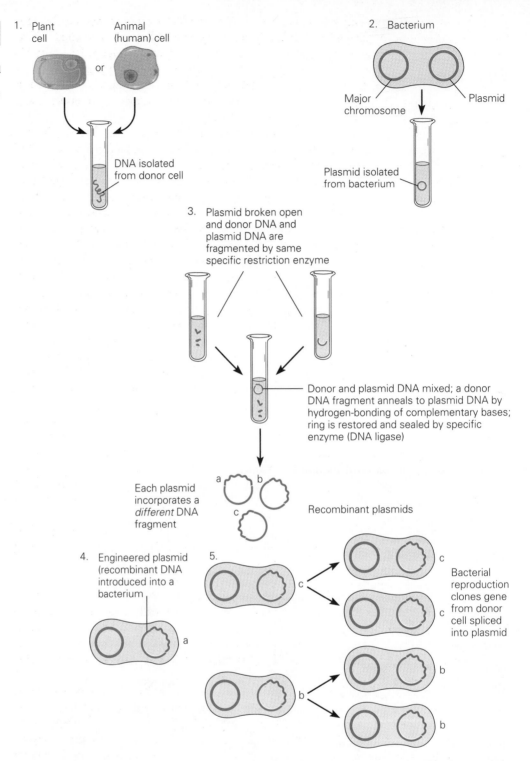

Figure 10.2

Molecular cloning. Plasmids are isolated from bacteria and cleaved by special enzymes into small fragments. DNA is simultaneously isolated from plant or animal donor cells. Plasmid DNA and donor DNA are incubated together, and a special DNA ligating enzyme restores the ring plasmid. The recombinant engineered plasmid is introduced into a bacterium that then reproduces clonally making more recombinant DNA.

1. Plant cell   Animal (human) cell
   or
   DNA isolated from donor cell

2. Bacterium
   Major chromosome   Plasmid
   Plasmid isolated from bacterium

3. Plasmid broken open and donor DNA and plasmid DNA are fragmented by same specific restriction enzyme

Donor and plasmid DNA mixed; a donor DNA fragment anneals to plasmid DNA by hydrogen-bonding of complementary bases; ring is restored and sealed by specific enzyme (DNA ligase)

Each plasmid incorporates a *different* DNA fragment

a   b
c
Recombinant plasmids

4. Engineered plasmid (recombinant DNA introduced into a bacterium
   a

5.
   c   c
       c
   Bacterial reproduction clones gene from donor cell spliced into plasmid
   b   b
       b

Molecular Cloning

Molecular cloning begins with DNA from a test organism, a cloning vehicle, a host bacterium, and a set of enzymes. The test organism is chosen because of special interest in its DNA (figure 10.2, step 1). For disease diagnosis in humans, for example, DNA might be obtained from a liver or bone marrow biopsy, or from cells in amniotic fluid. Alternatively, DNA might be taken from a plant that is believed to have a trait of agricultural importance such as, for example, resistance to an insect pest.

Second, a **vector** or **cloning vehicle** is required. Two types of vectors are commonly used, **plasmids** and **phages.** A plasmid is a small circular segment of extrachromosomal DNA, about five thousand base pairs long, found in the cytoplasm of bacteria (figure 10.2, step 2). Plasmids reproduce independently of the DNA of the bacterium and have long been known to carry genes for antibiotic resistance. Furthermore, plasmids are transmissible between bacteria, meaning that antibiotic resistance has now spread to virtually all disease-causing bacteria of humans.

Phages, more properly called bacteriophages, are viruses that attack bacteria (chapter 12). Like plasmids, phages move between bacteria, and they reproduce independently of the bacterium itself. The key trait of both plasmids and phages is their ability to incorporate short segments of foreign DNA into their own genomes. Thus, both can accept DNA from another organism, reproduce it along with their own DNA, and transmit or move that DNA to a host bacterium. Neither plasmid nor phage is considered to be alive, however, because neither is capable of independent self-reproduction. Each uses the enzymes and cellular apparatus of a bacterium to reproduce itself. Whether a plasmid or a phage is chosen for molecular cloning experiments usually depends upon the test species being examined. A phage will incorporate much larger segments of foreign DNA than will a plasmid and is thus often chosen for working with DNA from species with a large genome, such as humans.

The third ingredient in molecular cloning is a host bacterium such as *E. coli*. The bacterium is the key to amplification of DNA from other species. When the bacterium reproduces, it reproduces its own genome plus the DNA incorporated into the bacterial genome by a lysogenic phage (see chapter 12 to review aspects of phage biology). If the bacterium contains plasmids, they will have been reproduced independently in the cytoplasm; daughter bacteria will receive plasmids in their cytoplasm also. Thus, tiny amounts of DNA can be rapidly turned into very large quantities by bacterial reproduction.

The final key ingredient is one or more of a set of enzymes called **restriction endonucleases** or **restriction enzymes.** These enzymes were first identified as a key component of the defense mechanism of a bacterium against viral attack. Their name derives from their ability to restrict the entry of foreign DNA (such as that of a phage) into a cell by enzymatically cutting it into harmless fragments. Each species of bacterium produces its own set of restriction enzymes, against which its own genome is protected. Furthermore, some of these enzymes cut only at specific four- or six-base sequences of DNA.

Thus, a large group of these enzymes is available to use in the first step in molecular cloning—cutting DNA into manageable fragments at specific sites. Because the DNA of each species has a unique sequence, a particular restriction enzyme will cut DNA from one species into a set of fragments that differs from that of all other species.

In step 3 of molecular cloning (figure 10.2), purified test DNA from a source such as human bone marrow is incubated with a specific restriction enzyme, which cuts it into thousands of segments. Usually, each segment is either about 250 base pairs long (for an enzyme that recognizes a four-base sequence; any random four-base sequence of the four nucleotide bases will occur roughly every $4^4 = 256$ bases) or about 4,000 base pairs long ($4^6 = 4,096$, for an enzyme recognizing a six-base sequence).

At the same time, a cloning vehicle such as a plasmid is also cut in exactly one place with a restriction enzyme so that the circular DNA is "opened up" but not fragmented. Next, a small quantity of the test DNA is mixed with a large number of broken plasmids, and an enzyme called **DNA ligase** is added. DNA ligase, a naturally occurring enzyme, functions to ligate, or join, broken segments of DNA. The excess of plasmids over fragments ensures that the reformed plasmids will incorporate no more than one fragment of test DNA. This step results in many **recombinant plasmids,** each containing its own DNA combined with a fragment derived from another organism.

These plasmids are then incubated with their host bacterium (step 4), which incorporates them. The result is often referred to as a **DNA library,** meaning that the bacterial culture contains a set of bacteria that has incorporated fragments of DNA from the test species. All the DNA from the test species may have been incorporated, but a single bacterium will have only a short fragment of the entire test genome.

Figure 10.3

How a probe works.
(Agar plate) Dr. Jack
Griffith.

1. Make radioactive probe:

   (a) Isolate hemoglobin mRNA from red blood cells

   Red blood cells · · · m-RNA for hemoglobin

   (b) Use reverse transcriptase and radioactive DNA bases
   to make cDNA* (= probe for hemoglobin gene)

   cDNA

   m-RNA

2.
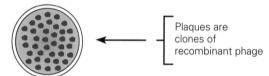

   (a) Grow recombinant phage in bacteria on agar plate
   (b) Make nitrocellulose replica of agar plate
   (c) Probe filter with cDNA* under conditions allowing cDNA* to hybridize with *any* complementary DNA sequences on filter; expose filter to X-ray film

   Phage plaque

3. Pick colony off original agar plate; grow phage on new bacterial culture

   Plaques are clones of recombinant phage

4. Isolate DNA from culture; it will contain the cloned hemoglobin gene

   hemoglobin-DNA (= gene)

Often, the next step is simply to allow the bacterial culture to incubate and reproduce itself, thus *amplifying* the quantity of bacteria *and* the test DNA. Once bacteria are in a library, essentially infinite quantities of DNA can be produced—cloned—for subsequent testing. This means that particularly interesting DNA from only one source can be used in laboratories all over the world, and that the problem of tiny concentrations of DNA is overcome by standard microbiological techniques. If a phage is used as a cloning vehicle, the phage colony itself is directly amplified by culturing in a medium containing its bacterial host.

Probing

The next part of the procedure is like searching for a needle in a haystack. The haystack, or DNA library, consists of a set of molecular clones, and the object is to select efficiently only the one set of interest from the tens of thousands of clones in the library. This is achieved by the use of a specific genetic **probe.**

A probe is a sequence of nucleotide bases complementary to any segment of a particular gene and long enough to hybridize uniquely with that gene. Construction of a probe is the key aspect of gene isolation and one of the most difficult parts of recombinant DNA technology.

Often, a probe is created by collecting large amounts of the messenger RNA specific to the gene to be isolated. For example, mRNA for human hemoglobin can be obtained in large quantities from red blood cells; few other mRNAs are found in red cells. Other tissues and cell types would be specific for other types of mRNA.

Next, reverse transcriptase (chapter 12) from retroviruses is used to construct a segment of DNA complementary to the sense strand of the gene of interest from the isolated mRNA. Radioactive nucleotide bases are added to the culture where the complementary DNA (or **cDNA**) is produced. Thus, genetic probes normally are radioactive cDNA segments.

The first step in the use of a probe in a phage library is to "plate" the phage containing the DNA library on an agar medium covered with a bacterial lawn, a continuous growth of bacteria (figure 10.3). The result is a set of phage colonies—clones—each originating from a single recombinant phage. Each phage clone is called a *plaque,* which looks like a clear hole in the bacterial lawn, where the bacteria have all been lysed by the replicating phage.

The problem is identifying which of these colonies, if any, contains the gene of interest. A nitrocellulose filter is then applied to the surface of the agar medium; the filter absorbs some of the phage from each colony and thus becomes a mirror image of the original plate. The original plate is saved for future use.

Next, the nitrocellulose filter is heated to separate, or *melt,* the two strands of DNA in the phage in each plaque. DNA melts at about 85°–100°C. The filter is now incubated in a solution containing the radioactive probe under conditions favoring *hybridization,* or rejoining, of the complementary but separated strands of DNA. In the phage clones containing the gene that matches the cDNA of the probe, hybridization will occur to the probe. Those colonies can be identified by exposing the agar plate to x-ray film, then matching the colonies that exhibit radioactivity to the original plate. Those clones can then be cut out of the original plate for further study and use (figure 10.3).

The most difficult part of these procedures is developing the probes. Many genes may not produce a protein or mRNA in sufficient quantities to allow production of a probe. Additionally, a single agar plate can grow only a few dozen plaques; screening enough plates to ensure coverage of the fifty thousand or more functional genes in the entire genome of a complex organism is an arduous task, which is not always feasible.

Analysis

Segments of DNA isolated by probing may be many thousands of nucleotide bases long. Further analysis is usually required. The exact goals of the analysis determine the methods, and an increasing array of techniques are available. Often, DNA that has been fragmented by restriction enzymes is placed into a supporting gel matrix and then subjected to a powerful electrical current. Called **gel electrophoresis** (from the Greek word *phorein,* meaning "to carry"), this process depends upon the fact that DNA fragments have differing electrical charges and thus move through the supporting gel matrix to different locations depending upon their charge, shape, and size (box 10.1). Finally, specific stains are applied to the gel to visualize the location of these DNA fragments.

The results in each lane are one or a series of bands representing DNA fragments of a single individual. Comparison of individuals (lanes on the gel) may reveal few or many differences between individuals, depending on the particular segment of DNA being examined and also on the degree of relationship between individuals.

The most complete but difficult analysis visualizes fragments of all the DNA in one lengthy segment, looking much like a supermarket bar code (figure 10.1). These types of electrophoretic gels can be used to determine the complete **sequence** of hundreds or even thousands of nucleotide bases in one region of a chromosome. Sequence analysis reveals the greatest amount of variation.

Sequencing DNA is painstaking, time-consuming, and relatively expensive work. Additionally, DNA sequences are difficult to interpret; there is often too much information. Base sequences are like alphabetic characters in a language whose words, grammar, and syntax are unknown. Therefore, simpler analyses are often useful. A very common type involves the use of restriction enzymes, electrophoresis, and a **probe** specific to a single locus. This analysis produces a relatively simple pattern on a gel that identifies variation between individuals at a single locus (figure 10.5b).

Different patterns exhibited by different individuals are called **restriction fragment length polymorphisms,** or **RFLPs** (pronounced RiFLiPs). This is because each band on the electrophoretic gel identifies a single fragment of DNA, and the location of a band usually allows identification of the fragment length. RFLPs are extremely useful as markers to identify individuals carrying mutations or alleles for deleterious conditions.

DNA probes can also be applied to chromosomes rather than to electrophoretic gels, a technique called **chromosomal** *in situ* **hybridization.** The probe then binds to (hybridizes with) the site on the chromosome to which the probe is specific. The probe is constructed to include either fluorescent or radioactive molecules, so that the location of the probed gene can be visually identified (figure 10.6).

BOX 10.1

Electrophoresis

Figure 10.4

(a–c) Description of gel electrophoresis for proteins and (d) example of phenotypes, genotypes, and electrophoretic patterns of the β-globin gene.

Tissue homogenates are placed in a gel made of starch, agar, polyacrylamide, cellulose, or some other substance providing a homogenous matrix.

Samples

(a)   Top view

The gel is placed in a solution with an appropriate substrate for the enzyme to be assayed and a salt that produces a colored band when reacting with the product of the reaction catalyzed by the enzyme

Staining box   Origin   Solution with substrate and salt

+   −

Gel   Top view

(c)

The gel with the samples in it is placed in an electric field for several hours.

Electrode

Sample slot   Gel   Contact wick

+

Buffer tank

Side view

(b)

Phenotype	Normal	Sickle-cell trait	Sickle-cell anemia
Genotype	SS	Ss	ss

Origin

Electrophoretic pattern

(d)

Electrophoresis is a technique that separates a mixture of molecules on the basis of differences in their size, shape, and electrical charge. It has been widely used in genetics to study variation in both DNA and proteins. Samples of DNA can be prepared for electrophoresis by precipitating the DNA from ruptured cells in solution and then fragmenting the DNA with restriction enzymes. Proteins are prepared by physically grinding a tissue such as liver or red blood cells to release the proteins into solution. Following centrifugation, the supernatant (liquid containing proteins) is used for electrophoresis, and the pellet of cell fragments is discarded.

The prepared sample (either DNA or protein extract) is placed in a supporting gel, often agar or potato starch, resembling tough jelly. A strong direct electrical current is then run through the gel, causing DNA fragments and proteins to move at different rates depending primarily upon electrical charge. After a few hours, the current is removed. A radioactive probe is then used to visualize specific segments of DNA on an autoradiograph. Or, the gel is stained to reveal specific proteins.

Often, the results can be given a direct Mendelian interpretation. For a single locus, a single band is a homozygote, while a single band with a different mobility (i.e., location on the gel) is a homozygote for a different allele. A heterozygote may have two bands, one corresponding to each homozygous type. In this way, for example, two hemoglobin components can be detected in the blood of a person who has sickle-cell trait, one for the normal hemoglobin and one for the sickle-cell hemoglobin (figure 10.4d).

Electrophoresis has made direct examination of genetic variation possible on a wide scale. Electrophoresis is routinely used around the world as a research tool and as a diagnostic technique for a rapidly increasing number of genetic diseases. The technique is applicable to virtually all organisms. Prior to development of electrophoresis, identification of genetic variation usually relied on long-term breeding experiments or other indirect methods.

Production of
autoradiograph

DNA samples
migrate from
top to bottom

X-ray film placed
on top of gel

(a)

Figure 10.5

DNA gel electrophoresis. (a) Agarose gel is placed in vertical or horizontal apparatus; radioactively-labeled samples are placed in wells and migrate from cathode to anode. The gel is dried and covered with x-ray film and the exposed film developed to show location of DNA fragments. (b) Gel electrophoresis permits visualization of restriction fragment length polymorphisms (RFLPs) after using the enzymes Taql and Ncol on six and five individuals, respectively, and a probe specific to the human muscle protein creatine kinase, whose gene is located on the long arm of chromosome 19; the lengths of DNA is thousands of base pairs (kb) are shown in the margins.

(b) Dr. Fielding Hejtmancik, Institute for Molecular Genetics, Baylor College of Medicine. Houston.

Taq I   Nco I

4.5kb —
3.8kb —

−3.6kb

−2.5kb

(b)

Figure 10.6

In situ hybridization. A fluorescent DNA probe specific to the gene producing an enzyme in human muscle (glycogen phosphorylase) identifies the site that the gene occupies on both chromosomes 11.

Lichter, P., et al., 1990 *Science,* 247:64–69. Photo by Peter Lichten and David C. Ward. Copyright 1990 by AAAS.

In situ hybridization is being used to produce detailed maps of human chromosomes. Knowledge of the exact location of disease-causing genes can have significant medical benefits. For example, in situ hybridization can guide researchers to important genes that should be examined in more detail, perhaps by sequencing. Genes that cluster around one site on a chromosome may give clues to conditions not previously known to be related. A linkage has now been identified that appears to relate the same segment of chromosome 21 to both Down's syndrome and Alzheimer's disease, conditions with very different physical expressions but that may prove to have a common cause.

The Polymerase Chain Reaction

DNA research and technologies, as practiced in the 1970s and 1980s, made astonishing advances, but were also technically difficult, expensive, and slow. Months were often required for a single experimental result—cloning to completion of analysis. On the other hand, techniques are rapidly being developed that should make DNA research simpler, cheaper, and faster.

Of particular importance is the **polymerase chain reaction (PCR)** (figure 10.7). PCR is a technique that allows a single DNA segment of as many as 6,000 base pairs to be enormously amplified. A target sequence of DNA is chosen for amplification. Short sequences of about 20 bases on either side of the target must be known; complementary DNA sequences known as *primers* are produced artificially by a DNA synthesizing machine. The target DNA is then heated to denature it (separate the two strands). The primers, extra nucleotide bases, and DNA polymerase are added to the mixture, which is then allowed to cool. As cooling proceeds, the primers attach to their complementary sequence on either side of the target sequence on both strands. Each strand then serves as a template for the formation of complete new DNA copies of the target.

At the completion of the cycle, the number of copies of the target sequence (and only the target) has been doubled. The process can be repeated indefinitely, producing highly concentrated DNA of a single gene. This contrasts markedly with a DNA library, where the entire genome has been fragmented and amplified. Because PCR produces such concentrated and pure DNA, analysis is far easier than with a DNA library.

Furthermore, PCR requires only a single copy of DNA as a starting point. Thus, DNA from a single cell from a human embryo can be amplified by PCR for analysis. A single copy of DNA from a single cell can be amplified 1 million times in twenty cycles and 1 billion times in thirty cycles, all in one afternoon. Even a single hair contains sufficient DNA for PCR amplification and subsequent analysis.

Furthermore, PCR can reconstruct genes from DNA fragments in a relatively poor state of preservation. Dried blood many years old may have enough DNA for use in PCR. Museum specimens contain sufficient DNA to allow study of extinct species such as the Tasmanian wolf and quagga. DNA has been studied from human mummies 7,000 years old and found to contain genes no longer known in any surviving populations. In the most extraordinary instance, DNA was extracted from magnolia leaves that had been buried in anoxic mud at the bottom of a cold lake in Idaho 17–20 million years ago. The leaves had not decayed; when removed from the encasing rock they were still green, and an 820 base pair sequence of a chloroplast gene was shown to be very similar in sequence to the same gene in present day living magnolias.

Safety

When recombinant DNA experimentation began in earnest in the early 1970s, many scientists and laypersons expressed serious concern about the potential dangers. The possibilities for a technological disaster seemed very real. Moving genes between species seemed to offer the opportunity for creation of new and possibly dangerous organisms. The use of *E. coli* as the amplification host for recombinant vectors seemed especially threatening. Humans are the natural host of *E. coli;* recombinant forms might produce toxins or initiate infections that could not be controlled and could lead to disease epidemics. Release of new genetic combinations, even new species, into the environment might have disastrous effects. With some justification, the public imagination ran wild.

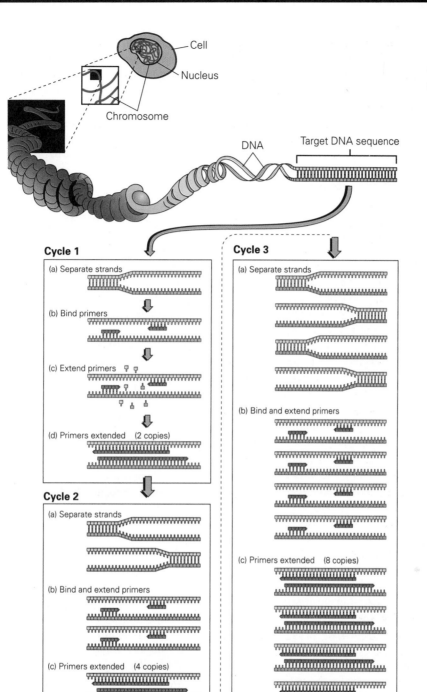

Figure 10.7

The polymerase chain reaction.

Cell

Nucleus

Chromosome

DNA

Target DNA sequence

Cycle 1

(a) Separate strands

(b) Bind primers

(c) Extend primers

(d) Primers extended   (2 copies)

Cycle 2

(a) Separate strands

(b) Bind and extend primers

(c) Primers extended   (4 copies)

Cycle 3

(a) Separate strands

(b) Bind and extend primers

(c) Primers extended   (8 copies)

Cycles 4-25

At least 10^5-fold increase in DNA

Fears culminated in Cambridge, Massachusetts, where the City Council passed laws to prohibit recombinant DNA research within the city limits, which include Harvard University and Massachusetts Institute of Technology. This placed some scientists in the equivocal position of seeing their research and careers threatened by the actions of a public who had only a limited knowledge of their work. However, the scientific community itself recognized the novel nature of recombinant DNA research and the potential for danger.

A series of scientific meetings in the early 1970s eventually led to a major international conference at Asilomar, California in 1975, which established a set of *voluntary guidelines* for scientists engaged in recombinant DNA research. These guidelines prohibited certain types of research entirely. Some potentially dangerous experiments could be conducted only in facilities designed for maximum containment of pathogens; only a few such facilities existed and were used for studies with the most dangerous disease organisms such as smallpox and anthrax. Many other types of experiments were considered acceptable if standard laboratory protocols were followed.

Businesses and private individuals could not be forced to comply with these guidelines. However, the guidelines were adopted by the National Institutes of Health, who fund most public research of this type in the United States. Failure to adhere to proper safeguards could result in an investigator losing future funding, as well as other penalties.

Experience in the past decade has led to relaxation of these guidelines. For several reasons, recombinant DNA research now appears to be considerably less dangerous than imagined in the early 1970s. Recombinant DNA itself is not dangerous. It may become so only after transfer to another living cell, which is not a simple procedure. Even in a species such as *E. coli,* which might be of potential danger to humans, laboratory strains are usually selected for special traits that restrict their infectiveness outside the laboratory. Such strains have been employed for decades. No disease epidemic in humans is known to have resulted from escape of an infectious recombinant agent from a laboratory.

Even more compelling is the information discovered since the early 1970s on the interindividual and interspecies transfer of DNA. It is now quite clear that genes are naturally transferred between individuals, even of different species and have always been throughout evolutionary history. New genetic combinations are always arising as a result of such natural processes. Such discoveries have been made primarily as a result of recombinant DNA research, and they argue convincingly that our initial fears of the dangers of recombinant DNA research were, largely but not totally, unfounded.

Nonetheless, some types of research inevitably carry an element of risk. By far the most hazardous may be military research to produce new **biological weapons** (box 10.2). Accidental as well as intentional release of biological warfare agents could have massive and highly unpredictable effects.

Professional ecologists have expressed concern over the **release into the natural environment of genetically modified organisms.** They have particularly noted the potential for unexpected harmful effects of genetically engineered crops and pest control agents. This view is more cautious than that of companies hoping to produce these agricultural products. These companies stand to make enormous amounts of money if their products achieve wide use, and they have predictably emphasized the probable safety of genetically modified agricultural products.

Recombinant organisms might in some way upset the "balance of nature" if, for example, they proved to be superior competitors to wild species, or superior predators, or if they carried diseases to which wild species were not resistant. While traditional plant breeding often led to exchange of genes between domesticated species and also accidental "leakage" of genes from domestic crops to closely related wild species via hybridization, recombinant techniques offer the potential for the creation of radically new phenotypes because genes can be moved between distantly related species. Experimental releases of recombinant organisms can be regulated to provide reasonable short-term levels of safety, but it is far more difficult to calculate the long-term risks to the natural environment.

Furthermore, the safety protocols for recombinant DNA research, still administered by the National Institutes of Health, are totally voluntary. Several serious violations of these protocols have occurred, and the perpetrators were reputable and well-intentioned scientists. If even the scientific community does not conform to safety guidelines, many people despair of compliance by businesses motivated by profits. In addition, governments have shown few limits on their behavior during times of war, and actions of unscrupulous individuals can seldom be restricted. Opponents of recombinant DNA research argue that these technologies should never be developed because they are certain to be misused (box 10.3).

On the other hand, the safety record thus far is good, and as noted previously, increasing knowledge seems to be diminishing our perceptions of danger. The techniques are simple enough and already so widely used that it would be virtually impossible to forbid recombinant DNA research entirely. The actual and potential benefits to be derived from this technology are so high that for most people the risks seem worth taking.

Cloning of Whole Organisms

Although molecular cloning is new, cloning of entire organisms is not. Clones (genetically identical copies) of various agricultural species of plants have long been produced by asexual (vegetative) reproduction. Navel oranges, lombardy poplars, dates (figure 10.8a), and many other cultivated species all comprise clones.

Some animals produce clones as well. Identical twins and triplets of any species are clones. Armadillos commonly give birth to identical quadruplets, making them extremely useful experimental animals (figure 10.8b).

In the 1950s the first clone of a complex animal was produced experimentally by a technique known as **nuclear transplantation** (figure 10.9). Nuclei from adult frog or tadpole cells were introduced into enucleated eggs. About 20% of eggs treated this way later developed into a frog that was an exact replica of the tadpole that donated the transplanted nuclei.

In 1978 a writer named David Rorvik published a book entitled *In His Image: The Cloning of a Man.* It purported to document the secret cloning of a wealthy man. Published as nonfiction, *In His Image* received a flurry of media attention. Rorvik, however, was unable to verify his claims, and in 1982 the publisher publicly admitted, following a court ruling, that no attempt had been made to confirm Rorvik's claims and that his book was a fraud and a hoax.

The successful cloning of mammals from somatic cells of embryos or adults has not been accomplished. Normal mammalian development requires specific contributions from both the male and female parent, unlike the situation observed in amphibians. Although it may be possible in the future, cloning of humans is generally regarded as ethically unacceptable.

Biotechnology

The uses of recombinant DNA technology seems as limitless as the imagination. In this section, we briefly discuss some of the applications that are presently affecting human life or that will soon do so.

Genetic Pharmacology

Molecular cloning is a powerful tool of the biochemical industry. Genes encoding natural chemical products—especially proteins and hormones—can be transferred to host bacteria that then produce limitless quantities of the chemical. In fact, the genes do not even have to be found. If the amino acid sequence of the protein can be identified, then an artificial gene encoding that sequence can be synthesized and inserted into a bacterial host. In this section we discuss medical and agricultural hormones and pheromones now in mass production using recombinant DNA techniques. All are commercially lucrative, and some raise significant ethical and social issues.

Human Insulin

The disease **diabetes** results from inadequate production in the pancreas of **insulin,** a hormone required for normal carbohydrate metabolism. Diabetics do not properly metabolize sugar, which then accumulates in the blood with many harmful effects. Diabetes is the fifth leading cause of death and the second leading cause of blindness in the United States.

The primary treatment for diabetes has long been shots of *bovine insulin,* purified from pancreases of cattle killed in slaughterhouses. Bovine insulin is not an ideal treatment, however. It is expensive to produce, and because it differs from human insulin at two of twenty-one amino acids in the A chain (figure 10.11a), it induces an immune (allergic) response in about 5% of all diabetics.

Since the early 1980s, recombinant DNA methods have been used to produce synthetic human insulin relatively cheaply (figure 10.11b). The amino acid sequences of both the A and B chains were determined; then geneticists constructed synthetic DNA sequences—artificial genes—encoding each chain.

BOX 10.2

Biological weapons

Biological weapons have been used for almost as long as organized warfare has existed. Bodies of smallpox victims were hurled into fortified cities by the Greeks to cause epidemics during sieges, and bodies of plague victims were similarly used in the Middle Ages. But recombinant DNA methods offer horrible new possibilities for efficient creation and widespread use of biological weapons.

Recombinant techniques could be used to produce either bacterial toxins such as botulinus toxin or highly virulent forms of existing diseases (table 10.1). Potential targets could include an enemy's crops, livestock, armies, or the general population. In World War II, scientists on both sides experimented with diseases such as anthrax, which infects humans, sheep, and cattle. Caused by a soil bacterium, respiratory anthrax kills most infected people within three days. Hardy anthrax spores survive in the soil for decades; an island off the coast of Scotland used for research in World War II is still uninhabited because of anthrax contamination.

Because biological weapons might backfire by killing friendly soldiers or civilians, a more devious strategy would be to use alleles that are highly specific to certain ethnic groups as "targets" for genetically engineered pathogens. Few alleles are unique to a race or population, but some highly significant differences certainly are. The allele for wet earwax (chapter 4) is found in about 85–95% of black and white Americans, but only in 2–15% of Koreans, Chinese, and Japanese. Similarly, populations of northern European origin (including white Americans) have frequencies of 70–99% for an allele for lactase production in adults; people with this allele can digest lactose, the sugar found in milk, as adults. Many other populations have very low frequencies of this allele and enzyme, and adults cannot digest milk. Thai and Vietnamese populations, for example, have frequencies of 0–5% for this allele, as do some African populations. Some populations in the Mediterranean area have high frequencies of alleles resulting in specific enzyme deficiencies (Glucose -6- phosphate Dehydrogenase). If carriers of these alleles inhale pollen or eat raw broad beans (*Vicia faba*), their red blood cells are destroyed resulting in a serious and often lethal hemolytic disease, favism. Thus, spraying an area with a high pollen dose (or a genetically engineered similar agent) could selectively disable specific persons.

Military scientists around the world are almost certainly engaged in certain types of recombinant DNA research. The budget for biological warfare research by the United States Defense Department increased from about $15 million in 1980 to as much as $100

Next, the artificial genes were inserted into plasmids of *E. coli*. These particular plasmids insert themselves and the artificial insulin genes into the *E. coli* genome immediately adjacent to the Z gene of the lac operon (see figure 9.2). Recall that the lac operon is normally turned off. However, it can be turned on simply by adding to the bacterial growth medium the inducer substance, lactose. This in turn induces transcription of bacterial mRNA, *as well as* the mRNA for the inserted A or B gene of human insulin. The system is extremely efficient; a single bacterium produces about 100,000 copies of one of the chains. As a final step, either the A or the B chain is purified from each separate culture, and the purified products then combined. Between 10% and 40% of the chains of each type bond to chains of the opposite type, forming entire molecules of synthetic human insulin.

Synthetic human insulin is identical to natural human insulin and therefore does not induce allergic responses in diabetics. It is also produced far more cheaply than bovine insulin. Additionally, the commercial success of synthetic insulin shows clearly the impossibility of separating so-called "pure" from "applied" research. When the lac operon was first proposed in 1961, it seemed little more than an interesting academic hypothesis that explained something about how nature functions, but had no obvious commercial value. In less

TABLE 10.1

Biological warfare agents that could be enhanced or produced by recombinant DNA technologies.

Disease	Pathogen	Symptoms
Anthrax	*Bacillus anthracis*	Inhaling spores causes fever, respiratory distress, death in 3–5 days
Cholera	*Vibrio cholerae*	Diarrhea, vomiting; sometimes fatal if untreated
Plague	*Pasteurella pestis*	Fever, shock, delirium; often fatal; killed ⅓ of population of Europe in Middle Ages
Tularemia	*Pasteurella tularensis*	Fever; often fatal; standard U.S. lethal biological agent until 1969
Smallpox	Pox virus	Fever, skin lesions; would have high mortality rate because vaccination is no longer common
Venezuelan equine encephalitis	VEE virus	Headache, fever, malaise; standard United States incapacitating agent until 1969
Botulism	*Clostridium botulinum*	Toxin produced by this bacterium is lethal

Reprinted by permission of the publishers from *Genetics: The Clash Between the New Genetics and Human Values* by D. Suzuki and P. Knutson, Cambridge, Mass.: Harvard University Press, Copyright © 1989 by New Data Enterprises Ltd. and Peter Knutson. Reproduced for Canada with the permission of Stoddart Publishing Co. Limited, 34 Lesmill Road, Don Mills, Ontario, Canada.

million by the end of the decade. Much of this is for defensive research, because the 1972 Biological Weapons Convention (signed by the United States, Soviet Union, and about half of the other countries in the world) commits the United States not to manufacture or stockpile biological weapons.

An important defensive use is the production of vaccines against possible biological warfare agents. Nonetheless, defensive technologies usually have offensive applications. A vaccine, for example, can be used to immunize friendly soldiers prior to a biological attack using the pathogen for which the vaccine provides protection.

than twenty years, knowledge of the regulatory mechanism for lactose metabolism in bacteria proved to be a key ingredient in an industrial process.

Human Growth Hormone

Pituitary dwarfism results from a deficiency of **human growth hormone (HGH),** produced in the pituitary gland of the brain. Through the mid-1980s, about 6,000 children per year were treated for this condition by administration of HGH obtained from pituitaries of human cadavers. Recombinant DNA techniques similar to those used to produce synthetic insulin are now producing synthetic HGH, identical to natural HGH, for the treatment of pituitary dwarfism.

HGH from cadavers has been used for decades, but proved to have a significant risk. In the early 1980s, three recipients of cadaver-derived HGH died of Creutzfeld-Jakob disease (CJD), almost certainly from highly infectious slow viruses transmitted from the pituitaries of CJD victims. As with insulin, recombinant production of HGH thus had an advantage over traditional treatment.

Figure 10.8

Examples of clones:
(a) date trees grown in
Palm Springs, California;
(b) mother armadillo and
identical quadruplet
offspring.

(a) © Edna Douthat/Photo
Researchers. (b) © Karl H.
Maslowski/Photo
Researchers.

(a)

(b)

Unlike insulin, however, synthetic HGH raises significant ethical concerns. The number of pituitary dwarfs is far smaller than the hundreds of thousands of victims of diabetes, meaning that the financial incentive to biotechnology companies is far smaller to produce HGH, unless other uses can be found. And they can, because HGH plays a significant role in metabolism, causing fat tissue to be metabolized while conserving muscle tissue. Thus, HGH may be beneficial to people trying to lose weight, an enormous market. People who are naturally deficient in HGH are shorter than normal, and they also prematurely develop wrinkles and other signs of aging. While HGH may not increase longevity, it is being used experimentally to try to limit or even to reverse wrinkling and other characteristics of aging.

Use of HGH in "cosmetic endocrinology" could thus vastly increase the potential market for the drug. But how widely should such a drug be used? Given the broad effects of HGH on metabolism, there may be many other effects that are unknown and undesirable or even dangerous if the drug is taken in large doses or over long periods of time.

Even more seriously, many parents may be tempted to give HGH to normal children because of the well-documented advantages of tallness. In general, tall people tend to have higher salaries than short people, to command positions of power in society, and to be perceived more favorably than short people. The taller candidate usually wins United States presidential elections. Even the English language imbues height with value: A proud person sits "tall in the saddle"; a broke person is "short"; a generous act is described as "big," while a mean-spirited person is "small."

But even if HGH can make people taller, should it be given routinely to otherwise normal children? What constitutes normal height? Americans are over fifteen centimeters (half of a foot) taller on average now than a century ago. Do we want a nation of basketball players?

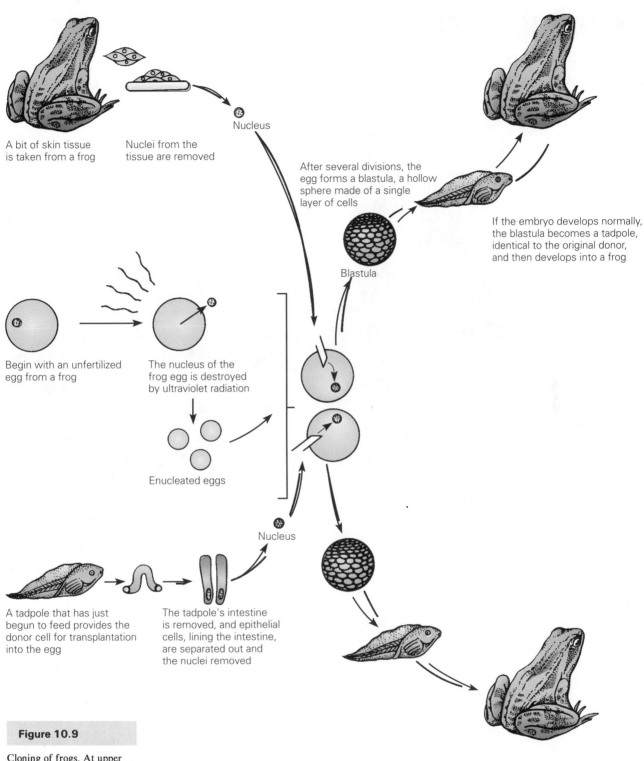

A bit of skin tissue is taken from a frog

Nuclei from the tissue are removed

Nucleus

After several divisions, the egg forms a blastula, a hollow sphere made of a single layer of cells

Blastula

If the embryo develops normally, the blastula becomes a tadpole, identical to the original donor, and then develops into a frog

Begin with an unfertilized egg from a frog

The nucleus of the frog egg is destroyed by ultraviolet radiation

Enucleated eggs

Nucleus

A tadpole that has just begun to feed provides the donor cell for transplantation into the egg

The tadpole's intestine is removed, and epithelial cells, lining the intestine, are separated out and the nuclei removed

Figure 10.9

Cloning of frogs. At upper left, the nucleus of a skin cell from an adult frog is used to produce a frog identical to the donor. With the use of multiple eggs and multiple nuclei, a clone of frogs is produced. At lower left, the nuclei are taken from the intestinal cells of a tadpole. The success rate of cloning from adult tissues is much lower than when using nuclei from tadpoles.

BOX 10.3

Jeremy Rifkin: an implacable opponent

Figure 10.10

Jeremy Rifkin. Rifkin leads a formidable coalition of opponents of DNA research and technology.

Bettmann Archives.

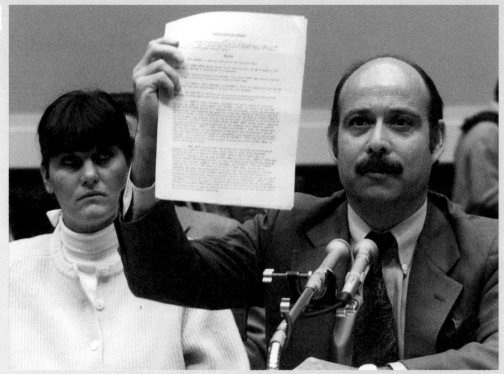

Fears of the consequences of inevitable misuse of recombinant DNA technologies have produced a significant reservoir of public distrust encompassing virtually all DNA research and its implementation. Motivations and justifications of opponents of recombinant DNA research vary widely, but the unquestioned leader of the opposition is Jeremy Rifkin (figure 10.10).

As leader of the innocently titled activist organization Foundation for Economic Trends, Rifkin has skillfully and energetically worked to impede DNA research and technologies through media campaigns and legal battles. He has questioned the ethics of DNA researchers, some of whom (he correctly observes) have substantial financial interests in biotechnology corporations. Rifkin has used court injunctions to stop or delay field tests of genetically modified agricultural species. He has fought running battles with the Recombinant DNA Advisory Committee of the National Institutes of Health in attempts to stop experimental medical applications of DNA technologies. Rifkin argues that even apparently benevolent therapies can have unintended consequences, leading down a "slippery slope" to such undesirable ends as forcing persons to undergo genetic treatments. He has also cautioned that DNA technologies could lead to workers being genetically modified to tolerate harmful chemicals in the workplace, rather than forcing companies to behave responsibly by limiting exposure of workers to these chemicals.

Rifkin seems to be opposed to new biological technologies of all varieties. He is cochairman, for example, of the National Coalition against Surrogacy. This group has lobbied against surrogate parenthood—not a DNA technique—and at least thirteen states now outlaw surrogate parenthood where the surrogate is paid for her services.

Rifkin has proven to be a resourceful opponent. His success almost certainly reflects a persistent unease among the public about implications and possible future uses of DNA technologies. In general, the scientific community has expressed little sympathy for his views. Nonetheless, Rifkin has ensured that the debate over practical and ethical aspects of DNA technologies has been intense and persistent.

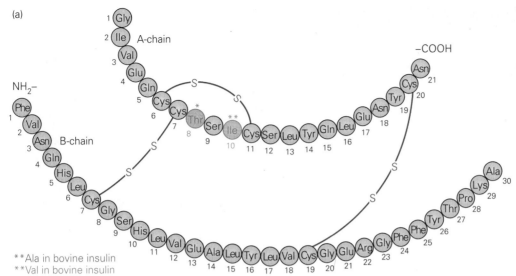

(a)

A-chain

B-chain

NH₂–

–COOH

**Ala in bovine insulin
**Val in bovine insulin

Figure 10.11

An important biotechnology product, human insulin. (a) The hormone is a protein composed of two polypeptide chains; the A chain has twenty-one amino acids, and the B chain has thirty; note the two amino acid differences to bovine insulin. (b) The production process begins when synthetic genes for A and B chains are introduced into the lac operon of separate batches of *E. coli*. Both cultures are induced to begin production by lactose; finally, the resulting A and B polypeptides are purified and allowed to combine to form active human insulin.

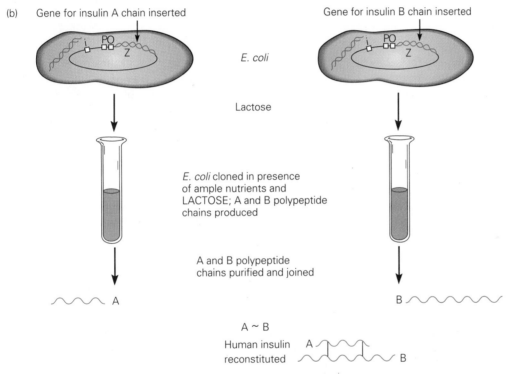

(b)

Gene for insulin A chain inserted

Gene for insulin B chain inserted

E. coli

Lactose

E. coli cloned in presence of ample nutrients and LACTOSE; A and B polypeptide chains produced

A and B polypeptide chains purified and joined

A

B

A ~ B
Human insulin reconstituted

A

B

What is to stop use of HGH to produce larger soldiers? What limits should be placed on pharmaceutical companies who have a vested interest in increasing the market for HGH? And, as always, the wealthy will have first access to HGH: Is it fair for the wealthy to use a drug that could have the effect of further increasing the disparity in wealth and power between the rich and poor?

Even more frustrating, it is not even certain that HGH will work as expected. The primary effect of HGH may not be to cause growth, but to regulate other aspects of metabolism. Malnourished children produce elevated amounts of HGH, as do athletes in training. It may be that HGH is most effective very early in life, before a child is found to be shorter than normal. The proper doses for differing ages have yet to be determined. In an extreme case, even administering drugs routinely to a child could lead the child to believe something is wrong with him or her, producing a "placebo effect" that actually leads to decreased growth.

Society surely has an obligation to provide proper treatment to pituitary dwarfs, whose lives are seriously affected by their condition. Obligations for the use or restriction of the same treatment on other people seem far less clear.

Genetic Engineering: A Miracle Made Reality

Figure 10.12

An advertisement by Ben & Jerry's Homemade, Inc., opposing use of recombinant bovine growth hormone because it may harm small farming operations.

From Marjorie Sun, "Market Sours on Milk Hormone," *Science,* Vol. 246:877, 17 November 1989. Copyright © 1989 by the American Association for the Advancement of Science. Reprinted by permission.

Bovine Growth Hormone

Very different issues are raised by a similar product of recombinant technology, **bovine growth hormone (BGH).** Synthetic BGH is virtually identical to natural BGH. Regular doses of BGH can increase milk production in dairy cows by 10–20%. Development of recombinant BGH has cost the pharmaceutical companies about $200 million dollars; thus, they have a large incentive to see wide-scale use of the hormone, and the opportunities for increased production must be tempting to many farmers. The U.S. Food and Drug Administration (FDA) has evaluated the hormone and supported its safety. What could be simpler?

As with HGH, several issues cloud the prospects for BGH. In the United States, a campaign led by Jeremy Rifkin (box 10.3) has cast doubt on the future use of BGH. Rifkin and others have claimed that BGH is a serious health risk, a claim *not* supported by experimental data but one that appears to have sufficiently raised public fears that major food companies and supermarket chains withdrew products produced from BGH milk until further FDA analysis was completed.

An issue on which Rifkin is probably on firmer ground is that of social and economic impacts of BGH. The supply of milk and dairy products is high, meaning that increased supplies resulting from BGH are likely to lower prices further, benefitting large industrial style farms at the expense of small farms. Some companies have opposed use of recombinant BGH on the basis of expected harmful effects on family farms (figure 10.12). In Europe, the debate on BGH centers on the social consequences of its introduction.

Pest Control

Another use for genetic engineering of valuable biochemicals lies in the area of insect pest control. Present practice relies primarily on pesticides, which have many disadvantages. Most are synthetic chemicals that are not species-specific, killing valuable as well as target pest species. Most also remain in the environment—like DDT—and accumulate in the tissues of rare species like hawks and sea birds.

Recombinant DNA manufacture of naturally occurring chemicals has none of these or other disadvantages. For example, species-specific control can be achieved by the use of **sex pheromones,** chemicals that insects use to find partners during the brief mating season. In many species of insect pest, the female releases a pheromone that attracts males and identifies her not only as a female, but also as a member of the proper species. These chemicals have been used as a form of pest control by saturating fields with them, overwhelming the pheromones released by females and so confusing the males that they cannot find females with which to mate.

Most sex pheromones occur in such tiny quantities that purifying them from natural sources is prohibitively expensive. However, once even a tiny amount of pheromone has been obtained, recombinant DNA methods can allow mass production for agricultural use at reasonable prices. Residues of such chemicals pose no long-term health hazard to other species because they are species-specific and also because as naturally occurring biochemicals they rapidly degrade in nature.

Diagnosis of Diseases

Identifying diseases or disease-causing organisms is of obvious medical importance, but often not a simple matter. Early and precise diagnosis allows improved treatment that saves lives and money. In this section, we discuss three ways in which recombinant DNA techniques for diagnosis can be used. Some of these are already widely available, and they can be expected to quickly become a routine tool of medical practice.

Genetic Diseases

Over four thousand genetically caused diseases have been identified in humans. In principle, RFLP analysis should allow diagnosis of any simple genetic disease, that is, one caused by defective alleles showing Mendelian segregation at a single locus. The primary requirement is development of a probe specific to the particular disease. Probes allowing diagnosis already exist for many serious genetic diseases, including Duchenne's muscular dystrophy, cystic fibrosis, sickle-cell anemia, and Huntington's disease (figure 10.13).

Homologous pair
chromosome 11

β→   ←β

7.6 kb

13.0 kb

↑   Sites cut by Hpa I

▢   Normal β^A-globin gene

▣   Sickle-cell Hb β^S-globin gene

(a)

7.6 kb—

13 kb—

(b)

Figure 10.13

RFLP diagnosis of sickle-cell hemoglobin. (a) A radioactive probe to the beta globin gene has been prepared. Total genomic DNA is obtained from both parents, their first child who has sickle-cell anemia and the fetus that the mother is carrying. Each DNA sample is digested with the restriction endonuclease, Hpa I. (b) The resultant fragments are separated by electrophoresis and probed with a radioactive probe to the beta globin gene. Comparison of the patterns of the normal (carrier) parents and the affected first daughter with that of the fetus shows the unborn child will be like the parents and carry the allele, but not be homozygous for the sickle-cell allele.

DNA testing can be done at any stage of life, including prenatally using cells obtained either by chorion villus sampling or amniocentesis. Because amplification of target DNA sequences by PCR is far faster than cell culture, DNA diagnosis is significantly quicker than for some previous prenatal tests.

The first probes used for RFLP analysis were often specific only to segments of DNA closely linked to specific diseases. This meant that tests had a significant possibility of giving a false result. Increasingly accurate probes have been developed, however, reducing the probability of incorrect diagnosis. The ideal probe is specific not only to the gene causing the disease, but to the short segment of the allele (even an alteration to a single base) that actually causes the disease. This can be a serious diagnostic limitation, as noted in chapter 4 for cystic fibrosis. The CF gene is very large, and the disease is caused by mutations at many different sites on the gene. A different probe is required to identify each mutation, and it may not be feasible to develop a probe for very rare mutations.

We previously noted also that diagnosis is not a cure. However, DNA techniques that allow diagnosis represent a significant first step on the road to a cure. Development of a probe allows mapping of the chromosomal location of the gene causing the disease. The gene can then be sequenced, which should allow identification of the gene product involved in the disease. Eventually, understanding of the genetics, biochemistry, and metabolism of the disease can be expected to lead to cures, although the latter steps may still require many years. Nonetheless, for many well-known genetic diseases such as Huntington's, the biochemical defect has been entirely unknown. DNA techniques are thus a major breakthrough in knowledge that may finally lead to treatment or cure. Additionally, where a single disease such as

Genetic Engineering: A Miracle Made Reality

BOX 10.4

Retrospective diagnosis: Lyme disease

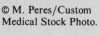

Figure 10.14

The Lyme disease vector—the *Ixodes* tick.

© M. Peres/Custom Medical Stock Photo.

The appearance of a new disease is both mysterious and frightening. There are many unknowns: How serious is the disease? Who is at risk? What will the long-term effects be? Can a treatment or prevention be devised? Is the disease of natural origin or the result of some unexpected human disturbance of nature?

Lyme disease, officially recognized and described fully only in 1975, raises many such questions. It produces a variety of symptoms, including severe malaise, arthritis, and temporary dementia. It is now known to be caused by the microorganism *Borrelia burgdorferi.* Lyme disease is transmitted to humans by a *vector* or intermediate organism, the tick *Ixodes* (figure 10.14). Diagnosis of the disease is often confirmed by the severe rash that spreads, concentrically, from the site of the *Ixodes* bite.

The disease was identified from cases in the northeastern United States, including Lyme, CT, the source of the name. It has spread to 47 of the 50 states, and the first diagnosis of cases in Australia was made in 1990. However, Lyme disease may have originated in Europe, where a probable case was described early in the twentieth century.

Is Lyme disease really new, or has it existed in the United States unrecognized but still occurring at a low rate for most of this century? The latter hypothesis is supported by a condition known as *Montauk knee* by long-time residents of Long Island, NY, possibly a reference to the arthritic symptoms of Lyme disease.

DNA analysis has now provided answers to some of the mystery surrounding the origins of Lyme disease. *Ixodes* ticks preserved in arthropod collections from Long Island and the northeastern United States have been examined by probes specific for *Borrelia,* the causative agent. Positive results for *Borrelia* have been obtained from ticks collected as early as about 1940, indicating that the disease has existed for many decades but has not been recognized. Scientists must now try to answer why the disease occurred in a relatively latent form for so long, only to increase in frequency so rapidly, beginning in the early 1970s.

CF is caused by many different mutations, diagnosis of the differing genetic causes of the disease can allow specific treatment for each of the forms.

Identification of Disease Organisms

Traditional diagnostic tests for disease-causing microorganisms have several important disadvantages. They usually require culturing on appropriate media. Culturing is slow, sometimes taking days. Cultures are often not specific to disease-causing strains; for example, some types of *E. coli* cause diarrhea, while others are normal inhabitants of the human gut. Additionally, some disease species cannot be cultured, or they occur at such low concentrations that culturing may not work. The HIV virus, which causes AIDS, often occurs at concentrations less than one viral particle per one milliliter of blood.

DNA probes solve all these problems. Culturing is not necessary, because target DNA from any organism, even in very low concentrations, can be amplified by PCR. Probes can be designed to be specific not only to any bacterial or viral species, but to particularly virulent strains. Identification is thus highly precise and takes only minutes or, at most, a few hours. They can even be used on tissues or whole organisms preserved for many decades (box 10.4).

Improved identification of microorganisms will also allow improved scientific study of the natural history of disease organisms. Daily we are assaulted by myriad microorganisms, many of them with the potential to cause disease. Until now, however, it was not possible to establish the frequency and distribution of these organisms in our surrounding environment. Such information can help us to understand how infectious organisms manage to circumvent our immune systems, which normally protect us from invasions by disease organisms.

Cancer

In chapter 11 we discuss genes that can lead to loss of normal regulation of cell replication or cancer. These genes, called **oncogenes,** can be activated to cause cancer by several mechanisms, including somatic mutation, chromosomal rearrangement, and viral activity. Probes specific to particular oncogenes can allow diagnosis of the onset and cause of particular cancers, especially leukemias. Additionally, probes can identify individuals carrying certain oncogenes, and therefore at risk for certain cancers, in families with histories of the disease. This can allow improved surveillance and early treatment if the cancer appears.

DNA Fingerprinting

Within a species, most sections of the genome show little or no variation between individuals. One region of the genome of humans and many other species, called **hypervariable minisatellite DNA,** is so variable that every individual is genetically unique. In 1984 British geneticist Alec Jeffreys and his colleagues discovered a cDNA probe specific to this region of the genome. RFLP analysis of this region is called **DNA fingerprinting** because, like traditional **forensic** (legal) fingerprinting, the pattern for every individual is unique (figure 10.1b).

The number of bands detected per individual using DNA fingerprinting or other RFLP analyses depends upon the number of endonucleases used. Three or four enzymes cut DNA into more pieces than does a single one because each endonuclease cleaves the DNA at different specific sites. The odds of two individuals (who are not identical siblings) having an identical set of bands depend upon the number of endonucleases used, but the odds are usually far less than one in one million, and often less than one in one billion. Because restriction sites are transmitted as Mendelian alleles, individuals share half their bands with each parent and half with each full sibling.

Parentage

Paternity or maternity testing with DNA fingerprinting is thus extremely accurate. Unlike testing with ABO or other blood groups, where possible parents can only be excluded but never confirmed, the odds of mistakenly identifying a parent of a child by DNA fingerprinting are negligible. The first official use of DNA fingerprinting was to confirm maternity in cases of children claiming British citizenship on the basis of maternal relationship.

The power of DNA fingerprinting is demonstrated by a German paternity case. A woman sued the alleged father for financial support for her three-year-old child. DNA fingerprints revealed that the man was not the father *and* that the woman was not the mother. This finding suggests that babies were switched or mistakenly labeled at birth, and the hospital has undertaken a search for all other children born at about the same time as the child in question.

Some people recommend that blood samples be taken from all children at birth and maintained in a permanent DNA database. Such samples could later be used for many identification purposes—lost children, paternity testing, verification of citizenship, etc.—but a DNA database raises the question of security: How can we be certain that any genetic information will not be misused? Decades later an employer or insurer, for example, could request information on genetic diseases. In fact, dried blood samples are already a part of the medical record of many people as a result of the Guthrie test for PKU, and dried blood can be used for DNA fingerprinting after many decades of storage.

Figure 10.15

(a) Restriction fragment length polymorphism (RFLP) patterns from three generations in one family. The genotypes are shown directly below pedigree symbols; M is a marker lane. Showing standard molecular weight markers. AN represents Allele Numbers, shown on far right of gel. Source: Nakamura, Y., *Science,* March 27, 1987, fig. 3, 235:1616–22. © 1987 by the AAAS.

Identification of Criminals

Criminals often leave tissue at the scene of a crime, especially in cases of murders and sexual offenses. Sufficient DNA can often be obtained from blood, semen, or even a single hair to produce DNA fingerprints that can later be compared with those of suspects. Vaginal swabs taken from rape victims usually have enough DNA from the rapist's sperm to produce a DNA fingerprint. Semen stains from clothing used as evidence in rape cases over a decade old have contained sufficient DNA for testing.

In many states and countries, courts have the authority to require blood samples for criminal and paternity investigations. By 1990, over one hundred criminal cases in the United States had used DNA fingerprints as evidence. Often, knowledge of the high certainty of DNA evidence is sufficient to lead a criminal to confess. Equally importantly, DNA evidence can unequivocally clear innocent suspects.

One of the first uses of DNA fingerprinting for major crimes involved the rape and murder of two girls in Britain in the late 1980s. DNA tests confirmed that the same man had committed both crimes, and also that the main suspect was innocent. With few other leads, police used traditional blood tests on over four thousand men living in the area of the crimes. All but four hundred could be excluded.

DNA fingerprints on these four hundred did not immediately identify the killer, but a man with the unlikely name of Colin Pitchfork was implicated because he boasted that he had paid another man to give blood for him. Pitchfork had not been a suspect, but DNA tests of his blood confirmed that he was guilty. The story of this first murder case solved by DNA fingerprinting is recreated in *The Blooding* by J. Wambaugh.

Technical difficulties have led to some problems in the forensic use of the DNA fingerprinting. In some cases, DNA evidence has not been allowed because scientists have yet to fully agree on proper controls. The technique is difficult, and improvements are occurring rapidly. Often, fingerprints reveal so many bands (e.g., fig. 10.1b) that 20% or more are likely to match between entirely unrelated individuals just due to chance. Full siblings might then be expected to show 70% or more matches, and even distant relatives might show 50% matches.

However, perfect matches are still expected only for tissues from the same person. In general, the principle of DNA fingerprinting appears to be well-established and likely to become a routine method for criminal identification.

A very different ethical question involves, as always, cost. At present, DNA fingerprinting is expensive, requiring well-equipped biochemistry laboratories and highly-trained specialists. Because the technique is equally effective in proving persons innocent as well as guilty, anyone suspected of major crimes should have equal access to DNA data to clear themselves, regardless of the cost. The law has yet to address this issue, but because of the accuracy of identification, the police may well consider DNA testing so efficient that it becomes mandatory in any serious case where tissue samples are available as evidence.

Agriculture and Conservation

Highly valuable livestock can be identified and their parentage confirmed using DNA testing. Such a system is being implemented worldwide, for example, for thoroughbred race horses. DNA tests can confirm maternity of calves and lambs in large flocks, where young are frequently separated from their mothers. High quality mothers sometimes acquire offspring of another female or lose their own.

Wild populations of animals are often highly inbred, particularly for some isolated populations. This means that DNA fingerprints can sometimes characterize populations, as well as individuals. Ivory tusks removed from elephant carcasses, for example, often have bits of flesh attached at the base that can be used for DNA testing. Fingerprints can then establish that the ivory was poached from northern African populations where elephants are in danger of extinction and cannot be taken legally, instead of from southern African populations where populations are still relatively large and some harvesting or hunting is legal.

DNA Sequencing

Scientific journals now publish page after page of DNA sequences from a wide array of species. The U.S. National Institutes of Health directs an international computer database that stores and disseminates for scientific research all known DNA sequences. The quantity of data is increasing so rapidly that NIH is struggling to keep pace with the flow of new sequences.

As noted earlier in this chapter, establishing DNA sequences for key genes is likely to be a first step in understanding the molecular processes that produce diseases. Interpreting DNA sequences and the biochemical pathways they govern will occupy geneticists for many decades. Nonetheless, sequence information is already changing our understanding of medically important genes and can be expected to lead to improved medical practice (box 10.5).

Techniques for identifying base sequences were established only in 1973. By 1977, the entire sequences for several small viruses were known, each about 5,000 base pairs long. By 1990, the largest organism whose entire sequence was known was the cytomegalovirus, with about 250,000 base pairs. Over 37 million base pairs from all species have been described. The cost of obtaining sequences is estimated to be about $3–5 per base pair.

Two important species are the next targets for obtaining complete sequences. One of these, *E. coli,* has had about 800,000 base pairs identified, of a total of almost 5 million. The complete sequence is likely to be known by the end of the century. The other species is *Homo sapiens.* The human genome is about 3 billion base pairs long, or about 1,000 times that of *E. coli.* So far the longest human gene whose sequence is fully known is only about 70,000 base pairs. But the number should increase rapidly.

The Human Genome Project

Proposals to sequence the entire human genome began to be discussed among scientists in 1985. The project was technically feasible, but major doubts were voiced because of the enormous costs, technical difficulties, and even the value of the information. Coordinating the efforts of the many laboratories that would be required would be a complex and enormous job. Even if the project succeeded, it was unclear to many if the information would be useful

BOX 10.5

New understanding of a textbook gene: the ABO locus

For many decades, the ABO blood group has been used (as in this book) to illustrate Mendelian patterns of inheritance in humans. Millions of biology students have pricked a finger with a needle for a drop of blood for agglutination testing (figure 4.11). Added interest in this locus derived from its use in paternity testing.

Since its interpretation as a Mendelian locus in the 1920s, however, the ABO locus has remained a genetic "black box." Knowledge of the functions of the ABO antigens or their precise molecular structures has been extremely limited. Since the 1960s, the ABO locus has been thought to produce an enzyme called *glycosyltransferase* that catalyzes the last step in production of the A and B antigens. It was uncertain why O individuals produce no red blood cell antigen equivalent to the A and B antigens.

DNA sequencing has now dramatically increased understanding of the ABO locus. In 1990, about 1,000 base pairs encoding the locus were sequenced. These sequences confirmed that glycosyltransferase was indeed the product of the ABO locus. They also revealed that small genetic differences can have significant phenotypic effects. Alleles A and B differ in only four out of nearly four hundred amino acids. Allelic subgroups A1 and A2 differ at only a few more sites. Allele O differs from A and B only at a single site: the base guanine at site 258 is deleted, causing a frameshift mutation (chapter 8) and a resulting protein that is not a functional glycosyltransferase. That is, the product of the O allele is unable to catalyze a reaction to produce any red blood cell surface antigen.

Knowledge of the molecular basis for agglutination reactions will probably not affect blood transfusions, but may ultimately prove important to understanding certain diseases related to ABO type. Frequencies of some diseases have long been known to significantly vary among people of differing blood types. In particular, a group of cancers known as adenocarcinomas are relatively rare in type O individuals and relatively frequent in type A. Does the absence of an ABO antigen somehow confer protection against cancer? Does the A antigen facilitate cancer? Molecular answers to such questions could help to cure or prevent cancer in the future.

or even interpretable. Much of the genome appears to consist of non-coding sequences that may have no function. Better, some argued, to target the money to those genes that are known, or could be identified, to be of medical importance.

Surprisingly quickly, however, proponents won their case, although significant opposition remains. In 1988, the National Institutes of Health established an Office of Human Genome Research with James Watson, one of the discoverers of the double helix, as director (figure 10.16). In 1989, the office was upgraded to the National Center for Human Genome Research (NCHGR). Working jointly with the Department of Energy, the NCHGR prepared the National Genome Plan. The resulting Human Genome Project officially began on October 1, 1990.

The Human Genome Project is ambitious, and many think its success will depend to a large degree upon the scientific and administrative skills and prestige of Watson. The enthusiasm of the project's supporters is obvious in their language: Deciphering the human genome has been called the Holy Grail of genetics; others call it the Rosetta Stone of biology; Watson himself says it "will provide the ultimate answers to the chemical underpinnings of human existence."

The first major goal of the Human Genome project is to produce physical and linkage maps of the human genome by 1995. A physical map would identify the chromosomal location of a large number of probes scattered at approximately equal intervals throughout the genome. One proposal calls for about 30,000 of these probes—landmarks on the DNA—

(a)

(b)

Figure 10.16

James Watson, co-discoverer of the structure of DNA and, more recently, science administrator and director of the U.S. Human Genome Project and the international Human Genome Organization (HUGO).
(a) In 1962; (b) in 1989.

(a) Bettmann Archives;
(b) Margot Bennett, "Plan for Genome Centers Sparks a Controversy," *Science,* 246:204–05, 1989.

spaced at intervals of about 100,000 base pairs. Genes of particular interest could then be pinpointed to a small section of a chromosome (figure 10.17). A linkage map relates loci in a very different way by their frequency of recombination.

The second major goal of the Human Genome Project is to sequence the entire genome within fifteen years, by September 30, 2005. The involvement of the U.S. Department of Energy is explained by their experience in managing large science projects and databases as will be required to achieve this goal. The Department of Energy also needs the information to assist with monitoring the genetic effects of radiation.

Another major goal is the study of the social and ethical consequences of such detailed knowledge of the human genome. Watson has promised that 3% of the entire budget of the Genome Project will be devoted to ethical and social studies.

And what will the budget be? The best estimates are about $3 billion over the fifteen year lifespan of the project. Critics have argued that such an enormous amount of money could be spent more usefully to aid victims of genetic diseases in many other ways. Supporters of the project note that $3 billion is in fact small by comparison with another comparable "big science" project, the U.S. moon landing program of the 1960s and 1970s. Even in medical terms, $3 billion in fifteen years is relatively modest: The Cystic Fibrosis Foundation spent $120 million alone in a five-year search for the single gene causing CF.

Most scientists think that the rewards of genetic knowledge and medical improvements likely to arise from the Human Genome Project will probably justify the expense many times over. But there is no guarantee. The project is so technically complex that it could prove unmanageable. Once the sequence is fully known, it may not be satisfactorily or usefully interpreted for decades or centuries. James Watson confronts an enormous scientific and leadership challenge.

HUGO

Sequencing the human genome must be an international project for at least two reasons. First, the magnitude of the task requires that a large number of laboratories work on specific components of the project, and many countries around the world have well-trained scientists eager to participate. Furthermore, while the United States appears prepared to provide the bulk of the funding, contributions from other countries will probably play a significant role.

Second, and more importantly, the human sequence is knowledge that rightfully belongs to all humanity. The uses of this information will be virtually limitless, and all countries deserve equal and fair access to it. A cloud of suspicion would accompany the attempt of any country to claim such knowledge as its sole property.

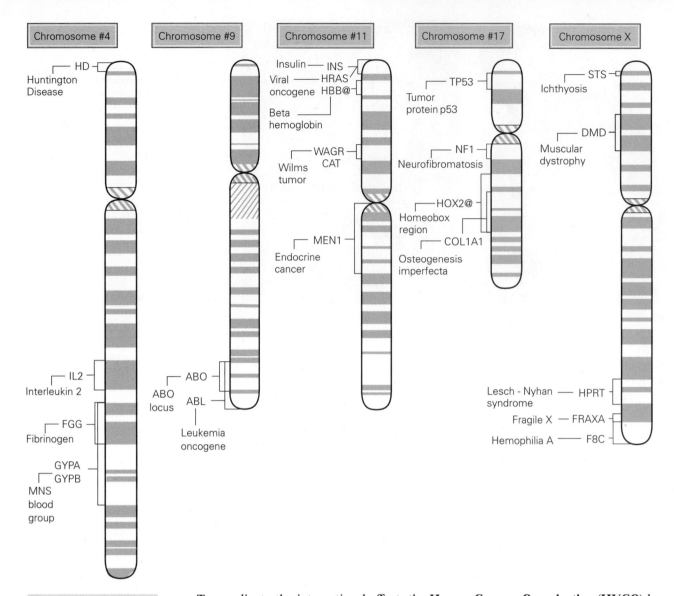

Figure 10.17

Chromosome location of some human genes. The Human Genome Project aims to produce physical maps that permit determination of the precise chromosomal locations for the four thousand known disease-causing genes.

From J. Claiborne Stephens, et al., The Human Genome Map 1990 in "Mapping the Human Genome," *Science,* Vol. 250, p. 237, October 12, 1990. Copyright © 1990 by the American Association for the Advancement of Science, Washington, DC. Reprinted by permission.

To coordinate the international effort, the **Human Genome Organization (HUGO)** has been formed, with—who else?—James Watson as director. The national genome projects of the United States, the United Kingdom, the U.S.S.R., Italy, and Japan plan to use HUGO as an international "umbrella" organization, and other countries are likely to join, as well.

Already, potential difficulties of cooperation have arisen. When Japan was slow to join the international effort, Watson warned that countries not contributing fairly could not expect to have the results of the project. Despite the importance of making sequence information universally available, he argued, the genome project should not be a form of subsidy to, for example, the Japanese pharmaceutical industry. Japan has since increased its financial commitment.

Germany may not participate because of its deep suspicion of human genetic research resulting from the history of eugenics in that country under the Nazi's.

Gene Therapy

The ultimate treatment for disease-causing genes is at the level of the gene itself. Defective genes such as the sickle-cell allele might be somehow removed or corrected or inactivated. More promisingly, present strategies aim at inserting one or a few genes into target tissues to produce a gene product that can lead to normal metabolism.

Such **gene therapy** seems most feasible for diseases caused by a single mutation. First candidates for gene therapy include sickle-cell anemia, thalassemia, hemophilia, and phenylketonuria, diseases whose abnormal gene products are well-known. Other conditions, such as Huntington's disease, cystic fibrosis, and muscular dystrophy, may also be subject to gene therapy when their abnormal gene products become better known.

Gene therapy treating somatic cells is generally regarded as a logical extension of traditional medical therapies. Diabetes, for example, has long been treated by insulin administration to supplement the small amounts produced by the patient. Inserting the insulin gene into pancreatic cells would allow additional insulin to be produced by the diabetic's own metabolic processes. Because only somatic cells are affected, the inserted gene would not be passed to children of the diabetic who also would carry the gene for diabetes.

Most people accept that somatic cell gene therapy poses no special ethical problems, so long as the germ cells are not affected. Opponents of somatic cell therapy argue that we can never be certain that this will always be the case.

Gene therapy poses two technical problems: (1) insertion of the desired gene into the genome of at least some of the host's cells, and (2) regulation of the transferred gene. Genes appear to be surprisingly easy to move; viruses can be used as vectors to insert a single gene into a target tissue such as bone marrow, and fragmented chromosomes can even be physically inserted intact into a host cell. Once inserted, however, turning the transferred gene on and off is tricky. Strategies for ensuring a gene is turned on include inserting the gene at a site where it will always be active or transferring the target gene and a second controlling gene as a unit. Feasibility of gene therapy has been shown many times in a variety of experimental animals. Rabbit and human hemoglobin genes, for example, have been transferred to and expressed in mice.

More dramatically, genes encoding rat growth hormone have been transferred to mice, producing giant mice (figure 10.18). An obvious practical use of this technique would be to produce larger livestock. Transfer of a gene for high production of bovine growth hormone might offer an alternative to the highly controversial administration of recombinant BGH, although it might also produce its own controversies. Genes for human blood clotting factors have been successfully introduced into sheep linked to a gene for β-lactoglobulin, a protein produced in milk and its regulatory sequences. The sheep has then produced human blood clotting factor in its milk, effectively becoming a factory for this medically important chemical.

The first known trials of gene therapy took place in the early 1980s on two victims of β-thalassemia, a serious type of anemia caused by defective β-hemoglobin. Bone marrow cells that produce red blood cells were removed from both victims and cultured with purified quantities of the normal β-globin gene; it was hoped that the growing cells would incorporate the normal genes. The marrow cells were then reimplanted into the bones of the victims, but neither showed evidence of accepting or expressing the normal genes. Unfortunately, the experimental treatment was conducted secretly, in violation of the ethical guidelines of the National Institutes of Health, who had sponsored the research. The physician's funds were withdrawn and his future research restricted.

Physicians W. French Anderson, R. Michael Blaese, and Steven A. Rosenberg began the first authorized gene transfer experiment in humans in 1989. A marker gene was inserted into a cancer patient's naturally occurring cancer-fighting cells called tumor-infiltrating lymphocytes (TIL), so that the movement of TIL cells could be followed through the body. The transferred gene was intended to have no health effect, so this was *not* a type of gene therapy, but rather an experiment designed to learn more about the body's natural anticancer mechanisms and existing therapies. Nonetheless, the experiment used the type of transgenic technologies required for gene therapy.

Unlike the earlier β-thalassemia trial, this experiment had the full approval of NIH and its Recombinant Advisory Committee. The TIL experiment was vigorously opposed by Jeremy Rifkin and colleagues, and NIH conducted fifteen separate rounds of scrutiny of the proposed experiment before approving it.

Figure 10.18

Experimental gene transfer.
(a) The mouse on the right is
normal. The mouse on the
left carries a transferred rat
growth hormone gene. The
genetically engineered mouse
had growth hormone levels
800 times greater than its
companion and was nearly
twice its weight. (b) The carp
on top has a transferred gene
for trout growth hormone
that caused it to grow larger
than normal carp (below).
(c) This transgenic mouse
carries a sheep hair gene
(intermediate filament
keratin protein). The foreign
gene is expressed in hair cells,
but its regulation is aberrant
leading to premature loss of
hair in waves along the
mouse's body.

(a) R. L. Brinster, R. E.
Hammer, and R. M. Evans in
Nature, 300: 611–615, 1982.
Photo courtesy of Ralph
Brinster; (b) Auburn
University, *Science,* 247:
1298, 1989; (c) George
Rogers, *Nature,* 9: 1485–
1493, Rogers and Powell,
1990.

(a)

(b)

(c)

The first NIH-approved case of gene therapy began in September, 1990. Physicians Anderson, Blaese, and R. Michael Culver inserted a gene for adenosine deaminase (ADA), an enzyme required for a functioning immune system, into T cells (see chapter 12) removed from a four-year-old girl suffering from a disease called ADA deficiency (figure 10.19). The girl lacked the normal ADA gene and thus was susceptible to many types of opportunistic infections. About one billion T cells cloned from her original cells with the newly inserted gene were then administered intravenously to the young girl. Several years will be required to confirm that the transgenic cells have been accepted by the girl and that the ADA enzyme is being expressed, thus curing her condition.

The experimental ADA therapy can be viewed as a means for inserting a complex drug delivery system into a person lacking an essential natural chemical, in this case ADA. The present alternative is administration of synthetic ADA at a cost of about $60,000 per year. In principle, any necessary drug could be produced in the recipient's body by gene therapy.

One promising method has already achieved initial success in experimental animals. Using recombinant technologies, small **organoids** have been produced at selected sites in the body as drug production and delivery structures. Cells containing transferred genes for (1) constructing blood vessels and (2) secreting specific proteins are attached to synthetic fibers and then inserted into the body (figure 10.20). A gene called *heparin-binding growth factor (HBGF)* directs formation of the small organoid and blood vessels that attach to the circulatory system. Then, drugs produced by the second transferred gene are secreted directly into the body's system.

T cells from
ADA-negative
patient

(a)

Mouse leukemia
retrovirus +
human ADA gene

(b)

Figure 10.19

Gene therapy for ADA
deficiency. (a) Lymphocytes
(T cells) from a patient
suffering from ADA
deficiency are removed and
incubated with (b) mouse
leukemia retroviral DNA in
which a normal human ADA
gene has been inserted.
(c) The T cells are cloned in
the laboratory until there are
roughly a billion transformed
cells, each containing the
normal human ADA gene.
(d) The patient's transformed
T cells are transfused back
and allowed to "recolonize"
her immune system,
hopefully effecting a cure.

(a and c) © David M.
Phillips/Visuals Unlimited.

(c)

T cells carrying
human ADA gene

Clone to ~ 10^9 cells

(d)

Transfer to patient

(e)

Figure 10.20

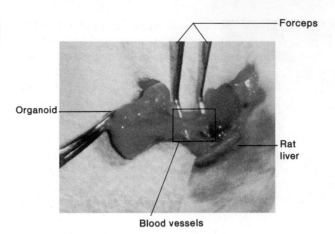

An organoid made from Gore-Tex fibers coated with collagen. The organoid is in a rat connected to the rat liver by blood vessels growing between the organoid and the liver.

John A. Thompson, "Gore Tex Organoids and Genetic Drugs," *Science,* 246:747–49, 1989.

Ethical Concerns Related to Recombinant Technologies

Scientists, philosophers, theologians, and the average citizen have expressed ethical reservations at every step in the development of DNA research and technologies. In many cases, a consensus has been reached that advances in DNA technologies pose no undue practical or philosophical threats to societal values. Significantly, the scientific community itself led in imposing controls and safeguards on recombinant research in its early stages.

A few people, including activist Jeremy Rifkin, have opposed virtually every aspect of recombinant technologies and even fundamental research. Some do so for religious or philosophical reasons, believing that it is arrogant and immoral for humans to interfere with the role of a divine creator by tinkering with the essence of life. Others fear that the technologies are inherently so unpredictable or unsafe that scientists are likely to create unforeseeable and even unimaginable environmental catastrophes. Some things, they argue, remain better unknown.

Most people probably acknowledge that any technology can be misused, but that at least some types of DNA research have the potential to significantly improve human life and that many DNA technologies differ little from conventional technologies in their levels of risk or interference with nature. Somatic cell gene therapy, for example, is generally accepted as similar in kind to traditional medical therapies. Modification of germ cells or cloning of entire human beings, on the other hand, is viewed by most as unethical and entirely unacceptable for both philosophical and practical reasons.

For most people, then, the key issue is the use of genetic knowledge. New knowledge that allows us to understand ourselves and our world and that can be used humanely to improve the human condition raises few ethical problems. Knowledge and technologies that can be used to harm people, or that are used to assign worth, are potentially objectionable.

At least three primary areas of concern exist over the misuse of DNA (and other genetic) knowledge: environmental damage, assigning relative worth to people, and money. The potential for DNA technologies to cause massive **environmental damage** is unquestioned. In the extreme case, recombinant technologies can be used to create biological weapons that could kill millions. Unintentional but nevertheless major alterations to natural ecosystems could occur through release of genetically modified organisms. Just as some physicians conducted gene therapy trials without NIH approval, so also did a well-meaning American scientist release, without the required approval of the Environmental Protection Agency (EPA), genetically modified bacteria intended to treat elm trees infected with Dutch elm disease. In both cases, the scientists took action because of frustration with the extensive ethical and safety review processes. No serious health or environmental consequences occurred in either case, but the cases demonstrate the willingness of individuals to circumvent regulatory procedures when they feel circumstances require it. And if well-intentioned people violate safety restrictions, so too will people of lower scruples.

American regulatory procedures for both gene therapy and environmental release of genetically modified organisms have also been evaded by conducting experimental tests outside the United States where restrictions are fewer or nonexistent. Genetically engineered animal vaccines, for example, were tested in New Zealand and Argentina by American institutions because of the long American review procedures and because legal liabilities are lower in those countries. On the other hand, EPA-approved field tests of genetically modified bacteria in California were vandalized by opponents of any environmental release. Feelings and motivations run high on both sides of the issue. Unreasonable regulations are certain to be broken, and opponents of DNA technologies may sabotage implementation of approved uses.

Security of DNA information is a major concern of people who fear that DNA technologies can be used to **value** or **assign relative worth** to people. As DNA probes for an increasing number of genetic diseases become available, employers and insurers may require DNA tests. Employment and insurance might then be denied to, for example, people with the sickle-cell allele. It is even imaginable that individuals of certain blood or tissue types might be discriminated against on the basis of predispositions to particular diseases.

In fact, discrimination on the basis of genotype is already accepted in extreme cases, such as aborting fetuses shown to have a serious genetic disease. For particular diseases, acceptability of abortion might change if a cure or treatment is found. PKU demonstrates a serious disease where a relatively simple treatment renders abortion of affected fetuses unacceptable.

The third primary concern related to DNA technologies, **money,** affects many likely developments. Who, for example, should have access to DNA fingerprinting in criminal trials? Only those who can afford it, or any who need it, and at government expense? Should gene therapy be available only to those who can afford it? Should pharmaceutical companies be allowed to use bioengineering technologies to produce chemicals such as growth hormone whose primary market may be for cosmetic rather than health purposes? Should pharmaceutical companies be subsidized to produce therapeutic chemicals for which only a tiny market exists, such as for rare genetic conditions? Should corporations be allowed to patent recombinant organisms they have designed and created? What rights do the public have to DNA technologies developed in universities using public funds and then taken over by private companies? And, what financial rewards are appropriate to the scientists themselves who develop new technologies?

Such questions often have no easy or generally acceptable answers. For certain, recombinant DNA technologies are here to stay. Their potential for good and evil is as limitless as the human mind. Quite likely, our lives will be improved in countless ways by new genetic technologies. We can only hope that reason prevails in their use and that risks remain minimal.

In the best of all worlds, DNA technologies will continue the long genetic tradition of teaching us about ourselves: Who are we? What does it mean to be a human? The more we understand ourselves and the meaning of *humanness,* perhaps the more likely we are to acquire the wisdom to act humanely and prudently in all aspects of life.

Summary

1. Genetic engineering, gene splicing, and recombinant DNA technologies all refer to modern ways of manipulating DNA.
2. Biotechnology is the industrial use of biological systems and processes.
3. Molecular cloning uses recombinant DNA techniques to amplify specific genes or gene sequences.
4. Either plasmids (for small pieces of DNA) or phages (for larger DNA) can be used as cloning vectors.
5. Restriction endonucleases are bacterial enzymes that cut DNA at specific sequences. They are used in gene cloning and also to produce characteristic restriction fragment length polymorphisms (RFLPs) in DNA. These RFLPs are used as markers for a variety of medical and forensic purposes.
6. DNA ligase is an enzyme used to produce recombinant plasmids; it rejoins broken segments of DNA.
7. DNA libraries are cultures of bacteria, with plasmids or phages, that carry genes or gene fragments from recombinant DNA studies.

8. cDNA probes are DNA sequences complementary to the sense strand of the gene to be studied. They are often produced using the enzyme reverse transcriptase and mRNA for the gene in question.

9. Segments of DNA isolated by a probe can be analyzed for restriction fragment length polymorphisms, or by chromosomal *in situ* hybridization, or by sequencing.

10. The polymerase chain reaction allows single fragments of DNA to be amplified many thousands or millions of times for analysis by standard techniques.

11. Beginning with the Asilomar Conference in 1975, voluntary and government guidelines have restricted the conditions under which certain types of DNA research have been conducted.

12. Environmental release of genetically modified organisms could cause serious ecological damage and is subject to close governmental regulation.

13. Cloning of many cells and genes can be done routinely; cloning of entire organisms is more problematic. Although some plants and frogs can be cloned, mammals cannot.

14. Genetic engineering methods have been used successfully to produce proteins and hormones of great medical importance, such as insulin and growth hormone.

15. DNA technologies can now allow direct testing for genetic diseases of humans, including prenatal diagnosis. Identification of a disease-causing gene is the first step to a possible cure. DNA testing can also improve identification of disease organisms.

16. Hypervariable minisatellite DNA is so variable that RFLP analysis can identify individual variations in a procedure called DNA fingerprinting. This technique has many forensic and legal applications, including testing parentage and identifying criminals.

17. The U.S. Human Genome Project aims to produce maps of the human genome by 1995 and a complete human sequence by 2005. The project is controversial because of the high cost, the potential misuse of the information, and the possibility that the information may not have immediate benefits to medical science. The Human Genome Organization (HUGO) is the international group coordinating national genome projects.

18. Gene therapy aims to cure genetic diseases by inserting normally functioning genes into the genome. Transferred genes can be viewed as delivery systems for therapeutic drugs.

19. The three primary causes for concern over misuse of DNA technologies are (1) the possibility of environmental damage, (2) the possible use of DNA information for assigning relative worth to people, and (3) the possibility of financial considerations distorting the fair use of these technologies.

20. Some people object to any research or development of DNA technologies on philosophical and practical grounds. Others believe that DNA technologies offer many significant benefits to human life and that the risks can be controlled and limited.

Key Words and Phrases

biological weapons

biotechnology

bovine growth hormone

cDNA

chromosomal *in situ* hybridization

diabetes

DNA fingerprinting

DNA library

DNA ligase

DNA sequencing

environmental release of genetically modified organisms

forensic

gel electrophoresis

gene splicing

gene therapy

genetic engineering

genetic pharmacology

Human Genome Organization

Human Genome Project

human growth hormone

human insulin

hypervariable minisatellite DNA

insulin

interferon

molecular cloning

nuclear transplantation

oncogenes

organoids

phage

plasmid

polymerase chain reaction

probe

recombinant DNA technologies

recombinant plasmids

restriction endonucleases or restriction enzymes

restriction fragment length polymorphism (RFLP)

satellite DNAs

sex pheromones

vector or cloning vehicle

Questions

1. Discuss the genetic potential of the cloning of single genes, cells, organs, and entire organisms. Distinguish between plants and animals.
2. Do you believe it is important to sequence the entire human genome? Why?
3. Should there be government regulation of research using recombinant DNA methodologies?
4. Should prenatal diagnosis be available to all pregnant woman? Or, should it only be available to those at risk for known genetic disease?
5. Why would geneticists use plasmids rather than phages for molecular cloning?
6. How can RFLPs be used in prenatal diagnosis?
7. How can RFLPs be used in paternity testing? How does this genetic information compare to data from blood groups?
8. Electrophoresis separates proteins (as well as DNA), and the variation is often given a Mendelian genetic interpretation. Why is this justifiable?
9. Consider the German parentage legal case where DNA tests revealed that a three-year-old child was probably given to the wrong mother, who is now suing a possible father for financial support. What are the legal *and* ethical issues involved? Assuming that DNA tests confirm that children were switched in the hospital: Should the children be returned to their genetic parents? Even after they have bonded with their "incorrect" parents and might suffer psychological damage if switched? Whose rights have precedence if one couple wants to switch children and the other does not?

For More Information

Arnheim, N., T. White, and W. E. Rainey. 1990. "Application of PCR: Organismal and Population Biology." *Bioscience*, 40:174–82.

Bains, W. 1989. "Disease, DNA, and Diagnosis." *New Scientist*, 6 May 1989.

Behringer, R. R. et al. 1989. "Synthesis of Functional Human Hemoglobin in Transgenic Mice." *Science*, 245:971–73.

Cantor, C. R. 1990. "Orchestrating the Human Genome Project." *Science*, 248:49–51.

Caskey, C. T. 1987. "Disease Diagnosis by Recombinant DNA Methods." *Science*, 236:1223–29.

Connor, S. 1988. "Genetic Fingers in the Forensic Pie." *New Scientist*, 28 January 1988.

Cooper, D. N., and J. Schmidke. 1986. "Diagnosis of Genetic Disease Using Recombinant DNA." *Human Genetics*, 73:1–11.

Culliton, B. J. 1989. "Gore Tex Organoids and Genetic Drugs." *Science*, 246:747–49.

Dodd, B. E. 1985. "DNA Fingerprinting in Matters of Family and Crime." *Nature*, 318:506–7.

Donis-Keller, H. et al. 1987. "A Genetic Linkage Map of the Human Genome." *Cell*, 51:319–37.

Drlica, K. 1984. *Understanding DNA and Gene Cloning: A Guide for the Curious.* John Wiley & Sons, New York.

Golenberg, E. M., et al. 1990. "Chloroplast DNA Sequence from a Miocene *Magnolia* Species." *Nature*, 344:656–58.

Gould, S. J. 1988. "Integrity and Mr. Rifkin." *An Urchin in the Storm.* W. W. Norton & Co.: New York. (This is a scathing review of the book *Algeny*, by Jeremy Rifkin.)

Hall, S. S. 1990. "James Watson and the Search for Biology's 'Holy Grail'." *Smithsonian*, 20: 41–49.

Higuchi, R. et al. 1988. "DNA Typing from Single Hairs." *Nature*, 332:543–46.

Juskevich, J. C., and C. G. Guyer. 1990. "Bovine Growth Hormone: Human Food Safety Evaluation." *Science*, 249:875–84.

Kolata, G. 1986. "Genetic Screening Raises Questions for Employers and Insurers." *Science*, 232:317–19.

Landegren, U., R. Kaiser, C. T. Caskey, and L. Hood. 1988. "DNA Diagnostics—Molecular Techniques and Automation." *Science,* 242:229–37.

Lewin, R. 1989. "DNA Typing on the Witness Stand." *Science,* 244:1033–35.

Lewis, R. 1987. "Genetic-Marker Testing: Are We Ready for It?" *Issues in Science and Technology,* 4:76–82.

Lichter, P., et al. 1990. "High-Resolution Mapping of Human Chromosome 11 by *in situ* Hybridization with Cosmid Clones." *Science,* 247:64–69.

Marx, J. L. 1982. "Building Bigger Mice through Gene Transfer." *Science,* 218:1298.

Miller, J. A. 1987. "Mammals Need Moms and Dads." *Bioscience,* 37:379–82.

Murray, M. 1987. "Battling Illness with Body Proteins." *Science News,* 131:42–45, (recombinant DNA technology).

Pershing, D. H., et al. 1990. "Detection of *Borrelia burgdorferi* DNA in Museum Specimens of *Ixodes damini* Ticks. *Science,* 249:1420–23.

Pursel, V. G., et al. 1989. "Genetic Engineering of Livestock." *Science,* 244: 1281–88.

Roberts, L. 1987. "Flap Arises Over Genetic Map." *Science,* 238:750–52.

Suzuki, D., and P. Knudtson, 1989. *Genethics: The Clash between the New Genetics and Human Values.* Harvard University Press, Cambridge, MA.

Wambaugh, J. 1989. *THE BLOODING,* Bantam Books, New York.

Watson, J. D. 1990. "The Human Genome Project: Past, Present, and Future." *Science,* 248:44–49.

Wickelgren, I. 1989. "Please Pass the Genes." *Science News,* 136:120–36.

Yamamoto, F., H. Clausen, T. White, J. Marken, and S. Hakomori. 1990. "Molecular Genetic Basis of the Histo-Blood Group ABO System." *Nature,* 345:229–33.

Chapter

Misinformation

Figure 11.1

Aneuploid phenotypes in humans: (a) Down's syndrome, trisomy-21; (b) Patau's syndrome, trisomy-13; (c) Edwards' syndrome, trisomy-18. The children in (b) and (c) are so badly affected; they will live only a few months.

(a) M. Coleman/Visuals Unlimited; (b) Kutay Taysi; (c) James W. Hanson, University of Iowa.

(a)

(b)

(c)

The hereditary mechanisms discussed in the preceding chapters are precise, but they are not perfect. During the S phase of the cell cycle in humans, for example, the entire diploid complement—about 3 billion base pairs—of each of the 10 trillion or more cells in the body is copied (replicated) from preexisting DNA in another cell. Mistakes in copying, which occur at a very low rate during replication, change the base sequence within a DNA molecule.

During meiosis, a second opportunity exists for alterations of DNA. The formation of chiasmata during prophase I of meiosis permits the equal exchange of chromosomal segments between homologues. However, nonequivalent types of chromosomal changes also occur: segments may be totally lost or moved to non-homologous chromosomes. Division of chromosomes between daughter cells is sometimes not equal, and sometimes the parent cell does not divide at all, ending up with two diploid sets rather than one. Similar mistakes also occur during mitosis. The result is that the genome of any cell is likely to differ in some way from that of its parent cell.

Heritable changes in the DNA and chromosomes are termed **mutations.** Mutation is derived from the Latin word *mutare* meaning to change. A genetic mutation may affect only one nucleotide base, or it may involve long segments of a chromosome. In extreme cases, it may affect one whole chromosome or even the entire diploid complement. Mutations occur at any time during the life of a cell, not just during meiosis or mitosis.

The results of mutation differ greatly depending upon whether they affect germinal or somatic tissues. Germinal mutations affect the gonads or gametogenic tissues. They cause the genetic complement of the gamete to differ from that of the parent, and germinal mutations are passed on to succeeding generations. Somatic mutations occur after fertilization in nonreproductive cells of an organism. The mutation is transmitted via mitosis to all descendants of that cell but, because the gonads are not affected, not to succeeding generations.

The effects of somatic mutations depend on when in development they occur and which genes are affected. Early mutations are likely to affect larger numbers of cells than late mutations. Mutations of some genes have minor phenotypic effects or none at all, but some mutations are lethal. Mutation of oncogenes, for example, can induce cancers.

Most mutations are harmful. Even the substitution or deletion of a single base may be lethal. However, many mutations appear to be recessive, so their potentially harmful consequences may be masked by a dominant homologue, thus having little effect. On the other hand, mutations also provide the ultimate source of genetic variation for continued evolution of life on earth. All living species result from the genetic modification of preexisting species. The millions of living species represent an accumulation of mutational events whose effects on survival and reproduction were positive.

Chromosome Mutation: Numerical Changes

The most spectacular mutations involve the addition of entire haploid sets, **polyploidy,** or the addition or loss of entire chromosomes, **aneuploidy.**

Polyploidy

Variations in the number of haploid sets are common in many species. Plants such as mosses spend most of their life cycle in a haploid phase, and for all sexually reproducing organisms, the gametic phase of the life cycle, however brief, is haploid. In bees and wasps, differences in ploidy determine sex: females are diploid, males are haploid. Some fish, salamander, and lizard species consist entirely of triploid females; no males at all are found in these species. Furthermore, tissues within an individual sometimes vary in ploidy. Endosperm of plant seeds is usually triploid, and human liver cells exhibit many degrees of ploidy—triploid, tetraploid, and even octoploid.

In plants, polyploidy is relatively common. Within a genus, many species often have a diploid number that is an exact multiple, three or greater, of the smallest haploid number for that genus (table 11.1). Furthermore, many common weeds such as dandelion, crabgrass, and wild oats are polyploid.

Polyploidy in plants originates in at least two ways. In some cases, polyploid cells arise from a failure of the normal mitotic apparatus. A cell duplicates its chromosomes and then completes prophase and metaphase, but the potential daughter cells fail to separate. The parent cell reenters interphase with two entire diploid sets: in effect, it has become tetraploid. Subsequent mitotic divisions produce a clone of tetraploid cells within the plant.

Polyploids also originate from *unreduced gametes.* If bivalents do not separate during anaphase I of meiosis, any gametes subsequently formed are diploid or unreduced. Union of a diploid gamete with a normal haploid gamete produces a triploid zygote; union of two diploid gametes produces a tetraploid zygote.

Commonly, polyploid cells and individuals are larger and more vigorous than normal. Tetraploid portions of a plant are often clearly distinguishable from the diploid portions. Many domestic fruits and grains originated as polyploids. They are valued highly because they often produce fruit (including grain) substantially larger than that of their diploid progenitors. In fact, the importance of polyploidy to modern society can hardly be overstated (box 11.1). Food production worldwide has more than kept pace with the exponentially increasing human population, primarily because of the development of new strains of basic grains—rice, wheat, and corn—many of which are polyploid.

Sexual reproduction is often problematical for polyploids because of difficulties in *pairing* of homologues. In a tetraploid, for example, all four homologues may synapse during meiosis, or three or two may synapse. Sometimes, balanced gametes are produced, each containing a complete diploid complement, but unequal results are also common. With triploids, pentaploids, and other uneven degrees of ploidy, gamete formation is even more likely to be unbalanced. Triploids are usually not fertile and must be propagated entirely by asexual means. The domestic banana is a triploid derived from a wild diploid species and surviving entirely by human propagation. The small black spots in the center of a domestic banana are remnants of the seeds that did not develop. An even more extraordinary example, the boysenberry, is heptaploid (7N).

BOX 11.1

Wheat—A polyploid that rises to the occasion

Without polyploidy, most humans would have a very different diet. Among the foods we would miss might well be bread, made from one of several species of wheat (*Triticum*). Wheat is the single most important grain crop; over 20% of calories consumed by humans come from wheat. Varieties of wheat are grown throughout the world, except in the tropics.

Triticum comprises a genus that shows a polyploid series (figure 11.2). The diploid species, *T. monococcus* or einkorn wheat, 2N = 14, was the first to be used for food about 10–12,000 years ago, about the time civilization began. Like most domesticated crops, seeds from wild plants were probably collected for food and subsequently brought in cultivation. Einkorn wheat is hulled and small compared to its polyploid relatives.

Emmer and durum wheat belong to the tetraploid species *T. turgidum,* 2N = 28. These wheats originate from a naturally occurring wild tetraploid variety, whose grains are larger than einkorn wheat. Emmer wheat is hulled, but durum wheat is particularly desirable because it is free-threshing, meaning that the hull is easily removed by threshing and, unlike hulled wheats, does not need to be ground for human consumption.

Timopheev's wheat, *T. timopheevi,* is another ancient hulled wheat that is now grown in very small amounts only in Georgia and in the Soviet Union. It is tetraploid (2N = 28), but has a genetic origin different from emmer wheats.

Bread wheats, *T. aestivum,* are the most agriculturally important varieties. Bread wheats are hexaploids, 2N = 42, derived from hybridization of tetraploid emmer wheat with a diploid relative, the wild grass *Aegilops squarrosa* (2N = 14). Bread wheats are especially desirable because they are free-threshing and have large grains, and particularly because they have a high content of proteins called glutens. Glutens are sticky and elastic, allowing wheat breads to rise without collapsing when leavened (mixed with yeasts, which produce carbon dioxide gas).

Wheat has long been the favored grain of most non-tropical cultures because of its superiority over other cereal crops in making bread. Bread wheat accounts for about 90% of wheat production in the world today. Durum wheat, the other important agricultural variety, grows well in warm Mediterranean climates and is used in macaroni and other pastas. Einkorn, Timopheev's, and emmer wheats make very poor bread. Their production is extremely limited, and they are used primarily for porridge, beer, and livestock feed.

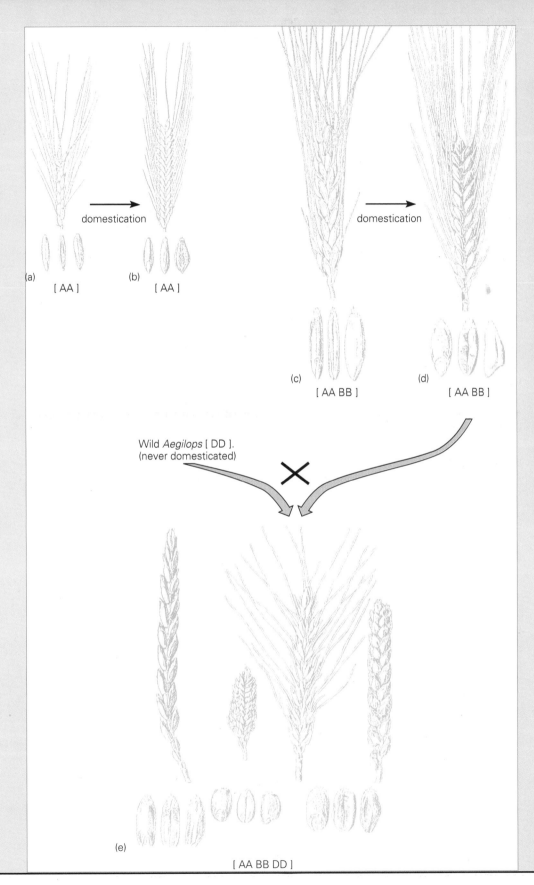

Figure 11.2

Origin and domestication of species of wheat. (a) Wild einkorn wheat, *Triticum monococcum;* diploid (AA), brittle ear and hulled grain. (b) Cultivated einkorn wheat, *T. monococcum;* diploid (AA), non-brittle ear, hulled grain. Widely cultivated in the past; relic today. (c) Wild emmer wheat *T. turgidum;* tetraploid (AABB), brittle ear, hulled grain. (d) Cultivated *T. turgidum;* tetraploid (AABB), non-brittle ear, hulled grain. Widely used in the past, durum wheat derived from this species. (e) Hexaploid (AABBDD) cultivated *T. aestivum.* Non-brittle ear, hulled grain; most variable wheats which are used extensively for making breads.

Taken from figures a, b, c, and d from E. Schiemann, *Weizen, Roggen, Gerste. Systematik, Geschichte, und Verwendung,* copyright © 1948 Gustav Fischer Verlag, Jena; and Table 4 in Zohary and M. Hopf, *Domestication of Plants in the Old World,* copyright © 1988 Clarendon Press. Used by permission.

(a) domestication

[AA]

(b)

[AA]

(c) domestication

[AA BB]

(d)

[AA BB]

Wild *Aegilops* [DD].
(never domesticated)

(e)

[AA BB DD]

TABLE 11.1

Diploid numbers found in certain genera of plants and animals

Genus	Chromosome Number	Ploidy
Chrysanthemum (N = 9)	18	2N
	36	4N
	54	6N
	72	8N
	90	10N
Dahlia (N = 16)	32	2N
	64	4N
Triticum (wheat) (N = 7)	14	2N
	28	4N
	42	6N
Gossypium (cotton) (N = 13)	26	2N
	52	4N
Nicotiana (tobacco) (N = 12)	24	2N
	48	4N
Rubus (blackberries) (N = 7)	14	2N
	21	3N
	28	4N
	35	5N
	42	6N
	49	7N
	56	8N
	65	9N
	84	12N
Hyla (treefrogs) (N = 12)	24	2N
	48	4N
Xenopus (frogs) (N = 18)	36	2N
	72	4N
	108	6N
Cnemidophorus (lizards) (N = 23) (many triploid all-female species)	46	2N
	69	3N

Figure 11.3

The cryptic species pair of grey treefrogs, *Hyla chrysoscelis* (2N) and *Hyla versicolor* (4N). These two species look so similar that they can be distinguished only by karyotype analysis or by differences in the calls of the males.

(a) © Joseph T. Collins/Photo Researchers; (b) © Thomas Gula/Visuals Unlimited.

(a)

(b)

Because of the increased productivity of many polyploids, horticulturalists now use chemicals to increase the frequency of polyploidization. Application of *colchicine* to growing tissues of plants stops the mitotic cycle prior to separation of the daughter cells, producing tetraploidy in the dividing cell. Successive colchicine treatments can produce octoploids (8N) and even higher levels of ploidy. The domestic strawberry, for example, is an octoploid.

Polyploidy is far less common in animals than in plants, probably because sex chromosomes are more common in animals than in plants. Nonetheless, surveys of chromosome numbers among animal taxa reveal that polyploid events have sometimes occurred. The gray treefrog, *Hyla versicolor,* is a tetraploid, presumed to have arisen from its morphologically indistinguishable diploid relative, *Hyla chrysoscelis* (figure 11.3). The two species were originally distinguished by slight differences in the male's call, and subsequent examination uncovered the chromosomal differences between them.

Polyploidy is apparently always fatal in humans. A few triploids and tetraploids have been born and even survived a few months, but all suffered obvious malformations. The frequency of conceptions involving polyploids, however, is not trivial. About 12% of all spontaneously aborted fetuses are known to be triploid or tetraploid, and the frequency may be considerably higher. In fact, more than 50% of all conceptions are estimated to be inviable due to genetic causes.

Polyploidy in humans results from at least two different causes: unreduced gametes and dispermy. The fertilization of an unreduced diploid egg by a haploid sperm gives rise to a triploid zygote, while the far-rarer union of two diploid gametes produces a tetraploid. **Dispermy,** the fertilization of an egg by two sperm, produces a triploid fetus.

Partially polyploid individuals are also known. Recall from chapter 6 that **mosaic** individuals consist of two or more discrete cell lines. Both 46,XX/69,XXX and 46,XY/69,XXY genotypes have been identified. Such mosaics may originate through the incorporation of a haploid polar body nucleus in a line of cells early in development. The earlier such an event occurred, presumably the larger the proportion of triploid somatic cells. Variation in the proportion of triploid cells might also result from differential growth rates of the two cell lines.

Aneuploidy

A substantial proportion, perhaps as high as 30%, of all human conceptions are aneuploid, differing by one or a few chromosomes from the normal forty-six. The possession of three rather than two homologues of a particular chromosome is referred to as **trisomy** for that chromosome. *Monosomics* have only one, and *nullisomics* have none, of a particular chromosome.

The primary cause of aneuploidy appears to be **nondisjunction** of homologues during meiosis (figures 6.11 and 6.13). Nondisjunction produces gametes with one chromosome too many or one chromosome too few. Subsequent fertilization of such a gamete gives rise to an aneuploid zygote. Nondisjunction also occurs during mitosis, producing individuals mosaic for various aneuploidies. As with polyploid mosaics, the effects of mosaic aneuploidy are more severe when the mosaicism originates early in development.

In most cases, however, aneuploidy has severe phenotypic effects. The genome appears to be in very delicate balance, and any alteration in dosage of chromosomes or chromosome fragments is almost always deleterious. With the single exception of **Down's syndrome,** the only aneuploidies viable to adulthood in humans are those involving the sex chromosomes discussed in chapter 6. The relatively high viability of sex chromosome aneuploids is probably explained by the dosage compensation mechanism. Because only one dose of most genes on the X chromosome is required, the loss of one X chromosome as in Turner's females, XO, is not necessarily lethal. Cytogenetic studies of spontaneously aborted fetuses, however, indicate that over 90% of XO fetuses do not survive to birth, suggesting that not all genes on the inactivated X chromosome are inactivated.

Dosage compensation also explains why XXX, XXXX, XXXXX, and XXY individuals are viable. The additional Xs are simply inactivated, although again phenotypic syndromes associated with these aneuploidies suggest that genes on the extra chromosomes are expressed to a limited extent. Nonetheless, triplo-X females are essentially normal phenotypically, and both triplo-X and tetra-X females may be fertile.

Because the tiny Y chromosome appears to carry only male determining genes, aneuploidies affecting it also are viable. XYY males are relatively normal phenotypically and are fertile. XYYY males are also viable.

Aneuploidies affecting the autosomes are always deleterious. Loss of an autosome is lethal; no monosomies other than Turner's survive past infancy. Trisomies surviving to birth have now been reported for at least chromosomes 8, 9, 13, 18, 21, and 22, but those involving 8, 9, and 22 are extremely rare. Trisomies of the smaller chromosomes (13–22) affect fewer genes and appear to be somewhat more viable. About 25% of all spontaneous abortions involve trisomic fetuses.

Trisomy 13, Patau's syndrome, occurs in a frequency of one in 5,000 to one in 10,000. Trisomy 18, Edwards' syndrome, occurs somewhat more frequently than Patau's. Affected individuals of both syndromes are severely phenotypically affected, and most survive only a few months or less (figures 11.1b,c).

Trisomy 21, Down's syndrome (figure 11.4), is relatively frequent, occurring in about one in 700 live births in the United States. It is the most common form of heritable mental retardation, affecting about 10% of all mentally retarded people. Affected individuals suffer numerous phenotypic abnormalities: They are short in stature, and the face, hands, and heart are malformed (figure 11.1a). They are subject to cataracts and other eye problems affecting the lenses, and they are 20–50 times more likely to develop leukemia than a normal person.

Nonetheless, Down's victims are relatively viable, despite the major genetic anomaly. Early this century life expectancy for a Down's child was only about nine years, but improved therapies have increased average life expectancy to more than thirty years; about 25% live longer than fifty years. Down's persons over thirty-five years old develop the same type of beta amyloid protein deposits found in the brains of persons with Alzheimer's disease, a form of senile dementia.

The trisomy affecting chromosome 21 is usually caused by meiotic nondisjunction, most commonly in the gametes of the mother. Furthermore, it has been known since 1876 that the incidence of Down's syndrome increases with the age of the mother (figure 11.5). At birth each ovary of a female contains several hundred thousand primary oocytes, in which meiosis has been arrested at prophase I. Meiosis does not resume until the oocyte is about to be ovulated. Apparently, the accuracy of meiotic division decreases with age, and after about age thirty the frequency of nondisjunction associated with chromosome 21 increases substantially. For women over age forty-five, the probability of producing a Down's child is about one in forty.

Higher than normal frequencies of Down's children have also been discovered for mothers under the age of seventeen. The birth rate for girls under age seventeen has increased dramatically, permitting statistical studies of birth defects in very young mothers. One theory that may explain the relatively high incidence of Down's syndrome for both older and younger mothers relates the frequency of nondisjunction to estrogen levels. Estrogen helps control the rate of meiosis in primary oocytes. As estrogen levels decrease with age, a slowing of the meiotic process may make nondisjunction more likely. Estrogen levels do not stabilize following puberty until about age twenty. Down's children produced by very young mothers thus may reflect fluctuations in hormone levels. If age or age-related factors increase the likelihood of nondisjunction, then trisomies involving chromosomes other than chromosome 21 should also be more common in older mothers, and perhaps younger mothers. In fact, the incidence of trisomies 13 and 18 *is* higher in older women. Other trisomies may also be common in these women, but because they are lethal, they are difficult to detect.

In the 1970s improved techniques for chromosomal banding and analysis showed that in as many as 25% of all cases of Down's syndrome, the extra chromosome is derived from the father rather than the mother. The probability of a father having an affected child increases with age, as it does for the mother. The rate of increase is relatively low until age fifty and steep thereafter.

Figure 11.4

Genetic relationships of Down's syndrome: (a) About 95% of all cases result from trisomy 21; the trisomy is shown in the box in this karyotype; (b) a map of chromosome 21 and some of the genes that may cause phenotypic abnormalities associated with Down's.

(a) From "The Causes of Down Syndrome" by D. Patterson. Copyright © Scientific American, Inc. (b) Adapted from "The Causes of Down Syndrome" by D. Patterson. Copyright © 1987 by Scientific American, Inc. All rights reserved.

(a)

Genes	Associated conditions
SOD-1	Premature aging; mental retardation
Gart	Mental retardation
Ets-2	Acute myelogenous leukemia
Alpha-A-Crystallin	Cataracts

(b)

This finding is of considerable value to genetic counselors. It means that the chromosomes of both parents should always be investigated in trying to pinpoint the source of the genetic problem. Older mothers who have produced Down's children have often borne an unnecessary burden of guilt, assuming they were the "cause" of their child's defect.

Figure 11.5

The incidence of Down's syndrome as a function of the mother's age. Note the slight elevation in frequency of Down's births for very young mothers and the marked increase for mothers over thirty-five years of age.

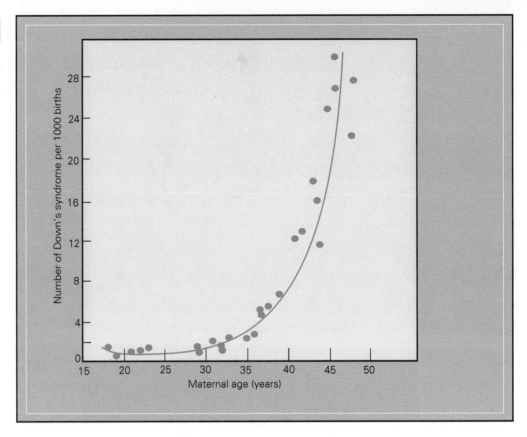

Figure 11.6

Three types of chromosomal deletions: (a) terminal deletions result from one break and cause loss of genetic material, shortening the chromosome; (b) interstitial deletions result from two breaks that do not include the centromere and, again, cause loss of genetic material; (c) two deletions, one near each end of the chromosome, may lead to formation of ring chromosomes if the two "sticky" terminal ends join.

Chromosome 21 is one of the smallest in the genome, containing only about 45 million nucleotide base pairs or 1.5% of the total nuclear DNA. Because of its small size and its association with several important genetic conditions, it is now one of the best known chromosomes. About 1,500 genes are thought to be located on chromosome 21. Only about twenty

Figure 11.7

Karyotype of a normal person following exposure to X rays. Note ring chromosome (a) and dicentric chromosome (b).

Michael Bender.

of these have been mapped thus far, and none of these is thought to be the gene causing Down's syndrome. Nonetheless, these genes already offer important clues about phenotypic aspects of Down's and related conditions (figure 11.4b).

Chromosome Mutation: Structural Changes

While DNA is a relatively stable molecule, it may break at any point along its length. Most breaks heal or are repaired with no change in base sequence. Some breaks, however, cause permanent alterations of chromosome structure. The phenotypic effects of such chromosomal mutations range from trivial to lethal.

Deletions

Segments of chromosomes can be lost or **deleted** in several ways. Following a single break, the free ends may not rejoin. The fragment lacking a centromere is subsequently lost during cell division, a *terminal deletion* (figure 11.6a).

Two breaks in chromosomes may also produce deletions. If the centromere is not located between the breaks, an internal segment may be lost; if the two terminal segments rejoin, an *interstitial deletion* may occur (figure 11.6b). If the centromere is located between the breaks, both terminal fragments may be lost. The free ends then join, forming a **ring chromosome** (figure 11.6c). Chromosome breakage often results from *mutagenic* (meaning change-producing) chemicals and radiation in the environment. Ring chromosomes are often found in the karyotypes of persons exposed to mutagens (figure 11.7).

Another cause of deletions is **unequal crossover** in meiosis. If the exchange of chromatid segments during prophase I is not precisely equal, a segment of DNA is deleted from one chromatid and added to its crossover partner (figure 11.8). Such an event produces four different types of gametes, one representing each parental type, plus one containing the deletion and one containing a duplication.

The phenotypic consequences of deletions are analogous to those of monosomy and presumably increase in severity in proportion to the size of the deletion. As with monosomy, most deletions are probably lethal. The individual is hemizygous for those genes that have been lost, and lethal or deleterious alleles are thus subject to expression. Many viable deletions have now been described for humans, all of which have some detrimental phenotypic effect. Chromosomal deletions are clearly associated with certain types of cancers (pages 297–98). No deletion has been discovered to be beneficial.

Figure 11.8

Unequal crossing-over produces chromosomes with duplications and deletions. Improper pairing in meiotic prophase I followed by crossing-over of non-sister chromatids results in two normal chromosomes, plus one with a deleted segment and one with a duplicated segment.

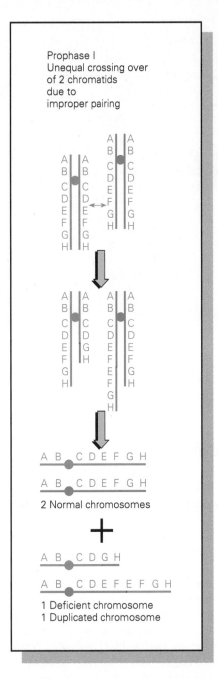

Probably the best-known deletion is one affecting the short arm of chromosome 5, producing the *cri du chat* (cry of the cat) *syndrome.* Affected individuals have microcephaly, wide-set eyes, severe mental retardation, and as infants a distinctive cat-like cry for which the syndrome is named. Survival of these individuals is relatively high, with some living until adolescence.

Duplications

The effect of unequal crossover in meiosis is not only the deletion of chromosome segments, but the addition of those segments to the homologous chromatid—a chromosomal **duplication** in one of the gametes (figure 11.8).

Such duplications are unknown in humans, in part due to difficulty in detecting them, but in laboratory species such as *Drosophila,* duplications and their phenotypic effects are reasonably well described. The effects of duplications would be expected to be comparatively less severe than loss of genetic material, although their rarity in humans suggests otherwise.

Figure 11.9

Chromosome inversions:
(a) production of inversions
by two breaks in a single
chromosome followed by
healing; (b) homologous
pairing in meiosis of
individuals who are
heterozygous for an inversion.
Notice that in both inversions
only two of the possible four
gametes will contain an entire
chromosome, one normal and
one inverted. The remaining
two meiotic products will be
missing some genes, have
additional genes, be missing a
centromere, or even have two
centromeres. Both of the
latter conditions interfere
with normal meiosis.

It is generally believed that duplications have a beneficial evolutionary aspect. Duplicated alleles already occur in the disomic condition and should therefore not be required for any phenotypic function. Duplicated genes are thus evolutionarily free to assume new functions, representing an important mechanism of evolutionary change.

Inversions

When two breaks occur in one chromosome, the result need not be a deletion. Instead, the broken ends may rejoin, but with the central segment reversed in orientation—an **inversion** (figure 11.9a). If the inverted segment contains the centromere, the inversion is *pericentric;* if not, it is *paracentric.*

Inversions do not change the amount of DNA in a chromosome; they simply change the order of some genes. In many cases this change has no phenotypic effect. Inversions may substantially alter linkage relationships and can thus be expected to show phenotypic manifestations. Such **position effects** have been demonstrated frequently in *Drosophila.* They have not, however, been clearly associated with inversions in humans, even though a small but significant percentage of humans probably have at least one inversion.

The most obvious effects of inversions are associated with meiosis. Synapsis of homologues heterozygous for an inversion is possible only with the formation of a loop in one of the homologues (figure 11.9b). Such an event is common and produces normal gametes if crossover does not occur. Should a crossover occur, however, some normal and some unbalanced gametes, containing deletions and duplications, result. Reduced fertility may be one consequence of chromosome inversions.

Another hypothesized effect of inversions is the protection of well-adapted linkage groups. The meiotic difficulties associated with inversions seem to favor the suppression of crossover in inversion heterozygotes. An inversion associated with well-adapted linkage groups may increase the probability of transmitting such a group of genes intact to future generations.

Figure 11.10

Chromosome translocations
involve moving entire
segments from one
chromosome to another.
Homologous chromosome
ends are numbered to help
keep track of moving
segments: (a) a simple
translocation occurs when a
portion of one chromosome
breaks and attaches to a non-
homologous chromosome
arm; (b) in a reciprocal
translocation, breaks occur in
two or more non-homologous
chromosomes, and the non-
homologous chromosomes
reciprocally exchange
chromosome segments.

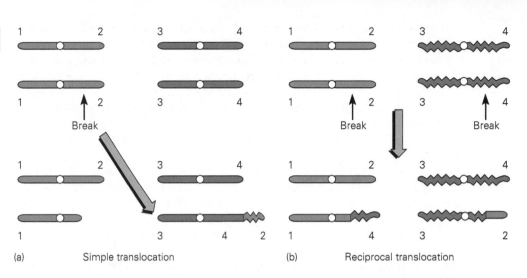

Figure 11.11

Meiosis in a cell that is
heterozygous for the
reciprocal translocation
depicted in figure 11.10b
results in different products
depending upon whether
segregation at anaphase I is
alternate (a) or adjacent (b
and c). (a) In alternate
segregation both gametes
have a balanced genome,
although half have normal
chromosomes and the other
half have translocation
chromosomes. (b and c) In
both types of adjacent
segregation, homologous
centromeres go to the same
(c) or opposite (b) poles,
resulting in all gametes being
duplicated and deficient for
portions of the genome.

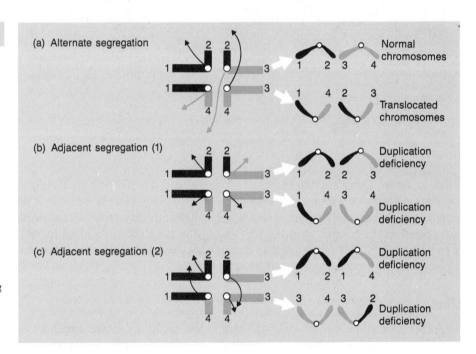

Furthermore, the meiotic difficulties characterizing inversion heterozygotes could provide a genetic basis for reproductive isolation between populations; that is, inversions may
play an important role in the process of species formation. If a particular inversion is highly
adaptive, it could become the only type of chromosome in the population, eventually leading
to speciation of that population. Such an hypothesis is supported by comparison of banding
patterns between closely related species. Species within the same genus often have identical
or very similar chromosome numbers, but banding patterns reveal that many inversions have
occurred in the course of their evolution. Chromosomal rearrangements, particularly pericentric inversions, appear to have played a significant role in the evolution of humans and
our Great Ape relatives (see box 2.1 and figure 2.11).

Translocations

A final type of chromosomal rearrangement involves transfer of genetic material between
non-homologous chromosomes, **translocation.** In a *simple translocation,* a break occurs in
one chromosome, and the acentric fragment then attaches to a member of a different pair
(figure 11.10a). If breaks occur in non-homologous chromosomes, the acentric portions may
be exchanged, a *reciprocal translocation* (figure 11.10b).

Because genetic material is not lost or gained, translocations often have little obvious phenotypic effects. Because they alter linkage relationships, however, the phenotype might be expected to change as a result of position effects. A more significant problem, associated with other chromosome changes as well, occurs during meiosis. When two pairs of chromosomes have undergone reciprocal translocation, synapsis can occur only if all four chromosomes form a single X-shaped configuration (figure 11.11). Subsequent gamete formation produces six different gametic combinations, only two of which are balanced. It is estimated that about 9% of couples with multiple recurrent spontaneous abortions are carriers of balanced translocations.

Translocations affect many different chromosomes in humans. The best known of these involves a translocation of the long arm of chromosome 21. Most commonly, it is translocated to chromosome 14, but it may also attach to chromosomes 13, 15, 21, or 22. No phenotypic effect is observed in the person in whom the translocation originates. However, if a gamete containing the translocated chromosome 21 unites at fertilization with a normal gamete, the resulting embryo has trisomy 21, Down's syndrome.

Such individuals, referred to as **translocation Down's,** have all the symptoms associated with the standard Down's. They account for about 5% of all Down's cases. Translocation Down's have the normal diploid number (46) with one chromosome having the extra translocated arm of chromosome 21. One of the parents of a translocation Down's, on the other hand, is aneuploid, having only 45 chromosomes. However, because that parent still has essentially two copies of chromosome 14 and chromosome 21 (one normal 14, one normal 21, and the translocation 14/21), the parent has a normal genetic complement and is thus unaffected. The parent is said to be a *translocation carrier.*

Identification of the type of Down's a child has is important in counseling parents considering further reproduction. Nondisjunction events that cause the standard Down's child are unrelated. Thus, the probability of having a future Down's child is unrelated to the previous occurrence of a Down's child, although it is related to parental age.

Parents of a translocation Down's, on the other hand, have a relatively high probability of giving birth to future Down's children. In effect, translocation Down's is an heritable disease. One of six gametic types produced by the carrier parent bears the extra chromosome 21. Because three of the gametic types are apparently inviable, the theoretical probability of producing a future Down's child is roughly one in three. Further, phenotypically normal children may also be translocation carriers for chromosome 21, thus perpetuating the Down's-producing chromosome in the family. Although parental age appears to have no affect in the origin of translocation Down's, recent studies suggest that the father is more likely than the mother to be the source of the translocation, in contrast to standard Down's.

Another translocation of considerable medical importance, the **Philadelphia chromosome,** abbreviated Ph1, is associated with 96% of the cases of chronic myelogenous leukemia. The Ph1 chromosome is the shortened chromosome 22, whose long arm has been translocated, most commonly to the long arm of chromosome 9 (figure 11.12). The Philadelphia chromosome is a somatic chromosomal mutation acquired sometime during the life of an affected individual, rather than an inherited germinal mutation. Discovered in 1960, the Ph1 chromosome was the first karyotypic abnormality associated with a specific cancer. At least 42 different translocations have now been associated with specific cancers.

Fragile Sites

When grown in culture, some human chromosomes display narrowings or gaps called **fragile sites** (figure 11.13). Discovered in 1965, over forty such sites are now known. Fragile sites occur at specific points in the karyotype, not randomly, and the sites are heritable. Their precise molecular structure is unknown, and they may represent transposable elements.

In the past, fragile sites were regarded as a novelty of little significance. In part, this derived from difficulty in demonstrating their existence. Fragile sites do not appear in karyotypes of all cells taken from an individual, but rather in cells that have been grown in certain culture media. Rapidly increasing knowledge, however, strongly suggests that fragile sites may be of considerable medical importance, and they are already known to be in one instance. Further, they appear to occur at surprisingly high frequencies; one site was found to occur in one in forty individuals. Considering all forty known sites, a significant proportion of the population carries one or more fragile sites.

Figure 11.12

Karyotype of a person diagnosed as having chronic myelogenous leukemia. The longer chromosome 9 (upper box) has a piece of chromosome 22 (lower box) translocated to it, and the resultant short chromosome 22 is known as the Philadelphia chromosome.

Victor B. Eichler.

The **fragile X syndrome,** identified in the late 1970s, is the one medically important phenotypic abnormality unquestionably related to a fragile site (position Xq27.3, figures 6.15 and 11.13b). Fragile X syndrome is the second leading cause (after Down's syndrome) of inherited mental retardation in the United States, partially explaining why mental retardation is more common in males than females. Understanding this condition has been complicated by the unusual pattern of expression: The fragile X site is inherited in standard X-linked fashion, but about 20% of males with the fragile X site do not express the disease. Penetrance and expressivity in heterozygous females are highly variable; about 50% of females with the site show no expression whatever.

These sites are called fragile because they are prone to breakage, at least in culture. Evidence linking specific break points to diseases is circumstantial, but highly suggestive in one respect. At least nine fragile sites occur in breakpoints associated with **oncogenes** or cancer-causing genes (discussed later in this chapter). This association is *statistically* highly significant. Fragile sites may prove to be a key element in understanding the causes of cancer.

Genetic counsellors do not normally inform patients of chromosomal or genetic abnormalities unless that information can be used meaningfully by that person to change her or his life. Even for conditions such as fragile X where a clear relationship to disease exists, counselling is difficult because of variable expressivity. A counsellor's advice for fragile X would be carefully matched to the person and family affected. Because no autosomal fragile site is known to cause disease, the most likely (and probably only) effect of informing a person that she or he carried an autosomal fragile site would be to cause unnecessary and pointless anxiety. Counsellors are thus recommended not to inform persons that they carry these sites.

Gene Mutation

The smallest type of mutation affects only one or a few DNA bases, a **gene** or **point mutation.** Point mutations cannot be observed by inspection of the chromosomes, and they may have small or large phenotypic effects. The basic types of point mutations are shown in table 8.3. A **frameshift mutation** involves the addition or deletion of one or a few bases, resulting in a shift in the triplet reading frame. A frameshift can render nonsensical all triplets downstream from the site of the mutation. Deletion of even a single base may thus have a significant phenotypic effect by causing the production of nonsense mRNA or no mRNA at all (e.g., box 10.4 and the O blood type).

A **substitution mutation** occurs when one base is replaced by another. If a simple base substitution produces a triplet that encodes a different amino acid than before mutation, the mutation is said to be **missense.** About 30% of all substitutions, however, do not change the amino acid encoded, particularly substitutions affecting the second and third bases in a triplet. These are known as **same sense mutations.**

(a)

Figure 11.13

Fragile sites. (a) Composite karyotype showing seventeen of the best known fragile sites; (b) fragile X chromosome, with fragile site marked by arrow. On the left is a scanning electron micrograph of the X; on the right is a diagrammatic representation of a G-banded X chromosome at metaphase (i) and prometaphase (ii).

From *Fragile Sites on Human Chromosomes.* (Oxford Monographs in Medical Genetics) by Grant R. Sutherland and Frederick Hecht. Copyright © 1985 by Oxford University Press, Inc. Reprinted by permission.

(b)

The effect of point mutations is likely to increase with the number of bases affected. It is generally assumed that most mutations are deleterious, although their expression is often masked because the mutations are also usually recessive. As with chromosomal mutations, the immediate effects are normally not beneficial, but the evolutionary effects may be positive. Point mutations are the source of all new base sequences and provide the ultimate reservoir of genetic variation, continually renewed, upon which natural selection acts. Point mutation is one of the bases for the diversity of life on earth.

Mutations apparently occur at random. That is, while we can estimate the *rate* at which mutations occur, we cannot predict which base pairs will be affected. No source of mutation is known to be specific in its action. In that sense, evolution is clearly an undirected phenomenon. The variation available to natural selection cannot be controlled, and specific mutations do not arise in direct response to environmental changes.

TABLE 11.2

Mutation rates of particular alleles expressed as autosomal dominants or X-linked recessives in males

Trait	Mutant Genes Per 100,000 Gametes Per Generation
Epiloia	0.8
Aniridia	0.5
Microphthalmus	0.5
Wardenberg's syndrome	0.4
Facioscapular muscular dystrophy	0.5
Pelger anomaly	0.9
Myotonia dystrophica	1.6
Myotonia congenita	0.4
Huntington's chorea	0.2
Retinoblastoma	0.4
Neurofibromatosis	13–25
Hemophilia	2.7
Sex-linked muscular dystrophy	5.5

Table from *Human Genetics* by Daniel L. Hartl. Copyright © 1983 by Harper & Row, Publishers, Inc. Reprinted by permission of HarperCollins Publishers Inc.

The Rate of Spontaneous Mutation

The background rate of spontaneous mutations is surprisingly constant throughout nature. On average, about one mutation can be expected in a gene in every 100,000 to 1 million gametes per generation. In humans, the rate of spontaneous mutations has been estimated on the basis of the rate of first appearance in a family of genetic diseases that are expressed when the controlling allele is present in only one dose, i.e., diseases caused by autosomal dominant alleles, or those caused by X-linked recessives in males (table 11.2).

Table 11.2 shows that the mutation rate differs among loci. The variation could be due to several factors, such as the number of bases in the gene or the relative mutability of particular base sequences. Estimates giving similar values of the spontaneous mutation rate have been made on many other species, including both plants and bacteria.

While 0.1–1.0 mutations per 1 million gametes per generation may seem a relatively trivial rate of occurrence, remember two points. First, most mutations are deleterious. If the rate of mutation were much higher, the mortality rate due to lethal genetic combinations might be unbearably high. Recall that more than 50% of all human conceptions are believed to be inviable due to genetic factors, even at this rate of mutation.

Second, humans are estimated to have 10,000 to 100,000 structural genes, plus an unknown number of regulatory genes. If the mutation rate is estimated to be one mutation per 100,000 gametes per generation and each person has 100,000 genes, we can expect that, on the average, each person will have one new point mutation per gamete. Since every person results from the union of two gametes, each would be expected to carry about two new mutations. Other evidence indicates that this estimate is conservative and that, on average, each person carries about three to five new, potentially harmful recessive mutations. By comparison, the rate of chromosomal mutation is considerably lower than the rate of point mutation: only about one gamete in one hundred is estimated to carry a chromosomal aberration.

Causes of Mutation

Many factors causing mutation are now known. As noted in chapter 8, mutability is a key requirement for the genetic material. One characteristic of the chemical structure of DNA is a certain amount of instability. Tautomeric shifts occur when hydrogen atoms in the nucleotide bases undergo slight alterations in position. When this happens, the base is unable to pair with its normal complementary partner, but instead pairs with another base; for example, guanine may pair with thymine, or cytosine with adenine. Normally, tautomeric shifts are only transitory states of a base and have little effect. If, however, a tautomeric shift occurs

BOX 11.2

Mutations are random

Until the time of Darwin, little more than a century ago, it was commonly believed that the universe had been designed to meet the needs of humans. One of the most significant philosophical as well as scientific contributions of modern genetics has been the demonstration that mutation, the source of all genetic diversity, is random. The needs of humans or other species are not a factor in the origin of genetic variation. Mutations are *not* produced in response to a need of an organism, although the result of some mutations is that organisms are better able to survive in changing environments.

An elegant experimental technique, *replica plating,* demonstrates that random mutations may be naturally selected by differing environments (figure 11.14). Large populations of *E. coli* are grown on normal medium (plate 1). Several additional plates of growth medium are prepared to which have been added a toxic chemical or drug such as penicillin (plate 2). A piece of velvet stretched over a block of wood is gently pressed against the growing colonies in the normal culture, inoculating the velvet with a mirror image of the colonies on the normal plate.

Next, the velvet is used as a stamp to inoculate the experimental plates. On the experimental plates, only those colonies grow that contain mutations making them resistant to penicillin. Because the mutation occurred *prior* to the inoculation of replica plates, all replica plates have penicillin-resistant colonies growing in the same location. Further examination of the resistant colonies on the original plate verifies that the mutation was present *before* introduction to the penicillin-containing cultures. Thus, the mutation occurred randomly—without respect to the environment. It became apparent only in the presence of a particular environmental change that selected it, i.e., that permitted its survival in preference to genetically different individuals.

at the time of DNA replication, the mispairing causes a change in the base sequence in the complementary chain that can be transmitted to future copies of the DNA—a substitution mutation.

Tautomeric shifts occur as a result of natural chemical properties of the nucleotide bases and help explain the normal spontaneous mutation rate. Many environmental agents also affect the rate of mutation. Additionally, variations in the mutation rate have been associated with biological factors such as age and sex. Achondroplasia, for example, occurs more frequently in matings involving older fathers.

Chemicals and radiation are the primary physical agents affecting mutation rates. Mustard gas, used widely in World War I, was one of the first chemicals shown to be mutagenic. Mustard gas, a powerful mutagen, reacts directly with DNA bases, causing both genic and chromosomal mutations. Other types of chemicals, **base analogues,** have structures similar to the normal nucleotide bases and are thus subject to accidental incorporation in DNA. Caffeine, for example, is a base analogue of adenine. If incorporated in place of adenine, it may pair with guanine instead of thymine, causing a base substitution during cell division (figure 11.15).

The early 1950s marked the beginning of large-scale production of modern chemicals. Now, over sixty-three thousand chemicals are used routinely in industry and commerce, and about 1,000 more are added each year. Literally thousands of these are known or strongly suspected of being mutagenic. Some are used only in medical or industrial applications, but many are commonly encountered in everyday life. Among the latter are such chemicals as pesticides and herbicides, caffeine, formaldehyde (used in household insulation), toners (used in photocopy machines), some components of dyes and cosmetics, and sodium nitrites (used as preservatives in meat products such as ham, bacon, and hot dogs). Over twenty-five hundred chemical additives are used in processing food.

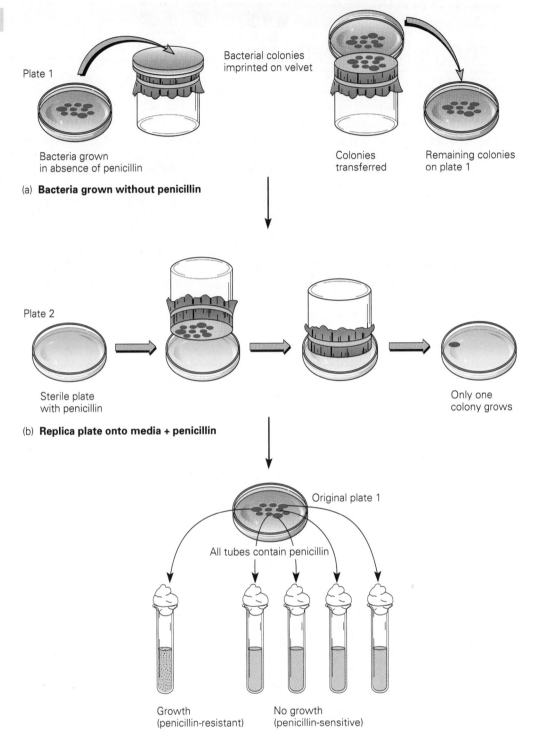

Plate 1

Bacterial colonies imprinted on velvet

Bacteria grown in absence of penicillin

Colonies transferred

Remaining colonies on plate 1

(a) **Bacteria grown without penicillin**

Plate 2

Sterile plate with penicillin

Only one colony grows

(b) **Replica plate onto media + penicillin**

Original plate 1

All tubes contain penicillin

Growth (penicillin-resistant)

No growth (penicillin-sensitive)

(c) **Isolation and test of colonies**

A second major category of mutagenic agent is radiation. Two basic types of radiation are of health importance to humans: *electromagnetic* (including **ultraviolet radiation, X rays,** and gamma radiation) and nuclear (including alpha and beta particles, protons, and neutrons). In 1927, H. J. Muller demonstrated that the mutation rate could be increased 1,500 times in the progeny of male *Drosophila* subjected to high doses of X rays. Moreover, genetic damage is proportional to the dose administered, and the effect is cumulative. In other words, one large or **acute dose** has roughly the same mutagenic effect as many smaller or **chronic doses** administered over many days, as long as the total dose is the same.

Original DNA

Figure 11.15

How caffeine (CF) acts as a base analogue of adenine and causes mutation. Note that the CG base pair (II) replaces the original AT (III, IV) base pair at the top of the DNA molecule, due to interference of caffeine. The DNA molecule replicates in the normal fashion, except that the caffeine that replaced the A mispairs with G, rather than with T, as A normally does.

Two different DNAs   Two original DNAs

UV irradiation

Thymine dimer

Figure 11.16

Formation of thymine dimers in DNA. Exposure to ultraviolet radiation causes adjacent thymine molecules in a single strand of the DNA to dimerize and break their bonding with opposite adenine molecules. Normally enzyme systems monitor the DNA, excise these dimers, and replace them with thymines that bond properly.

Leland G. Johnson, *Biology,* 2d ed. Copyright © 1987 Wm. C. Brown Publishers, Dubuque, Iowa. All Rights Reserved. Reprinted by permission.

Exposure to some types of radiation, like many chemicals, is unavoidable. Some portions of the world have high levels of background radiation from uranium and other minerals, and most of us are routinely exposed to medical and dental X rays. The ultraviolet (UV) component of sunlight is probably the single most important natural mutagen in our environment. UV light easily penetrates the epidermis and breaks the DNA of underlying tissues. In particular, where two thymine molecules occur adjacent to one another, UV light causes them to bond to one another rather than to the adenines of the complementary strand, forming **thymine dimers** (figure 11.16). If these dimers are not excised and the DNA repaired, mutations occur, as DNA with thymine dimers cannot replicate normally.

Figure 11.17

A person with xeroderma
pigmentosum.

Lester V. Bergman and
Associates, Inc.

DNA Repair

Given the mutability of DNA and the presence of many mutagens in the environment, it is
perhaps not surprising that cells possess mechanisms to repair DNA. One repair process
involves removal (excision) of the mutant segment by certain enzymes, followed by its replacement with the proper set of nucleotide bases. Thymine dimers caused by exposure to
UV light are normally repaired in this way.

Excessive exposure to UV radiation, however, may produce so much mutation that the
repair system is overwhelmed. In such cases, skin tumors eventually form. About 400,000
new cases of skin cancer occur annually in the United States due to UV radiation. Fortunately, these growths are normally benign, and reduced exposure to sunlight reduces the risk
of recurrence.

The role of DNA repair mechanisms is well illustrated by the genetic disease **xeroderma
pigmentosum** (meaning dry, heavily pigmented skin; figure 11.17). Individuals homozygous
for this autosomal recessive allele cannot produce enzymes necessary for excision of thymine
dimers from DNA and subsequent DNA repair. Affected individuals are thus extremely
sensitive to sunlight. Even the slightest exposure causes dry, flaking skin, heavy pigmentation, and eventually numerous disfiguring skin cancers. Death usually ensues by age thirty
due to metastases of inoperable skin cancers. This unhappy condition illustrates both the
high frequency of natural mutations and the vital importance of the DNA repair processes.

Mutations and Human Health

Genetic mutations, both point and chromosomal, cause major human health problems. In
some cases, this information allows prevention or reduction in the incidence of certain genetic
conditions. In the long term, increasing knowledge should prevent or treat many serious health
problems caused by mutations.

In this section we consider aspects of mutation related to aging and developmental abnormalities. In the next section, we consider cancer. We conclude the chapter by briefly considering the politics of dangerous substances.

Aging

Numerous theories have been advanced to explain the process of aging, and several are based
largely on genetic factors. The **mutation theory of aging** holds that aging is a degenerative
process reflecting the accumulation of deleterious somatic mutations throughout life. The
Second Law of Thermodynamics states that, in general, the level of disorganization (entropy
or randomness) increases throughout the universe. The high level of organization in living

organisms is maintained only by the many homeostatic and repair mechanisms of the body. As somatic mutations accumulate, the genetic repair mechanisms gradually are unable to fix all of them, and the result is a net deterioration of physical and mental function, culminating in death.

Another hypothesis relates the decline in cellular function to the activities of *free radicals* in the cells. Free radicals are groups of atoms that are highly reactive with other molecules in the cell. Formed by ionizing radiation or as by-products of normal cell chemistry, they are usually inactivated by enzymes. Free radicals are so active that some combine with DNA, apparently altering its function permanently. Free radicals could thus provide a mechanism for the mutation theory of aging.

The **commitment theory of cellular aging** suggests that development is a genetically programmed process designed, like all products of natural selection, to maximize the genetic contribution of an organism to future generations. That is, the developmental sequence characterizing a species has evolved because individuals with that sequence produced more offspring. Thus, species differ in their maximum duration of life because they are *genetically programmed* for differing lengths. Bacteria may live only a few minutes, mice only a few years, humans, elephants, whales, and tortoises a hundred years or more. Organisms die at the appropriate time not because of genetic failure, as in the previous hypotheses, but because of their genetically directed developmental program. Scientists have not been able to alter significantly the maximum life span of any species, suggesting strongly the existence of genetically controlled limits that differ between species.

Leonard Hayflick's studies of the growth of cells in culture support this view. Human fibroblast cells in laboratory culture will not continue to divide indefinitely. The number of doublings that occur in culture decreases with the age of the cell donor. The limit for cells taken from a human embryo is about fifty doublings, but the limit for cells taken from a thirty-year-old human is only about twenty doublings. Species with different maximum life spans also possess different potentials in cell culture. The Galápagos tortoise, one of the longest lived of all animals, may live more than 150 years. In culture, embryo fibroblasts of the Galápagos tortoise undergo 90–125 doublings. A mouse, on the other hand, usually lives only two or three years, and its embryonic fibroblasts are capable of only 14–28 doublings. The **Hayflick Limit** is one of several possible mechanisms controlling the aging process. It describes a phenomenon—the limited growth of cells in culture—and thus provides a clue to the genetic processes related to aging. However, the exact meaning of the Hayflick Limit is still an object of scientific debate.

The **autoimmune theory of aging** is another important hypothesis. This theory holds that decreasing function of one of our primary protective systems, the immune system, leads to deterioration in cellular function with age.

Note, however, that these theories are not mutually exclusive. All are supported by strong but circumstantial evidence, and all may be correct to some extent. Maximum length of life does seem to be a genetically governed trait of a species, but somatic mutation could well be one of the mechanisms enforcing that maximum.

Progeria and *Werner's syndrome* are two rare autosomal genetic disorders characterized by rapid aging and thus a greatly shortened life span (figure 11.18). Individuals with these conditions age about ten times more rapidly than normal, begin to go bald, have wrinkled skin, and even arthritis and heart disease by the time they are seven or eight years old. Death usually results from a heart attack or stroke before the child is fifteen. Cell cultures taken from individuals possessing these disorders are capable of far fewer doublings (2–10) than normal, suggesting a change in the normal mechanism controlling mitosis.

Developmental Abnormalities: Mutagens and Teratogens

Aging is a normal and predictable part of the developmental process. As we have seen in chapter 9, development is a highly integrated and complex process of phenotypic changes guided largely by a precisely programmed sequence of genetic events. While medical science will continue to study the causes of aging, we should bear in mind that aging is a natural and generally desirable process. None of us, for example, would wish to retain the phenotype of a five-year-old forever!

Figure 11.18

Two children 8 and 9 years old at time of photo. Both are victims of progeria, a rare genetic disease characterized by rapid aging.

AP/Wide World Photos.

The precision of development, however, makes the normal developmental sequence vulnerable to disturbance. If mutation affects a key regulatory gene, the medical consequences can be serious. Similarly, mutations affecting "dosage" of particular genes (in reality, loss of normal controls on expression of those genes) also can cause major developmental anomalies. Down's and Turner's syndromes are examples of the effects of dosage alteration due to mutation.

Environmental agents such as chemicals and radiation can adversely affect human health through *mutagenesis.* **Mutagenic** agents cause genetic mutation. Many environmental mutagens are probably also **teratogenic** (from the Greek *teras,* meaning monster), producing physical defects or deformities in developing embryos.

Mutagens and teratogens are *not* the same thing. Substances that cause mutations need not cause deformities, and vice versa. Genetic mutation probably often has no significant health effects. It is likely, however, that mutagenesis is an important agent of teratogenesis, and as we have seen, developmental abnormalities are often associated with genetic damage.

The first trimester (the first three months) of gestation is a particularly vulnerable time for a developing fetus because during this period the basic tissues and organ systems are established. Chemicals that normally have little or no harmful effect on humans may cause severe damage to a fetus during this period; it is for this reason that pregnant women are urged to stop all use of tobacco, alcohol, and drugs. During pregnancy, even such drugs as aspirin should be taken only on the advice of a doctor.

Perhaps the most notorious teratogen in recent decades has been the drug *thalidomide,* widely used as a sleeping pill and tranquilizer in Europe in the early 1960s. Children of women who used thalidomide during the first trimester of pregnancy were born with abnormalities in the bones of the arms and/or legs (figure 11.19). Phocomelia is a rare genetic disease with the same phenotypic effects. The sudden high frequency of children born with this phenotype led to the discovery that thalidomide produces **phenocopies** of phocomelia. As soon as the teratogenic effects of thalidomide were recognized, the drug was removed from the market.

A drug whose teratogenic effects on the developing fetus have been widely recognized is alcohol, the single most widely used dangerous chemical in our society. Although the effects of alcohol addiction or misuse are well known, it is less widely realized that alcohol is a potent teratogen for the developing fetus. The Bible mentions the harmful effects of alcohol

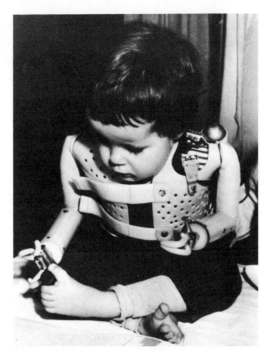

Figure 11.19

A three-year-old girl whose mother took thalidomide during the first trimester of her pregnancy. Absence or shortening of arms are characteristic abnormalities induced by thalidomide.

© Omikron/Photo Researchers.

on pregnant women, and British doctors at the turn of the century observed that "maternal intoxication" was a major source of fetal damage. The public, however, has only slowly begun to perceive the dangers of alcohol use by pregnant women.

Alcohol is a small molecule that easily crosses the placenta, reaching concentrations in the fetus equal to those in the mother's bloodstream. Its extreme danger to the developing fetus was explicitly recognized medically in 1973 with the description of **fetal alcohol syndrome (FAS).** Alcohol (specifically, ethanol) causes abnormal development of the head and nervous system. The effects are greatest during the first trimester of pregnancy, particularly during the first month. Substantial damage may be done to a fetus even before the mother realizes she is pregnant. Symptoms of FAS include head and facial deformities (figure 11.20), microcephaly, mental retardation, reduced growth and development rates, and ten-times higher probability of perinatal death. About one in 750 births in the United States is severely affected due to FAS. Mothers who drink alcohol have an increased incidence of stillbirth.

The extent of fetal damage increases with maternal alcohol consumption. Even two drinks per day can cause damage. Recent primate studies have shown that maternal alcohol consumption interrupts placental exchange of oxygen between mother and fetus for up to an hour or more. Such oxygen deprivation may be responsible for the mental retardation characteristic of FAS. Symptoms of FAS have been detected in 2% of the offspring of mothers admitting to two or fewer drinks per day during and just before pregnancy; 11% of infants of mothers having two to four drinks per day; and 19% of infants of mothers having more than four drinks per day. FAS is the leading cause of mental retardation in the United States. It has been suggested that women should be required to abstain from alcohol during pregnancy.

Cancer

Cancer (figure 11.21) is a disruption of developmental and regulatory systems, arising from the loss of normal controls on cellular growth, differentiation, and migration. Cancer occurs when a cell or group of cells begin uncontrolled growth, forming tumors or **neoplasms** (meaning new forms). Cells within the neoplasm do not differentiate properly, if at all. They are *invasive,* infiltrating tissues and destroying surrounding cells, and they often spread or **metastasize** from the point of origin to other locations in the body, where they lodge and continue uncontrolled proliferation. Without medical intervention, cancers are normally fatal.

Figure 11.20

Fetal alcohol syndrome:
(a) many facial abnormalities
result from exposure to
alcohol during gestation;
(b) the effects are similar in
children of all races.

(b) Dr. Ann Pytkowicz
Streissguth, University of
Washington: *Science* 209 (18
July 1980), 353–61, © 1980
by the AAAS.

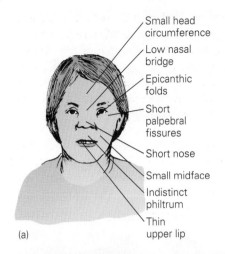

Small head
circumference

Low nasal
bridge

Epicanthic
folds

Short
palpebral
fissures

Short nose

Small midface

Indistinct
philtrum

Thin
upper lip

(a)

(b)

Figure 11.21

Human cervical cells:
(a) normal; (b) cancerous.
(Note: cancerous cells have
much larger nuclei, less
cytoplasm, and more of them
are in stages of division.)

(a and b) © Cabis/Visuals
Unlimited.

(a)

(b)

Cancers usually kill by malnutrition and infection. Cancerous cells grow far more rapidly than surrounding cells, robbing the surrounding cells of vital nutrients. In spite of good nutrition, cancer victims become progressively emaciated. Malnutrition makes the body susceptible to disease. Cancers invade the lymph nodes and overwhelm their protective responses. Cancerous cells infiltrate the major organs, blocking circulation, and eventually causing failure of life-sustaining processes.

Three main categories of cancers are identified by the tissues affected. (1) **Carcinomas** affect a variety of organs, accounting for about 85% of all cancers. They include cancers of the skin, glands, gastrointestinal tract, respiratory tract, and urinary and reproductive systems. (2) **Leukemias** and other cancers of the blood-forming tissues account for about 10–12% of all cancers. They are a major source of mortality among children. (3) **Sarcomas,** the rarest of all cancers, affect the bones and muscles. About 2–3% of all cancers are sarcomas.

Cancer is a general term, applied to over two hundred different diseases that probably have many different causes. Circumstantial evidence has long suggested a genetic role in cancer. For over a century some cancers have been known to "run in families" and thus be at least partially hereditary. Chromosomal mutations were often found in cancerous cells, although these could have been the result rather than the cause of cancer. And, some substances known to be mutagenic have also been shown to be **carcinogenic,** or cancer-causing. Nonetheless, as recently as 1970, few scientists believed that genetics played a central role in causing cancer.

That view has changed dramatically, however. Normal cell growth is now understood as a dynamic equilibrium involving two sets of genes, **proto-oncogenes** that promote cell proliferation and **tumor suppressor genes** that limit it. Cancers represent a serious imbalance in this genetic equilibrium, in which genes promoting proliferation escape their usual controls. Usually, this results from a series of mutational events affecting several genes.

Proto-oncogenes and Oncogenes

In 1976 J. Michael Bishop and Harold Varmus identified genes that can cause cancer. These genes, found in all of us, exist in two states within the cell. In their normal state they appear to promote cell growth, but their activity is tightly controlled. About fifty different types of these normal genes, called **proto-oncogenes,** have now been identified. Under certain circumstances, however, these genes undergo mutation and are transformed into **oncogenes,** or cancer-causing genes. Activated oncogenes appear to cause 10–30% of all cancers. Bishop and Varmus were awarded the Nobel Prize for Physiology or Medicine in 1989 for this key breakthrough in understanding both normal growth processes and genetic malfunctions that cause cancer.

Exactly how oncogenes cause cancers is the subject of intense research, but a variety of mechanisms appear likely. For example, the translocation of segments of chromosomes 9 and 22 that creates the Philadelphia chromosome results in the proto-oncogene *c-abl* from chromosome 9 being attached to the proto-oncogene *c-bcr* on chromosome 22. Together, these two genes function as a single new oncogene, producing an abnormal enzyme that in some way induces chronic myelogenous leukemia.

Alternatively, chromosomal deletions removing or disabling the regulatory portion of a proto-oncogene could result in higher than normal levels of the growth-promoting substance produced by the proto-oncogene, leading to the uncontrolled growth characteristic of cancers. *Gene amplification* does not normally occur in mammalian cells, but could also cause production of abnormally high levels of growth-promoting enzymes. In some human breast cancers, a proto-oncogene is amplified between two and twenty times.

Moreover, *viruses* appear to cause about 20% of human cancers by oncogenic action. Viruses are extraordinarily simple entities, consisting only of a small genome—five to several hundred genes—enclosed in a protein coat. Their life cycle can be quite simple (see chapter 12). Viruses inject their own genome into the cytoplasm of normal cells. The viral genes insert themselves into the host genome and then direct the host cell's metabolic machinery to produce more viruses, which are ultimately released from the host cell, continuing the viral life cycle.

Viral activity appears capable of causing cancer in several ways. First, insertion of the viral genome may induce mutations in proto-oncogenes, leading to oncogenic activity. Second, the powerful viral regulatory apparatus may induce hyperactivity in the proto-oncogene. Third, viruses themselves contain oncogenes that induce cancer when inserted into the host genome. In fact, oncogenes were first detected in viruses, and the discovery that human and other animal cells contain comparable and nearly identical genes (called proto-oncogenes) was a major advance in unravelling the complex story of cancer. It is now accepted that viral oncogenes were originally derived from the genome of their hosts, a sort of primeval gene transfer.

Viruses have long been implicated in cancer. Cancers were experimentally induced by injecting viruses into laboratory animals as early as the 1950s. In 1981, **human T-cell leukemia virus (HTLV)** was identified as a cause of adult leukemia. Leukemia, one of the three major types of cancer, is characterized by production of abnormally large numbers of white

Figure 11.22

Child affected with Burkitt's lymphoma.
Science VU/Visuals Unlimited.

blood cells that have reduced immune activity. Leukemic white blood cells do not function as normal antibody-producing cells. Instead, they accumulate in the circulatory system, disrupting normal circulation and in general disrupting normal body function. Leukemia is the most common childhood cancer, with about 3,000 new cases of childhood leukemia diagnosed annually in the United States.

A different type of virus, the **Epstein-Barr Virus (EBV),** is the cause of **Burkitt's lymphoma,** a common cancer of African children. Burkitt's lymphoma causes tumors of the jaw and facial bones (figure 11.22) and affects the spleen and lymph nodes.

Tumor Suppressor Genes

Cell proliferation is normally limited by a variety of genes called **tumor suppressor genes** or **anti-oncogenes.** Two of these genes are now reasonably well-known, and many others are suspected. The loss of function of these genes appears to be an important step in the development of many cancers.

Hereditary retinoblastoma is a malignant cancer of the retina in children. It is controlled by the *RB-1* locus on the long arm of chromosome 13, whose normal product acts to suppress tumor formation. Development of the cancer appears to be a two step process, each step affecting one copy of the gene at the *RB-1* locus because the normal gene product acts in a dominant fashion. The child is born heterozygous at the *RB-1* locus, receiving a normal allele from one parent but a mutant allele from the other parent, the first step. The normal allele is sufficient to suppress cancer, which only develops when the normal allele is mutated or lost by deletion, the second step. The *RB-1* gene appears to serve a general tumor suppression function, as chromosomal mutations at this locus are also associated with other cancers.

Cancer more commonly affects adults than children, and most adult cancers appear to have complex causes, often involving tumor suppressor genes. The p53 gene located on the short arm of chromosome 17 produces a protein known to activate transcription of another gene, thought to be a tumor suppressor. Deletions and mutations of the p53 gene are now known to be involved in cancers of the lung, colon, rectum, and genitalia.

The complexity of cancer is shown by a model for the development of colorectal cancer (figure 11.23). Cancer appears to result from a sequence of changes that include activation of the *ras* oncogene, loss of three chromosomes that each may carry a tumor suppressor gene (including chromosome 17 that has the p53 gene), and loss of methylation on DNA that may change gene expression. Each of these changes alone is not sufficient to cause cancer, which occurs only when the entire sequence has been completed.

This multi-step model of cancer is consistent with clinical data showing that cell abnormalities progress through a clearly defined series of non-malignant stages called *adenomas* before culminating in a malignant carcinoma. It also agrees with molecular studies showing that colorectal tumor cells exhibit numerous allelic deletions and additions (figure 11.24), and that more severe cancers contained significantly more mutations than less severe cancers.

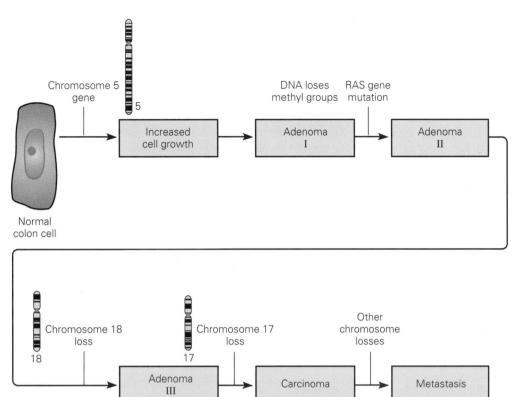

Figure 11.23

A model for the development of colorectal cancer.

Tumor suppressor genes may serve many functions. It is likely that they play a role in DNA repair, inhibiting or fixing the many chromosomal mutations that occur as a part of cell division. They may contribute to normal differentiation, as normally differentiated cells seldom become malignant, in contrast to tumor cells that are usually poorly differentiated. And, tumor suppressor genes clearly limit cell proliferation. Research on tumor suppressor genes is proceeding rapidly. Because tumor suppressor genes appear to be the body's natural anti-cancer protection, understanding these genes may be a major step in the development of treatments and even cures for cancer.

Cancer and Society

It is often claimed that cancer is a peculiarly modern affliction and that cancer rates have increased throughout the last century because of the increasing exposure of humans to cancer-causing agents as a result of world-wide industrialization. This is a difficult hypothesis to evaluate for many reasons, and it is likely to be only partly true. In particular, claims of increasing rates of cancer cannot be fully verified. Diagnostic techniques are far better now than even a decade ago, and precise rates of particular cancers in the past century cannot be accurately known. Furthermore, cancer is still largely a disease of the old. Average life expectancies have nearly doubled worldwide in the last century, so it is hardly surprising that cancer and other degenerative diseases seem to be more common.

On the other hand, as we saw earlier in this chapter, the rate of genetic mutation can be significantly increased by exposure to many types of chemicals and radiation. We know that these agents may interfere with developmental pathways, resulting in teratogenic effects. Any increase in mutation rate could reasonably be expected to affect proto-oncogenes and tumor suppressor genes as well as the rest of the genome, thus having carcinogenic, or cancer-causing effects. Evidence is now clear that some environmental agents, often very common ones, cause cancer.

However, society's response to knowledge about cancer-causing agents is complex, especially when dangerous substances are extremely useful. Furthermore, establishing carcinogenicity, even for extremely toxic agents, is not always easy. For the remainder of this chapter we consider aspects of the identification and use of potentially carcinogenic or mutagenic agents.

Misinformation   **299**

Evidence for DNA deletions associated with colorectal cancer. (a) RFLP patterns from normal (N) and cancerous (C) tissue from a single patient. The DNA of this patient was screened with three restriction endonucleases and probed for three different genetic markers. In each case, note the absence of a band in the cancer tissue indicating an allelic loss. (b) Histogram of allelic loss based on RFLP mapping for each chromosome. Short arm loss (hatched) and loss from long arm (open) are scored separately.

(a) Vogelstein, B. et al., *Science,* 244:207–211, 1989; (b) From B. Vogelstein, et al., "Allelotype of Colorectal Carcinomas," *Science,* Vol. 244:209, 14 April 1989. Copyright © 1989 by the American Association for the Advancement of Science. Reprinted by permission.

(a)

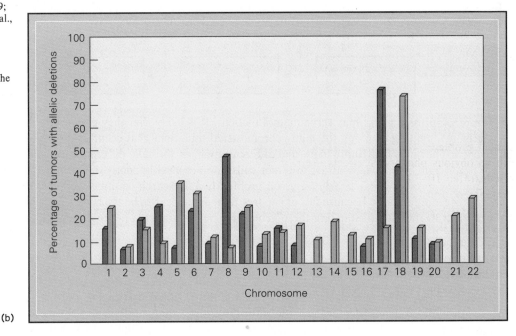

(b)

Measuring Mutagenicity and Carcinogenicity

Assays for chromosome damage are difficult and expensive. Because everyone experiences low levels of normal chromosome damage due to age, exposure to sunlight, radiation, and other background mutagenic sources, assays for chromosome damage in a population exposed to a potential mutagen or carcinogen must be carefully conducted. First, the normal level of mutation must be determined in a *control population,* and then the test population must be examined for damage beyond that found in the control group. Such tests should also be *blind:* the geneticist examining the karyotypes should not know to which test group a particular karyotype belongs.

While mutagenicity does not necessarily reflect carcinogenicity, the extent of chromosomal damage is a widely accepted measure of the extent of *exposure* of a test population to potentially harmful environmental agents. Chromosomal damage, however, is meaningful only on a population basis. A particular individual can have many chromosome abnormalities that are totally unrelated to exposure to a particular mutagenic agent. Furthermore, aberrant chromosomes in a test subject cannot be interpreted to mean that the individual will contract a genetic disease.

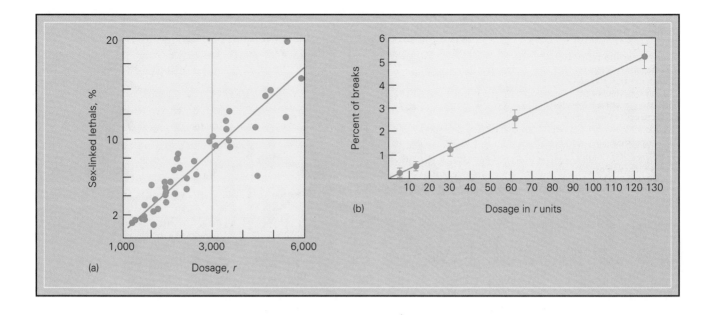

(a) Dosage, *r*

(b) Dosage in *r* units

On the other hand, a high level of genetic abnormalities in a test population may mean that the population is at risk for subsequent development of particular cancers. Because most chromosome abnormalities cannot be directly linked to cancer, the use of information on chromosomal anomalies poses serious ethical questions.

Is it appropriate, for example, to inform a person with karyotypic anomalies that he or she has suffered genetic damage, when the anomalies cannot be directly linked to a specific medical problem? Remember that chromosomal abnormalities are found in many people with no obvious phenotypic effects. Remember, also, that many cancers have long **latent periods,** often 10–30 years, between the time of exposure to a carcinogenic agent and the development of cancer. Many people would die of other causes prior to onset of cancer. For others, the psychological trauma of anticipating a disease, perhaps unspecified, that might or might not develop, would constitute mental torture as debilitating as the disease itself, despite the fact that the person was otherwise healthy during the latent period. Imagined harm can often be more severe than actual physical harm.

Thus, the tests for mutagenicity and carcinogenicity must be conducted with extreme experimental rigor, and the results must be interpreted carefully and explained with sensitivity to the needs of the subjects.

Sister chromatid exchange (SCE). Currently, two types of karyotypic data provide information concerning the extent of exposure to mutagenic agents. The first is the observed level of chromosomal mutations: deletions, translocations, ring chromosomes, and so forth. For example, the number of mutations resulting from chromosome breakage increases in direct proportion to the dose of radiation an individual receives (figure 11.24).

A second indicator of exposure is the frequency of **sister chromatid exchange (SCE).** SCE is essentially a crossing-over between homologous sites on the sister chromatids of a duplicated chromosome (figure 11.25). SCE is not considered to be a mutation, because the exchange of material between identical chromatids is presumably reciprocal, leaving the chromatids genetically equivalent. SCE is not known to have any harmful medical effects, but a high frequency of SCE is associated both with certain pathological conditions and with exposure to certain chemicals. Furthermore, SCE is a more sensitive measure of exposure to mutagens than is the level of chromosome mutations; the level of SCE increases at doses that do not produce chromosome abnormalities. Thus although the genetic significance of SCE is not precisely known, it is a common and easily measured consequence of exposure to mutagens.

Figure 11.25

Increasing doses of X-radiation increase the number of gene and chromosomal mutations in experimental material: (a) increased frequency of sex-linked lethal mutations in fruit flies with increasing dosage of X rays (summary of several studies); (b) increased frequency of chromosome breakage in grasshopper cells exposed to increasing dosage of X rays.

Figure 11.26

Sister chromatid exchange (SCE) in (a) Chinese hamster ovary cell and (b) human lymphocyte, indicating possible exposure of these cells to mutagenic agents.

(b) Wolff, Sheldon and Rodin B. *Science* 200 (5 May 1978) 543–45, © 1978 by the AAAS.

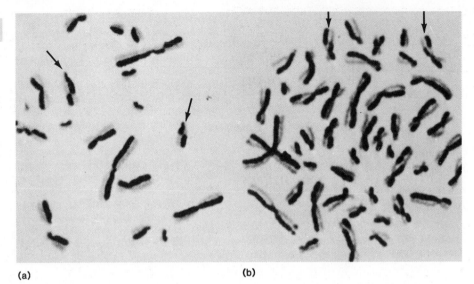

(a)   (b)

The Ames test.   Both chromosome mutations and SCEs are correlated with past exposure to toxic environmental agents. Obviously, it would be useful to assay for toxicity and, especially, carcinogenicity *prior* to introduction of chemicals into the environment. This is not a trivial problem. The American Chemical Society now lists over 6 million chemicals in its registry. About 63,000 chemicals are used routinely in medicine and industry, and about 1,000 more, never before known on earth, are synthesized in laboratories every year and made available for new uses. Which will cause genetic damage? Which will cause cancer? Deciding which chemicals to test, how to conduct such tests, and how much to spend on testing are major questions for business, government, and society.

Prior to 1974, tests for carcinogenicity in humans used laboratory mammals such as mice—an expensive, time-consuming process. Testing one chemical for carcinogenicity using mice or rats costs about $250,000 and requires three years. Maintaining just one mouse in a laboratory colony for a year costs $300–500. Tests with laboratory mice and rats are now used only on chemicals that are seriously suspected of being carcinogens; but these tests suffer basic limitations: chemicals that cause cancer in mice may not cause cancer in humans.

In 1974, Bruce Ames of the University of California–Berkeley developed a test using the bacterium *Salmonella typhimurium* to assay for the mutagenic effects of chemicals. The Ames test is rapid and inexpensive, requiring only two days and a small fraction of the cost of tests using animal models. It is now widely employed as a first estimate of the potential health hazards of new chemicals.

The **Ames test** uses a mutant strain of *Salmonella* that can only grow in a medium containing the amino acid histidine. To perform the test, this histidine-deficient strain of *Salmonella* is added to a medium containing *no* histidine. The chemical to be tested for mutagenicity is then added, along with an extract of enzymes from rat liver. The liver enzymes are added because some metabolic pathways in mammals differ from those of bacteria, producing metabolic by-products unique to mammals that may themselves be carcinogenic. Human metabolism may also produce unique by-products, but human liver extracts are not widely available (for obvious reasons). A set of synthetic human enzymes was developed to make the Ames test more directly applicable to human metabolism.

This concoction is incubated for two days and then examined for bacterial growth (figure 11.27). The only bacteria to grow should be those that have undergone a mutation permitting them to produce their own histidine. The more bacterial colonies present, the greater the number of mutations that have occurred. Because mutations occur at random, most mutations will affect genes other than the one or ones controlling histidine synthesis. Thus, the observed mutations represent only a small fraction of the total; the Ames test provides a conservative estimate of the total mutational effect of a chemical.

(a)

Smokey ash brown

Natural black

Moonhaze

A

Moonlit mink

Frivolous fawn

Wildfire

B

(b)

The results of one medically and commercially important Ames test are shown in figure 11.28. From 1973–1977, the chemical tris (or, more precisely, tris-2, 3-dibro-mopropyl-phosphate) was routinely added to children's clothing, especially pajamas, as a fire retardant. To test for mutagenicity, a series of Ames tests using varying amounts of tris was conducted: the larger the dose of tris used, the more mutant colonies of *Salmonella* that were found. Tris is absorbed directly through the skin, so protecting children from fire was also exposing them to doses of a mutagenic and potentially carcinogenic chemical—a truly modern dilemma. Tris metabolites were also detected in the urine of children who had worn tris-impregnated pajamas, even after the pajamas had been washed many times. Needless to say, tris is no longer used in children's clothing.

The Ames test is a direct measure of mutagenicity, not carcinogenicity. However, mutagenicity on the Ames test correlates strongly with carcinogenicity in humans. A positive result for mutagenicity on the Ames test is generally accepted as an indication of carcinogenicity. Over 90% of all chemicals known to be carcinogenic in humans have given positive results on the Ames test.

Are more direct measures of carcinogenicity possible? Perhaps, but as with tests using mammalian models, such tests might be prohibitively expensive. The only direct test would use human subjects, which is ethically unacceptable. A further complication is the long latent period shown by most cancers. Is it more reasonable to use the Ames test to estimate toxicity of a new chemical, or to expose thousands of persons, then wait decades to see if they develop cancer? Present practice uses the Ames test as a first step in evaluation. If a chemical that may be widely used shows mutagenicity on the Ames test, subsequent tests on laboratory mammals can be ordered to confirm the Ames result.

Some Examples of Common Chemical Carcinogens
and Mutagens: Recreational Drugs

Mutations may be either germinal or somatic. In the long term, germinal mutations are the most troubling because their effects could persist for many generations. However, somatic mutations are increasingly implicated as a cause of cancer. Cancer is now the second leading cause of death in the United States, behind heart disease. In 1988, over 800,000 persons were diagnosed as having cancer, and over 485,000 Americans died of cancer, more than all who died in World War II plus the Vietnam War. Cancers of the lung, breast, and colon account for more than 50% of the cancer-related deaths in the world. While significant progress has been made in diagnosing and treating many cancers, the incidence of some cancers has increased over the past twenty-five years, perhaps due to the increasing levels of toxic chemicals in the environment.

Figure 11.27

The *Salmonella* assay of the Ames test. (a) Series of plates with (A) control with no mutagen, indicating spontaneous back mutation rate; known mutagens: (B) furylfuramide, (C) aflatoxin, and (D) 2–aminofluorene. (b) Plates exposed to three different hair dyes (A) without liver enzymes and adjacent control plate; (B) with liver enzymes and adjacent control plate. Note increased mutation in B.

(a and b) Bruce N. Ames.

Figure 11.28

Results of an Ames test on the mutagenicity of the flame retardant, tris, compared to other compounds. All tests used *Salmonella* strain TA100. Note magnitude of revertants (mutations) when exposed to tris.

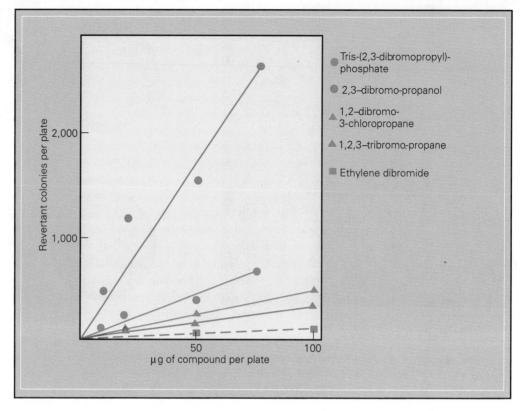

Tobacco. The most dramatic increase in cancer since World War II has involved cancer of the lung, primarily because of the increased rate of cigarette smoking. The scientific evidence concerning the harmful effects of tobacco is abundant and unequivocal. In 1982, the Surgeon General of the United States issued a report saying, "Cigarette smoking is clearly identified as the chief preventable cause of death in our society and the most important public health issue of our time."

There are 350,000 smoking related deaths in the United States each year. This is the equivalent of two fully loaded 747 jumbo jets crashing and killing all aboard every day each year. Of the more than 111,000 deaths from lung cancer, 85% could have been prevented had the victims not smoked. A smoker is 1,000 times more likely to contract cancer than a nonsmoker.

Death is not the only cost to society. Medical care for smoking-related illnesses costs about $13 billion annually, and lost productivity and wages cost $25 billion. Society bears a massive economic burden due to the activities of smokers. The health problems caused by cigarette smoke are not limited to smokers themselves. Children of mothers who smoke during pregnancy often have low birth weight. Children who live with a smoking parent are at increased risk of respiratory problems. The nonsmoking spouse of a smoker is estimated to have about 30% higher risk of contracting lung cancer than the spouse of a nonsmoker. About 20% of all deaths from lung cancer among nonsmokers are estimated to be caused by environmental tobacco smoke. Passive smoke is estimated to kill between 20,000–40,000 persons annually.

Cigarette smoke contains over 6,800 different chemical compounds, many of them highly toxic. Included on the list are hydrogen cyanide (which causes emphysema), vinyl chloride, arsenic, the radioactive metal polonium, and nicotine. Urine from smokers is highly mutagenic on the Ames test (figure 11.29), probably explaining the high rate of bladder cancer among smokers.

The substance of most concern in tobacco is nicotine. Nicotine, an alkaloid, is an extremely potent natural poison. A few drops of pure nicotine absorbed through the skin are fatal. A grim fact: before World War II more tobacco was grown in the United States for use in insecticides than for cigarettes. Today, nicotine is used by veterinarians as an external parasiticide.

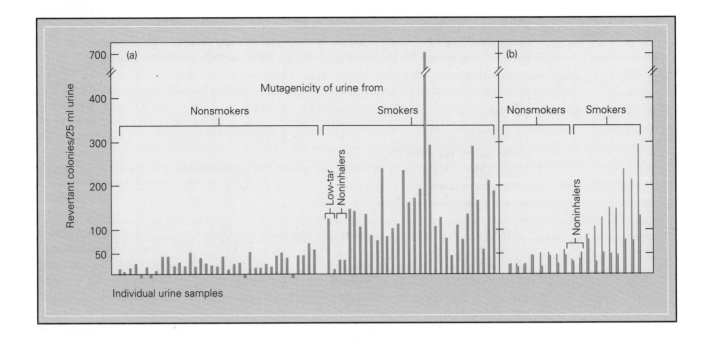

The evidence relating smoking to cancer is as strong as the evidence for any toxic substance. The tobacco industry, however, is so powerful a political force that the scientific evidence and the immense costs of smoking to society are ignored. Despite the Surgeon General's conclusion that cigarette smoking is the "chief preventable cause of death" in the United States, the government does little to protect the public from the harmful effects of smoking: warning labels are required on cigarette packages and television advertising of cigarettes is restricted. Advertising of cigars, pipe tobacco, and chewing tobacco is not limited. In fact, the government provides a massive financial subsidy to the tobacco-growing industry.

The industry attempts to defend itself by perpetuating the Big Lie, a technique described in George Orwell's *1984*. Tell a lie so big and outrageous that it cannot possibly be true, and tell it often enough, and soon the public will come to believe it. Says the chairman of the Tobacco Institute, "Scientific research has not been able to establish [the causal] link [between smoking and cancer]." In fact, no causal link between carcinogen and cancer is more firmly proven.

Marijuana.   *Cannabis sativa* (marijuana, pot, grass, or hashish) is the most widely used drug among adolescents. The health effects of marijuana have been seriously debated during the past two decades.

In 1982, the National Academy of Science's Institute of Medicine released a comprehensive review of all data relevant to the health effects of marijuana. The evidence is unequivocal that smoking marijuana leads to respiratory problems similar to those caused by smoking cigarettes. There is not yet evidence, however, to relate marijuana smoking to increased risk of lung cancer. Two reasons may explain the latter finding. First, lung cancer has a long latent period. As marijuana has been widely used in the United States only since the 1960s, sufficient time may not yet have elapsed for the increased incidence of lung cancer among marijuana users to become evident. Second, marijuana users probably smoke fewer marijuana cigarettes than tobacco smokers smoke tobacco cigarettes. Although marijuana smoke has about 50% more carcinogenic material than tobacco smoke, marijuana users may be subjected to smaller doses of carcinogens, thus reducing the relative risk of lung and other cancers. In such a case, a proportionately larger amount of data would be required to establish a carcinogenic role for marijuana.

Marijuana use may also produce temporary sterility, particularly in males. Marijuana suppresses testosterone production, in turn inhibiting sperm production. It interferes with both DNA and protein synthesis. In laboratory mice, marijuana use decreases the size and weight of the prostate and testes. Spermatogenesis is impaired, with many ring chromosomes

Figure 11.29

Relationships between number of revertant colonies on Ames test (number of mutations), as determined by exposure to urine from smokers and nonsmokers. Dark color represents urine collected in the evening; light color represents an early morning specimen: (a) single samples from individuals and (b) paired morning and evening samples from other individuals.

and translocations visible in gametes. Additionally, among offspring of mice subjected to marijuana treatment, both reduced fertility and meiotic chromosome abnormalities were detected, suggesting that marijuana causes germinal as well as somatic mutations.

Furthermore, the active ingredient in marijuana crosses the placenta and can affect a fetus. It is also secreted by the mammary glands and may thus affect nursing infants. Evidence indicates that marijuana suppresses the activities of the immune system. Marijuana impairs motor response and short-term memory, and it reduces learning capacities. The effects on the nervous system last four to eight hours past the feelings of intoxication, considerably longer than the effects of alcohol.

A 1988 study showed smoking one marijuana cigarette was as unhealthy as smoking five regular cigarettes. The evidence is strong that marijuana has many potentially harmful effects on health and may also be mutagenic.

Caffeine. Caffeine is a stimulant of the central nervous system. It occurs naturally in coffee, tea, and other plants, and it is an ingredient in soft drinks and aspirin. Although caffeine has no analgesic or pain-relieving effects, it is used in aspirin for its stimulating effect on the central nervous system, increasing alertness and apparently making a person feel better. Caffeine is an habituating drug, but withdrawal is achieved far more easily than with narcotics.

Caffeine is a base analogue (figure 11.15), suggesting that it is likely to be mutagenic. It has been shown to be mutagenic in bacteria. In eukaryotes, it breaks chromosomes only during replication, and its mutagenic effects have yet to be shown as an important health problem. Caffeine crosses the placenta easily, reaching the same concentration in the fetus as in the mother. Recent studies have given conflicting results concerning potentially harmful effects on fetuses. Rats fed the caffeine equivalent of two cups of coffee per day produced offspring with slowed rates of skeletal growth, and a ten-cup-per-day equivalent reduced litter sizes.

Lysergic acid diethylamide (LSD). In 1967 chromosome studies of users of LSD revealed translocations and chromosome breakage. The subjects of these investigations, however, had used the drug illicitly; therefore, these findings must be considered at best circumstantial and not the result of controlled experiments. For example, the amount of chromosome damage prior to use of LSD is unknown. Drugs obtained illicitly often include many impurities, which may also be mutagenic. Furthermore, drug users often mix drugs, and the combined effects of several toxic chemicals are unpredictable and often extremely severe.

Recent controlled studies still leave the mutagenic capacities of LSD in doubt. Moderate doses of pure LSD have not produced chromosomal mutations in humans. Large doses of LSD are weakly mutagenic in *Drosophila* and fungi; LSD is probably not significantly mutagenic in concentrations that would normally be taken by humans. There is no evidence that LSD is carcinogenic.

Radiation, Mutation, and Cancer

In 1921, when almost nothing was known of the nature or causes of mutations, **H. J. Muller** said, "Beneath the imposing structure called Heredity there has been a dingy basement called mutation." For the remainder of his professional career, Muller studied mutation. His work on the mutagenic effects of radiation, for which he received the 1946 Nobel Prize for Physiology or Medicine, comprised the first detailed examinations of a mutagenic agent.

X-radiation was discovered in 1895 by Wilhelm Conrad Roentgen. He and his colleagues used X rays indiscriminately to examine the skeletons of patients (figure 11.30). Within a few years some of these scientists and physicians lost their hands and, in some cases, their lives as a result of radiation burns. Despite the early awareness of the potential dangers of X rays, their use was widespread. Until the 1950s, many shoe stores in the United States used fluoroscopes, a form of X ray, to check the fit of new shoes. The devices were so interesting that many children were encouraged by their parents to X ray their feet for entertainment while the rest of the family shopped.

Figure 11.30

One of the first X rays taken, this was done in 1898 by Wilhelm Roentgen, who discovered X rays in 1895.

The Bettmann Archives.

Other forms of radiation were only slowly recognized as dangerous. Radium, a radioactive isotope formed by the decay of the element uranium, was discovered by Marie Curie in 1898. Like X rays, radium was treated more as an innocuous novelty than as a dangerous substance. Between 1915 and 1930, thousands of persons in the United States drank radium as a medicinal substance. In Europe, radium-containing chocolate bars were marketed commercially. As recently as 1953, an American company sold a contraceptive jelly that contained nearly a microgram of radium in a two gram tube. Today, some persons still travel to radon mines in the Rocky Mountains to "take the cure." Radon is a gas formed by the disintegration of radium. As humorous as some of these examples may seem, the consequences were not always trivial. Cancers of the jaw bone, face, and throat were developed by many women who painted radium on the dials of wrist watches. These women often held the paint brushes in their mouths, receiving large cumulative doses over many years.

Aware of some of these dangers, Muller demonstrated in 1927 that X-radiation increases the mutation rate in *Drosophila*. Within a year, X rays had been shown to increase the mutation rate in plants as well. These observations led Muller to introduce the concept of **genetic load.** Genetic load is a measure of the amount of deleterious mutant genes in a population. Muller believed that increased radiation exposure could significantly increase the genetic load carried by humans. Since most mutations are recessive, their harmful effects would at first be hidden, but at increased frequencies the harmful alleles would eventually be expressed, having serious long-term consequences for our species.

By 1930, radiation genetics was recognized as an important aspect of genetic research. Since Muller's work, the role of radiation in causing both mutations and cancer has been well established.

Types of Radiation

Two types of high energy radiation are genetically important (figure 11.31). *Electromagnetic radiation* is similar to visible light, moving in waves. Examples include microwaves, ultraviolet light, and X rays. Some forms of electromagnetic radiation, such as radio waves, appear to be of little significance genetically, but UV light and X rays contain much higher energy and can cause considerable damage. Because of its ultraviolet component, sunlight is the most potent natural carcinogen, causing about one-third of all tumors in humans. Fortunately, almost all of these are benign skin tumors that can be removed by surgery. This type of tumor is treated with great success, so it is usually omitted from cancer statistics.

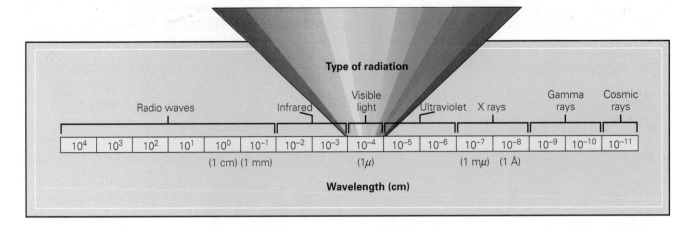

Figure 11.31

Spectrum of electromagnetic waves from long radio waves to very short cosmic rays.

Particulate radiation is even higher in energy than electromagnetic radiation. Radioactive elements emit high energy particles (protons, neutrons, and electrons) that can penetrate the human body. Electromagnetic and particulate radiations release energy as they penetrate living tissue, causing reorganization of the tissue's atomic structure. One effect is the alteration of DNA, genetic mutation.

Various units are used to measure amounts of radiation. A **rad** describes the amount of energy absorbed by living tissue. A **rem** (roentgen equivalent man) measures the biological damage done by radiation. For small doses, rads and rems are roughly equivalent. Very small doses are often measured in millirads or millirems, $\frac{1}{1000}$ of a rad or a rem. The exposure and absorption of radiation are seldom distributed equally over the entire body, as illustrated by the women painting radium on the faces of watches. The **genetically significant dose (GSD)** is the amount of radiation absorbed by the gonads, which might produce germinal mutations.

Sources and Effects of Radiation

Exposure to radiation may come from natural or human-made sources (table 11.3; figure 11.32). Natural or background radiation comes from sunlight and other cosmic radiation, radioactive isotopes in geological formations, and radioactive substances in food. Doses of cosmic radiation vary with altitude. The earth's atmosphere filters cosmic radiation efficiently, but at high elevations the atmosphere is much thinner than at sea level, increasing the exposure to radiation. A person living at sea level is exposed to about 30 mrem per year, but in mountainous regions exposure may be as high as 110 mrem per year.

Ultraviolet radiation is filtered from sunlight by the ozone in the stratosphere, a layer of the atmosphere found between ten and thirty miles above the earth's surface. In the early 1970s, scientists focused public attention on the potential depletion of this ozone layer by the increasing release into the atmosphere of a group of chemicals known as chlorofluorocarbons (CFCs), widely used in the past few decades as coolants in refrigerators and propellants in aerosol sprays. CFCs interact with stratospheric ozone, converting it to molecules of oxygen, which do not filter ultraviolet radiation from sunlight.

These early warnings were greeted with skepticism, but in 1978 the use of CFCs in spray cans was banned in the United States, though not elsewhere. Convincing evidence is now accumulating that a significant reduction in stratospheric ozone has begun and that CFCs are at least partly responsible. In particular, a so-called "ozone hole," or thinning in the concentration of ozone, was detected in the atmosphere over Antarctica each austral spring during the 1980s. By 1986, the levels of ozone over Antarctica were reduced by about 50%, although they recovered to about normal levels during austral summer. Reductions in stratospheric ozone are occurring elsewhere in the world.

The consequences of worldwide reductions in ozone could be severe. Ultraviolet radiation causes skin cancer and may contribute to other health problems, such as cataracts. Decrease in the ozone concentration by only a few percent could dramatically increase the incidence of skin cancers in humans. Domestic livestock could be expected to be adversely affected, as well.

TABLE 11.3

Radiation doses and units

Rad = amount of radiation absorbed by tissue
Rem = amount of biological damage done by radiation (rem = roentgen equivalent man)
GSD = genetically significant dose; the amount of radiation reaching the gonads

Source of radiation	Average dose in mrem/year (range)
Natural	
Cosmic	45 (30–130)
Terrestrial	60 (30–115)
Internal radionuclides	25
Total	130 (Unavoidable)
Human made	
Global nuclear fallout	4
Nuclear power	0.003
Medical diagnostic	72
Radiopharmaceutical	1
Occupational	0.8
Miscellaneous	2
Total	80
Grand total	210

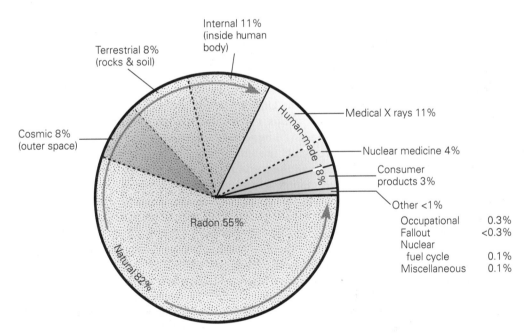

Figure 11.32

Sources of ionizing radiation exposure in the U.S.

From E. Marshall, "Academy Panel Raises Radiation Risk Estimate," *Science,* Vol. 247:23, 5 January 1990. Copyright © 1990 by the American Association for the Advancement of Science. Reprinted by permission.

An international agreement in 1987 initiated a program for reducing CFC use and production among developed nations, but it is a long-term plan and does not aim for total elimination of CFCs. The program also does not apply to the developing nations of the world, most of which lie in tropical regions and have a great need and demand for refrigeration. If dramatic increases in skin cancer occur, society will be forced to balance the production of these useful chemicals against their harmful effects. Whether the depletion of ozone could be reversed is unknown.

Figure 11.33

Mutation rate in a variety of
organisms as a function of
genome size and the exposure
to radiation.

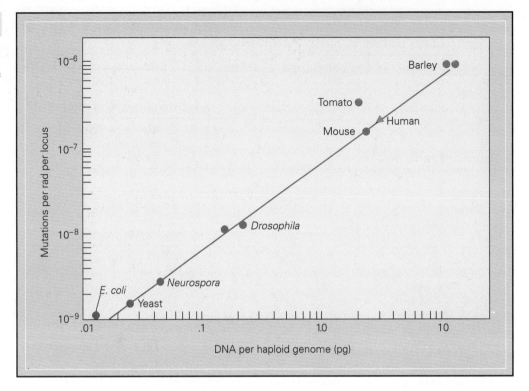

The amount of radiation derived from natural terrestrial sources varies substantially. Some mountainous regions of China have levels as high as 230–440 mrem per year; and in one region of India background radiation approaches 3,000 mrem per year. As these doses are 1/100 that received in Hiroshima and Nagasaki (see the following section) it is deemed an acceptable risk.

In general, radiation derived from human-made sources is less than natural radiation (table 11.3). Radiation from global nuclear fallout has decreased since the abolition of above-ground testing of nuclear weapons by the United States and the Soviet Union. France and China, however, continue above-ground tests. In some cases, populations immediately downwind from nuclear test sites have received large doses of radiation, and patterns of global weather circulation ensure that eventually the entire earth receives increased radiation from nuclear testing. Radioactive isotopes of the metal strontium are a byproduct of nuclear explosions. These isotopes fall on vegetation, where they are consumed by cattle, then transmitted to humans in cow's milk. Strontium replaces calcium in bones and thus accumulates in human tissue.

Medicine is the primary artificial source of radiation. X rays are an important diagnostic tool, and radiation is used to combat cancers. As the efficiency and precision of x-ray machines have improved, exposure levels have decreased.

The harmful effects of excessive exposure to radiation are well documented. In high doses, radiation causes burns, nausea, loss of hair, anemia, leukemia, sterility, cancer, and genetic mutation. In general, the severity of these effects increases in direct proportion to the dose received. At low doses, the effects are more difficult to document, and considerable controversy exists concerning the existence of a safe lower limit of radiation. Is a certain small amount of radiation exposure tolerable? Or does any exposure, however small, significantly increase the risk to health? While available data are inconclusive, some evidence indicates that, in general, the mutation rate is roughly constant for all organisms, when adjusted for size of the genome (figure 11.33). Thus, because mammals have a comparatively large amount of DNA, the total number of mutations induced by a constant amount of radiation would be greater in humans than in bacteria and other organisms with small amounts of DNA.

TABLE 11.4

Comparison of LD50s among different mammalian species

Organism	LD50 (rads)
Human	450
Monkey	600
Sheep	250
Dog	250
Rabbit	800
Rat	700
Mouse	800
Guinea pig	450

From V. P. Bond, T. M. Fliedner and J. O. Archambeau, *Mammalian Radiation Lethality.* Copyright © 1965 by Academic Press, Orlando, FL. Reprinted by the permission.

At present, the United States government has established 170 mrem per year as the upper limit of acceptable exposure to human-made sources such as X rays and radioactive material. It is estimated that if every person in the United States received the maximum allowable amount (170 mrem) there would be thirty-two thousand additional cancer deaths annually.

For comparative purposes, toxicity is often expressed in terms of an **LD50**: the dose of an agent that is lethal to 50% of a test population or species. For humans, the LD50 of radiation is estimated to be 450 rads. Mammals appear to be among the most susceptible species (table 11.4).

The genetic damage caused by radiation includes chromosome breakage and mutation, sister chromatid exchange, and point mutation. In combination with other types of radiation damage, these genetic changes may lead to cancer, particularly cancer of the blood and blood-forming organs, i.e., leukemia. Leukemia is the most common form of cancer associated with exposure to X-radiation. In the first half of this century, X raying fetuses was a common medical practice. The children of mothers X rayed during pregnancy have a 50% higher incidence of leukemia than children of mothers not X rayed. Leukemia has a relatively short latent period, and one of the first effects noted among survivors of the bombings of Hiroshima and Nagasaki was an increased incidence of leukemia.

Other cancers may appear later, of course. Long-term medical tests have been conducted on residents of the Marshall Islands, who received large radiation doses after a test nuclear explosion on the Bikini Islands in 1954. Following the explosion, the wind shifted and un-expectedly carried the fallout to the Marshall Islands. Some of the residents received doses estimated at 175 rads, only slightly less than *half* the LD50. All became ill, but subsequently recovered. Ten years after exposure, over 50% of the islanders still had visible chromosomal mutations in the white blood cells, and five years later a slight increase in incidence of benign tumors of the thyroid was detected.

The Marshall Islanders received a single massive *acute dose,* but because the effects of radiation are cumulative, a series of small or *chronic doses* may be equally damaging. About five hundred pounds of radioactive plutonium are processed annually in Rocky Flats, Colorado. Thus, the residents of Rocky Flats are chronically exposed to radiation. The incidences of both leukemia and lung cancer among the residents of Rocky Flats are about double those found in a nearby control community.

Some of the best data relating to the effects of chronic exposure to low doses comes from regions of high background radiation. In the region in China where the dose of background radiation is as high as 440 mrem annually, studies of large samples of the population detected no significant increase in frequencies of chromosomal damage, incidence of hereditary diseases and congenital malformations, spontaneous abortion rate, or frequencies of various cancers. Only a higher than normal incidence of Down's children suggested the possibility of genetic damage caused by high background radiation.

On the other hand, studies of populations in India living in regions of background radiation about ten times greater than that in China (as high as 3,000 mrem annually) detected significantly increased incidences of chromosomal aberrations, Down's syndrome, and severe mental retardation. These problems can reasonably be attributed to the high levels of radiation.

Hiroshima and Nagasaki

The survivors of the atomic bomb explosions at Hiroshima and Nagasaki have been the subjects of the most comprehensive epidemiological studies ever conducted. Since 1945, more than $100 million have been spent to follow the medical histories of over 100,000 persons. Initially, the survivors were examined for the immediate medical effects of the explosion and radiation exposure. During the 1960s, however, karyological techniques for humans were developed that permitted expansion of the studies to genetic effects. (Remember, the correct diploid number for humans was not known until 1956.)

The studies of survivors and their descendants are continuing. Specific doses of radiation have been calculated for each survivor on the basis of their location at the time of the explosion, and the dose estimates have been correlated with the probability of developing certain cancers. The mean dose was 132 rem per survivor.

The incidence of cancer is clearly related to the dose of radiation among the survivors of the nuclear explosions. Survivors have sustained rates of cancer far higher than normal, so great in fact that they constitute a cancer epidemic. Furthermore, the data clearly show the variation in latent period associated with different cancers. During the late 1940s and the 1950s, heavily exposed persons developed leukemia at a rate thirty to forty times higher than normal, but the rate decreased almost to normal by the early 1970s. In the late 1950s cancers of the lung and thyroid gland were in high frequency. In the 1960s and 1970s cancers of the breast and salivary glands increased. By 1980, rates of multiple myeloma and cancer of the colon had increased. Exposure to a single high dose of radiation has obviously altered the probability of contracting many types of cancer throughout the entire *lifetime* of the survivors. In 1987 a reassessment of the radiation received by survivors of the bombings was completed. These results indicated that the elevated cancer rates seen in this population were caused by *smaller* doses of radiation than was previously believed. This means cancer risk from radiation is higher than previously estimated and consideration is being given to lowering the maximum allowable radiation exposure by 30–70%.

Genetic studies begun decades after the explosion also reveal residual effects. Karyotypes of survivors show that chromosome fragments and ring chromosomes are common, in spite of the expectation that these abnormalities would not have survived or been transmitted to descendant cells for so many years. Although, again, the genetic changes cannot be proven to have caused the cancers, the associated mutagenic and carcinogenic effects of radiation are abundantly clear.

There are several notable positive aspects of the data from Hiroshima and Nagasaki. First, no latent pathologies other than cancers have yet been shown to occur at higher than normal levels. Second, the harmful effects of the radiation seem limited to the survivors themselves and do not extend to their children. Incidences of death, congenital abnormality, and point and chromosomal mutations are not above normal in the children of the survivors. It was initially feared that genetic mutations resulting from radiation would be passed on for generations, but such does not appear to be the case.

Such findings, however, should not be interpreted to mean that germinal mutations did not happen, or that reproduction of survivors has not been deleteriously affected. The high incidence of somatic mutations almost certainly means that germinal mutations did, in fact, occur. The long-term effects of germinal mutations may have been limited because gametes possessing numerous deleterious mutations were probably selected against. That is, affected gametes either were not viable, or they were so seriously affected that they were relatively unlikely to participate in a successful fertilization. Because most mutations are recessive, however, it is possible that higher than normal levels of germinal mutations have been transmitted but not expressed in offspring of survivors. These mutations would increase the genetic load of the population as a whole. Increased levels of harmful mutations would be

(a)

(b)

Figure 11.34

(a) Atomic bomb explosion over Hiroshima, August 6, 1945. (b) Peace Park, Hiroshima, August 6, 1988. A memorial service marking the 43rd anniversary of the bombing of Hiroshima. Note the symbolic skeleton dome, gutted by the bomb, in the background.

(a) AP/Wide World Photos; (b) Bettmann Archives.

expected to be expressed as higher numbers of "genetic deaths" in succeeding generations. Although no evidence yet indicates harmful effects on children of survivors, long-term harmful effects on the Japanese gene pool cannot be ruled out.

Finally, it is worth noting that many survivors (called *hibakusha,* meaning explosion-affected persons) have not married or reproduced. Failure to marry has resulted from disfigurement, injury, or cancer, but in some cases fear of producing genetically damaged children may also have been a reason.

Chernobyl. On April 26, 1986, the largest accidental release of radioactive material occurred in the Soviet Union. There were roughly 1,000 workers present at the Chernobyl nuclear power plant in the northern Ukraine at the time of the accident. More than 135,000 people were evacuated from the area around the plant. Emission of highly radioactive material, including the highly volatile Iodine-131, and Cesium-134 and-137, were released over a nine-day period until the reactor was safely contained (figure 11.35). An immediate consequence of the accident was 237 cases of acute radiation sickness, including 31 reported deaths (of rescue and fire-fighting personnel) at Chernobyl.

There were no such effects outside of the immediate Chernobyl region. There is no expectation that this accident will produce adverse genetic effects outside of Chernobyl, despite the dispersal of a radioactive plume across much of the Northern Hemisphere. The projected impact on the health of persons most exposed to radioactive fallout from Chernobyl is less

Misinformation   **313**

Figure 11.35

Histogram indicating estimated daily release of radioactivity (in millions of curies) from the nuclear reactor at Chernobyl, until the release was contained.

From J. F. Hearne, "Nuclear Power After Chernobyl," *Science,* Vol. 236:676, 8 May 1987. Copyright © 1987 by the American Association for the Advancement of Science. Reprinted by permission.

than that caused by a lifetime of cigarette smoking. Projections one year after the accident suggest that 21,000 Europeans may die of radiation-induced cancer over the next fifty years. This is twice the number of expected Soviet victims—because of the vagaries of wind patterns. Globally, the Chernobyl accident can be expected to contribute 39,000 new cancer deaths—but this number is inconsequential compared to the 630 million cancer deaths expected. The major global impacts of this accident, therefore, appear to be economic (costs are estimated to go over $15 billion) and social.

The Politics of Toxic Substances

Of the thousands of chemicals in our daily lives, many are mutagenic, carcinogenic, or teratogenic. Many of these (caffeine, tobacco, and alcohol, for example) are widely used. Other common substances are equally toxic, although their harmful effects are less publicized: formaldehyde (used in household insulation and biological preservatives), toners (used in photocopy machines), hair dyes, and many medical drugs, for example.

Governments throughout the world regulate toxic substances to varying extents. The decision to regulate depends not only on scientific evidence, but on social and political considerations as well. Many toxic substances have beneficial uses, and their regulation involves balancing the positive and negative aspects of their use. No better example exists than X rays. Radiation can cause cancer, but medical benefits warrant continued use. Not only are X rays a major diagnostic tool, but radiation is an important weapon against cancer. Rapidly dividing cancer cells are more seriously affected by X rays than are normal, slowly dividing cells, so a concentrated dose of well-aimed radiation may selectively kill a growing cancer. Similarly, caffeine is toxic only in amounts far greater than most people ingest in coffee, tea, or soft drinks, so their positive effects seem to outweigh any gain to be derived from restriction of caffeine in beverages.

In reality, of course, political decisions may ignore health considerations. Powerful economic interests underlie the continued tolerance of substances such as tobacco, the single greatest cause of cancer in the United States. Its health and economic cost to society is immense, but tobacco is not restricted. Even worse, the tobacco industry is supported by large government subsidies. Similarly, powerful lobby groups were able to forestall action by the Environmental Protection Agency to restrict use of formaldehyde, a highly carcinogenic substance.

We are not suggesting that use of all toxic substances should be severely limited. That would be unwise and probably also impossible. We do believe, however, that the interests of governments and various political and economic groups do not always coincide with the health interests of the public. The carcinogenic effects of atomic radiation were poorly understood throughout the 1940s and 1950s. As a result, not only the unfortunate survivors of Hiroshima and Nagasaki, but also many American, British, New Zealand, and Australian military personnel and civilians, as well as Pacific Islanders, were exposed to nuclear test radiation. These exposures occurred primarily due to our ignorance of the genetic and medical risks. Even after knowledge advanced, the government was often slow to assume responsibility for its actions or even to admit them.

As genetic knowledge increases, the dangers of many environmental agents will become even clearer. Appropriate regulation of such agents requires a knowledgeable citizenry to ensure that decisions are founded on scientific data relating to health risks, and not based on political and economic expediency. The American political system appropriately allows open debate of public issues such as the regulation of toxic or potentially toxic substances. Both sides of arguments may be forcefully expounded in public. In some cases, such as the regulation of sodium nitrite, even professional scientists and toxicologists differ in their opinions concerning the need for regulation. Nitrites indeed increase the risk of cancer in laboratory mammals, but when used as meat preservatives, they reduce the danger of transmitting food poisoning and other bacterial diseases. Which danger should we choose?

In other cases, propagandizing relies not on evidence, but on irrelevant claims that cloud consideration of the issue at hand. For example, the evidence is clear that cigarette smoke has many serious affects on the health of smokers and nonsmokers—and no beneficial effects. The Tobacco Institute therefore cannot cite positive aspects of cigarette smoking and is forced to rely on a fraudulent issue: that smoking does not affect others and should therefore be a matter of choice. In addition to concern regarding the health risk created by cigarette smoke, many nonsmokers find smoke foul smelling and irritating to their eyes. Some nonsmokers are allergic to cigarette smoke—not surprising, considering the thousands of compounds spewed into the air by burning cigarettes. Smoking clearly affects nonsmokers, yet the industry obscures the issue by focusing on the freedom of choice of smokers.

A second example illustrates the same point. In the late 1970s, *dioxin*, one of the most toxic chemicals known, was implicated as a cause of spontaneous abortions, birth defects, and cancer. Dioxin is so toxic that no laboratory experiment has been able to determine a dose so small that it is not toxic—even one part per billion. Because dioxin occurs as a contaminant in several herbicides commonly used by wood product companies, these companies and the chemical manufacturing companies were subjected to considerable public scrutiny and adverse publicity concerning the use of dioxin-containing chemicals in the environment. One major corporation conducted an advertising campaign observing that chemicals are the basis of life, while showing some of the chemicals used by the company to increase the productivity of managed forests.

One such advertisement promoted the virtues of an all-purpose chemical called "dihydrogen oxide"—H_2O, water. The implication was that if this corporation valued water so highly, then all other chemicals it used must be equally benign. Dihydrogen oxide belongs in the same list of good things as baseball, hot dogs, apple pie, and mom. In other words, the corporation was saying, "Trust us—we know what's best." Such commercials are clever and often persuasive. Only an educated and informed public can evaluate how much truth is contained in such propaganda.

Summary

1. Mutations are heritable changes in DNA. They may be somatic, affecting somatic tissues and transmissible only to descendants of a particular cell within the body, or they may be germinal, affecting the gonads and transmissible to succeeding generations. Germinal mutations may produce long-lasting effects on the species, while somatic mutations often have harmful effects on the health of the bearer.

2. Most mutations probably have harmful effects on survival and reproduction. However, mutations also provide the ultimate source of genetic variation and thus play a fundamental role in evolution.

3. Variations in ploidy are common in plants and some animals. The gamete stage of life history is ordinarily haploid for all species, and haploidy may also be a mechanism of sex determination. Polyploids often possess characteristics of great commercial value, and many domestic crops are either artificial or natural polyploids.

4. Polyploidy in animals is relatively infrequent. Triploid and tetraploid human fetuses occur, but never survive.

5. Aneuploidy, the deficiency or excess of one or a few chromosomes, is relatively common and occurs as a result of meiotic or mitotic nondisjunction. In humans, monosomy for the X chromosome, Turner's syndrome, is the only viable monosomy. Trisomies are relatively more viable than monosomies, but they are usually associated with physical and mental abnormalities. Only trisomies involving the sex chromosomes and trisomy 21, Down's syndrome, survive to adulthood.

6. Some mutations affect only portions of one or two chromosomes. Deletions and duplications involve loss or gain of a segment of a chromosome. Inversions occur when a segment of a chromosome is reversed in position within the chromosome. Translocations occur when segments of one chromosome attach to a non-homologous chromosome. Most mutations of these types are deleterious, and a few have been linked to specific diseases.

7. The effects of chromosomal mutation depend upon the genes affected. For example, Down's syndrome may be caused either by an extra chromosome 21 (trisomy) or by an extra long arm of chromosome 21, translocated to another chromosome and inherited from one parent (translocation Down's). Thus, the condition is caused by a gene or genes carried on the long arm of chromosome 21. Similarly, Turner's syndrome may be caused by loss of one X chromosome or by deletion of only the short arm of one X.

8. The Philadelphia chromosome (Ph1) is a mutant form of chromosome 22 resulting from a translocation with chromosome 9. Discovered in 1960, it was the first chromosomal mutation related to a specific cancer, chronic myelogenous leukemia.

9. Fragile sites are heritable narrowings or gaps located at precise points on chromosomes. Over forty are known, and they appear to represent sites at which chromosomes may be especially subject to breakage. Some fragile sites occur near or at sites where oncogenes are located. The only fragile site thus far clearly associated with genetic disease is that causing a form of X-linked mental retardation, the fragile X syndrome.

10. Point or gene mutations affect only a small segment, sometimes only a single base, of a chromosome. Point mutations may be either frameshift or substitutional.

11. Mutations are random. Neither their location nor their result can be predicted. However, the rate with which they occur is relatively constant—about 0.1–1.0 mutations in every one hundred thousand generations of each gene. Each human probably carries about three to five lethal new mutations, which are not expressed because they are recessive.

12. The natural rate of mutation is affected by many different environmental agents. Various forms of radiation and chemicals are mutagenic, causing significant increases in the mutation rate.

13. Cells are adapted to a certain level of background mutation. Most mutations are probably fixed immediately by mechanisms of DNA repair.

14. Nonetheless, mutations are probably involved in many health problems, especially aging and cancer.

15. The mutation theory of aging hypothesizes that aging results from accumulated deleterious mutations, producing a steady degeneration of bodily function. The commitment theory of aging claims that development is under the direct control of genetic programs, which include maximum limits on life span.

16. Teratogens are substances that interfere with normal embryonic development. During the critical first trimester of pregnancy, chemicals that might be otherwise harmless can cause severe abnormalities in development.

17. Alcohol has been shown to cause teratogenic changes in the developing embryo, resulting in a suite of features in the newborn known as fetal alcohol syndrome (FAS). Mental retardation and stunted growth are key features of FAS.

18. Cancer cells are characterized by uncontrolled growth, lack of contact inhibition, and ability to spread throughout the body, metastasis.

19. Based on the tissues involved, three types of cancers are recognized: carcinomas, cancers of the skin, glands, reproductive tract, etc.; leukemias, cancers of the blood-forming tissues; and sarcomas, cancers of the bones and muscles.

20. Proto-oncogenes, normal cellular genes involved in growth, can be converted to oncogenes, cancer-causing genes, in a variety of ways. These include point mutations, chromosome rearrangements, viruses, radiation, and chemicals.

21. Tumor suppressor genes form part of the body's normal defense mechanisms against cancer by restricting cell proliferation. Mutations of the *RB-1* and *p53* tumor suppressor genes are known to be involved in development of some cancers.

22. Cancer probably occurs when the normal balance of proliferation-stimulating (proto-oncogenes) and proliferation-inhibiting (tumor suppressor) genes is disturbed. This appears to be a multi-step process requiring several mutations and other changes to regulatory mechanisms.

23. Many mutagenic substances are also carcinogenic. Assaying for potential carcinogenicity can be done by scoring cells for sister chromatid exchanges (SCEs) or by performing the bacterial Ames test.

24. Many familiar compounds have been implicated in mutagenicity and carcinogenicity (including tobacco, marijuana, caffeine, and LSD).

25. X rays and ultraviolet radiation are capable of causing genetic damage, including chromosome breakage, SCE, and point mutations.

26. Both acute and chronic doses of radiation cause the same total amount of point mutational damage.

Key Words and Phrases

acute dose	base analogue	commitment theory of cellular aging
Ames test	Burkitt's lymphoma	
aneuploidy	cancer	deletion
anti-oncogene	carcinogenic	dispermy
autoimmune theory of aging	carcinoma	Down's syndrome
	chronic dose	duplication

Epstein-Barr Virus (EBV)

fetal alcohol syndrome (FAS)

fragile sites

fragile X syndrome

frameshift mutation

gene mutation

genetically significant dose (GSD)

genetic load

Hayflick Limit

human T-cell leukemia virus (HTLV)

inversion

latent period

LD50

leukemia

metastasis

missense mutation

mosaic

H. J. Muller

mutagenic

mutation

mutation theory of aging

neoplasm

nondisjunction

oncogene

phenocopy

Philadelphia chromosome (Ph1)

point mutation

polyploidy

position effect

progeria

proto-oncogene

rad

rem

ring chromosome

same sense mutation

sarcoma

sister chromatid exchange (SCE)

substitution mutation

teratogenic

thymine dimer

translocation

translocation Down's syndrome

trisomy

tumor suppressor genes

ultraviolet radiation

unequal crossover

xeroderma pigmentosum

X rays

Questions

1. The tortoiseshell color pattern in cats is the result of the random inactivation of one or the other X chromosome carrying the genes for coat color (a type of mosaicism—see chapter 5). If a male tortoiseshell cat is produced by a mating of a rust female and a black male cat, can you tell in which of the parents the nondisjunctional event occurred? How?

2. What possible advantage might there be to producing a triploid watermelon, in addition to having a larger and more vigorous plant and fruit?

3. Two parents each have blue eyes. Assume that eye color is controlled by a single pair of alleles, with B for brown eyes dominant to b, which produces blue eyes. Their first child has blue eyes, but their second has one brown and one blue eye. Explain the likely genetic origin of this mosaicism.

4. Why are chromosomal aneuploidies normally a developmental disaster?

5. How can nondisjunction lead to mosaicism?

6. Of what importance are gene mutations; chromosome mutations?

7. Distinguish between nonsense, missense, and same sense mutations.

8. What are the major characteristics of cancer cells?

9. How can mutations cause cancer?

10. Describe the Ames test. Of what value is this test?

11. Describe some of the mutagenic and carcinogenic effects of tobacco; of marijuana.

12. What are the genetic effects of ultraviolet radiation and x-radiation?

13. If a dose of one thousand rems produced sex-linked recessive lethal mutations in 12% of a total of one thousand exposed fruit fly gametes, how many mutations would you predict would occur after exposing one thousand gametes to four thousand rems? How many mutations would you predict if one hundred gametes were exposed to one thousand rems?

14. Why are most mutations harmful?

15. What is the health significance of the Philadelphia chromosome?

16. What are possible mechanisms for converting proto-oncogenes to oncogenes?

17. Which is genetically safer, an acute dose of radiation or a chronic dose? Explain your answer.

For More Information

Ames, B. N., R. Magraw, and L. S. Gold. 1987. "Ranking Possible Carcinogenic Hazards." *Science,* 236:271–80.

Ames, B. N., M. Profet, and L. S. Gold. 1990. "Dietary Pesticides (99.99% all Natural)." Proc. Nat. Acad. Sci. U.S.A. 87:7777–81.

Angier, N. 1987. "Light Cast on a Darkling Gene." *Discover,* 8:85–96.

Arehart-Treichel, J. 1979. "Down's Syndrome: The Father's Role." *Science News,* 116:381–82.

Benowitz, S. I. 1985. "Retinoblastoma: Unmasking a Cancer." *Science News,* 127:10–12.

Bishop, J. M. 1987. "The Molecular Genetics of Cancer." *Science,* 235:306–11.

Cairns, J. 1981. "The Origins of Human Cancers." *Nature,* 289:353–57.

Croce, C. M., and G. Klein. 1985. "Chromosome Translocations and Human Cancer." *Scientific American,* 252:54–60.

Epstein, S. S. 1974. "Environmental Determinants of Human Cancer." *Cancer Research,* 34: 2425–35.

Galloway, J. 1989. "Cancer is a Genetic Disease." *New Scientist,* 18 March 1989, pp. 54–58.

Goldman, M. 1987. "Chernobyl: A Radiobiological Perspective." *Science,* 238:622–623.

Harman, D. 1981. "The Aging Process." Proc. Nat. Acad. Sci. U.S.A., 78:7124–28.

Harris, H. 1986. "Malignant Tumours Generated by Recessive Mutations." *Nature,* 323:582–83.

Hayflick, L. 1980. "The Cell Biology of Human Aging." *Scientific American,* 242:58–66.

Kolata, G. 1986. "Two Disease-Causing Genes Found." *Science,* 234:669–70.

Kolata, G. 1987. "Human Cancer Gene Sequenced." *Science,* 235:1323.

Kolata, G. 1987. "Oncogenes Give Breast Cancer Prognosis." *Science,* 235:160–61.

Lave, L. B., and G. S. Omenn. 1986. "Cost-Effectiveness of Short-Term Tests For Carcinogenicity." *Nature,* 324:29–34.

Marx, J. 1989. "Many Gene Changes Found in Cancer." *Science,* 246:1386–88.

Marx, J. 1990. "Marijuana Receptor Gene Cloned." *Science,* 249:624–26.

Muller, H. J. 1955. "Radiation and Human Mutation." *Scientific American,* 193:58–68.

Nicolson, G. L. 1979. "Cancer Metastasis." *Scientific American,* 240:66–94.

Patrusky, B. 1982. "What Causes Aging?" *Science 82,* 3:112.

Patterson, D. 1987. "The Causes of Down Syndrome." *Scientific American,* 257:52–60.

Roberts, L. 1987. "Atomic Bomb Doses Reassessed." *Science,* 238:1649–51.

Roberts, L. 1987. "Radiation Accident Grips Goiânia." *Science,* 238:1028–31.

Rosenberg, S. A. 1990. "Adoptive Immunotherapy for Cancer." *Scientific American,* 262:62–69.

Sager, R. 1989. "Tumor Suppressor Genes: The Puzzle and the Promise." *Science,* 246:1406–12.

Smith, R. J. 1980. "Government Says Cancer Rate Is Increasing." *Science,* 209:998–1002.

"Smoking and Cancer." 1982. *Morbidity and Mortality Weekly Report,* 31:77–80.

Stanbridge, E. J. 1990. "Identifying Tumor Suppressor Genes in Human Colorectal Cancer. *Science,* 247:12–13.

Sutherland, G. R., and F. Hecht. 1985. *Fragile Sites on Human Chromosomes.* Oxford University Press, New York.

Upton, A. C. 1982. "The Biological Effects of Low-Level Ionizing Radiation." *Scientific American,* 246:141–48.

Vogelstein, B., et al. 1989. "Allelotype of Colorectal Cancers." *Science,* 244:207–11.

Wilson, R. 1987. "A Visit to Chernobyl." *Science,* 236:1636–40.

Yunis, J. J. 1983. "The Chromosomal Basis of Human Neoplasia." *Science,* 221:227–36.

Zohary, D., and M. Hopf. 1988. *Domestication of Plants in the Old World.* Clarendon Press, Oxford.

Chapter

12

© Neg. 333218
Courtesy Department
of Library Services,
American Museum of
Natural History

Immune genetics

Figure 12.1

Different types of human blood cells: (a) red blood cells; (b) platelets; (c) macrophages; (d) scanning electron micrograph of human lymphocyte.

(a) © Stanley Flegler/Visuals Unlimited; (b) © Robert Caughey/Visuals Unlimited; (c) © Robert Becker/Custom Medical Stock Photos; (d) © David M. Phillips/ Visuals Unlimited.

(a)

(b)

(c)

(d)

Living organisms are continually subject to invasion by pathogenic organisms and toxic substances. As a consequence, various defenses against biological and physical assaults have evolved. Some defenses are general, or nonspecific, resisting a broad array of potentially harmful agents. Others are specific, defending against only one pathogen or toxin. The function of specific defense mechanisms is closely intertwined with aspects of the hereditary system. In this chapter, we explore the immune system and some of the applications of immune genetics to human health. We conclude with a consideration of viral diseases such as AIDS, whose deadly effects largely depend upon the ability of viruses to avoid and subvert the normal immune defense mechanisms of the body.

Nonspecific Defense Mechanisms

Nonspecific defense mechanisms protect the body from many different kinds of pathogenic and toxic agents. These nonspecific defenses are often called **natural resistance.** They represent a set of defenses with which all individuals are born, and they function more or less continuously throughout one's entire life.

One important category of natural resistance is the physical protection provided by the external surface of the body. The dry, tough skin is a barrier to almost all pathogens and toxins. The linings of the respiratory, digestive, urinary, and reproductive systems are also considered to be external to the body. All are lined with a warm, moist epithelium; the epithelial surfaces have mechanisms to remove or kill most infective environmental agents. The lungs, for example, are lined with ciliated epithelium. The moist surface traps dust and bacteria, and the action of the cilia sweeps the pollutants out of the lungs to the throat, where they are expelled.

Many fluids of the body surfaces are toxic to bacteria, often because of high acidity or alkalinity. Urine, for example, cleanses the urinary tract. Tears wash particulate matter from the eyes and also contain lysozyme, an antibacterial enzyme. Sweat cleanses and protects the skin. Ingested microorganisms that manage to reach the stomach are almost certain to die in the highly acidic gastric juice (pH 2.0–4.0). Lactic acid in the vagina discourages the growth of most microorganisms. The normal flora and fauna of the gut, still technically outside the body, are also effective antipathogens.

If external barriers are penetrated, internal mechanisms of general defense respond. *Phagocytic* (meaning cell-eating) cells attack and destroy infective bacteria. Connective tissue may rapidly grow to surround sites of bacterial and parasitic invasion, isolating the infection from the rest of the body. Many chemicals in the body inhibit bacterial and viral growth. Polypeptides such as histones and spermine act against bacteria, and interferon is an antiviral agent. All of these factors contribute to the general *inflammation* response against infection.

Specific Defense: Immunity

Specific defense mechanisms are directed against particular pathogens such as bacteria, viruses, parasites, fungi, and the toxins that these organisms may release. The diseases tetanus and botulism, for example, are caused by bacteria of the genus *Clostridium*. Their potentially lethal effects result from potent toxins released by the bacteria, rather than directly from the growth of the bacteria themselves. The resistance of an individual to a particular microorganism or its product is referred to as **specific immunity.** Because specific immunity normally occurs only after recovery from initial infection by that microorganism, specific immunity is also called **acquired immunity.**

Specific immunity is achieved by three extraordinary and essential aspects of the immune system. First, the immune response is highly *specific,* meaning that it recognizes and responds individually to each type of invading organism or toxin, and it is highly *diverse,* meaning that it is capable of identifying an almost limitless range of pathogens and toxins. Second, the immune response distinguishes *self* (the tissues of the body) from *non-self* (all foreign substances). Third, the immune system *remembers* non-self substances that it has previously encountered, allowing it to respond very rapidly to second and all subsequent challenges from foreign substances.

Specific immunity results from the interactions within the body of **antigens,** foreign substances that elicit the production of antibodies when introduced into the body, and **antibodies,** the proteins produced by the body in response to foreign antigens. Occasionally, the *self/non-self* recognition mechanism breaks down. The body may fail to identify certain antigens as foreign, thus not mounting a defensive response; or the body may perceive its own tissues to be foreign and produce antibodies against itself. In either case, the medical consequences are serious.

Following invasion of the body by antigens such as proteins, bacteria, parasites, viruses, and toxins, several types of specific cells respond to neutralize these antigens. **Lymphocytes,** a class of white blood cell, are especially active (box 12.1). **T cells,** one type of lymphocyte, attack foreign antigens directly, a process known as **cellular immunity. B cells,** a second type of lymphocyte, produce *antibodies* that interact with and destroy foreign antigens, a process known as **humoral immunity. Natural killer (NK)** cells are large lymphocytes that attack viruses and developing tumors. **Macrophages,** very large phagocytes, act directly to engulf foreign antigens and interact with T cells. The total *immune response* reflects the many cooperative actions of these cell types to destroy or neutralize specific foreign antigens.

Immune Function

Acquired immunity is an essential protective mechanism of the body, yet there is no anatomically distinct immune system. Instead, immune function depends chiefly upon the actions of B cells, T cells, NK cells, and macrophages. These cells all originate in bone marrow, but they exist and function within the two major circulatory networks, the **cardiovascular system** and the **lymphatic system** (figure 12.2).

The cardiovascular system consists of the heart, whose pumping action continually circulates materials throughout the body; arteries, vessels that carry blood away from the heart; veins, vessels that return blood to the heart; and capillaries, vessels that connect the arteries and the veins and whose internal diameter permits passage of only a single red blood cell at

BOX 12.1

Blood

The average person has about five liters of blood. The liquid portion, *plasma,* accounts for roughly 55%; the remaining 45% consists of red blood cells, white blood cells, and platelets (which participate in forming blood clots). In a healthy individual each milliliter of blood contains about five million red cells, ten thousand white cells, and 300,000 platelets (figure 12.1).

Red blood cells are produced in the bone marrow. As they mature, they lose their nuclei, becoming bags of hemoglobin functioning in gas exchange. Red cells survive about four months in the circulatory system, eventually deteriorating and being destroyed by the spleen.

White blood cells come in many forms and fulfill many different functions. Most white blood cells, **leukocytes,** have a life span of two weeks or less. Only the antibody producing leukocytes, B cells, survive as long as one hundred days. One major class of white blood cells include the *macrophages* (figure 12.1). Like giant amoebae, macrophages move to a site of infection, engulfing and ingesting foreign matter. Macrophages are present in large numbers early in an infection.

Lymphocytes, a second major class, comprise about 40% of all leukocytes. Depending upon their developmental history, lymphocytes differentiate to one of three basic types. About 65–75% of lymphocytes are processed by the thymus gland, acquiring a surface antigen that identifies them as T cells. Lymphocytes called B cells originate in the bone marrow and mature in the blood. B cells produce antibodies following exposure to an antigen. About 10% of lymphocytes are natural killer (NK) cells, which lead in the body's attack against invading viruses and developing cancers.

a time. Nutrients and other essential materials in the blood diffuse through the walls of the capillaries into the interstitial fluid surrounding all cells in the body; waste products from the tissues diffuse back into the capillaries for ultimate removal from the body.

The lymphatic system, on the other hand, provides only one-way flow of materials from the tissues to the cardiovascular system. Tiny vessels of the lymphatic system drain fluid called *lymph* from all body tissues. Lymph is the liquid portion of the blood, without blood cells and large proteins. Lymph passes through a series of increasingly larger lymphatic vessels until it reenters the cardiovascular system near the heart. Located throughout·the lymph system are lymph nodes, which filter bacteria and other foreign particles from the lymph fluid. The tonsils are masses of lymphatic tissue surrounding the throat, forming a protective ring that defends the digestive and respiratory tracts from pathogenic invasion. Decades ago, tonsils were commonly removed during childhood because almost all children experience bouts of tonsillitis, inflammation of the tonsils. Because the protective importance of the tonsils is now better understood, the tonsils are less frequently removed.

The association of lymphocytes (B cells, NK cells, and T cells) and macrophages with the two circulatory systems is fundamental. The circulatory systems penetrate all portions of the body. Essentially all living cells in the body are no farther than the width of one cell from a capillary. Thus, lymphocytes and macrophages can be transported anywhere in the body within seconds to begin their protective action. Furthermore, bacteria, toxins, and other foreign substances are not permitted to circulate freely throughout the body after they penetrate the skin. Instead, they are trapped in nearby lymph nodes, where they are attacked by lymphocytes and macrophages. Concentrations of lymph nodes occur in the neck, armpits, and groin, regions that become swollen and tender during times of illness and infection.

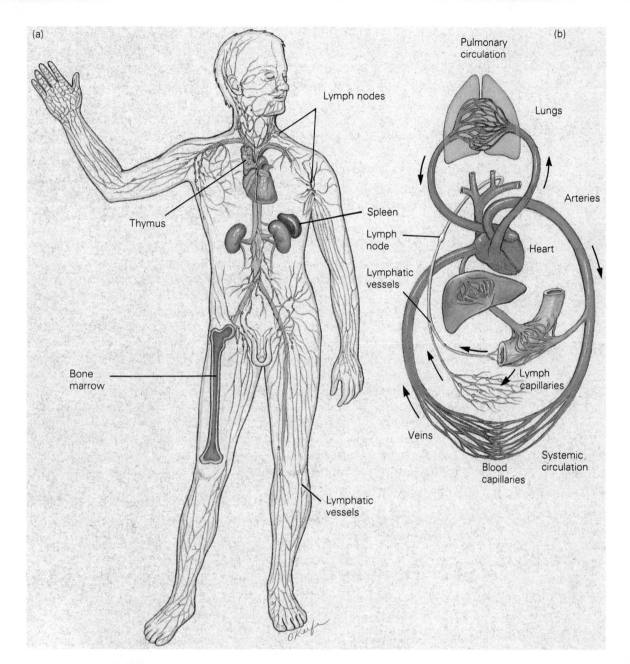

Discriminating Self and Non-self

Immunity depends upon the ability of the body to distinguish *self* from *non-self*. Exactly how substances are identified as foreign is still uncertain, but this capacity is essential to normal immune function.

The developing fetus cannot distinguish its own tissues from foreign substances, particularly those of its mother. While in the uterus, the fetus is dependent upon its mother's immune mechanisms. During gestation, the fetus learns to identify its own tissues immunologically. The self/non-self distinction is not possible until the individual has catalogued its own tissues.

Experiments on laboratory mammals have shown that exposure of fetuses to foreign antigens prior to birth results in those antigens not being recognized as foreign after birth. Skin grafts between mice of different genetic strains are ordinarily rejected. However, if a

Figure 12.2

The organs and vessels of the (a) cardiovascular and (b) lymphatic systems.

From "The Immune System" by Niels Kaj Jerne. Copyright © July 1973 by Scientific American, Inc. All rights reserved.

mouse of one strain is exposed in the uterus to skin of a second strain, postnatal skin grafts are often successful. It appears that during gestation, the body is identifying those tissues present as self.

Immune Development

A nursing newborn infant receives a large dose of antibodies from its mother in the form of *colostrum,* a thin, yellow substance, rich in antibodies, secreted by the mother's mammary glands immediately before and after birth. Thus, the newborn infant is provided short-term immune protection by the mother's antibodies, while the infant's own immune system matures. Bottle-fed infants miss out on this valuable protection. By about six months of age, the infant's own immune mechanisms are fully functional.

Both lymphocytes and macrophages are produced in the bone marrow and released into the blood (figure 12.3). Lymphocytes originate from *stem cells,* and subsequent differentiation depends upon their fate after entering the bloodstream. Approximately 65% of the lymphocytes enter the **thymus gland,** where they undergo differentiation to become T cells (T refers to the role of the thymus in their development). Following maturation in the thymus, T cells are released back into the blood. The thymus gland, a soft, reddish organ that is part of the lymphatic system, lies directly over the heart and just behind the upper part of the breastbone (figure 12.2). The thymus is relatively large in infancy and childhood. Its decrease in size, beginning at puberty and continuing to death, is believed to be related to a general decline in function of the immune system with age. By age seventy, the immune system is only about 25% as effective as in one's youth. Many deaths of elderly persons are due to insufficient immune response to infections such as pneumonia.

The remaining lymphocytes do not enter the thymus, but undergo differentiation in the bloodstream to become B cells (about 25%) and NK cells (about 10%). (B does not refer to blood in this case, but the association is a good memory aid for distinguishing T cells and B cells.) Macrophages also originate in the bone marrow and mature in the blood. Following maturity, T cells, B cells, NK cells, and macrophages settle throughout the circulatory system. Many reside in the walls of the smaller blood vessels, but a large proportion are concentrated in the major lymphatic organs, the lymph nodes and the *spleen.* The spleen, a large, abdominal organ with a flattened oblong shape, is a reservoir for blood and an important site of defense against infection (figure 12.2).

Immune Response

When a foreign substance penetrates the body surface, the lymphocytes and macrophages engage in a complex set of interrelated responses. Some macrophages and B cells engulf (phagocytize) the foreign cells or proteins. These are digested, and short segments of the foreign substances called **antigenic determinants** are "presented" on the surface of the B cells and macrophages. Acting in concert with proteins produced by the genes of the HLA complex (see chapter 4), some T cells bind to the presented antigenic determinants and become activated.

The ability of the immune system to respond to almost any foreign antigenic determinant derives from the enormous, genetically determined variation—*diversity*—in T cells. The diversity of response is further enhanced by the highly variable HLA proteins, controlled by dozens of alleles at several loci, that interact with the T cells and foreign antigens. The genetically mediated variation in the T cell and HLA systems ensures that the immune system can respond specifically to millions of different antigenic determinants.

Bacteria and other foreign substances usually have many antigenic determinants; blood cells, for example, have several thousand. Each T cell (and associated HLA protein) circulating in the blood is *specific* for a single antigenic determinant, so when a new foreign antigen is presented, only a very few T cells will respond. Different T cells, however, will respond to different antigenic determinants for the same invading bacterium.

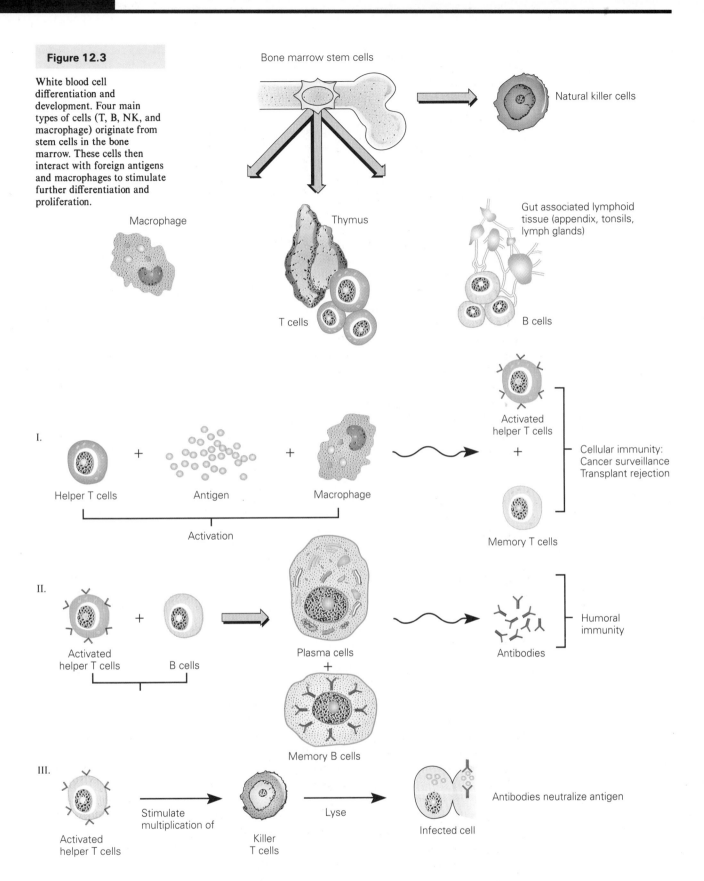

Figure 12.3

White blood cell differentiation and development. Four main types of cells (T, B, NK, and macrophage) originate from stem cells in the bone marrow. These cells then interact with foreign antigens and macrophages to stimulate further differentiation and proliferation.

Bone marrow stem cells

Natural killer cells

Macrophage

Thymus

T cells

Gut associated lymphoid tissue (appendix, tonsils, lymph glands)

B cells

I.

Helper T cells + Antigen + Macrophage

Activation

Activated helper T cells + Memory T cells

Cellular immunity: Cancer surveillance Transplant rejection

II.

Activated helper T cells + B cells

Plasma cells + Memory B cells

Antibodies

Humoral immunity

III.

Activated helper T cells

Stimulate multiplication of

Killer T cells

Lyse

Infected cell

Antibodies neutralize antigen

Activation of T cells that have bound to antigenic determinants has several important consequences. First, the T cells begin to rapidly multiply. Each T cell has the capacity to respond to only one presented antigenic determinant. Different T cells respond to different determinants from the invading bacterium (or other antigen). As each T cell responds, it begins dividing (by mitosis) to produce an enormous clone of T cells, each clone specific to only one determinant of the invading organism. This process is known as **clonal selection.** Second, the selected T cells differentiate further. Some become **helper T cells** that secrete chemicals called **interleukins,** which enhance and stimulate growth and division of both T and B cells. Others become **killer** (or **cytotoxic**) **T cells,** that secrete chemicals directly on to the surface of invading cells and thus kill them. A third group of T cells are transformed into **memory cells.**

At the same time, a parallel response is being mounted by the B cells. Stimulated by interactions with helper T cells and the effects of interleukin, B cells also undergo a process of clonal selection and differentiate further to become **plasma cells.** The primary activity of plasma cells is the production of specific antibodies or, more properly, **immunoglobulins.** Immunoglobulins attach to the antigens to which they are specific, inactivating toxins or destroying foreign cells. Some B cells, like some T cells, survive weeks or years as **memory cells,** preserving the antigenic determinant so that, should a subsequent infection by the same foreign agent occur, the body can rapidly respond. This provides the basis for resistance to specific diseases, once an individual has survived a first infection (figure 12.3).

Immunoglobulins

About five to ten days after an infection begins, immunoglobulins specific for the invading antigen are abundant enough to be detected in the blood. This short period is the time required for B cells to proliferate and mature to plasma cells, following stimulation by helper T cells. During this process, the B cell nucleus is squeezed off to one side as the entire cytoplasm fills with endoplasmic reticulum. This extensive membrane system is the site of massive antibody production, the main function of plasma cells.

An individual B cell produces only one type of antibody in response to the antigenic determinant brought by helper T cells. The maturing B cell reproduces mitotically, generating a clone of identical plasma cells, all producing identical antibodies (box 12.2). Thus, upon exposure to a single bacterial infection, several different antigenic determinants on the surface of the invading bacteria elicit proliferation of plasma cells and subsequent antibody production. Each plasma cell can produce roughly two thousand identical antibody molecules per second. The five- to ten-day period of immunological response reflects the normal time required for common illnesses such as colds and influenza to run their course.

Types of Antibodies

Five types of antibodies (immunoglobulins) are known, each playing a different role in the immune reaction. *Immunoglobulin G (IgG)* comprises about 80% of the antibodies in blood. IgG is active against bacteria, viruses, and toxins. IgG's chemical structure permits it to easily cross the placental barrier and provide the primary specific defense of the fetus.

About 7% of plasma antibodies are *Immunoglobulin M (IgM),* a relatively large immunoglobulin. The antibodies to the ABO blood group are IgM. Because IgM molecules are too large to cross the placenta, ABO incompatibility of mother and fetus is not a problem.

About 12% of plasma antibodies are *Immunoglobulin A (IgA).* IgA, however, is the most abundant immunoglobulin in the secretions of exocrine glands. Large quantities of IgA are found in saliva, tears, milk, nasal mucus, urine, and fluids of the digestive system. IgA appears to provide an external defense to viral infection.

Two other types of immunoglobulins occur in small quantities, less than 5% of the total plasma antibodies. Neither is known to have protective functions for the body, and they may actually be harmful. *Immunoglobulin D (IgD)* occurs in relatively high concentrations in the serum of some persons with myeloma, a type of bone cancer. Recently IgD has been shown to be involved in differentiation of memory B cells. *Immunoglobulin E (IgE)* is the causative agent of allergies, the bane of millions. IgE also appears to be involved in fighting parasitic infections.

BOX 12.2

Antibody structure and diversity

Figure 12.4

An antibody molecule, IgG. Each antibody molecule is composed of four polypeptide chains (two heavy and two light). The heavy chain consists of 446 amino acids; the light chain consists of 214 amino acids. Each chain has a variable region (V) and a constant region (C). The variable regions function in antigen recognition.

Antibodies are large proteins composed of four polypeptide chains, synthesized on the endoplasmic reticulum of plasma cells. R. R. Porter and G. M. Edelman received the 1972 Nobel Prize for Medicine or Physiology for their description of antibody structure.

Two large polypeptides called *heavy chains,* each 446 amino acids long, join for slightly more than half their length to form the Y-shaped framework of an antibody (figure 12.4a). Two *light chains,* each 214 amino acids long, are linked to the heavy chains by bonds between the sulfur atoms of sulfur-containing amino acids. Amino acids 1–114 of each heavy chain are a *variable region* that functions in antigen recognition. The remaining portion of all heavy chains is invariant. Each light chain also has a variable region (amino acids 1–108, functioning in antigen recognition), as well as a constant region.

The key to antibody diversity is somatic mutation and rearrangement (by recombination) of portions of at least five separate genetic loci (with hundreds of alleles) encoding the variable regions of both heavy and light chains of immunoglobulins. B cells, which have been stimulated by activated helper T cells, undergo a process of differentiation resulting in each individual B cell clonally expanding to produce antibodies specific to one particular antigenic determinant. The nearly limitless diversity of the mammalian immune system is directly attributable to the cutting and joining of the gene sequences encoding the variable regions of the immunoglobulins (the antigen binding sites). This process allows production of hundreds of millions or even billions of different types of antibodies from a relatively small amount of genotypic variation.

Clonal Selection and Antibody Diversity

The human immune system has the incredible capacity to produce unique antibodies to nearly every foreign substance it encounters. An individual produces millions of different types of antibodies in a lifetime. How is this vast antibody diversity generated and controlled? Evidence now suggests that hundreds of gene fragments encode the variable regions of the heavy and light antibody chains, and that somatic mutation and recombination are important genetic events in the origin and maintenance of antibody diversity (box 12.2). The vast number of antibodies we each produce are the products of cutting and splicing a much smaller set of genes in the maturing plasma cells. Thus each plasma cell is committed to produce a single unique antibody, against one of many determinants on the surface of the invading antigen.

Antigenic determinants on invading pathogens or toxins are almost certain to be matched by at least one or a few of the antibodies on the surfaces of circulating B cells, because of the enormous diversity of these antibodies. Only B cells that are specific to the invader begin proliferation, resulting in a clone of plasma cells that, because of further somatic mutation, produce large quantities of antibody highly specific for the invader. This view of the production of antibody diversity is known as the **clonal selection theory.** The apparently limitless capacity of the immune system to respond precisely to a continually changing set of environmental challenges reflects the extraordinary versatility of the genetic mechanism. Antibody production (a highly variable phenotypic character) is a finely tuned survival response resulting, like all phenotypic traits, from a complex interaction of the genotype and environment.

Acquisition of Specific Immunity

One may acquire immunity actively or passively. An individual's immune defenses are not fully developed until about six months after birth. During the preceding fifteen months of life, immune protection is provided by antibodies produced by the mother. Any infections or toxic agents can gain access to the fetus while *in utero* only by penetrating the mother's defense mechanisms. If that is achieved, the fetus is still protected by maternal IgG that freely crosses the placenta.

Passive Immunity

Passive immunity is the immune protection resulting from the transmission of antibodies from one individual to another. The example of antibodies passed from a mother to her nursing child is probably the most common form of passive immunity. As noted earlier, colostrum, the first portion of mother's milk, is exceptionally rich in antibodies. Not surprisingly the incidence of colds is significantly lower among nursing infants than among bottle-fed infants. The fact that the mother has been exposed to a much wider variety of antigens in her lifetime than has the newborn means that her immune system responds faster and more efficiently than the immune system of her infant. By receiving antibodies from its mother, a nursing infant is better protected from infectious diseases (box 12.3).

Artificial passive immunity is now commonly employed in medical practice. Persons at risk for specific diseases are given injections of *gamma globulins,* the antibody-containing fraction in blood. Mumps, for example, is a mild disease in children but a serious disease for adults. Prior to the development of a mumps vaccine, adults exposed to mumps were given injections of gamma globulin from persons who had previously had mumps and thus had anti-mumps antibodies.

Gamma globulin injections provide only temporary immunity. If long-term protection is needed, successive injections are required. In rare cases, bone marrow transplants can be used to transfer the entire antibody-producing machinery from one person to another. The recipient benefits from a lifetime of passive immunity.

Active Immunity

Throughout life, an individual's immune mechanism produces antibodies in direct response to substances encountered in the environment, i.e., **active immunity.** Upon first exposure to a foreign antigen, some B cells are stimulated to differentiate into plasma cells, initiating the immune response. Within two weeks, antibody production is well underway. Some T cells and some B cells survive for years as *memory cells*. Upon subsequent exposure to the same antigen, the immune response is immediate. The rapid response of the memory cells is the basis for immunity to measles, mumps, and other childhood diseases *after* one bout of illness.

Vaccination against disease is a common method for inducing active immunity without having to suffer any disease symptoms. The antigen is prepared by either making an extract of the disease-causing organism, or by genetically crippling the disease organism so it cannot reproduce and using this defective strain as the antigen. Subsequent injections of the same vaccine (*boosters*) restimulate the plasma cells producing higher concentrations of anti-

BOX 12.3

Behavior and passive immunity

Maternal fondling and kissing of babies is a trait that may have been favored by natural selection throughout mammalian (including human) evolution. Mothers of many species continually lick and groom their young, just as human mothers kiss their infants (figure 12.5). This behavior exposes the mother's immune system to the bacteria and other organisms on the infant's face and body. If the child is exposed to any antigens not yet encountered by the mother, kissing behavior ensures the mother's exposure to these antigens and ensures that her system makes the antibodies most needed by the child. The mother's antibodies will then be passed in her milk to the nursing infant.

Maternal exposure to bacteria and other antigens carried by her newborn has a second beneficial effect. The mother has previously been exposed to many of these antigens and has memory cells that recognize them. In effect, she has been vaccinated against most common environmental agents. Thus, the mother's immune response to "old" antigens is swifter and stronger than the initial response of her infant to "new" antigens. The normal fourteen-day lag in immune response time could be fatal to a fragile infant. Kissing behavior may thus provide a nursing mammalian infant with double immunological protection against its new environment.

bodies and memory cells. Upon later exposure to the live disease-causing organisms, there is immediate antibody production that destroys the organism *before* it can cause disease.

Induced active immunity by vaccination is one of the greatest success stories of modern medicine. Immunization programs have effectively eliminated such diseases as smallpox, tetanus, typhoid, typhus, polio, and diphtheria in the developed world—diseases that have killed thousands or millions throughout history.

Health, Genetics, and Immunity

Many diseases result from a weakening or breakdown of the immune system. In other cases, normal immune function results in an inappropriate response directed against an individual's own tissues rather than an invading organism or toxin. Increased knowledge has permitted advances in medical treatment of diseases associated with the immune system.

Figure 12.6

Rejection of a skin graft, showing necrosis of the tissue.

L. Brent, from Roitt, Ivan: "Essentials of Immunology" 3rd ed. © Blackwell Scientific Publications.

We now review some important genetic aspects of immunological medicine. Particular attention is given to viruses, organisms with a special capacity for subverting immune mechanisms, as illustrated by the disease AIDS.

Tissue and Blood Typing

In chapter four we introduced the ABO and MN blood groups associated with red blood cells, and the HLA group associated with white blood cells. All three groups, and many others as well, are the result of proteins called *cell surface antigens* occurring on the blood cells. These groups are of value for forensic purposes, such as paternity testing and identification of criminals from blood and other tissues left at crime scenes. Additionally, patterns of variation among ethnic and cultural groups have often provided important anthropological evidence for understanding the history of human populations.

These blood groups are clearly of medical importance, as shown by the necessity for matching ABO types for transfusions, or by the association of specific ABO and HLA types with diseases (see chapter four). In the next sections we consider another important medical aspect of HLA type and introduce the Rh group.

HLA Type and Transplants

The immune response poses a major barrier to *transplantation* of tissues and organs from one person to another. Except in a few special cases, the recipient of a transplant experiences an immediate tissue rejection response. This reaction primarily involves the T cells, which isolate and ultimately kill the transplanted tissue (figure 12.6).

A few sites in the body are not susceptible to transplant rejection. These **immunologically privileged sites** include corneas, bone, cartilage, heart valves, and testes. It is thought that because the cornea has little vascularization and little contact with blood and lymph, circulating T cells that monitor the body for foreign tissue never "see" the cornea. Thus, corneas can be transplanted with little fear of rejection, and such operations save the sight of many persons every year. In spite of the ease of cornea transplants, not enough corneas are available to meet present demand.

Most tissues in the body, however, are subject to continual immune surveillance and to a rejection reaction following tissue or organ transplants. The *human leukocyte antigens* (HLA) govern the tissue rejection response. Recall that HLA type depends upon at least 5 loci and over 110 alleles, generating millions of unique genotypes and a very low frequency of any one of them. Although blood transfusions are performed routinely as a result of the limited number of ABO genotypes, such is not the case with tissue and organ transplants.

The extraordinary HLA diversity in humans explains the immunological barrier to tissue and organ transplants. Even parent/child or sibling/sibling transplants are seldom successful, as HLA genotypes within a single family can all be distinct. In figure 4.13, for example, considering only the major loci, the parents share no HLA alleles and produce four children, all of differing HLA genotype. Thus, if a child in the family needs a transplant, *none* of the family members is a suitable donor. If crossover occurs and the minor loci are considered, the possible genetic variation makes an HLA match within any pedigree extremely unlikely.

A notable exception to the rarity of HLA matching is the case of monozygotic twins. These individuals are genetically identical, therefore, all tissues have the same surface antigens. Tissue rejection is no barrier to transplants between identical twins.

Tissue matching is a serious medical problem. Only rarely is a healthy organ with the right genotype available for transplant at the right time. Physicians now accept a partial match in serious cases. If a match at the A and B loci can be achieved, the use of immunosuppressant drugs can limit the rejection response due to other genetic variation (box 12.4). Researchers are examining many drugs and other techniques that can inhibit the rejection of transplanted tissues. One method that may prove useful for transplanting tiny amounts of tissue involves enclosing the donor cells in an ultra-fine mesh bag. Diabetics, for example, need only a small number of insulin-producing cells. Experiments suggest that it may be possible to transplant such cells into the pancreas of a diabetic, enclosing donor tissue in a mesh fine enough to keep antibodies out, but large enough to permit the entry of nutrients and the exit of insulin.

Autografts. Sometimes tissue from a person's own body is transplanted to a new site, a procedure known as an *autograft* ("auto-" means self). An autograft has the special advantage of a perfect tissue match—self to self—and therefore tissue rejection due to HLA mismatch is not encountered. Autografts of skin are a standard procedure for certain types of plastic surgery and for treating severe burns, where areas of skin are so badly damaged that they must be replaced. Autografts of large veins from the leg are now used to replace cardiac veins and arteries damaged or occluded by atherosclerosis. In some cases, an amputated or congenitally damaged thumb or finger can be replaced by autografting a toe to the hand.

Fetal-cell surgery. *Parkinson's disease* is characterized by gradual and severe loss of voluntary motor function; victims suffer tremors, lose the ability to speak, and eventually become totally incapacitated. The disease is caused by deterioration of the brain's capacity to produce dopamine, a chemical required for communication among brain cells. The only available treatment is administration of the drug L-dopa, which is similar but not identical to dopamine. Victims usually respond to L-dopa for a short time, but gradually its beneficial effects diminish. Some victims suffer serious side effects from the drug. A new therapy for Parkinson's disease is being developed experimentally using laboratory animals: defective dopamine-producing brain cells are replaced with fetal brain cells. Transplants of fetal cells are not subject to the tissue rejection response of adult tissue and may be accepted as the patient's own tissue. If successful, this procedure would effectively be a permanent cure of this disease.

As noted in box 7.2, however, fetal-cell surgery confronts society with serious ethical problems. If transplanting fetal tissues proves widely successful for treating Parkinson's disease and other conditions, the number of spontaneously aborted fetuses that might be available for such transplants will be far exceeded by the potential demand. How will decisions on the use of fetal tissues be made? Who should have access to fetal tissues? Should *in vitro* fertilization be used to generate excess embryos to supply the transplant market? Might mothers be under increased pressure to abort children with only minor defects diagnosed prenatally? Rapidly advancing medical technology will force society to answer these and other questions in the near future.

Cyclosporin: immunosuppression and biodiversity

Figure 12.7

Surgeon removing a heart from a sterile pan during a heart transplant operation.

© Jim Olive/Peter Arnold Inc.

Each year in the United States, over 1,500 people undergo heart transplant surgery, about 2,000 people receive a new liver, and more than 9,000 receive a new kidney. Between 80% and 90% of these people survive at least one year after surgery, far better than the 65% first year survival rate in 1980.

Improved survival following transplants is largely due to the use of new drugs that effectively suppress the body's immune response. The most widely used of these drugs is *cyclosporin,* a derivative of a soil fungus that occurs naturally in Norway and was discovered in 1970. Cyclosporin works by reducing T-cell secretion of interleukins, natural hormones that stimulate T-cell proliferation following transplants or other invasions by foreign cells and toxins.

Because cyclosporin controls the rejection response so successfully, transplant surgery is now a practical, almost routine treatment for certain serious illnesses. Like penicillin, cyclosporin is a drug of humble biological origins that has greatly improved medical practice.

Cyclosporin also exemplifies the practical importance of biological diversity. Most plant species have never been investigated for their medical importance, yet many probably produce chemicals that can greatly benefit humanity. At present rates of extinction, many of these potentially beneficial chemicals will never be known.

The Rh Blood Group

Karl Landsteiner received the 1930 Nobel Prize for Physiology or Medicine for his discovery of the ABO blood group, but this extraordinary scientist was still far from retirement. In 1940, Landsteiner and Philip Levine used RBCs from a rhesus monkey to produce antibodies in rabbits. When tested against blood from humans, the rabbit antibodies to rhesus monkey RBCs caused agglutination of the blood of 85% of the Caucasian test population. These individuals were thus designated *Rh positive*. The remaining 15% whose blood showed no agglutination reaction were designated *Rh negative*. These data indicated that 85% of the test group had an antigen on their RBCs similar to that present on the rhesus monkey RBCs, accounting for the agglutination reaction.

Rh blood type is an heritable character, but it follows a complex pattern that is still not fully understood. We can understand medical aspects of the Rh blood group if we use a normal Mendelian model for one locus with two alleles. Allele R determines the presence of the Rh antigen and is dominant to r. Thus, RR and Rr individuals are Rh positive; rr individuals are Rh negative.

The Rh blood group has considerable medical importance. Human serum contains no naturally occurring antibodies to the Rh antigen, unlike the ABO system. If an Rh negative person is exposed to Rh positive blood (as might occur with a transfusion, for example), he or she develops anti-Rh antibodies. The antibodies are not produced rapidly enough that they can attack the RBCs of the first transfusion, but should subsequent transfusions of Rh positive blood occur, the anti-Rh antibodies will be present in large quantities, causing agglutination and a shock reaction. (This is exactly the principle behind immunization against various diseases.) Donor blood and recipient blood is routinely cross matched to avoid such a reaction.

However, an analogous process sometimes results from Rh incompatibility between mother and developing fetus. If an Rh negative female mates with an Rh positive male, any child they conceive has the possibility of being Rh positive, like the father. If the father is homozygous Rh positive, all their children will be Rh positive; if the father is heterozygous Rh positive, half the children will be Rh positive. (The reciprocal mating, Rh positive female with Rh negative male, has no important medical consequences.)

Normally, large molecules or cells cannot cross the placenta between mother and fetus. However, at the time of birth the placenta may tear, allowing entry of some of the child's blood into the mother's system. If the mother is Rh negative and the child Rh positive, the mother is essentially being immunized against the Rh antigen, just like the rabbits that Landsteiner injected with blood from rhesus monkeys. She will develop large quantities of anti-Rh antibody.

Rh antibodies are chemically different from anti-A antibodies and anti-B antibodies and are small enough that they *do* pass across the placenta. Should the Rh negative mother conceive subsequent Rh positive fetuses, they will be at severe risk from the mother's anti-Rh antibodies. The antibodies attach to the surface of the RBCs of the fetus and destroy them, causing the condition **erythroblastosis fetalis** (literally, "destruction of the red blood cells of the fetus"), a type of hemolytic anemia. These infants are called *blue babies* at birth. The bluish tint of their skin is due to the destruction of the RBCs, reducing the blood's capacity to carry oxygen. Destruction of the RBCs also causes *jaundice,* a condition resulting from abnormally high levels of *bilirubin,* a pigment released into the blood from the breakdown of hemoglobin. High levels of bilirubin may cause palsy and mental retardation in the newborn, and if RBC destruction is severe enough, the infant may be stillborn.

Fortunately, understanding this condition has led to a variety of effective therapies. Newborn infants with erythroblastosis fetalis can be given a complete transfusion, replacing all their blood (which contains the mother's anti-Rh antibodies) with blood of their own type *without* those antibodies. In rare cases, such transfusions are given prior to birth. Careful exposure of the infant to fluorescent light degrades the excess bilirubin rapidly enough that it causes no damage.

Preventive therapy is also given an Rh negative woman having a child with an Rh positive man (and thus at risk of producing a blue baby). Immediately after the birth of her first and all subsequent Rh positive children, the mother is given massive injections of anti-Rh antibody. These antibodies destroy the child's RBCs *before* they trigger an immune response in the mother. Thus, the Rh negative mother does not develop anti-Rh antibodies in her system that could harm future children.

ABO Incompatibility

Because antibodies to the ABO blood group occur naturally in the serum, and because maternal-fetal incompatibilities are more common for the ABO group (85% are Rh positive), you might expect hemolytic anemia to occur frequently for the ABO and other blood groups. Although some infant anemias are due to incompatibility of blood groups, hemolytic anemia occurs far less frequently than would be expected. Anti-A antibodies and anti-B antibodies are so large that they normally do not cross the placenta from mother to fetus, in contrast to the smaller anti-Rh antibodies.

An unexpected occurrence is the reduction in hemolytic anemia due to Rh incompatibility as a result of simultaneous ABO incompatibility. For example, suppose a mother of blood type O, Rh negative carries a fetus of type A, Rh positive. As fetal RBCs leak across the placenta, the mother's natural anti-A antibodies attach to and destroy them *before* the mother's system can begin production of anti-Rh antibodies. Thus, that child and future children are protected from hemolytic anemia due to Rh incompatibility.

Malfunctioning Immune Systems

We have learned most about immunity through studies of individuals with defective immune systems. About one person in fifty thousand is born without a thymus gland. These children cannot process T cells and thus have no tissue-mediated immunity. They are extremely susceptible to infectious disease and have a high mortality rate. They do produce B cells and can mount a weak humoral response, even without helper T cells. Some athymic children can be cured by transplanting donor thymus tissue into their abdominal region (the tissue may come from anyone, since without T cells, tissues are not rejected). In some cases the transplanted thymus grows and functions normally, eventually processing leukocytes into mature T cells and bestowing normal immunity on these children.

Other children have a thymus but suffer from a sex-linked recessive condition known as **agammaglobulinemia** (meaning without gamma globulins). These children do not produce B cells and thus cannot produce antibodies. Simple bacterial infections are life threatening to these children because antibodies are the primary defense against bacterial disease. They can, however, mount an immune reaction to viruses, indicating that viral immunity is a function of the T cells.

Agammaglobulinemia can be treated by a bone marrow transplant, which provides a source of normal B cells. This surgery is successful only with precise HLA matching. The T cells of the recipient are fully functional and will reject unmatched tissue. In addition, the donor bone marrow is also immunologically functional and can mount an immune reaction to the recipient, called *graft-versus-host disease*. Incompatible grafts can be fatal.

Bubble Babies

Rarely, a baby is born with no immune capability at all, the result of a sex-linked recessive gene. Victims of *severe combined immune deficiency (SCID)* have no thymus gland, no lymph nodes, and no plasma cells. With no immune protection, the boy (why not girl?) succumbs to the first infection contracted.

In 1971, a boy with SCID was born to a Texas couple who had already lost one son to this disease. At birth, David was isolated in a sterile plastic bubble. Initially doctors had expected to find a suitable bone marrow transplant for David, but his HLA type occurs in the population with a frequency of only one in 32,000. After twelve years of searching, a suitable donor was not identified. In 1984, successes with a new experimental technique for preventing graft-versus-host disease in bone marrow transplants led David and his family to opt for a transplant using bone marrow from his sister.

His sister's HLA type was not a perfect match with David's, so her marrow was pretreated to destroy all mature T cells, which play a major role in graft-versus-host disease. The marrow was subsequently transplanted, but after several weeks, David died from complications of the transplant (see page 342). New methods of gene therapy, however, make the potential for curing HLA-mismatched SCID patients appear promising (see chapter 10).

Allergies and Immunity

An *allergy* is immunological hypersensitivity to a foreign antigen or *allergen*. Much evidence is accumulating on the genetic basis of allergy, and some allergies (for example, ragweed, table 4.5) have been associated with particular HLA genotypes. Over 35 million Americans have some type of allergy. The effects may be innocuous or annoying, as with seasonal hay fever or contact dermatitis. Alternatively, some allergies are life threatening, as in the case of allergies to insect stings or hypersensitivity to penicillin or other drugs. Allergies to common foods, such as milk and wheat, may require major alterations in life-style and habits. Virtually any substance is capable of inducing an allergic response: some women have been found to be unable to conceive because they are allergic to their husband's semen.

IgE antibodies are the culprit in an allergic reaction. These antibodies are highly specific, combining first with the offending allergen and then attaching to special cells in the body known as **mast cells.** Mast cells are found in many tissues in the body, but are most abundant in the dermis (the layer just under the dry outer skin) and the tissues surrounding the gut, respiratory tract, and blood vessels. IgE triggers mast cells to release copious amounts of **histamine.** Histamine causes blood vessels to dilate and fluid to leak from the circulatory system into surrounding tissues. It also stimulates the copious production of mucus. Food allergies and hives are other results of histamine release, caused by an outpouring of fluid into the dermis, which results in fluid-filled, itchy pustules.

Antihistamine drugs relieve allergy symptoms by blocking histamine receptors on nerve endings. They do not stop histamine release by the mast cells, thus accounting for the limited effectiveness of the drugs. Present research is seeking a class of drugs that would occupy the binding sites on the mast cells to prevent histamine release altogether.

The life-threatening allergic response of some persons to bee and wasp stings results from a massive histamine release. So much plasma pours out of the blood-stream that the victim's blood pressure drops rapidly. Without medication, the victim may enter shock and die.

Autoimmune Diseases

In some instances, the normal mechanism for distinguishing self from non-self fails, causing one of a group of **autoimmune** diseases. Autoimmune disease occurs when antibodies against one's own tissues, *autoantibodies,* are produced.

In normal individuals a very low level of autoantibodies is always present, but with no harmful effects. As an individual gets older, the production of autoantibodies tends to increase, perhaps contributing to the aging process. Although the normal function of autoantibodies is unclear, they may provide a mechanism for destroying cells in the body that spontaneously transform into cancerous cells, before these cells can cause disease.

Autoimmunity has been implicated in many diseases. The symptoms vary with the tissues affected. In multiple sclerosis, autoantibodies are produced to the myelin sheath of the brain and spinal cord. Autoantibodies to muscle protein occur in myasthenia gravis; in rheumatoid arthritis, autoantibodies are produced to IgG itself. Rheumatic fever is a complication that sometimes follows a childhood streptococcal infection. Antigens on the strep bacterium resemble those on heart tissue. The antibodies developed to defend against the strep bacterium may function as autoantibodies, causing permanent heart damage.

Systemic Lupus Erythematosus

Systemic lupus erythematosus (SLE) is an autoimmune disease affecting about one-half million Americans, mostly women. One symptom of SLE, a red rash across the nose and cheeks, accounts for its name, lupus erythematosus (meaning red wolf). Victims of SLE produce antibodies to their own DNA. These autoantibodies attack DNA, destroying cells in the kidneys, lungs, heart, and brain. About one-fourth of SLE victims suffer severe kidney damage and another one-third suffer severe brain damage. SLE is a chronic condition with bouts of remission and relapse. Nine of ten patients are women. It has been observed that, in general, women have a more active immune system than men, but also have a higher frequency of autoimmune diseases.

SLE is treated with immunosuppressant drugs and *plasmapheresis* therapy. In plasmapheresis, a patient's blood is slowly removed. All antibodies, including the damaging autoantibodies, are filtered from the blood, which is then returned to the patient. Temporarily lacking antibodies, the patient is susceptible to infection, but plasmapheresis often induces remission of the disease.

Twin studies indicate a genetic predisposition for SLE. If one identical twin gets SLE, the chance of the second twin contracting SLE is 50–60%. In fraternal twins, the concordance of the disease, however, is only 2–3%.

(a)

(b)

(c)

(d)

Viruses and Immunity

Viruses (figure 12.8) are remarkable yet simple entities, which are composed of a protein coat enclosing a core of nucleic acid, either DNA or RNA. They may have as few as five genes and never more than several hundred. Viruses probably infect all living things and are generally species specific. That is, a tulip virus infects only tulips, not roses or humans or bacteria. Additionally, many viruses are organ specific, growing only in one type of tissue. Leukemia viruses, for example, grow only in white blood cells.

Viruses reproduce only inside a living cell. The virus subverts the regulatory apparatus of an infected cell, forcing it to make viral protein (for the coat of the virus) and nucleic acid, at the expense of the host cell's normal metabolism. Such infections generally end with the death of the cell and the release of hundreds of new viruses.

Many familiar diseases are caused by viral infections. DNA viruses cause smallpox, warts, mononucleosis, fever blisters or cold sores, shingles, genital herpes and some cancers. Among the common RNA viruses are those causing polio, rabies, mumps, measles, encephalitis, and influenza. For the remainder of this chapter, we consider aspects of the life history of viruses and focus on the ways that viruses interact with the immune system to produce certain diseases, including AIDS.

Life Cycles of Viruses

Viral infections have different outcomes. The virus may immediately use the host's metabolic machinery to replicate itself many times. Following replication, the progeny viruses are released, killing the host cell. This is known as a **lytic infection** (figure 12.9).

Alternatively, the virus can enter the host cell in a **latent** or relatively inactive form. Viral DNA attaches to and becomes an integral part of the host's DNA. When normal cell division occurs, the viral DNA is replicated along with the host DNA. Thus, each descendant cell bears a copy of the latent virus genome within its own genome. This type of infection, known as **lysogenic infection** (figure 12.10), may persist for years within a host cell and its

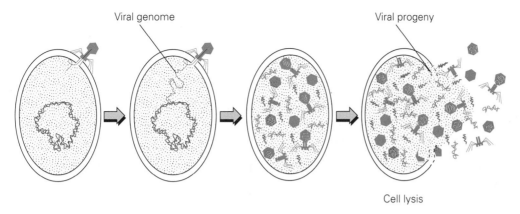

Viral genome

Viral progeny

Cell lysis

Figure 12.9

The life cycle of a lytic virus. The virus infects the cell by injecting its genome into the cell. The viral genes use the cell's nucleic acid and protein-synthesizing apparatus to make viral protein and nucleic acid. The viral particles are assembled and then the cell ruptures, releasing all the viral progeny.

Viral DNA

Integration of viral DNA

Induction of lysis (uv)

Bacterial DNA

Cell lysis and release of hundreds of viruses

Figure 12.10

The life cycle of a lysogenic virus in a bacterium. Following injection of the viral genome into the cell, the viral DNA integrates into the bacterial DNA. As the bacterium replicates, the viral genome is copied also, thus spreading throughout the bacterial population. Later, often in response to shock or stress (here ultraviolet radiation), the viral DNA is excised from the bacterial genome. The virus then enters a lytic phase, culminating in the death of the bacterium and the release of viral progeny.

descendants without producing pathogenic effects. While in its latent phase, a lysogenic virus sometimes appears to confer benefits upon the host cell. For example, the host may be protected from infection by other viruses. Infected cells may have growth advantages over other cells.

Following infection, the host cell grows normally until subjected to an environmental stress such as radiation, heat, or toxic chemicals. Such an assault can cause the lysogenic virus to enter a lytic cycle. The viral DNA then excises itself from the host chromosome and causes the host to produce hundreds of new viruses. After a short period, the host cell lyses, releasing infective viral progeny. Thus, the life cycle of a lysogenic virus terminates in a lytic phase.

Finally, some animal viruses are able to infect and remain latent within cells without integrating their genomes into that of the host cell. These viruses may cause little harm to the host cells and replicate with the host until some stimulus triggers them to become virulent or cause cellular transformation.

Reverse Transcriptase

RNA viruses cannot insert their RNA genome into the DNA of the host cell. For an RNA virus to be integrated into the genome of a host cell, it must first make a DNA transcript of itself. In addition to their RNA genome, some RNA viruses (**retroviruses**) carry an enzyme known as **reverse transcriptase.** Upon entering the cell, this enzyme builds viral DNA from the viral RNA genome, in direct contrast to the information flow predicted by the central dogma (chapter 8). Following the action of reverse transcriptase, the DNA transcript of the viral RNA genome is incorporated into the DNA of the host cell. The remainder of the life cycle of the RNA virus is similar to that of the DNA virus (figure 12.11).

Immune Genetics   **339**

Figure 12.11

The life cycle of an RNA tumor virus involves an extra step over that of a lysogenic DNA virus. The viral RNA and reverse transcriptase are injected into a cell. A DNA copy of the entire viral genome is then made and this DNA is incorporated into the host's genome. Then by normal transcription, viral RNA, more enzyme, and viral protein are made, producing mature viral progeny.

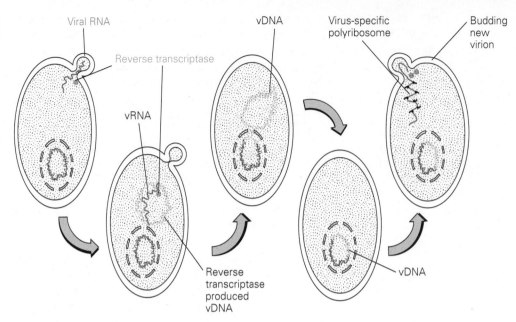

Color = Viral DNA made from vRNA using reverse transcriptase ●

A large group of retroviruses have now been identified. These viruses have a normal lysogenic life cycle, inserting a DNA copy of their genomes into infected host cells. Additionally, they have been found to cause cancer in many species.

Transduction

Viruses sometimes move segments of DNA from one cell to another, a process known as **transduction.** Transduction can be considered a form of bacterial genetic recombination. When a bacterium lyses due to a viral infection, some of the viral progeny may include segments of the DNA of the bacterial host. When the virus subsequently infects another bacterium, the DNA from the first host can be transferred to the new host.

Transduction is one mechanism responsible for the widespread drug resistance of many bacteria. In bacteria, genes controlling drug resistance are often carried on pieces of extrachromosomal DNA known as **plasmids.** Transducing viruses are responsible for spreading these drug resistant genes among a wide variety of bacteria, particularly in hospitals. While drug resistant strains of bacteria pose an obvious medical problem, genetic engineers use transduction for spreading beneficial genes between valuable domestic species. Genes for resistance to disease or insects could be transmitted among plants by such a technique.

Transduction, however, is only one way in which genes can spread through a population. More often, a gene for drug resistance is disseminated because its carrier survives and transmits the gene to its descendants, i.e., through the action of natural selection. Transduction differs from natural selection in spreading drug resistance because (1) only one, or a few, genes are transmitted at one time and (2) genes may be transmitted across species boundaries. It is this latter trait that makes transduction of particular importance to genetic engineers.

Viral Diseases

As noted earlier, viruses are the cause of many serious diseases. Viral diseases are particularly pernicious for two reasons. First, unlike bacteria, viruses are not usually susceptible to treatment with antibiotic drugs. Viruses must simply be allowed to run their course, treating the symptoms but relying upon the body's own immune mechanisms to fend off the disease.

Second, after the first infection, **latent viruses** can retreat to tissues of the nervous system and remain hidden there for years. They may cause no further symptoms or they may recur with varying severity from time to time, as with oral herpes (which produces cold sores in the mouth) and genital herpes. The presence of latent viruses can be detected by the presence of viral antigens in the host's tissues or by viral DNA sequences in some of the host's cells.

We now briefly examine a few viral diseases whose pathology appears to depend upon their ability to disrupt or evade the immune mechanism.

Mononucleosis and EBV

Burkitt's lymphoma (figure 11.22) is a relatively common cancer in African children but is rare elsewhere in the world. Tumor cells from virtually all persons with Burkitt's lymphoma are infected with the **Epstein-Barr virus (EBV),** strongly implicating the EBV as the cause of the cancer. Surprisingly, about 80% of the world's population are infected with EBV, although the cancer is limited primarily to Africa.

Commonly an EBV infection produces only a mild cold, and the affected individual develops antibodies to the virus, a response similar to that of any viral influenza. In the United States, a small fraction of persons exposed to EBV develop **infectious mononucleosis.** First described in 1920, mononucleosis is characterized by fever, tonsillitis, swollen lymph glands, and an enlarged liver and spleen. It is *not* highly infectious and usually runs its course in one to four weeks, depending upon the patient's health and genotype. Only about 1% of all patients develop complications such as pneumonia or kidney failure. Mononucleosis is also characterized by an increase in the white blood cells known as mononucleocytes, causing some scientists to consider mononucleosis a form of self-limiting leukemia.

EBV enters the body through the mouth or respiratory tract. Mononucleosis develops in four to seven weeks. An EBV infection normally produces permanent immunity, but the virus persists in a lysogenic form in the B cells. It is sporadically released and thus spreads through the population. Approximately 35–80% of students entering college have antibodies to EBV. Each year about 10–15% of those students without antibodies to EBV catch mononucleosis from their peers, who probably had no more than a cold from their initial infection with EBV.

Mononucleosis may be a disease whose incidence has actually increased with improved hygiene. Prior to the twentieth century, essentially all persons may have been exposed to EBV (and numerous other viruses) at an early age. Such early exposure, at a time when the immune system is most active, probably resulted in a mild disease, in contrast to the potentially serious illness experienced by college students, who may not be exposed to the virus until well past puberty.

Cancer

Worldwide, about 20% of all cancers are now known to be caused by viruses. In particular, three DNA viruses are clearly associated with specific cancers. The precise method by which cancers are induced remains to be determined in most cases, but it seems likely that viral infections can be one of several steps that may be required for cancer to develop (figure 11.23).

Several strains of the **human papilloma viruses** cause cervical and other genital cancers. A protein product of the papilloma virus appears to combine with the RB-1 protein of the hereditary retinoblastoma locus. Recall from chapter 11 that the RB-1 protein functions as an anti-oncogene or tumor suppressor. The papilloma protein immobilizes RB-1, leading to cancerous transformation of the affected cells.

The Epstein-Barr virus causes lymphomas and nasopharyngeal cancers. Like papilloma viruses, EBV appears to produce proteins that lead to cellular transformation. The ability of EBV to induce cancer was tragically confirmed during an attempt to cure SCID victim David, the "bubble boy" who had spent twelve years in a germ free environment waiting for a compatible bone marrow donor. With modern technology, doctors attempted to give David an immune system by pretreating marrow from his sister to prevent graft-versus-host dis-

Figure 12.12

These policemen wore masks to protect themselves during the 1918 influenza virus that killed 20 million people, more than were killed in World War I. The value of the masks is uncertain, but no other direct protection was available.

Culver Pictures Inc.

ease. David developed mononucleosis only eight days after the transplant from his sister, while still in a sterile environment. David died four months later of a B cell cancer and autopsy showed all of his own B cells were bursting with EBV. He probably received the virus from his sister's bone marrow transplant. Unintentionally, the attempted transplant proved EBV induces B cell cancer—and very rapidly in immune compromised patients.

Several hundred million people in southeast Asia are infected by the **hepatitis B virus.** An infected person has a risk of liver cancer about 100 times greater than an uninfected person. How the hepatitis B virus causes cancer is uncertain. Insertion of the viral genome into host cells might interfere with regulatory segments of a proto-oncogene, leading to activation of a cellular oncogene. Alternatively, the rapid rate of cell division required to repair damage from a hepatitis B infection offers increased opportunities for mutations that could lead to cancer.

Influenza

Three major types of **influenza** viruses (figure 12.8b), designated A, B, and C, are recognized by unique characteristics of their surface antigens. Types A and B are responsible for flu epidemics, while type C usually produces only minor symptoms. Influenza viruses, unlike other viruses, regularly change the antigenic determinants on their protein coat. These slight alterations, known as **antigenic drift,** permit the virus to escape recognition by the immune system, making humans susceptible to repeated bouts of flu.

Major alterations in the antigens of influenza viruses, **antigenic shifts,** occur less frequently, about every ten years. These changes can produce widespread, potentially severe flu epidemics. In only six weeks in 1918, a new flu virus was responsible for the deaths of at least 20 million persons and affected perhaps one hundred times that number. After major antigenic shifts, the Asian flu (1957) and Hong Kong flu (1968) epidemics also affected large portions of the world's population.

Influenza viruses are so insidious and ubiquitous because of their ability to undergo antigenic drift and antigenic shifts. These viruses accomplish antigenic drift by rearranging their genes, as well as by deletions, insertions, and point mutations in their genomes. These genetic changes can alter the antigenic determinants of a specific strain of virus, making the strain immunologically unrecognizable to antibodies produced in response to earlier infections. Such genetic changes are highly adaptive for viral survival. In effect, the virus is engaged in a continual arms race with the immune system of the host, the virus changing just enough to elude the host's antibodies.

Antigenic shifts and antigenic drift require that new flu vaccines be produced annually. Samples of influenza virus from around the world are monitored regularly in an attempt to identify altered strains that might be particularly virulent. Once such strains are identified, new vaccines are prepared.

Herpes

Herpes viruses (figure 12.8a) are DNA viruses that are found in all animal populations. Five distinct types of herpes infect humans: (1) type 1 herpes, causing facial cold sores; (2) type 2 herpes, responsible for genital herpes; (3) EBV, causing infectious mononucleosis; (4) varicella-zoster virus, producing chicken pox and shingles; and (5) cytomegalovirus, which results in serious mental retardation in an estimated 1% of all newborn infants in the United States. The effects of herpes viruses have been known to humans for at least 2000 years; the Roman Emperor Tiberius banned kissing at public ceremonies in an attempt to stop an epidemic of oral herpes.

Oral herpes infects at least 80 million Americans. Genital herpes, a far more serious problem, affects an estimated 20 million Americans. About 600,000 new cases are reported annually. Recent evidence shows that both oral and genital herpes viruses can be transmitted to either part of the body during oral sex. The initial manifestation of genital herpes is fluid-filled blisters on the genitalia. The extremely painful symptoms persist for several weeks.

Neither oral nor genital herpes is curable. As the blisters heal, the virus retreats to cells in the nervous system, where it remains in a latent condition. Some victims never experience another outbreak of the sores. However, many herpes victims have a recurrence of symptoms once or twice a year, while in others recurrence is far more frequent. Furthermore, *all victims* harbor the virus and are subject to an outbreak at times of stress.

Herpes viruses may have an even more serious effect on human health. Herpes viruses are known to cause cancer in frogs and chickens. Such a relationship has yet to be proven in humans, but women with genital herpes are eight to ten times more likely to develop cervical cancer than noninfected women.

Multiple Sclerosis and Measles

First described about 140 years ago, **multiple sclerosis (MS)** afflicts about two million people worldwide. MS is an autoimmune disease in which an individual's own antibodies attack the myelin tissue lining the central nervous system. Myelin is essential for transmission of information throughout the nervous system. Onset of the disease most commonly occurs between the ages of twenty and forty; twice as many women as men develop the disease. As the disease progresses, scar tissue forms in the myelin, preventing the flow of messages throughout the nervous system. Gradually the victim loses motor control and balance, eventually becoming paralyzed.

Viruses are strongly implicated in MS. Essentially all MS victims have had measles. Measles viral nucleic acid can be identified in one of four MS patients studied. The characteristic measles viral antigen has also been detected in the central nervous systems of MS patients, but not in other tissues of the body. Thus, MS may be initiated by entry of the measles virus into the central nervous system, where it subsequently flourishes. Normally, the central nervous system is strictly protected from infections of any kind. Persons who contract MS may have some genetic defect that permits the measles virus access to their central nervous system. Recent studies report a deficiency of interferon, which normally aids in the fight against viral infection, in patients with MS.

Measles is a **membrane budding virus.** When it replicates, it packages its RNA into the viral coat at the host cell wall (figure 12.13). During packaging, cell protein may be accidentally incorporated into the new virus. Although the nervous system is normally immunologically privileged (i.e., isolated from the body's immune mechanisms), viruses accidentally bearing proteins from the nervous system could enter the circulatory system as a result of membrane budding. The immune system does not recognize the protein (myelin, in the case of MS) as self and thus begins producing antimyelin antibodies, initiating the disease.

This scenario, analogous to the postulated relationship of EBV and mononucleosis, is reasonable, but has not been proven conclusively. Evidence from experiments with mice, however, is compelling. If samples of myelin are removed from a mouse, ground up, and then injected into the same mouse, symptoms identical to MS develop. Thus, it seems reasonable

Figure 12.13

Membrane budding viruses fuse with the host cell membrane and inject the viral genome into the host cell. When new viruses are produced, portions of the host cell membrane are incorporated into the viral protein coat. Such viruses thus serve to move pieces of host membrane to distant parts of the body, perhaps inducing autoimmune diseases.

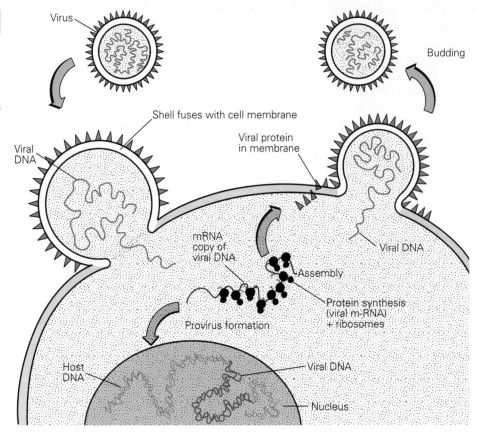

that the measles virus or other membrane budding viruses may provide a vehicle for exposing a person to their own myelin, as if the myelin were foreign tissue, which triggers the immune response and results in MS.

Age is another factor involved in the development of MS. Most persons have measles, but do not develop MS. Persons who contract measles after their teens are at higher risk of developing MS than those who had measles before age thirteen. Because the immune system is most efficient in juveniles, persons contracting measles early in life may mount an immune response sufficient to rid the body of the virus entirely. If, however, measles is not contracted until the late teens or twenties, the immune system may not cope with the virus as effectively. A latent reservoir of the virus could thus persist in the body, making the person prone to later development of MS.

Insulin Dependent Diabetes and Mumps

Diabetes is an inability to metabolize carbohydrates properly due to a deficiency in the production of insulin. One form, *maturity onset diabetes,* can normally be controlled by diet, exercise, and oral drugs. This type of diabetes has a strong genetic component. Over 85% of the victims have at least one diabetic parent, and twin studies show high concordance for this condition. If one twin contracted this type of diabetes, the other twin did also.

Insulin dependent diabetes is a severe disease, requiring daily insulin injections. It normally affects children, only 11% of whom have a diabetic parent. In insulin dependent diabetes, the white blood cells appear to participate in an immune reaction that destroys the insulin-producing cells of the pancreas. A relationship of this form of diabetes with mumps (caused by a virus) was first suggested by a Norwegian doctor in 1864. Like the measles virus, the mumps virus is a membrane budding virus. If the mumps virus is able to persist in the pancreas, the viruses could incorporate pancreatic antigens that stimulate an autoimmune response, just as in MS.

Figure 12.14

Cumulative number of cases of AIDS in the United States and those countries who reported to the World Health Organization by year of diagnosis.

Data from P. Piot, et al., "AIDS: An International Perspective," *Science,* Vol. 239, pp. 573–579; World Health Organization, Jan. 3, 1990; U.S. Center for Disease Control, December 1990.

Insulin dependent diabetes is also related to HLA types; about 60% of the victims are either type B8 or type BW15 (table 4.5). These particular HLA genotypes may make individuals more susceptible to viral infection or less able to rid their bodies of the offending virus, thus triggering diabetes. Both MS and insulin dependent diabetes clearly illustrate the complex interactions of latent viruses, genotype, and chance in the development of serious disease.

AIDS

The 1980s will be remembered in medical history as the decade that saw the onset of the extraordinary disease AIDS. No disease has ever been studied so intensively and so immediately once it was recognized as an epidemic of worldwide consequence. In less than a decade, AIDS has become one of the major causes of death by infectious disease throughout the world. Its rate of increase is geometric and shows no sign of slowing down (figure 12.14).

AIDS is the acronym for **acquired immune deficiency syndrome.** Acquired means that it is an infectious disease caught from another person. Immune deficiency means that the disease affects the immune system, drastically reducing its ability to fend off other infectious organisms or parasites. Syndrome means that the single condition has many effects.

AIDS was first clinically identified in 1981, when several physicians in the United States noticed an increased incidence of the rare cancer Kaposi's sarcoma in young homosexual males. Kaposi's sarcoma was previously associated primarily with patients who had been treated with drugs that suppressed the immune system, often in conjunction with transplants. Once the condition was clinically recognized, the number of cases increased rapidly, and in 1982 the name AIDS was given to the condition.

The HIV virus. Medical researchers throughout the world immediately realized the potential threats posed by AIDS and established research programs on both the medical and social aspects of the disease. In 1983 the cause of AIDS was identified as a virus, and in 1986 this virus was officially designated the **human immunodeficiency virus**—HIV. The entire genome of HIV has now been sequenced.

HIV is a retrovirus, incorporating itself into the host's genome and then releasing up to fifty thousand progeny at a time, time after time. As part of the host's genome, HIV is well protected from drugs. To destroy the cause of the disease, the drug would have to destroy the host's cells. Many experimental drugs are now undergoing human trials. The most widely used drug, AZT, interferes with the replication of the AIDS virus.

Development of an effective vaccine to prevent AIDS has proven to be extremely difficult because of the high levels of variability of the virus. At least three types of HIV virus have been identified, differing by more than 25% of their nucleotide sequences. Even within a single type, levels of variability are extremely high. Strains of a single type isolated within one individual may differ by 2–3% in their nucleotide sequences. These differences arise from genetic mutation. Unlike DNA viruses, retroviruses have no "proofreading" and repair mechanisms associated with replication, and for some reason HIV is even more mutable than other retroviruses. On average, HIV viruses undergo about one nucleotide substitution per 2,000 base pairs per replication, or a total of 5–10 mutations per genome per replication, a mutation rate about one million times higher than that of DNA viruses.

The HIV virus has a long and variable latent period. About half the people with the HIV virus or antibodies to it develop AIDS disease within five years. The disease appears to have 100% mortality; victims die, on average, about four hundred days after first diagnosis, most of that time being spent in hospitals. Roughly 80% of AIDS victims die within two years of diagnosis.

AIDS is lethal because the HIV virus attacks the body's primary immune defenses. The main target is helper T cells, but macrophages and B cells are also affected. This weakens the entire immune mechanism, and AIDS victims generally die as a result of secondary infection from one or more of over thirty types of disease organisms—other viruses, bacteria, protozoa, or parasites, none of which would normally be fatal.

Epidemiology. The HIV virus can live in a wide variety of body fluids, but the disease is contracted primarily by contact with the semen or blood of an infected person. The primary modes of infection have been sexual contact, sharing of dirty needles by drug addicts, and blood transfusions. Some health workers have contracted the disease by accidental needle sticks. Additionally, mothers may transmit the virus to their children either *in utero* or via maternal milk.

Contrary to early claims, AIDS is not a "gay disease"; AIDS is not restricted to any particular societal group. All segments of society have been exposed to the HIV virus, and anyone exposed to it may develop AIDS. Certain behaviors are especially risky. Sharing needles while using drugs greatly increases the risk of contracting AIDS, as does promiscuous sexual behavior, particularly without the use of a condom. Hemophiliacs as a group are at risk because of their need for frequent blood transfusions, although screening of donated blood has reduced the risk.

In North America, male homosexuals have comprised a major "at risk" group, partly because of high levels of promiscuity and partly because of the efficiency of transmission of the HIV virus by anal intercourse. During anal intercourse, the walls of the rectum are prone to tearing, and the virus can be transmitted from infected semen directly to the recipient's circulatory system.

Prevalence of AIDS is now increasing in the heterosexual community. Levels of transmission of the virus from males to females during vaginal intercourse are lower than for anal intercourse, presumably because the vaginal lining is a relatively effective barrier to the virus. Transmission of the virus from females to males occurs at even lower rates, for unknown reasons. On the other hand, transmission of the virus is enhanced by open lesions on the genitals, as may occur when either partner has other sexually transmitted diseases such as herpes, gonorrhea, or syphilis.

The United States has been more seriously affected by AIDS than any other country in the developed world. Retrospective analysis of samples stored in blood banks has established that AIDS has existed in the United States since at least the late 1950s, but with very low rates of incidence. Since identification of the disease in 1981, death rates from AIDS have increased exponentially (figure 12.14). By the end of 1987, about 25,000 people had died, and by the end of 1990, over 90,000 AIDS patients had died. By 1990, over 1,000,000 Americans were estimated to be infected; during the 1990s, this may mean about 75,000 new cases and perhaps 50,000 deaths annually. This contrasts with about 55,000 killed during the Viet Nam war, and about the same number killed in automobile accidents each year.

It is difficult to know if the number of infections, cases of AIDS, and deaths will continue to increase. Education programs appear to be effective in reducing the numbers of new infections in some gay communities, but infection rates may still be increasing in other at risk groups such as intravenous drug users and their children. Development of effective vaccines to prevent infections or treatments to cure infections seem unlikely before the turn of the century. But the rates of infection in North America in 1990—about 1/75 for men and 1/700 for women—mean that AIDS will be a source of tragedy for tens of thousands of Americans in the 1990s.

Worldwide, the number of infected persons may be as high as twelve million (one person in four hundred), and twenty million may be infected by the turn of the century. About 750,000 people are believed to have AIDS now, and by the year 2000 that number may be as high as six million. Roughly 60% of all cases occur in Third World countries. In central Africa, the most severely affected part of the world, one person in fifty is infected with the HIV virus. In contrast to North America, both sexes appear to be affected equally. In parts of countries such as Zaire, infection rates are as high as one person in five. For these impoverished countries, AIDS has already overwhelmed government health systems, a tragedy that will only become worse in the next decade.

Social and ethical issues. Any consideration of AIDS involves highly emotive topics such as death, sexuality, and drug use. These are further confounded by the enormous amounts of money now committed to the battle against AIDS. The combination of death, sex, drugs, and money has led to intense public debates and confrontations involving politicians, health care workers, prison workers, school officials, pressure groups such as the homosexual community and civil rights activists, and research scientists (box 12.5).

For such high risk groups as male homosexuals, the battle against AIDS has been a political struggle for funding to support research and medical needs for the afflicted. For homosexuals, AIDS has also been a continuing struggle against the stigmatization that homosexuals have long faced. Stigmatization is a risk confronting anyone with AIDS. People have been fired from their jobs because they have AIDS, and children with AIDS are sometimes not allowed to attend school or are harassed by parents of other children if they do attend.

Civil rights groups have fought against proposals for mass screening of the population for AIDS antibodies, believing that the results might be used for discriminatory purposes. Remember that the presence of HIV antibody only means exposure to the HIV virus, and

BOX 12.5

Scientists and public controversy (or, scientists really are human)

Figure 12.15

Leading protagonists in scientific controversies about AIDS: (a) Dr. Robert Gallo; (b) Dr. Luc Montagnier; (c) Dr. Peter Duesberg.

(a) Courtesy of Dr. Robert Gallo; (b) Courtesy of Luc Montagnier; (c) *Science,* 239, 1988; photo by Tom Alleman.

(a)

(b)

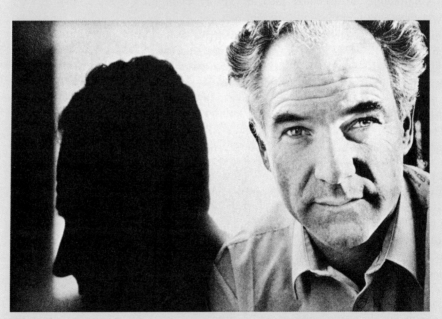

(c)

If you think that science is conducted by shy, introverted women and men in white lab coats working diligently to cure dangerous diseases out of the public eye, then you didn't pay attention to what might be called the "AIDS Wars" of the 1980s. From the time of identification of AIDS in 1981, scientists and physicians have been subjected to the harsh glare of media attention as the foot soldiers in the war against AIDS. From the very start, it was obvious that AIDS was both a terrible tragedy for the afflicted and also a growth industry for science, medicine, and the pharmaceutical industries.

Enormous amounts of money are involved. One year of care for a single AIDS patient costs about $38,000, a course of drugs about $7,000. A huge research establishment has been constructed to identify the cause(s) of AIDS, to develop vaccines and drugs, and ultimately to cure the disease. In 1989 the U.S. government spent about $1.3 billion for research and education related to AIDS, about the same amount spent on cancer plus heart disease, conditions that affect and kill far more people. For scientists, this money offers the opportunity to develop and supervise research programs, to train the next generation of scientists, to achieve international reputations and in many cases to acquire considerable wealth. Such high stakes produced two highly publicized controversies involving top scientists during the 1980s.

The first involved Dr. Robert Gallo of the U.S. National Cancer Institute and Dr. Luc Montagnier at the Pasteur Institute in France (figure 12.15a,b). Both led teams racing to identify the organism causing AIDS, and both claimed to find it first. After lengthy squabbling in the media and courts, an agreement was reached by which they agreed to share both credit for the discovery and the large income to be derived from the diagnostic test resulting from the discovery. Nonetheless, fuelled at least partly by the media, the controversy over priority has continued to simmer.

A second controversy has centered around Dr. Peter Duesberg, a professor at the University of California, Berkeley and one of the discoverers of retroviruses in the 1960s (figure 12.15c). Duesberg claims that the HIV virus has not been proven to be the cause of AIDS and that it is likely that other organisms or factors may be equally or more important in causing the disease. Duesberg argues that an "HIV industry" with vested interests is obstructing search for other possible causes. The HIV industry includes drug companies, research institutes, and scientists, all of whom built entire research programs around HIV and have large financial interests in HIV being the causative agent of AIDS. To a large extent, Duesberg has been a voice in the wilderness, as his views have been rejected by almost all the scientific community, but his cause has been actively taken up by some journalists both in print and on television.

Such sometimes unseemly controversies, in fact, permeate the world of science, but they seldom make "good copy" for the news media. That they do in these instances reflects the intense public interest in AIDS, as well as the enormous professional stakes (and sometimes financial interests) of the scientists involved. After all, the discoverer of the organism causing AIDS may one day be awarded a Nobel Prize (the ultimate in scientific prestige, and worth hundreds of thousands of dollars), and the scientist who finds a cure almost certainly will.

These controversies also may shatter the "ivory tower" stereotype of science and scientists held by many of the public, but that stereotype probably never reflected much truth. What is revealed instead is that scientists display the same range of behaviors as members of any other profession. Science is an activity engaged in for all the usual reasons—ambition, power, prestige, financial reward—but also for most out of a genuine desire to discover new things and to use that knowledge to benefit humanity. Gallo, Montagnier, and Duesberg have risen to the top of their professions almost certainly because they are ambitious men, but that is not reason to doubt their commitment to gaining knowledge useful in the battle against AIDS. As part of settlement of their priority dispute, Gallo and Montagnier agreed that income derived from the diagnostic test should be used to support foundations for AIDS research.

On the other hand, discerning truth in such controversies can be extremely difficult. If the truth were obvious, the controversy probably would not exist. While most scientists accept HIV as the cause of AIDS, many are also willing to acknowledge that Duesberg has some valid points to his argument. If HIV alone is the cause, for example, why does AIDS affect both sexes equally in Africa, but predominantly males in North America? Why, Duesberg asks, do different risk groups die from different secondary causes—Kaposi's sarcoma for one group, pneumocystis pneumonia for another, and so on? Such differential expression of disease is uncommon for other disease organisms. The truth about AIDS, whatever it may be, appears likely to be as complex and challenging as the truth about the scientists who study it.

not that the individual has the disease. In fact, the presence of HIV antibody could even mean that the person has effectively fought off an HIV infection and is thus immune. Furthermore, the latent period of the virus is still unknown; it is not even certain that all infected people will actually get the disease. Civil rights groups argue that HIV antibody tests could be used to penalize people who may be unaffected by AIDS, or who might die of other causes before AIDS strikes.

Research scientists, on the other hand, have argued that understanding the dimensions of the threat posed by AIDS requires detailed information on the frequency of the disease in particular demographic groups and thus that widescale screening for HIV antibodies is necessary. Testing for the presence of AIDS antibodies is now mandatory for members of the military and foreign services, and for obtaining some insurance policies. Visas to visit the United States are not granted to persons testing positive for the HIV antibody—a strange restriction considering that infection rates are relatively high in the U.S., and thus far more danger exists that the disease will spread from the U.S. to the rest of the world than vice versa.

The large federal budget now devoted to AIDS in the United States—over one billion dollars annually—reflects the results of a very effective political lobbying effort. Cancer and heart disease affect far more people every year, but do not receive proportionally equivalent funding.

(A bizarre comparison is with government policies related to tobacco. Recall that smoking is the single greatest preventable cause of death in the United States, with smoking-related diseases killing about 350,000 Americans annually—about four times the number killed by AIDS in the 1980s. Nonetheless, political lobbying by the tobacco industry has succeeded in achieving government policies and subsidies that actually *support* the tobacco industry and thus increase death rates in America. Lethal tobacco products are openly advertised, while advertisements for potentially life-saving products such as condoms have been severely restricted.)

For less developed countries, the AIDS epidemic has added a particularly terrifying new item to the list of horrors engendered by poverty. In Romania, hundreds of babies were infected with HIV because the government refused to admit the existence of AIDS and restricted funding for such essentials as sterile needles and syringes. Some countries are so poor that meaningful programs to prevent spread of the disease or to care for the afflicted are simply not possible. In some central African countries where a large portion of the population is infected by HIV, total government spending on health (not just AIDS) is about $7 per person per year. With such tiny amounts of money, even educational items such as posters are barely affordable.

The cost of a year's treatment for a single AIDS victim in the United States (about $38,000–see box 12.5) is an enormous fortune for citizens of countries such as Zaire. The average citizen of a country such as Zaire often cannot even afford condoms to protect themselves from HIV, much less treatment after contracting a serious illness. As the wealthy nations of the world develop treatments, preventatives, and possibly vaccines and cures for AIDS, their dissemination to poor countries will be a measure of the sense of humanity and fairness of the governments of the wealthy countries.

Summary

1. The body is protected from invasion by dangerous foreign substances by two different categories of defenses. Nonspecific defenses, such as the skin and various secretions, are effective against a broad range of harmful agents. Specific defenses or immune mechanisms develop after exposure to a particular foreign antigen.

2. Lymphocytes, a type of white blood cell, differentiate to either T cells or B cells, the primary agents of the immune response. T cells provide cellular immunity by directly destroying foreign antigens. B cells differentiate further to plasma cells, which in turn produce antibodies. Antibodies destroy foreign antigens in a process called humoral immunity.

3. Macrophages interact with helper T cells, assisting in the immune response.

4. Specific immunity does not develop until after birth. Prior to birth, the immune system is active in identifying its own tissues as self.

5. Antibodies are produced against antigenic determinants, portions of foreign molecules that are recognized as non-self.

6. Following an immune response, some T cells and some B cells survive for long periods as memory cells. In the event of further infection by the same agent to which the cells initially responded, memory cells facilitate an extremely rapid second response.

7. Five types of antibodies or immunoglobulins are known. IgG, IgM, IgA, IgD, and IgE.

8. The clonal selection theory of antibody diversity attributes the specificity of the immune system to recombination among multiple genes, as well as somatic mutation in differentiating plasma cells.

9. Specific immunity is acquired after birth. It may be acquired passively, by receiving antibodies from another person; or it may be acquired actively, by developing one's own antibodies. Vaccinations induce actively acquired immunity.

10. Transplant rejection is primarily a function of T cells. It can be prevented by precise matching of the human leukocyte antigen (HLA) type of the donor and recipient. The HLA system is genetically extremely diverse. When HLA types cannot be precisely matched, immunosuppressant drugs are used.

11. Allergies are a form of immune hypersensitivity and appear to have a genetic basis. Particular allergens elicit overproduction of IgE and histamine release, hallmarks of the allergic response.

12. Autoimmune diseases occur when individuals produce autoantibodies to their own tissues. Systemic lupus erythematosus (SLE), multiple sclerosis, rheumatoid arthritis, and rheumatic fever are important autoimmune diseases.

13. Viruses are extremely simple organisms, consisting only of an outer protein coat surrounding an inner core of nucleic acid. Viruses cannot replicate independently of living cells.

14. Lytic viral infections cause immediate death to the host cell accompanied by the release of progeny viruses. In lysogenic viral infections, the viral genome is incorporated into the host cell genome and is replicated each time the cell divides. Upon appropriate stimulation, the virus ends the latent phase and enters a lytic cycle, killing the host as it reproduces.

15. The life cycle of an RNA virus involves an additional step for a lysogenic infection: the RNA must be copied into a DNA strand that can then be incorporated into the host cell genome. Reverse transcriptase is an enzyme made by RNA viruses that permits the synthesis of a DNA molecule from an RNA template.

16. Transduction is the transfer of DNA from one cell to another by a virus. Resistance to antibiotics has been transferred among medically important bacteria by this process.

17. Viral involvement has been shown in mononucleosis and multiple sclerosis. These diseases appear to develop when the victim is exposed to EBV or measles viruses after their teens. The seriousness of the disease is related to several aspects of the victim's genotype, including their HLA genotype.

18. About 20% of all cancers are caused by viruses. Among the best known cancer-causing viruses are the human papilloma viruses, Epstein-Barr virus, and hepatitis B virus.

19. The herpes viruses are an example of latent viruses that may persist for decades in a relatively inactive form in the victim's nervous system.

20. Influenza is a persistent human disease because the flu viruses are capable of antigenic drift and antigenic shifts. These two genetic abilities permit the virus to escape recognition by the immune system and prevent populations from developing immunity to these potentially deadly diseases.

21. AIDS is an insidious infectious disease caused by HIV, a retrovirus that is so genetically variable that it is difficult to produce an effective vaccine for it. AIDS can be prevented by appropriate behavior, but no cure is presently available. The AIDS virus infects cells of the immune system, leaving its victims unable to defend themselves against normally nonfatal infections.

Key Words and Phrases

acquired immune deficiency syndrome (AIDS)

acquired immunity

active immunity

agammaglobulinemia

antibody

antigen

antigenic determinant

antigenic drift

antigenic shift

autoimmune disease

B cell

Burkitt's lymphoma

cardiovascular system

cellular immunity

clonal selection theory

Epstein–Barr virus (EBV)

erythroblastosis fetalis

hepatitis B virus

herpes virus

histamine

HLA

human immunodeficiency virus (HIV)

human papilloma virus

human T-cell leukemia virus (HTLV)

humoral immunity

immunoglobulin

immunologically privileged site

infectious mononucleosis

influenza

insulin dependent diabetes

interleukins

killer (cytotoxic) T cells

latent virus

leukocyte

lymphatic system

lymphocyte

lysogenic infection

lytic infection

macrophage

mast cell

membrane budding virus

memory cell

multiple sclerosis (MS)

natural killer (NK) cells

natural resistance

passive immunity

plasma cell

plasmid

retrovirus

reverse transcriptase

SCID

specific immunity

systemic lupus erythematosus (SLE)

T cell

thymus gland

transduction

Questions

1. Distinguish between active immunity and passive immunity.
2. What are autoimmune diseases and how might they be treated?
3. Why is the HLA system potentially more useful than the red blood cell antigens in paternity suits?
4. Distinguish between T cells, B cells, NK cells, macrophages, lymphocytes, and leukocytes.
5. Of what importance is the lymphatic circulatory system?
6. Why is Immunoglobulin E (IgE) called the allergy antibody? How does it produce the uncomfortable allergy symptoms?
7. The testes are an immunologically privileged site. If testicular material were surgically removed from a laboratory rat, minced finely, and reinjected under the skin, as with a vaccination, what do you predict would happen?
8. What success would you predict for transplants of testes between unrelated males; between twin brothers?
9. Distinguish between transformation and transduction.
10. Distinguish between lytic viruses and lysogenic viruses.
11. What is the significance of the enzyme reverse transcriptase?
12. How are viruses implicated in some autoimmune and degenerative diseases?
13. What is the relevance of antigenic drift to the difficulty of producing effective influenza vaccines?

For More Information

Ada, G. L., and G. Nossal. 1987. "The Clonal-Selection Theory." *Scientific American*, 257:62–69.

Barnes, D. M. 1986. "Nervous and Immune System Disorders Linked in a Variety of Diseases." *Science*, 232:160–61.

French, D. L., R. Laskov, and M. D. Scharff. 1989. "The Role of Somatic Hypermutation in the Generation of Antibody Diversity." *Science*, 244:1152–57.

Gallo, R. C. 1986. "The First Human Retrovirus." *Scientific American*, 225:88–98.

Gallo, R. C. 1987. "The AIDS Virus." *Scientific American*, 256:46–56.

Gonda, M. A. 1986. "The Natural History of AIDS." *Natural History*, 95:78–81.

Hamburger, R. N. 1976. "Allergy and the Immune System." *American Scientist*, 84:157–64.

Henle, W., G. Henle, and E. T. Lennette. 1979. "The Epstein-Barr Virus." *Scientific American*, 241:48–59.

Hunkapiller, T., and L. Hood. 1986. "The Growing Immunoglobulin Gene Superfamily." *Nature*, 323:15–16.

Janeway, C. A. 1988. "Frontiers of the Immune System." *Nature*, 333:804–6.

Janeway, C. A. 1989. "A Primitive Immune System." *Nature*, 341:108. (This article deals with natural killer cells.)

Jaret, P. 1986. "Our Immune System: The Wars Within." *National Geographic*, 169:702–34.

Kolata, G. 1982. "Alzheimer's Research Poses Dilemma." *Science*, 215:47–48.

Lewin, R. 1981. "New Reports of a Human Leukemia Virus." *Science*, 214:530–31.

Marx, J. L. 1975. "Cytomegalovirus: A Major Cause of Birth Defects." *Science*, 190:1184–86.

Marx, J. L. 1987. "Antibody Research Garners Nobel Prize." *Science*, 238:484–85.

Piot, P. et al. 1988. "AIDS: An International Perspective." *Science*, 239:573–79.

Rapp, F. 1978. "Herpes Viruses, Venereal Disease, and Cancer." *American Scientist*, 66:670–74.

Scholitssek, C., and E. Naylor. 1988. "Fish Farming and Influenza Pandemics." *Nature*, 331:215.

Smith, K. A. 1990. "Interleukin-2." *Scientific American*, 262:50–57.

Thompson, L. 1988. "AIDS Diary." *Discover*, 9:36–38.

Time. 1982. "The New Scarlet Letter and Battling an Elusive Invader (Herpes)." August 2:62–66, 68–69.

Tonegawa, S. 1985. "The Molecules of the Immune System." *Scientific American*, 253:122–31.

Treichel, J. A. 1982. "Monoclonal Antibodies Correct Severe Immune Deficiency." *Science News*, 122:244.

Varmus, H. 1987. "Reverse Transcription." *Scientific American*, 257:56–65.

Wallis, C. 1987. "Back to Normal: Hope for Parkinson's Victims." *Time*, April 13:39.

Weiss, R. 1987. "Cell Grafts Proceed, Value Uncertain." *Science News*, 132:341 (re: Parkinson's treatment).

Weiss, R. 1989. "The Viral Advantage." *Science News*, 136:200–203.

Yancopoulos, G. D., and F. W. Alt. 1988. "Reconstruction of an Immune System." *Science*, 241:1581–83.

Young, J. D., and Z. A. Cohn. 1988. "How Killer Cells Kill." *Scientific American*, 258:38–44.

© Dr. A. Jeffreys,
Nature 314, 67–73.

Genetic diversity

Figure 13.1

The different races of humans: (a) African black (Negroid); (b) European white (Caucasoid); (c) American Indian (Mongoloid); (d) Australian aborigine (Australoid).

(a) © Margaret Thompson/ Anthro Photo; (b) © Rick Smolan; (c) © Shostak/ Anthro Photo; (d) Neg. #333218, Courtesy Department of Library Services, American Museum of Natural History.

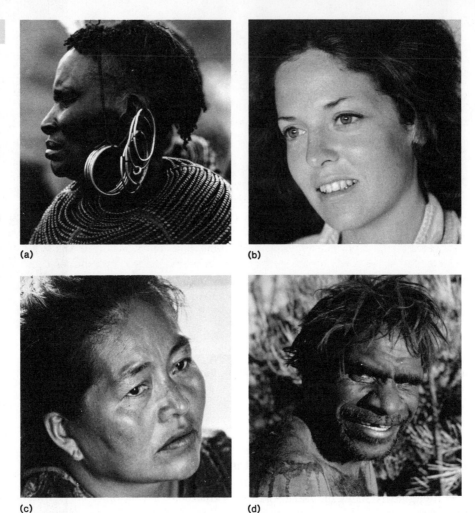

(a)

(b)

(c)

(d)

Genes are instructions for survival. A set of genes whose information matches (is *adapted* to) requirements of an individual's environment allows that individual to survive and reproduce. Maladapted genes lead to early death without reproduction.

Genes do not have an independent existence. They occur as integrated sets of information, analogous to a computer program, called the **genome.** In sexually reproducing species, there is a larger unit of integration, as well—the **gene pool.** The gene pool consists of all the genetic information, all the alleles, found in a group of interbreeding individuals. Thus for a population of N diploid individuals, the gene pool consists of all 2N alleles. All humans have the potential for interbreeding and thus all belong to one gene pool. The human gene pool, however, consists of many smaller gene pools determined by religious, social, and geographic factors. Because of this subdivision we recognize a number of human **races** (figure 13.1). In genetics, races are characterized by the frequency of alleles in a given population and by any alleles that may be unique to that particular gene pool. In reality, all humans have the same genes. We differ only in the particular alleles found at specific loci.

Each species may comprise a large gene pool, which is commonly subdivided into numerous smaller local gene pools. The total genetic variation found within the gene pools of individual species and also found in all species together is referred to as **genetic diversity.** Genetic diversity is the basis of all life on earth. Without genetic differences between individuals and between species, changing environments would quickly exterminate life.

Additionally, life is interdependent. Animals depend on plants and other animals for food. Plants depend on animals for dispersal and fertilization. Without other species, human life would not exist. Thus, humans need to understand the processes behind the present diversity of life on earth. How is genetic diversity generated and maintained? How can we ensure that the genetic basis for our continued existence will not be progressively depleted? How can we use genetic resources to sustain us in the future?

The Hardy-Weinberg Law

When the principles of Mendelian inheritance were rediscovered at the turn of this century, it was not understood how recessive alleles could persist in populations. Why, if brown eyes are dominant to blue eyes, do all persons not have brown eyes?

The answers to questions of this nature were provided in 1908. An English mathematician, G. H. Hardy, and a German physician, W. Weinberg, independently developed a mathematical model that showed how the frequency of recessive alleles in a population could be estimated knowing only the frequency of persons homozygous for the recessive allele.

The *Hardy-Weinberg equilibrium equations* apply to diploid, sexually reproducing populations and have five key underlying assumptions:

1. Genetic mutation must not occur.
2. Individuals must not migrate from one population to another.
3. Individuals must mate randomly with respect to genotype.
4. Reproduction and mortality must be the same for all genotypes—that is, *natural selection must not occur.*
5. The population must be very large.

In a mathematical sense, the Hardy-Weinberg equilibrium equations describe the fate of genetic variation in an "ideal" population that is not subjected to disturbing influences. Such ideal conditions virtually never occur in nature, yet many populations seem to conform to the expectations described by the Hardy-Weinberg equations. On the other hand, where populations differ from Hardy-Weinberg predictions, we can use the five underlying assumptions as a guide to understanding the process of genetic change and population divergence.

The Hardy-Weinberg equations describe frequencies of both alleles and genotypes in a population. The following notation is commonly used for a single locus with two alleles (A,a):

The frequency of allele A = p
The frequency of allele a = q

The frequencies of each allele vary between 0 and 1, and because there are only two alleles, the *first Hardy-Weinberg equation* states:

$p + q = 1$

In words, the sum of the frequencies of alleles A and a comprise 100% of the alleles at that locus in the population.

However, the population consists of diploid individuals, meaning that alleles always occur in pairs, with each pair constituting the genotype of a single individual. Thus, the basic data will always be the frequencies of the various genotypes in a population. Because the individuals are assumed to be mating randomly, the frequencies of the genotypes can be calculated as the product of the frequencies of alleles involved in each combination. The frequency of genotype AA is p $\times$ p, or p^2. Thus,

The frequency of AA = p^2
The frequency of aa = q^2
The frequency of Aa = 2pq

The frequency of heterozygotes, Aa (2pq), results from the fact that a heterozygote may be formed in either of two ways, receiving the A from the mother and a from the father, or vice versa (table 13.1).

Because a locus with two alleles can have only three genotypes, we know that their frequencies must also total one. Hence, the *second Hardy-Weinberg equation* states:

$p^2 + 2pq + q^2 = 1$

In practice, allele frequencies are calculated from data on genotype frequencies. For example, suppose that MN blood group data were determined for a large population of humans, with the following results:

Genotype	Number
MM	360
MN	480
NN	160

Allele frequencies can be quickly calculated, remembering that homozygotes have two copies of an allele and heterozygotes only one:

$p = [(2 \times \text{number of MM}) + \text{number of MN}]/(2 \times \text{number of individuals})$
$= [(2 \times 360) + 480]/(2 \times 1000)$
$= (720 + 480)/2000$
$= 1200/2000 = 0.60$

The frequency of the other allele, q, can be calculated the same way or more simply as:

$q = 1 - p$
$q = 1 - 0.60$
$q = 0.40$

Allele frequencies provide precise data for comparing populations, something we have already seen when comparing blood group data between human populations, for example. The interesting part of the Hardy-Weinberg calculations results from using calculated allele frequencies in one generation, as we have just done, to predict the allele frequencies in the next generation. Using the above data, we can calculate for the next generation:

$p^2 \quad = (0.60)^2 = 0.36 = \text{frequency MM}$
$2pq \quad = 2 \times (0.60) \times (0.40) = 0.48 = \text{frequency MN}$
$q^2 \quad = (0.40)^2 = 0.16 = \text{frequency NN}$

These frequencies are precisely those found in the original data. Continuing such calculations would show that, as long as these ideal circumstances apply, the allele frequencies *never* change. Hence, the Hardy-Weinberg Law states: *In a sexually reproducing diploid population not subjected to disturbing influences such as mutation, migration, non-random mating, natural selection, and small population sizes, neither allele nor genotype frequencies change in succeeding generations.* In other words, in a mathematically ideal world, evolution does not occur!

Geneticists know that in reality disturbing influences such as migration, mutation, and natural selection do occur, and that population sizes in nature are often very small. Moreover, biologists accept evolution as the central process in biology. Quite clearly, evolution does occur.

The Hardy-Weinberg equations are not applicable to non-sexual or to non-diploid species, or even to all loci in diploid sexual organisms. For example, Y-linked loci and mitochondrial loci are effectively haploid and thus not suitable for Hardy-Weinberg analysis. Modifications of Hardy-Weinberg calculations are required for X-linked loci.

TABLE 13.1

The Hardy-Weinberg equations and a Punnett square showing the derivation of the frequencies of each genotypic class

Let the frequency of the dominant allele **A** be represented by **p**
Let the frequency of the recessive allele **a** be represented by **q**
By calculation $[(p + q)^2 = 1]$ or by using the Punnett's square below, the frequencies of the three possible diploid types can be obtained:
The frequency of homozygous dominants (AA)   = **p²**
The frequency of homozygous recessives (aa)   = **q²**
The frequency of heterozygotes (Aa)   = **2pq**

Equation 1: $p + q = 1.00$
Thus,   Equation 2: $p^2 + 2pq + q^2 = 1.00$

All possible ♂ gametes
(their frequency)

	A(p)	a(q)
A(p)	AA p^2	Aa pq
a(q)	Aa pq	aa q^2

All possible ♀ gametes (frequency)

For example, consider a population where the frequency of the homozygous recessive genotype was 1/10,000. How many persons are carriers of the a allele?

$$q^2 = 1/10,000 = .0001$$
$$q = \sqrt{q^2} = .01$$
$$p = 1 - q = .99$$

The frequency of carriers in the population is $2pq = 0.0198 \simeq 2\%$

The importance of the Hardy-Weinberg Law is that it provides a basis for interpreting population differences and evolutionary change in the real world. Analysis of gene frequencies in populations over many generations often shows little or no change (a fact that is unsurprising in light of the Hardy-Weinberg Law). But in those cases where allele frequencies are changing, or where genotype frequencies are not those predicted, we can reasonably infer that one or more of the five disturbing influences is acting upon that population. Further study can often reveal the precise forces governing the evolutionary process.

The science of population genetics is an elegant mathematical development of the Hardy-Weinberg equations. The field of **ecological genetics** relates patterns of phenotypic and genetic variation within and among natural populations to principles derived from population genetics, particularly the five assumptions of the Hardy-Weinberg equations.

For the remainder of this chapter, we introduce aspects of genetic diversity and evolution as they relate to immediate human concerns: agriculture, conservation, and human population differences.

Agricultural Genetics

Domestication of wild species marked a crucial turning point in the origins of human civilization. Domestication increased the quantity and predictability of food and other important supplies. In turn, some members of developing societies were freed to focus their attention on art, commerce, engineering, law, medicine, and science. Today, domestic species are managed so productively that only a small fraction of the population in developed nations is

Figure 13.2

Measurements of maize (corn) demonstrating artificial selection for larger varieties by Peruvian Indians in the central Andes. The horizontal axis in these graphs represents time, beginning about A.D. 450 and continuing until A.D. 1530. On the vertical axes are various dimensions of maize kernels (average t standard deviation) excavated from prehistoric households at the indicated times. Note the increasingly larger maize varieties cultivated through time.

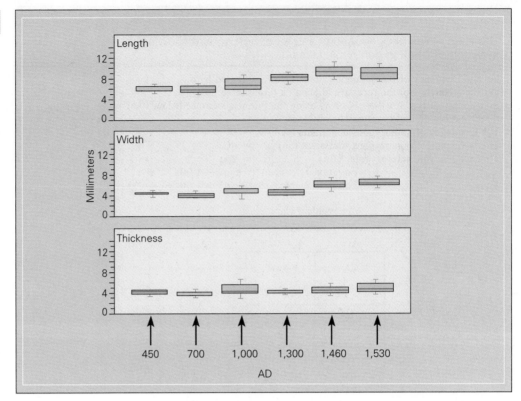

directly involved in agriculture. Much of the success in removing most members of society from the burden of food production is the result of genetic changes in domestic species during the past few thousand years.

Agricultural productivity, that is the *quantity* of domestic species, largely depends upon genetic *quality*. The essence of nature is variety. Some individuals are docile, easy to manipulate. Others produce more, healthier, or larger offspring. In the past, individuals with particularly desirable traits were chosen as parents of succeeding generations, and the desirable traits thus increased in domesticated populations. In other words, early humans exercised **artificial selection** (figures 1.2, 1.5, and 13.2).

Artificial selection has been used successfully for at least ten thousand years. As the science of genetics has developed in this century, artificial selection has become a precise and sophisticated technology.

Artificial Selection

Artificial selection remains the primary basis for crop and livestock management. It has been so successful that almost all domesticated species are quite different, phenotypically and genetically, from their wild progenitors. Traits that permit survival in the wild only complicate management in captivity. Wild cattle and sheep, for example, have long legs for speed and horns for defense, but domestic cattle and sheep are bred for short legs, and their horns are polled (figure 13.3).

The variety of traits selected artificially is astonishing. Food species have always been bred for features such as size, flavor, and disease resistance. But harvesting and distribution factors favor selection for some unexpected traits. Tomatoes, for example, have been developed with tough skins to withstand mechanical harvesting machines, extremely slow ripening to give long shelf life, and even cubical shape for efficient packaging.

The specialized requirements of medical research have often been met with unusual strains of experimental animals. Nude mice (figure 13.3), lacking a thymus gland, are used widely in cancer and immunological studies. A strain of pigs weighing an average of 140 pounds (65 kilograms) has been developed to provide a same-size model for surgical and other experimental studies relating to humans.

(a)

(b)

(c)

Figure 13.3

Examples of artificially selected organisms (a) cattle: guernsey, highland, zebu; (b) tomatoes; and (c) mice: diabetic obese mouse, nude mouse, and C57 black mouse.

(a) © Phillip E. Keenan/ Visuals Unlimited, © Link/ Visuals Unlimited, © Arthur R. Hill/Visuals Unlimited; (b) © William Ferguson; (c) © Dwight Kuhn, © 1991 Keith/Custom Medical Stock Photos, "Stanton Short—The Jackson Laboratory."

In spite of extraordinary success, artificial selection has serious limitations. Plant and animal breeders cannot produce new varieties on demand. Development of particular varieties is often a protracted, difficult process, with no guarantee of success.

Several factors limit the success of artificial selection. First, artificial selection is slow. A new mutation appears in only one individual. It must then be bred for many generations before a large number of individuals acquire the same trait. The time required to spread a new mutation depends upon the fecundity and generation time of the species, as well as the sex of the mutant individual.

Second, new genetic variation occurs at random. The effectiveness of artificial selection depends totally upon the particular alleles present in a population. Selection cannot change a population that is genetically uniform. Furthermore, most mutations are deleterious, so most new variation is useless to plant or animal breeders, who are forced to wait opportunistically for the rare, useful mutation to appear. If a particular trait never appears, or if it cannot be perpetuated, or if it is not recognized, then it can never be developed in a domestic species.

Third, the most rapid way to develop a strain of organisms bearing a desirable trait is through inbreeding. Consanguineous mating, however, leads to **inbreeding depression,** or decreased vigor, fertility, and survival. Inbreeding depression is the genetic price paid for the benefits of rapid selection.

Plant and animal breeders will continue to rely upon time-proven techniques of artificial selection. Modern genetic and reproductive technologies, however, are now providing solutions to the limitations of artificial selection. Coming decades should see the manipulation of captive gene pools become an increasingly precise and effective science.

Creating Genetic Variation

Artificial selection operates only on existing genetic variation, and selection is increasingly effective at higher levels of variability. *Mutation* is the ultimate source of genetic variation, so geneticists now accelerate the occurrence of new genetic variants using mutagenic agents such as *chemicals* and *radiation*.

A major problem with the application of mutagens is the unpredictability of the results. Valuable mutations usually occur only after many harmful ones. Individuals having desirable mutations may possess other mutations that are harmful.

The use of mutagenic substances is most profitable on species such as plants that produce large numbers of offspring. Each kernel on a stalk of corn, for example, represents a different individual. Therefore, a large number of individuals can be quickly assayed for mutations. Similarly, mutagens can be effective in generating commercially desirable results in bacteria, which have only a single copy genome. Thus, any mutations will be immediately expressed and reproduced extremely rapidly.

Polyploids, resulting from induced mutation, are important contributors to the world's food supply. As described in chapter 11, many commercially important strains of wheat, berries, and other food plants are polyploid. Polyploidy increases the overall size of organisms, especially the size of edible fruit and seed.

Inducing polyploidy in animals has had limited success. Polyploidy is achieved by subjecting recently fertilized eggs to temperature extremes. If these eggs successfully develop, the adults are polyploid and often sterile. Such techniques have been successfully applied to cultivation of salmon, oysters, and abalone.

Combining Genes in New Ways

Most species possess large amounts of genetic variation. New techniques of cell and gene manipulation are now permitting the transfer of specific genes between individuals as well as across species boundaries.

Grafting. Cuttings from different individual plants will often grow together, fusing to form a single new individual. Such **grafting** has long been a successful method of combining the best characteristics of different strains. Many cuttings can be taken from a single individual, so a newly arisen desirable mutant can be effectively cloned by appropriate grafting. All seedless oranges, for example, were propagated by grafting from the original mutant branch, as the seedless fruit is sterile. Much of the French and California wine industries now rely upon the grafting of the best French wine grape stock to insect-resistant American rootstock.

Grafting is also a method of "bulking up," that is, rapidly producing large numbers of individuals of a desirable genetic type. In this regard, grafting parallels the techniques of superovulation and embryo transfer now widely practiced on domestic livestock (box 7.1).

Hybridization. Interbreeding either different genetic strains within a species or individuals of different species is called **hybridization.** By definition, a *species* consists of all individuals that can freely interbreed in nature and produce fertile offspring. Commonly, members of different species cannot and do not interbreed, or if they do, the offspring are not viable. Occasionally, however, members of different species not only interbreed, but also produce viable offspring. Often, interspecific *hybrid* offspring have undesirable traits or reduced fertility, but in a few cases the results are spectacularly successful. Interspecific hybridization is more common in plants than in animals.

The best known interspecific hybrid is the mule, a cross between a male donkey and a female horse. Like many hybrids, the mule exhibits **hybrid vigor,** an increase in size, physiological vitality, or other desirable qualities often found in hybrid individuals. The mule is valuable as a work animal precisely because of these traits. Like many interspecific hybrids, however, mules are usually sterile.

Hybrid vigor can also result from the cross of different strains *within* a species. Most grain crops are the F1 progeny from hybridization of inbred strains. Their hybrid vigor is expressed as increased productivity. The F2 generation of these lines does not show hybrid vigor; their productivity is reduced. Hybridization has thus changed farming practices permanently. Farmers no longer save the best seed to produce next year's crop. Modern grain agriculture depends upon the breeding programs of seed companies that supply hybrid seeds annually.

Hybrid corn was developed in the United States in 1917. Since then hundreds of different hybrid combinations have been produced, each uniquely suited to a specific set of growing conditions. Often a single highly productive hybrid is grown over a broad geographical area, **monoculture.** While the productivity achieved with hybrid strains has been remarkable, monoculture increases the potential vulnerability of a crop to genotype-specific diseases.

Protoplast fusion.   Recent developments in *tissue culture* techniques also involve hybridization. In plant tissue culture, a small segment of plant tissue is grown in a petri dish in the laboratory. With proper chemical stimulation, the cells can sometimes be induced to grow into new plants. A technique known as **protoplast fusion** produces hybrid cells from the union of cells from different species grown in tissue culture. The hybrid cells exhibit properties of both parent species, qualities that in combination may produce an agriculturally superior plant.

Potential applications of tissue culture and protoplast fusion are almost limitless. For example, potatoes resistant to potato blight (the disease that caused the nineteenth-century Irish potato famine) have now been produced by protoplast fusion of potatoes with tomatoes in laboratory culture. Tomatoes belong to the same family as potatoes, but are resistant to the blight. Genes for resistance are incorporated into the hybrid offspring, which then produce normal, but blight-resistant, potatoes.

Protoplast fusion is still in the developmental stage. Mutagenic agents are used to induce fusion of cells, but the results of this process are unpredictable. It is not possible to control which cells fuse and which do not, nor to control which characteristics are expressed in the hybrid. The technique is so rapid, however, that many combinations can be tried. If a successful hybrid is produced, individuals can be cultured from only a few cells. One successful fusion can result, potentially, in a thousand or more offspring.

Gene transfer.   Recombinant DNA technologies provide a means of bypassing many of the limitations of artificial selection. Principles and techniques for the transfer of specific desirable genes into domestic plants and animals are essentially the same as those for *gene therapy* of humans as discussed in chapter 10, and as noted in that chapter, significant successes are already occurring, as in the case of genes for human blood clotting factors that have been transferred into sheep, which then mass produce these medically important chemicals.

The advantages of gene transfer are clear. Only a single gene or set of genes is introduced, while the remainder of the genome is unaffected. The desired gene(s) can be taken from any other individual, including members of other species, or the gene(s) can be constructed artificially, one base at a time, in the laboratory. Breeders no longer need to wait for the random appearance of a valuable gene by mutation. Subsequently, selective breeding can spread the gene throughout a population once the gene has been incorporated into the germline.

Domestic crop species are being subjected to transformation experiments with the goal of increasing their productivity. A major factor affecting productivity is the availability of nitrogen. Although nitrogen is an essential nutrient for plant growth, most plants cannot directly use abundant atmospheric nitrogen. Instead, plants use fixed nitrogen, i.e., nitrogen that has been combined with oxygen to form nitrates and nitrites by nitrogen-fixing bacteria. Nitrates are a primary component of fertilizers. A few plants, particularly legumes such as beans, clover, and alfalfa, are essentially self-fertilizing because they have nodules on their roots that house colonies of nitrogen-fixing bacteria. Such plants actually enrich, rather than deplete, the soils on which they grow. Plant geneticists are attempting to isolate and purify the nitrogen-fixing genes of these bacteria. If successful, gene-transfer techniques could be used to introduce these genes into domestic crop species.

Other successful experimentation with gene transfer has involved producing frost resistant strawberry plants by infecting them with "ice-minus" bacteria (normal bacteria from which the gene promoting ice formation has been removed); producing a genetically engineered vaccine (made by removing part of a natural viral gene) to protect swine against the deadly disease pseudorabies; adding herbicide resistance genes to normal food crops; transferring into tobacco plants a bacterial gene that triggers the production of a protein that kills caterpillars that feed on tobacco leaves; and adding extra copies of the gene for the hormone somatotropin to cattle to boost milk production (by 10–40%).

The Price of Agricultural Success: New Genetic Problems

Modern agriculture bears little resemblance to farming practices used prior to this century. Single crops are grown over vast areas, and massive doses of chemicals are applied to all crops. In addition to increased production, the new techniques also create a set of problems never before encountered by farmers.

Pesticide Resistance

The first widely-used pesticide was the chemical DDT. Following the discovery of its insecticidal properties in 1939, DDT was sprayed indiscriminately worldwide for control of mosquitoes, houseflies, and other insect pests. At that time malaria, for which mosquitoes are a vector, was the single greatest cause of human death in the world.

DDT was extremely effective against mosquitoes and houseflies, and for a few years the populations of these insects remained at low levels. Within a decade, however, DDT-resistant populations began appearing. While DDT killed most insects in a population, a few rare individuals possessed genetic traits that made them resistant to DDT. The large population sizes of insects make it likely that a few resistant individuals will exist in any population. The rapid reproductive rate of most insects means that in only a few generations large populations of pesticide-resistant insects will replace the non-resistant populations. Control of these populations consequently requires the application of increasingly larger doses of pesticide to keep pest numbers at a low level, the **pesticide treadmill.**

The result of several decades of use of DDT and other pesticides is that many insect populations in the world are largely resistant to the common pesticides employed against them, and malaria is still a major cause of human death in many parts of the world. For example, between 1970 and 1980 the number of documented pesticide-resistant insects increased from 224 to 428. One response to DDT resistance has been the rapid development of other pesticides. In a sense, an evolutionary arms race is required if pesticides continue to be a primary agent of pest control.

All species have the potential to develop resistance to toxic chemicals. A spectacular example is the Norway rat, a serious pest that consumes human food supplies and transmits disease. Following World War II, the anticoagulant warfarin was used to kill rats in cities. Warfarin was mixed with a food bait that included ground glass. Rats eating warfarin suffered internal hemorrhaging, causing death.

About ten years after the introduction of warfarin, however, warfarin-resistant populations of rats began to appear independently around the world. As with insect pests, genetic mutations had conferred resistance to warfarin giving rise to numerous resistant populations. In fact, some warfarin-resistant rats actively metabolize warfarin, using it to their physiological benefit.

Drug-resistant Diseases of Livestock

Moderate doses of antibiotics increase the productivity of pigs and cattle. As a result, antibiotics are routinely fed to healthy livestock, especially when they are moved into feed lots, confinement areas where they are fed massive amounts of grain to increase their weight prior to slaughter. The drugs may have a prophylactic effect in preventing the spread of disease when animals are kept in such large concentrations. However, drug-resistant strains of disease organisms are known to have resulted from these practices, just as antibiotic-resistant strains of human diseases have developed.

Fears that humans might become ill as a result of eating livestock that have been fed antibiotics were realized in 1984 when eighteen people became severely ill with salmonellosis. It was discovered that all eighteen were infected by an antibiotic-resistant strain of *Salmonella* that they had ingested in hamburger meat, which was traced to a single farm where cattle had been fed low doses of antibiotics. It is estimated that 30% of all chickens, 80% of all swine, and 60% of beef cattle marketed in the United States have been fed antibiotics for some part of their lives. The residual effect of the antibiotics on human consumers is uncertain. The FDA estimates that nearly 50% of all antibiotics produced in the United States are fed to livestock.

Monoculture and Pest Susceptibility

Today fewer than fifty species of plants (of a total of about 250,000 plant species worldwide) are used as food sources for most of the human population. The traditional diversity of cultivated strains of these few species has now been replaced by the planting of single, genetically uniform strains over vast geographic areas. Monoculture has many practical advantages over older methods of farming; however, as noted earlier, crops grown in monoculture are dangerously susceptible to epidemic disease.

One serious epidemic has already affected United States agriculture. In 1970, southern corn leaf blight destroyed about 15% of the corn crop, causing about $1 billion in damages. The blight-susceptible crop was saved only by quick action and good luck. Blight-resistant strains were introduced the following year. Damage from future epidemics, however, may not be controlled so successfully. Maintenance of substantial genetic variation in crop species provides the best long-term insurance against such potential calamities.

Genetic Techniques of Pest Control

In nature, potential pest organisms seldom reach destructively high numbers. Most species are limited by interactions with predators, diseases, parasites, and other factors.

Biological methods offer alternatives to the use of chemical pesticides. Some biological methods use genetic technology to circumvent many problems associated with the use of pesticides. Pesticides, for example, kill both pest and beneficial insects indiscriminately. The loss of beneficial insects may be more serious than the damage done by pests. Furthermore, pesticides such as DDT persist in the environment for years, accumulating in the tissues of living animals, including humans. The damage caused by chemical residues is serious, and the long-term effects are unknown. In contrast, biological control methods usually affect a single target species, for a relatively short period of time.

The Sterile Male Technique

Release of large numbers of sterile males into pest populations can significantly reduce their numbers (figure 13.4). Larval pest insects are grown to adulthood in artificial growth chambers. Adult males are then given large doses of radiation, rendering them sterile. The artificially reared, sterile males are released into nature. When fertile wild females mate with sterile males, no offspring are produced. Some females also mate with fertile wild males, but sterile males greatly outnumber fertile males, so most females mate with sterile males. The **sterile male technique** is most successful in limited pest populations where sterile males are released for several consecutive years. Note, again, that basic research has practical application. In this case, the work of H. J. Muller, who identified the mutagenic and sterilizing effects of radiation, provided a basis for the control of noxious insects.

The sterile male technique has been used successfully against a parasite of domestic cattle, the screwworm fly. The screwworm fly lays its eggs in the open wounds of mammals, particularly cattle. The eggs develop into larvae that feed on the wounds. Severe infection may cause death. In the 1950s, a sterile male program was begun against the screwworm, then causing an estimated $20 million damage to the American cattle industry. Since then, losses to screwworm have been dramatically reduced; in some portions of the United States this pest has been totally eliminated.

Figure 13.4

The sterile male technique. Large numbers of the insect are raised and exposed to sterilizing amounts of radiation. At the appropriate time, large numbers of sterile males are introduced into the natural population. These males mate with the normal fertile females, but none of her eggs are fertilized, so these females contribute no offspring to the next generation.

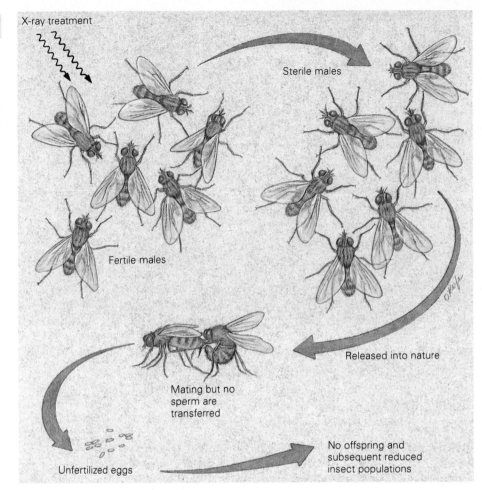

Chemicals

All organisms produce chemicals that regulate their own lives. In particular, *hormones* control development and reproduction. Some insect hormones can be used to help suppress pest populations. Hormones, however, usually occur in minute quantities, far too small for commercial use. Genetic engineering techniques (chapter 10) provide a method for mass production of insect hormones, much like production of artificial human insulin.

Pheromones, hormones released into the environment as mate attractants, are now used for pest control. Male insects searching for a mate, for example, often follow pheromone trails laid down by females. Scientists bait insect traps with pheromones, attracting males who are then killed. This technique requires many traps over large areas. Many males are still likely to find mates before entering the traps.

A second, more devious, approach involves aerial spraying of pheromones throughout an infested region. Mass application of pheromones confuses the males—females appear to be everywhere. Pheromone trails laid down by females are obscured, reducing the probability of a male finding a mate.

Growth hormones can also control pest species. *Juvenile hormones* suppress sexual maturation. *Anti-juvenile hormones,* on the other hand, produce premature maturation and consequent sterility. Aerial application of either of these hormones greatly reduces the reproductive potential of pest populations.

The advantage of biological control of insects by hormones is obvious. Unlike DDT or other pesticides, hormones are naturally occurring chemicals that are not toxic to humans or other animals. Their primary effect is to disrupt normal mating behavior and reproduction of a target species without the harmful secondary effects of traditional pesticides. The market for insect pheromones and other agricultural chemicals produced by recombinant DNA technology is substantial.

Figure 13.5

Potatoes for sale at an outdoor market in Ecuador. Several different varieties are grown by South American Indians in the Andes, and all are marketed throughout the growing season.

Source: D. Donne Bryant.

The Green Revolution: Mixed Verdict

The **Green Revolution** refers to a series of programs begun during the 1960s to increase agricultural productivity in underdeveloped countries. The development of genetic strains of food crops with high productivity has been a major goal of the Green Revolution. Several international institutes were established for this purpose. For example, the Centro International de Mejoramiento de Maiz y Trigo (CIMMYT) developed strains of wheat with relatively large seeds and short stems, thus increasing productivity. Nearly all wheat grown in Mexico and about 75% of that in India are the short-stemmed or dwarf varieties. Crop yields in India and Mexico have increased three and five fold, respectively. High productivity strains of maize developed by CIMMYT are now an important food crop in Mexico and Africa.

Geneticists at the International Rice Research Institute (IRRI) in India have developed high-yield strains of dwarf rice. Today, about 70% of all rice planted in the Philippines and 90% of the rice grown in Colombia are these dwarf varieties.

Public attention has focused on the genetic basis of the Green Revolution, but the new strains are successful only when modern techniques of cultivation are used. The crops must be grown in monoculture with massive applications of fertilizers, pesticides, and herbicides—a major change from traditional agriculture. In the Peruvian Andes, for example, native farmers grew fifteen or more different strains of potato in a single garden (figure 13.5). The variety of genotypes provided protection against soil depletion and loss of the entire crop to diseases or pests. Native farmers maintained their own seeds from year to year, whereas the hybrid strains must be purchased from the seed companies for each planting.

In short, traditional agriculture is labor intensive and self-sustaining. The Green Revolution requires capital-intensive techniques feasible only on a large scale because of large investments required for seeds, machinery, petroleum, and agricultural chemicals.

Although the goal of the Green Revolution—more food for more people—is laudable, results have been mixed. Productivity has increased greatly in many countries, but the imposition of modern technologies upon underdeveloped societies has sometimes caused significant social and economic stress.

Because small farmers cannot afford initial capital investments for the new techniques, they have often been forced to sell their land to large agribusiness corporations. Thus, the distribution of land has shifted from small farmers to large corporations. Profits made from increased productivity have been channeled into foreign investments rather than reinvested in the local economy. The small farmers, forced off the land, have often moved to urban ghettos, with little prospect of employment. The benefits of technology may increase productivity, but the social consequences are considerable.

Some biological consequences of the Green Revolution have also been unpredictable. The massive applications of pesticides to protect crops grown in monoculture have selected for pesticide-resistant strains of insects, including mosquitoes. In some underdeveloped countries the incidence of mosquito-borne malaria has increased dramatically with the introduction of modern farming practices.

Finally, increased agricultural production has not reduced the problems of world hunger. Increases in food production are being more than matched by increases in world population, most of which are occurring in the poorest countries. Only a few countries are net exporters of food. In addition, displaced small farmers who no longer grow their own crops cannot afford to buy food. Concentration of agricultural wealth in a few large corporations only serves to increase the disparity in wealth among sectors of the population, a factor that has added to political instability in many parts of the world.

Conservation

The results of biological evolution are extraordinary. Since life originated on earth over 3.5 billion years ago, species diversity has increased regularly to produce a fantastic array of living species. No one knows for sure how many species presently exist, but the number of species has recently been estimated to be between ten and thirty million. Most of these live in the tropics and have never been studied—or even described and named—by scientists. Tens of thousands, possibly millions, of species face an uncertain future because of the activities of humans.

Extinction and Genetic Resources

For the past several centuries, human numbers have increased exponentially. Since World War II, the rate of growth has actually increased and shows no signs of abating. Human populations now double about every thirty-five years. In 1987 the human population passed the five billion mark and by the turn of the century is likely to number between seven and eight billion.

The unprecedented human population explosion has resulted in massive loss of habitat for all other species. Very few places on earth are undisturbed by human activity. As the amount of space remaining for wild plants and animals has decreased, many species have become extinct. Many other species have seen their natural ranges fragmented. Large populations—and large gene pools—have been reduced to tiny remnants (box 13.1).

The potential disaster posed by massive species extinction is difficult to imagine, but predictions are dire. Of the possible thirty million species now alive, about one million are predicted to become extinct by the turn of this century. If extreme conservation measures are not implemented soon, half of all living species are likely to become extinct by the end of the next century.

The importance of this issue has been recognized by many organizations. The International Union for the Conservation of Nature publishes a series of Red Data Books that list the threatened, endangered, and possibly extinct species in countries throughout the world. The genetic aspects of extinction are explicitly recognized in the World Conservation Strategy of the United Nations, which makes the conservation of genetic resources one of its three primary goals.

Concern for diminishing genetic resources is well founded. Many people feel a strong ethical responsibility for the human role in species extinction, but compelling practical reasons underlie the need for biological conservation. Genetic resources comprise fundamental life support systems for humans. The many genetic varieties of plants and animals are the basis for all human food and agriculture; they are also the source of about one-fourth of all medicines and drugs. Plants play a major role in regulating weather and climate. Fungi and bacteria quickly recycle the remains of dead organisms. The loss of genetic diversity diminishes the quality of human life and ultimately threatens human existence.

Plant Genetic Resources

Human food supplies are narrowly based. Fewer than fifty plant species and about ten animal species are the major food sources for most of the world's population. Obviously, many other species could provide food, but many potential food sources are threatened with extinction before they can be properly evaluated.

Plants are also essential to medical practice. Traditional plant and herbal remedies for disease are still used by about 75% of the world's population. Plant-derived medicines are a major component of pharmaceutical practice in developed countries, where such essential drugs as morphine, codeine, atropine, and digitialis are all produced by plants, even in this age of biotechnology. Quinine, derived from a number of species of South American plants, is still the most effective anti-malarial preventive and treatment, despite centuries of use and many attempts to replace it with synthetic drugs. Only about 5,000 species of plants have been thoroughly tested for their potential medicinal uses, perhaps 2% of all known species.

Modern agricultural practices are a serious source of erosion of genetic resources. Most crop species are comprised of a large number of **land races,** locally adapted varieties well known to farmers in a particular region. Traditionally, farmers grew a number of these land races (figure 13.5). Now, farmers plant only one hybrid variety supplied by a seed company. The result is that many land races have become extinct, reducing the genetic resources on which future plant breeders can draw to produce new varieties. Some people also worry that the seed companies—often large multinational firms—have acquired all supplies of some land races. From the companies' viewpoint it represents good business practice, but farmers and the public have lost control of a resource formerly held in common. **Plant patenting laws** now allow plant breeders to own exclusive rights to specific varieties of plants and animals.

Loss of land races reduces the genetic base—the survival resources—of agricultural species, but some domestic crops have always had a narrow genetic base. No better example exists than coffee. The coffee plant originated in Africa, from where a few plants were taken by the Dutch to the East Indies (Indonesia) in the seventeenth century. These few plants formed the basis for large coffee plantations, which provided much of the wealth flowing into Holland from their colonial empire.

In the nineteenth century, the British surreptitiously took a single coffee plant from the East Indies to the Kew Gardens in London. A cutting of this plant was subsequently taken to South America, where it formed the basis for the South American coffee industry, now the largest in the world.

At each step in its history, the domestic coffee plant was subjected to a **genetic bottleneck,** reducing its total genetic variation drastically. The dangers of low variation were realized in the 1970s, when a disease epidemic struck the South American coffee plantations. Only the import of disease-resistant coffee plants from Indonesia—an input of genetic variation—was able to restore the South American plantations to production.

An equally dramatic example of the survival value of genetic diversity is shown by the history of the European wine grape, *Vitis vinifera.* In the late nineteenth century, an infestation of a root-destroying insect, *Phylloxera,* reached plague proportions and totally devastated most European vineyards. Their grape varieties were the finest in the world, the result of centuries and even millenia of captive propagation and selection for desirable characteristics.

Many of these unique and irreplaceable varieties, as well as the European wine industry itself, were saved by grafting cuttings onto rootstock from native American species of grapes, which are resistant to *Phylloxera.* Thus, the best wine grapes in the world survive as a genetic mosaic. Natural genetic diversity in wild species provided crucial information that saved an entire agricultural industry.

BOX 13.1

The blight of extinction

Many people give little thought to the meaning of extinctions, but few examples are more striking than that of the American chestnut tree, *Castanea dentata* (figure 13.6a). Only a century ago the chestnut comprised about 25% of the trees in the vast hardwood forests of eastern North America (figure 13.6b). A high quality hardwood, the chestnut was commercially valuable as timber for fence posts, telegraph poles, ship masts, and railway ties, and the nuts were used for candies and sweets.

About the turn of this century, the fungus *Cryphonectria parasitica*—called chestnut blight—arrived in the northeastern United States, probably imported inadvertently on Asian chestnut trees, a species closely related to the American chestnut. By 1950, the disease had spread throughout the entire range of the chestnut, killing virtually all individuals, perhaps as many as four billion trees. Today, the species survives only in a few rare individuals planted outside the natural range, thus escaping blight (figure 13.6a), or in a few remaining stumps that every year send up sprouts that later succumb to the blight.

Scientists and foresters have tried many schemes to save the species, but efforts thus far have not succeeded. In particular, forests have been searched unsuccessfully for individuals that are genetically resistant to the blight. If the genes conferring resistance to the blight

(a)

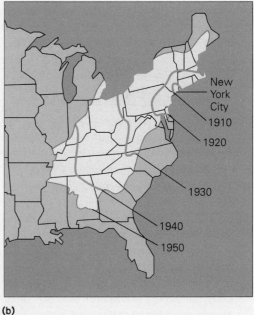

New York City
1910
1920
1930
1940
1950

(b)

Kew Gardens (figure 13.7), perhaps the most famous botanical gardens in the world, are a major tourist attraction for visitors to London. Most visitors are unaware that they were founded entirely as a repository for plants from around the world that might support the British Empire. Kew Gardens formed the center of a major program to obtain by any possible means plants that might be of value to the Empire. The British fully realized the commercial value of genetic resources, their exploitation of which allowed them to dominate colonial economies and accumulate immense wealth for their country.

In a sense, Kew Gardens may be considered a forerunner of plant gene banks, which are now being developed in many countries to preserve vanishing plant species and races. The *International Board for Plant Genetic Resources* (the *IBPGR*), based with the United Nations Food and Agriculture Organization in Rome, coordinates the activities of national can be isolated from Asian chestnuts, it might be possible to use gene transfer methods to introduce them to the American species. A slower method of achieving the same result is also being tried: to hybridize American and Asian chestnuts, then backcross the blight-resistant hybrids repeatedly to American chestnuts.

The commercial effects of loss of this species are large, but the ecological effects are far more difficult to measure. What other species, for example, have been brought to the brink of extinction because they require chestnuts for an important part of their life cycle, either as shelter or food?

Sadly, the American chestnut is not a rare example of a species harmed by diseases moved around the world by human activities. One need look no further than the American elm tree (figure 13.6c), another ecologically and commercially important member of eastern hardwood forests, for a second major species that is now rapidly succumbing to recently introduced disease. As with the chestnut, strains of the American elm resistant to its parasite (Dutch elm disease) have not been found. Even if scientists succeed in rescuing the elm and chestnut, which is by no means certain, it will be a century or more before these species could again become important members of the American forests.

(c)

(d)

plant germ plasm banks in many countries. These gene banks maintain collections of seeds to preserve differing varieties of agriculturally important species and their wild relatives. In the United States, the National Seed Storage Laboratory at Fort Collins, Colorado, administered by the U.S. Department of Agriculture, is the primary storage facility for plant germ plasm.

Although sound in principle, the reality of the task confronting plant germ plasm collections is immense. The collections for a single species may include millions of seeds. Seeds can survive for many years if stored in proper conditions, but they do not last forever. The collections must be renewed continually by planting and growing the seeds and then collecting the next generation of seeds. In addition, millions of wild species are largely ignored because most collections deal primarily with agriculturally or commercially valuable species.

Figure 13.7

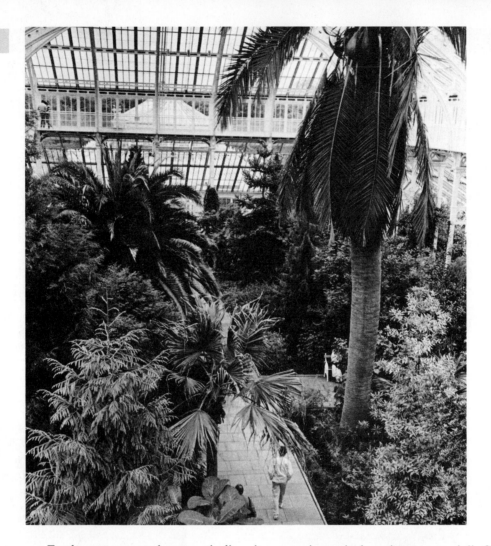

Furthermore, most plant genetic diversity occurs in tropical continents, especially South America, Africa, and Asia. Tropical areas of especially high plant diversity—called *Vavilov centers* (figure 13.8), after the famous Russian geneticist who identified them—are the site of origin of most domestic crop species. They also occur mainly in underdeveloped countries with rapidly increasing populations. Natural diversity is being greatly reduced in these countries due to human population pressures. These countries have urgent conservation problems with only limited resources, if any, to devote to the problem.

The IBPGR organizes assistance programs for the conservation of plant genetic diversity in underdeveloped countries, but such efforts sometimes have been highly controversial. The IBPGR is supported financially by seed companies, who are also represented on its board of directors. Other support has come from independent organizations such as the Kellogg Foundation, again with links to agro-industry and based in the developed world. Thus, many underdeveloped countries view germ plasm collections as a form of modern imperialism, which tries to steal from them their biological heritage. National germ plasm collections in developed countries are viewed as a contemporary equivalent to the Kew Gardens, where genetic resources of economic value are stored for national purposes.

Seeds stored in national germ plasm collections that participate in IBPGR programs are supposed to be available to and shared among all nations in the world, yet material is often not shared. The United States, for example, does not freely exchange seeds with communist nations. Following the Nicaraguan revolution in 1979, many seeds stored in the national collection in Nicaragua were removed for their own safety and stored in the United States. Subsequently, and despite many requests from the Nicaraguan government, the U.S. government has not allowed these seeds to be returned to Nicaragua. Some of the seeds represent varieties no longer found in nature that could greatly benefit Nicaraguan agriculture.

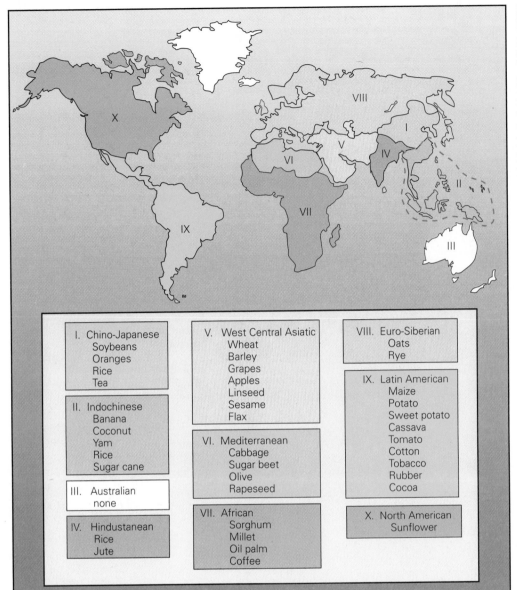

Figure 13.8

Regions of genetic diversity and their associated crops. These sites are called *Vavilov centers,* in honor of the Russian geneticist who identified them. They are the sites of origin of most domestic crop species.

From Jack Kloppenburg, Jr. and Daniel Kleinman, "The Plant Germplasm Controversy," *BioScience* 37:190–98. Reprinted by permission of the American Institute of Biological Sciences and the authors.

I. Chino-Japanese
Soybeans
Oranges
Rice
Tea

II. Indochinese
Banana
Coconut
Yam
Rice
Sugar cane

III. Australian
none

IV. Hindustanean
Rice
Jute

V. West Central Asiatic
Wheat
Barley
Grapes
Apples
Linseed
Sesame
Flax

VI. Mediterranean
Cabbage
Sugar beet
Olive
Rapeseed

VII. African
Sorghum
Millet
Oil palm
Coffee

VIII. Euro-Siberian
Oats
Rye

IX. Latin American
Maize
Potato
Sweet potato
Cassava
Tomato
Cotton
Tobacco
Rubber
Cocoa

X. North American
Sunflower

Thus, politics intrude on the conservation of genetic resources. Developed countries claim they are acting with the best motives, observing correctly that only they have the money and expertise to deal with such a major task. Underdeveloped nations point to the fact that multinational seed companies are likely to be the primary beneficiaries of germ plasm conservation programs, and also that their biological resources are now being plundered just as their mineral resources were in previous centuries. Meanwhile, rapid continuing development of tropical continents, particularly deforestation of the Amazon basin and remnant tropical forests in Africa and Asia, poses an ever-increasing threat to remaining plant genetic resources.

Wildlife Conservation

Wild animal species face the same problems confronting plant species: reduced and fragmented habitat, and decreased population sizes—conflict with humans at every turn. These problems are compounded by the large home ranges required by many species; caribou, for example, have migratory routes of several thousand miles. The problems are especially acute for large mammals, the species we often value most. Predicted extinction rates for large mammals in the next century are extremely high because so little habitat is available for nature reserves to meet their needs.

Figure 13.9

Zoos play a major role in contemporary conservation. (a) The San Diego Zoo is a world leader in the use of genetic and reproductive technologies to aid endangered species; (b) Przewalski's horses, the forerunners of the domestic horse, are bred in a carefully managed program at the San Diego Zoo, participating with other zoos to limit the harmful effects of inbreeding in the few survivors.

San Diego Zoo.

(a)

(b)

Genetics is now an important consideration for a wildlife manager in many ways. Rarity is partly a genetic problem. Small populations are subject to inbreeding and its harmful effects. As populations of animals decline and become fragmented, managers may be required to transfer individuals between populations to counter the effects of genetic drift and thus aid in maintaining normal levels of genetic variation within wild populations. Inbreeding depression resulting from loss of genetic variation may in some instances be a factor that contributes to extinction.

Rapidly improving genetic techniques have provided wildlife managers with important information for making management decisions. For example, genetic analysis of African rhinoceroses revealed that two disjunct and diminishing populations showed little genetic differentiation. Managers thus decided to establish a new hybrid population that might benefit from combining the small genetic variation of both types.

Figure 13.10

(a) The Leopard frog, *Rana pipiens*. (b) Distribution map of twenty species of leopard frogs in the *Rana pipiens* complex. Numbers refer to individual taxa.

(a) © Cleveland P. Hickman, Jr./Visuals Unlimited.

(a)

(b)

Chromosomal and biochemical genetic analyses have revealed the existence of many new **cryptic species,** that is, pairs or groups of species that externally resemble each other but genetically are quite distinct, such as the diploid/tetraploid pair of tree frogs shown in figure 11.3. Until the 1960s, for example, the "leopard frog" was commonly used in teaching and research across the United States. It was thought to be a single species ranging throughout North America. Genetic analysis has now shown that the leopard frog is really a cryptic species complex of about twenty species, some of which are relatively rare (figure 13.10).

About 30% more species of lizards are known from California now than in the 1950s, partly because of the discovery of many cryptic species. The rarity of many of the newly discovered species makes them of major conservation importance (box 13.2).

For some species, survival in nature may be impossible. Przewalski's horse (the forerunner of the domestic horse—figure 13.9b), the European bison, and Pere David's deer survive only in captivity, and their return to nature is only a dream for the future. Zoos have become a "modern Ark" that are the last hope for such species. Because the total capacity

BOX 13.2

Taxonomy is destiny

Figure 13.11

A very rare species: (a) the Brothers Island tuatara, *Sphenodon guntheri,* pictured on the cover of the issue of *Nature* where it was first described; (b) North Brother Island (population 12), four hectares, the only site where this species occurs.

(a) Dr. Allison Cree.

Conservationists often cannot save species that are not identified. Thus, taxonomy—the naming and cataloguing of all living species, and phylogeny—the study of species relationships—constitute an essential foundation of conservation practice.

The importance of taxonomy to conservation is demonstrated by the scientific history of the tuatara, an iguana-sized reptile known only from about thirty islands surrounding New Zealand. Originally thought to be a type of lizard, in 1867 the only known species (*Sphenodon punctatus*) was identified as representing an order of reptiles thought to have been extinct for eighty million years. As the sole representative of an order—far more important than a species—the tuatara was given full legal protection by the New Zealand government in 1895. Some tuatara populations continued to go extinct when rats invaded their islands, but tuatara were generally thought to be secure because the species occurred on a relatively large number of islands.

In 1990, however, genetic analysis revealed a morphologically cryptic second species, *Sphenodon guntheri* (figure 13.11a). Found only on tiny (four hectare) North Brother Island (figure 13.11b), roughly 230 individuals survived.

Any species consisting of a single small population is at high risk—a single catastrophe, natural or human-made, could cause extinction. Thus, the New Zealand Department of Conservation and the San Diego Zoo began a captive-breeding program aimed to establish new populations in zoos and in nature, giving the Brothers Island tuatara a more secure future.

The key point is that conservation depended on taxonomic identification of tuatara. Tuatara were given legal protection only after scientists identified them as a very special reptile, reflected in their classification in a unique order, and not as just another of the several thousand lizard species in the world. When the taxonomy was subsequently revised to identify a second species, new management programs were begun to save it. Failure to recognize a species taxonomically may greatly reduce its chances of survival: its taxonomy can be its destiny.

of zoos worldwide is limited, populations maintained in zoos will always be subject to intense inbreeding. ISIS, the International Species Inventory System, is a computer database that records the lineages of animals held in zoos worldwide, assisting zoos to coordinate breeding programs to minimize effects of inbreeding.

Genetics assists many aspects of breeding programs. In some cases, the pedigrees of captive animals are unknown, or individuals have been misidentified. Genetic analyses have assigned individuals to the correct species or subspecies, preventing unwanted and potentially disastrous hybridizations. Genetic studies also played a key role in breeding a pair of rare giant pandas at the National Zoo in Washington, D.C. After several years of mating without conception, the female panda was allowed to breed both with her usual mate and with a male imported from the London Zoo, who had previously fathered young. The female conceived successfully, raising the question of paternity: Would the London male have to be imported every year at great cost? Paternity analysis proved that the Washington male was the father, and he has subsequently fathered other young.

More technologically sophisticated reproductive help for endangered species is under development. Ultra-cold gene banks store sperm and embryos from endangered species, although ova cannot yet be successfully stored. Artificial insemination has been used successfully on giant pandas in Chinese zoos. The San Diego Zoo (figure 13.9a) has been a leader in applying techniques such as sperm banking, superovulation, *in vitro* fertilization, embryo transfer, and detailed genetic analysis to endangered species.

Such techniques are primarily applicable to mammals, but not, unfortunately, to most of the millions of other animal species. Most endangered species will not receive the conservation attention they deserve. As with plants, animal genetic diversity will continue to dramatically shrink in coming decades.

Human Genetic Diversity

Human genetic diversity has been shaped by the same evolutionary forces affecting all other species. Humans have been subject to reductions in genetic variation similar to those affecting other species. European colonialist expansion in the past five centuries has resulted in the extermination of many isolated populations. In some cases, indigenous peoples were subject to programs of genocide. In others, small local populations were literally consumed by hybridization as a result of intermarriages. In many other cases, local populations were not resistant to European diseases such as smallpox and syphilis. Contact with Europeans exposed the narrow genetic base of these populations to new diseases, and extinction ensued.

We will examine briefly key aspects of the processes guiding human population and evolutionary genetics.

Human Origins

Fossils show that life has been evolving on earth for over 3.5 billion years. Vertebrate animals have existed for over 500 million years, and mammals—warm-blooded animals with hair and mammary glands—for perhaps 150 million years. Primates (monkeys and their relatives) appear in the fossil record about 50–60 million years ago.

About 5–8 million years ago, three lineages diverged from a common ancestor: gorillas, chimpanzees, and hominids (the ancestors of humans). Living chimpanzees and humans differ only at about 1.6% of their nuclear DNA base sequences, while chimpanzees and gorillas differ at about 2.1%. Humans and chimps are thus each other's closest surviving relatives (figure 13.12). Mitochondrial DNA evolves more rapidly then nuclear DNA but reveals the same relationships: chimps and humans differ at 9.6% of bases in one mitochondrial gene, and chimps and gorillas at 13.1% of the bases of the same gene. Chromosome analyses also confirm the very close relationship of humans, chimpanzees, and gorillas (box 2.1). Further evolution within the hominid lineage produced at least four or five species within the genera *Australopithecus* and *Homo* (figure 1.9).

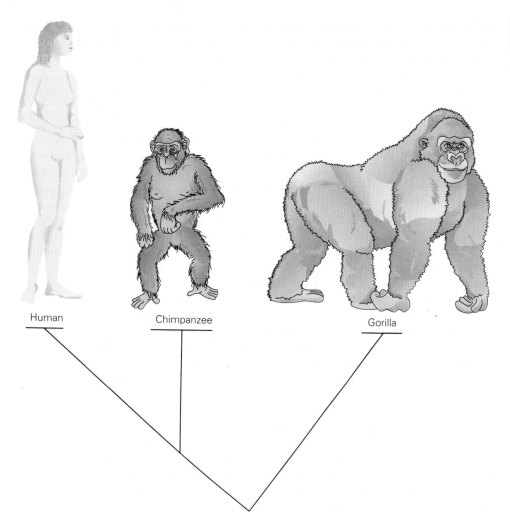

Figure 13.12

The relationship of humans to our closest surviving relatives (see Figure 1.9 for evolution within the hominid lineage). In 1967, Allan Wilson and Vincent Sarich provided biochemical evidence that chimpanzees were more closely related to humans than they were to gorillas. An academic "war" has raged since that time over the pattern of relationships depicted. An expanding body of DNA sequences from both nuclear and mitochondrial genes strongly support the close association of chimps and humans.

Human   Chimpanzee   Gorilla

For unknown reasons, all hominid species became extinct except *Homo sapiens,* wise man. Analysis of mitochondrial DNA variation indicates that all surviving human populations derive from a single female ancestor—*Eve* (chapter 1)—who lived about 200,000 years ago. Perhaps only in the past 100,000 years—two one-thousandths of 1% of the history of the earth—have humans been an abundant and ecologically important species. Human activities now dominate the earth.

Human evolution has thus been extremely rapid, especially the rate of increase of our key trait, intelligence. Precisely why intelligence evolved so rapidly is a subject of intense debate among anthropologists.

Throughout most of human history, human populations have probably been small. Migration, hybridization, and extinction of populations were probably common events. Presumably, natural selection and the other evolutionary factors that led to the development of land races in plants and animals also produced the diversity of human types. Human evolution has also been shaped by the diversity of human **cultures**—the sets of beliefs, values, behaviors, and institutions unique to human populations and ethnic groups.

Individual, Population, and Racial Differences

Daily experience guides our views of differences among humans. We are all able to distinguish by physical appearance alone virtually every person we ever meet. We can also sometimes identify apparent degrees of relationship; familial resemblances, for example, are often striking.

Most individuals can also be identified as belonging to a particular *race*. White Europeans, for example, are easily distinguished from black Africans. The apparent simplicity of this example is misleading because race is a complex concept with no objective definition, despite many attempts by scientists. In this century, some anthropologists have suggested that only three or four races of humans exist, while others have described more than one hundred races.

Many people acknowledge four primary groups of humans distinguishable by skin color and a variety of anatomical differences (figure 13.1): Mongoloid, generally yellow-skinned populations such as Chinese, Japanese, Malayans, Eskimos, and American Indians; Caucasoid, white- to brown-skinned populations originally found from Europe to Asia and northern Africa; Negroid, blacks of African descent; and Australoid, the aborigines of Australia. Some groups, such as the Polynesian peoples of the Pacific, however, do not fit easily into this classification.

The common practice of characterizing races by skin color is almost certainly a gross oversimplification that conceals the large amount of variability with each race and the limited differences between races. Within Negroid populations are the very tall Dinka and the very small pygmies and San tribespeople. Apache Indians, Cantonese Chinese, and Vietnamese differ greatly in appearance (figure 13.13), as do some Swedes from some Sicilians.

The concept of races probably reflects the obvious phenotypic effects of only a few loci. Alleles of some genes have marked phenotypic effects that correlate with commonly designated racial groups. The most obvious example is the set of genes controlling skin color; as few as four loci may encode the differing levels of pigmentation that lead to apparently obvious racial differences in skin color. Remember, too, how many environmental factors affect skin color: age, health, emotional state, exposure to the sun, exercise, and diet.

Genetic data have shed considerable light on human diversity. Differing types of genetic analysis—chromosome, protein, and DNA—all show a consistent pattern of extremely small differences between commonly designated racial groups but relatively high levels of individual variability. Each human is genetically unique (except identical siblings), and each population contains most of the genetic variation found within *Homo sapiens*. Populations and races differ mainly in the frequencies of the alleles. No alleles are known at any single locus that unequivocally distinguish all members of one race from all members of another, however, race might be defined.

Chromosomal variation among humans, for example, occurs only at the level of individuals, either in terms of abnormalities such as trisomy 21 (Down's syndrome) causing severe health problems or in variants such as fragile sites of no known medical importance. No chromosomal polymorphisms are known to distinguish human races or populations.

Electrophoresis (box 10.1) has been widely used to assess human variation at over fifty protein loci. At about 66% of these loci, the same allele is found in all individuals; those loci are said to be **monomorphic.** The remaining 34% of loci are **polymorphic,** that is, have two or more alleles. A particular individual, of course, possesses only a small fraction of this variation. On average, most individuals are heterozygous at about 3–5% of the loci examined by electrophoresis.

Comparisons *within* races show relatively high levels of differences between individuals. For example, on average, two randomly chosen American blacks differ at about 12% of their loci; two Japanese at about 12% of their loci; and two American whites at about 14% of their loci. Variation within each race is about equally large.

Comparisons *between* races, however, reveal only *limited additional variation.* Beyond the level of differences normally found within races, for example, American blacks and American whites differ at only an additional 1.6% of their loci; Japanese and American blacks at 1.5%; and Japanese and American whites at only 0.5%. The average genetic difference between members of the same race is almost as great as between members of different races, and in some cases members of the same race differ genetically more than members of different races.

(a)

(b)

Figure 13.13

Examples of variation among
persons considered as
belonging to the Mongoloid
classification of humans.
(a) A Vietnamese woman
from Danang; (b) a Noanama
man from Columbia;
(c) Apache men and women;
(d) a Chinese girl from
Sichuah; and (e) a mother
and child from Thailand.

(a) © Robert Alexander.
Photo Researchers, Inc.;
(b) © Borys Malkin/Anthro
Photo; (c) © Etter/Anthro
Photo; (d) © Dan Hrdy/
Anthro Photo; (e) © Dashaw/
Anthro Photo.

(c)

(d)

(e)

These patterns are revealed also in detailed analysis of frequencies of polymorphic alleles (encoding protein variation) among many human populations. Regional groupings of similar frequencies often emerge, but they often do not conform to racial distributions. In other cases, large frequency differences exist between geographically contiguous populations of the same race, while populations distant from one another may have similar frequencies, regardless of race (figure 13.14, table 13.2).

Figure 13.14

Distribution of ABO blood
group alleles in populations
around the world:
(a) frequencies of allele A;
(b) frequencies of allele B;
(c) frequencies of allele O.
(All numbers are
percentages.)

From *Human Polymorphic
Genes: World Distribution* by
Arun K. Roychoudhury and
Masatashi Nei. Copyright
© 1988 by Oxford University
Press, Inc. Reprinted by
permission.

(a)

(b)

(c)

TABLE 13.2

Variation in allele frequencies of cerumen (ear wax) type (chapter 4) among human populations. Allele W, wet type, is dominant to allele w, dry type.

Population	Allele Frequencies	
	W	*w*
Europe		
Finland	.46	.54
Germany	.82	.18
Asia		
Turkey	.48	.52
Iran	.42	.58
Afghanistan	.61	.39
India		
Non-tribes	.52	.48
Tribes	.61	.39
Maylasia		
Malays	.48	.52
Aborigines	.74	.26
China		
Mongolians	.06	.94
Chinese (Northern)	.02	.98
Chinese (Taiwan)	.14	.86
Taiwan		
Aborigines	.46	.54
Japan		
Japanese	.08	.92
Ainu	.29	.71
Korea	.08	.92
North America		
Alaska		
Aleut (Eskimos)	.30	.70
United States		
Whites	.89	.11
Blacks	1.0	0
Papago Indians	.23	.77
Middle America		
Mexico	.24	.76
(Maya Indians)		
Panama	.79	.21
(Cuna Indians)		
South America		
Peru	.12	.88
(Quechua Indians)		
Oceania		
Australia		
Aborigines	.82	.18
Melanesia	.47	.53
Micronesia	.39	.61

Source: Data from A. K. Roychoudhury and M. Nei, *Human Polymorphic Genes: World Distribution,* copyright © 1988 Oxford University Press, Oxford.

Average genetic differences between populations and species can be summarized by a statistic called **genetic distance (D).** Genetic distance estimates the number of different alleles per locus between the two groups on the basis of the frequencies of alleles in both populations. Genetic distances range from 0 to infinity. For genetically identical populations, D = 0. When about 40% of the loci compared have different alleles, D = 0.5; if 63% of loci differ, D = 1.0; and if 92% differ, D = 3.0.

Figure 13.15

(a) Genetic distances among
human races compared to
genetic distances between
humans and our nearest
evolutionary relative, the
chimpanzee. (b) The ranges
of genetic distance commonly
found between closely related
taxa. Genetic distances are
estimated on the basis of
proteins assayed by gel
electrophoresis.

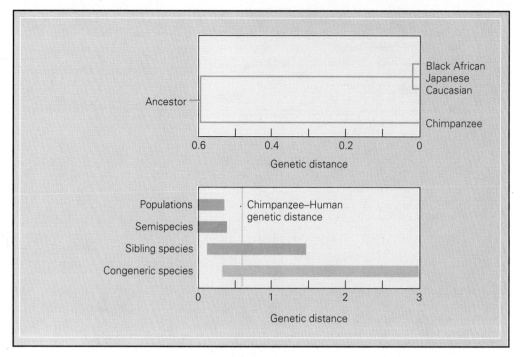

Estimates of genetic distance between humans and their nearest relatives, chimpanzees, vary from 0.39–0.62, depending upon the proteins examined (figure 13.15a). These values are relatively small compared to genetic distances found between other closely related species such as domestic dogs and coyotes. In fact, such small genetic distances are usually found between species within the same genus. The average protein in chimpanzees and humans differs by less than 1% in amino acid sequence.

On a finer scale, the average genetic distance between races and populations of humans is only about 0.01–0.02, tiny in comparison to the interspecific comparison (figure 13.15a,b). Interracial genetic distances are so small because, as noted earlier, genetic differences between races are normally *quantitative,* not qualitative. That is, populations differ in the frequencies of alleles, but most populations contain the same alleles found in other populations, regardless of race.

The genetic differences between human races are so small, and individual variation so great, that a prominent population geneticist observed: "If the holocaust comes and a small tribe deep in the New Guinea forests are the only survivors, almost all of the genetic variation now expressed among the innumerable groups of . . . people will be preserved." This statement does not deny racial differences, but it places them in proper perspective. The genetic differences that underlie characteristic racial phenotypes occur at a small minority of all loci. Conversely, even a small population contains most of the genetic variation found in the total human gene pool.

Direct analysis of DNA sequence variation provides a similar view of human individual and populations differences. On average, the nuclear DNA sequences of any two humans are 99.8% identical. This means that pairs of humans differ from one another at about one nucleotide base in five hundred.

Human population differences have been examined using rapidly evolving mitochondrial genes. The largest average difference found between any pair of groups was less than 0.6%. Differences between members of different races were only slightly greater than differences between members of the same race. The tree of population relationships (figure 13.16) shows that: (a) all surviving human populations trace their ancestry to a single female—*Eve* (chapter 1)—who probably lived in Africa about 200,000 years ago, and (b) each "race" has multiple origins. Within each race, some populations are more closely related to populations other than those of their own race. This almost certainly reflects the widescale intermingling of populations of all races over the past 200,000 years. "Racial purity" is a concept with no factual basis for humans.

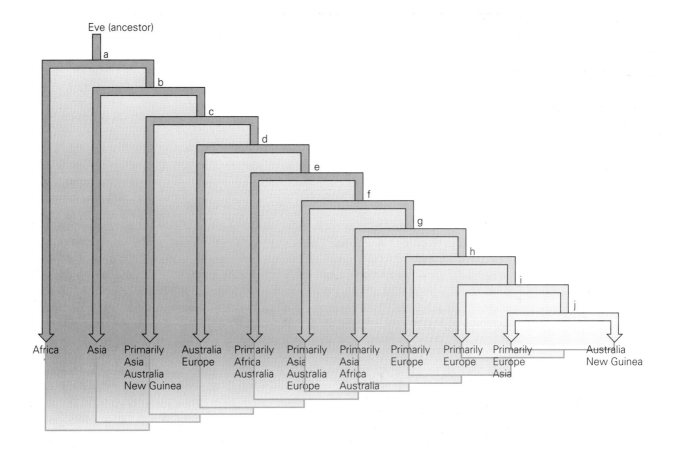

Eve (ancestor)

Africa | Asia | Primarily Asia Australia New Guinea | Australia Europe | Primarily Africa Australia | Primarily Asia Australia Europe | Primarily Asia Africa Australia | Primarily Europe | Primarily Europe | Primarily Europe Asia | Australia New Guinea

Racism

Figure 13.16

Human population relationships as shown by mitochondrial DNA similarities.

Source: Data from R. L. Cann, M. Stoneking, and A. C. Wilson, "Mitochondrial DNA and Human Evolution," *Nature* 325:31–36, 1987.

The belief that one race, population, or group (almost inevitably one's *own* race or group) is superior to another or all others—*racism*—is probably as old as human history. As we have stated previously, the science of genetics provides no support for racist contentions, but racism is still common in many, perhaps most, societies.

Racism has provided the motivation or justification for many of the major events in western history in recent centuries. The Civil War in the United States and World War II are obvious instances where race was a central issue, but many others could be cited.

The Divine Right of Kings, justifying the monarchy in England, for example, accepted the inherent (the word "genetic" had not yet been invented) superiority of royalty. At the other end of the spectrum, criminals and the poor, were believed to be inferior people. Many early colonists of America were poor criminals expelled from England on the belief that criminality was an inherited condition that could not be redeemed by imprisonment or other punishment. Therefore, they believed, better to improve England by permanently removing the offenders to America.

Following the American Revolution, English jails rapidly overfilled, creating a crisis for the legal system. An alternative was soon found—transportation (usually permanently) to the other end of the world, Australia. Between 1787 and 1868 about 160,000 people were transported to Australia for the protection of England from the "criminal class." America and, to a greater extent, Australia were founded by people thought to be inferior. Both these countries comprise a social experiment that, according to the genetics of the eighteenth and nineteenth centuries, should have produced nations of criminals, not independent democracies.

The advent of Darwinism in the mid-nineteenth century did not change things. To many, Darwinian evolution was not a demonstration of the relatedness of humanity, but rather a justification for the struggle between races. Entire theories of history were written based on racial conflict—as a virtue, because history had clearly (they claimed) demonstrated that white Europeans were the pinnacle of evolution and therefore deserved to dominate other races. The ongoing demise of native populations around the world due to European imperialism was seen as a natural process and thus desirable. (At the time of British settlement of Australia, no religious missionaries were sent to the Aboriginals because they were believed to have no souls and therefore not to be capable of salvation. Needless to say, the extermination of Aboriginal tribes would not trouble those who believed Aboriginals to have no souls—and who imperiously took over their lands.)

Late nineteenth century scientists often took the lead in developing racist explanations of human history. When the fossil record did not yield examples of the missing link between humans and apes, biologists and anthropologists often identified them in existing racial groups—Africans, Australian aboriginals (no soul, remember) and Orientals. To white male European scientists of the time, white male Europeans were obviously the pinnacle of evolution. Females and other races were evolutionarily less-developed life forms. Leading scientists donated their brains to museums as examples of the best evolution had produced.

Such views pervaded science of the time. For example, in 1866 J. L. H. Down identified a condition they called "mongolian imbecility" and now known as Down's syndrome. Because Down thought that victims of the syndrome resembled Mongols, he claimed that it represented a congenital reversion to a more primitive (i.e., Asian) human type. In the early 1900s, the condition was attributed to a recessive Mendelian gene, which could reveal a vestige of early human history. Only since the 1950s, when the disease was identified as chromosomal and not racially linked, has the name "Down's syndrome" become preferred, although the term "Mongoloid idiot" is still occasionally heard.

The sad fact is that from the time of Darwin—roughly 1860—until 1945, the end of World War II, many scientists, including some geneticists, used genetic theories to propose and justify racist social policies, some of which were implemented to terrible effect. Such scientific views have now been entirely discredited, partly due to the horrors of the racist policies and practices of Nazi Germany, partly due to advances in population genetics, and partly due to the demonstration of high levels of individual genetic variation, as described in the previous section.

Eugenics: The Past

In 1883 the English scientist Francis Galton (figure 13.17a) invented the word **eugenics** (from the Greek words meaning "good in birth") to refer to the science of improving the genetic qualities of the human population. The ideas of Galton, a cousin of Charles Darwin, were discussed with particular reference to human intelligence, which Galton believed (quite incorrectly) to be determined mainly by heredity.

Galton's ideas were extremely influential. By the early twentieth century, societies were founded in the U.S., Canada, and many countries in Europe, devoted to furthering eugenic proposals such as compulsory sterilizations for genetic defectives, immigration and marriage restrictions, and even breeding programs. Many of these programs were based on prejudice or bad science, but they were adopted anyway.

Professional geneticists also took up the cause. In 1904, a bequest from Galton established the Eugenics Record Office, later the Galton Laboratory for National Eugenics, at the University of London. In the United States, the Eugenics Record Office at Cold Spring Harbor, Long Island, New York, was established in 1910. Both institutions collected enormous amounts of information on families, trying to document the inheritance of many types of traits—mental and physical diseases, criminality, personality, and so on.

The founder of the Cold Spring Harbor Eugenics Record Office, and director for many years, was Geneticist Charles B. Davenport (figure 13.17b). Some aspects of his work were highly regarded, but many of Davenport's opinions of human inheritance were highly speculative, unsupported by data, and simply wrong. Davenport believed that mental illness (it was called "feeblemindedness" at the time) was inherited, as were poverty and criminality. Prostitution, he claimed, was caused by a dominant gene for "innate eroticism." "Thalassophilia," the love of the sea exhibited by naval officers, he attributed to a sex-linked recessive gene because it was expressed mainly in males!

(a) (b) (c)

Figure 13.17

Prominent early eugenicists:
(a) Francies Galton;
(b) Charles B. Davenport;
(c) Harry H. Laughlin.

(a) The Bettmann Archives;
(b and c) Cold Springs
Harbor Laboratory.

Davenport did not distinguish between races and nationalities, and he was greatly concerned with the immigration of peoples from southern and eastern Europe in the early 1900s, believing that it would cause future American generations to be "darker in pigmentation, smaller in stature, more mercurial . . . more given to crimes of larceny, kidnapping, assault, murder, rape, and sex-immorality."

Publications from the Eugenics Record Office were used with great effect by its director, Harry H. Laughlin (figure 13.17c), Davenport's student and protégé. Laughlin lobbied Congress for various eugenic laws, and in 1924 his efforts were rewarded with enactment of the Immigration Act, restricting immigration of southern and eastern Europeans (figure 13.18). Later, Laughlin lobbied for increased restrictions on Jews trying to escape from Nazi Germany. In 1936 the University of Heidelberg awarded Laughlin an honorary doctorate.

In fact, Nazi eugenicists were greatly inspired by American eugenics legislation, particularly state laws. Miscegenation—interracial marriage—was illegal in southern states. By 1930 twenty-nine American states and one Canadian province had enacted compulsory sterilization laws, and by 1939 over 30,000 Americans had been sterilized under these laws. Sterilization was sometimes by vasectomy, sometimes castration, and it was used on the feebleminded, habitual criminals, and sex offenders. (One of the first estimates of the diploid number for humans was made from cells in sectioned testes removed from three men in a Texas mental institution. All three were castrated because of "masturbation associated with insanity." The chromosome number was incorrectly estimated at forty-eight.) Harry Laughlin recommended a program to sterilize hundreds of thousands of genetically inferior Americans annually, with a goal of fifteen million sterilizations by 1980.

The worst horrors of eugenics, of course, occurred in Nazi Germany under the name "racial hygiene." In 1934, shortly after the Nazis came to power, 181 Genetic Health Courts were established to rule on eugenic matters. Mandatory sterilization laws were enacted, and marriage and sexual relations between Germans and Jews were prohibited. In 1939, "mercy killings" of 5,000 deformed and retarded children began, soon followed by over 70,000 adults judged to be defective, especially the mentally or chronically ill and the aged.

Apparently, physicians voluntarily conducted these initial mass killings once they were authorized to do them. Physicians, trained in eugenics, were not ordered to do them, but did them out of professional conviction. But the military soon used their techniques for mass killings of millions of Jews, gypsies, homosexuals, and others regarded by the Nazis as a genetic threat to the supposedly pure Nordic or Aryan stock of Germany.

The abhorrent actions of the Nazis permanently stained the word "eugenics." Following World War II many eugenics societies ceased activities entirely. Others directed their activities toward theoretical genetics or voluntary eugenic measures directed at individual health, not improvement of the gene pool. Nonetheless, it is worth remembering that explicitly racist laws and policies, often supported with flawed eugenic logic, characterized many developed

Genetic Diversity   **387**

Immigration into the United
States: (a) Ellis Island, New
York, the site of entry of
millions of immigrants to the
U.S.; (b) prospective
immigrants being examined
for health defects that might
limit their admission;
(c) numbers of immigrants,
by region of origin, showing
the effect of the Immigration
Act of 1924.

(a) AP/Wide World Photos;
(b) The Bettmann Archives;
(c) chart by Dale D. Glasgow
© National Geographic
Society. Used by permission.

(a)

(b)

500,000

Northern Europe
Southern Europe
Central & Eastern Europe
Asia
Africa
Americas

1890s 1900s 1910s 1920s 1930s 1940s 1950s 1960s 1970s 1980s

Arrivals estimated
from INS data.

(c)

nations long after 1945. The legal basis of segregation in the United States only began to be demolished in 1954. The White Australia policy, severely limiting immigration of non-whites, persisted through the 1960s; half-caste children were removed from their Aboriginal families for mandatory incorporation into white families because Aboriginals were thought to be a doomed race. Strict racial segregation or *apartheid,* is still the national law of South Africa.

Undercurrents of the history of eugenics also permeate current scientific debates. The IQ controversy exemplifies often fierce ongoing debates over the role of genetics in human behavior. Some opposition to genetic engineering plans for humans—gene therapy, for example—reflects enduring revulsion over the misuse of genetics in the first half of this century.

Eugenics: The Present

Many geneticists led the battle against the bad science and the racism of eugenics proponents earlier in this century. These geneticists believed that eugenics could improve the human condition, but that the focus must be on the welfare of the individual and not the supposed needs of the state.

Since World War II, genetic knowledge and technologies have been applied to humans, but with the aim of improved medical care and concern for the individual. Genetic counseling, for example, is an attempt to make genetic knowledge available to those who need it, informing individuals and families of choices available to them, but not coercing them into particular courses of action. PKU screening is mandatory in some states, but with the goal of helping affected individuals rather than improving the gene pool.

Genetics may be used in many ways to improve human life. Historically, medical practice has been **euphenic** (good appearance), aiming to manipulate the phenotype to nullify the harmful effects of a gene. Examples of euphenic therapies include eyeglasses, injections of insulin for diabetics, phenylalanine-free diets for persons with PKU, and removal of the cancerous eye for persons with retinoblastoma. In theory, improved understanding of genetic diseases could lead to euphenic treatment for each one.

Some people fear that euphenic medical practice may be **dysgenic**—that is, it may lead to an increase in the frequency of deleterious genes in the population. Insulin therapy, for example, allows diabetics to live relatively normal lives and produce many more children than previously. Genetic screening programs for diseases such as Tay-Sachs may mean that affected fetuses—homozygous recessives—are aborted, but that parents produce many children who are heterozygous carriers. Should we fear dysgenic results of such practices?

Probably not. The effect of any gene depends upon the environment in which it is expressed. Contemporary medical practice will continue to provide treatments (an environment) in which individuals bearing genes harmful to survival in primitive conditions will have normal phenotypes. If—and it is not a certainty—an increasing proportion of the population develops diabetes, the provision of insulin and other treatments will not be an unsupportable burden, particularly in comparison to systematic programs to eliminate such genes, i.e., persons, from the population. New medical treatments are likely to develop far more rapidly than deleterious genes can accumulate in the human gene pool.

Furthermore, every individual already harbors at least four or five deleterious recessive genes. Removal of all deleterious genes would involve exterminating that population. An equally intractable problem, of course, lies in defining deleterious. What is deleterious? In what context? Who decides?

Active eugenic programs aiming to improve the gene pool are practiced on domestic plants and animals, but many aspects are not applicable to humans for obvious ethical reasons. Eugenics programs can take two directions. *Positive eugenics* aims to increase the frequency of specific desirable genes through carefully designed breeding programs. Using a prize bull to father an entire herd of dairy cattle, for example, is a positive eugenics program.

The sperm bank for Nobel prizewinners (chapter 7) is a small-scale voluntary positive eugenic scheme whose population effects will be negligible; it represents no threat to human rights, whatever the motivation of participants.

Negative eugenic programs aim to diminish the frequency of undesirable genes by limiting or preventing reproduction of individuals carrying those genes. Compulsory sterilization laws represent a once widespread attempt at negative eugenics. While most of these laws have been repealed or changed, sterilization of mentally ill or retarded persons is still legal in some states and countries if requested by the family or legal guardian to prevent pregnancy or in some way benefit the affected person.

Therapeutic abortion of genetically defective fetuses can also be viewed as a negative eugenic measure. However, the ethical and legal basis of such abortions is to prevent unnecessary suffering of either the affected fetus or the mother, and *not* to benefit society or the gene pool.

Limiting reproduction of individuals affected by genetics disease can rapidly change allele frequencies only for diseases caused by dominant genes that are expressed early in life. Changing the frequency of Huntington's disease, for example, is difficult because of the late onset of the disease. Even with the newly available DNA probe for early diagnosis, not all persons at risk will want to be diagnosed, and coercion for such testing is considered unethical.

Population genetics theory shows that the frequency of deleterious recessive genes can be reduced only slowly by preventing reproduction of affected individuals. These **alleles** can never be eliminated, because at low frequencies virtually all the alleles are carried in the heterozygous condition. For example, for a condition affecting only one person in ten thousand, $q^2 = 0.0001$, the frequency of the allele $(q) = 0.01$, and it is carried by $2pq = 2\%$ of the population. If all persons with this allele were not allowed to reproduce, it would require five hundred generations to reduce the frequency to 0.0017. Thus, negative eugenics is not an effective way to deal with harmful recessive alleles.

Despite its history and its many limitations, eugenics is an important idea that is deeply imbedded in contemporary medical practice and in the science of genetics itself. Future eugenic developments are likely to be accepted only when they respect, and seek to improve, the life of the individual. Where they involve coercion of individuals for the supposed benefit of societies, memories of past excesses should provide an important barrier to their acceptance. Only well-informed citizens can adequately deal with the thicket of emotional, ethical, and legal issues raised by genetic advances that will confront us all in coming decades.

Summary

1. The Hardy-Weinberg equilibrium equations permit estimation of allele and genotype frequencies in populations.
2. Ecological genetics relates patterns of phenotypic and genotypic variation within and between populations.
3. Human life depends upon all other organisms with which we share the Earth.
4. Humans have long practiced artificial selection in breeding high-quality plant and animal species.
5. New genetic combinations of domestic species are created by mutagens, grafting, hybridization, protoplast fusion, and gene transfer.
6. Inadvertent selection by humans has resulted in the emergence of pesticide-resistant and drug-resistant strains of plant and animal disease organisms.
7. Genetic knowledge has led to varied methods of pest control, including sterile male releases and pheromone "trickery."
8. The interplay of politics and agriculture is exemplified by the Green Revolution of the 1960s.
9. It is imperative for humans to conserve the remaining genetic diversity of plants and animals around the world.
10. Genetic bottlenecks have greatly diminished the evolutionary potential of many species and driven others to extinction.

11. Various governmental agencies around the world are now conserving valuable plant and animal germ plasm.
12. The gene pool consists of all alleles contained in a group of interbreeding individuals. About one-third of all human genes are polymorphic, having more than one allele.
13. Races can be characterized by allele frequency differences. Over 85% of the allelic variation in human races is seen within populations, not between races.
14. Electrophoresis is a rapid way of visualizing allelic differences in structural genes. These allelic differences between populations can be expressed as a genetic distance.
15. Dysgenic factors lead to an increase in the proportion of harmful genes in a population. Modern health practices are unlikely to have significant dysgenic effects.
16. Euphenics involves manipulation of the phenotype to ameliorate genetic conditions. Insulin injections for diabetics are a good example.
17. Eugenics involves encouraging persons with favored genes to reproduce (positive eugenics) or discouraging persons with deleterious genes from reproducing (negative eugenics).
18. The history of eugenics—attempts to improve the human gene pool—is a shameful example of the misuse of science. Bad science was used to justify racism, widespread human rights abuses, and in Nazi Germany the murders of thousands of persons considered to be threats to the genetic well-being of the nation.

Key Words and Phrases

artificial selection

biological evolution

cryptic species

culture

dysgenics

ecological genetics

electrophoresis

embryo transfer

eugenics

euphenics

gene pool

gene transfer

genetic bottleneck

genetic distance (D)

genetic diversity

genetic engineering

genome

grafting

Green Revolution

Hardy-Weinberg equilibrium equations

hybrid vigor

hybridization

inbreeding depression

land races

monoculture

monomorphic

pesticide treadmill

pheromones

plant patenting laws

polymorphic

population genetics

protoplast fusion

race

racism

sterile male technique

superovulation

Questions

1. What are some of the problems associated with using mutagens to produce new genetic variants? Are plants or animals more likely to yield favorable results? Why?
2. To what domestic mammals other than cattle could superovulation and embryo transfer be most profitably applied? What are your criteria for selecting these species?
3. Until the twentieth century all farmers saved seed from one harvest to plant the succeeding year's crop. In advanced countries, farmers tend to purchase their seed annually from seed companies. What changes have caused farmers to purchase seeds annually? Can you think of any disadvantages arising from seed companies controlling seed stock?
4. Suppose a seed company developed a perennial variety of corn or wheat (all present varieties are annuals). What dilemma might such a development pose to a seed company? Why would a perennial corn be advantageous to the farmer?
5. Compare and contrast the advantages and the disadvantages of crop monoculture.

6. What characteristics shared by organisms such as bacteria, insects, and rats result in their ability to evolve rapidly in response to attempts to limit their populations with chemical pesticides and antibiotics?

7. Could the sterile male technique be applied to control rats or other noxious mammals? What problems might arise from such attempts?

8. Many triploid (or other polyploid) plants are of commercial value because they produce exceptionally large fruit or seeds. Induced triploid animals have, in some cases, increased commercial value due to their lack of reproductive tissues. Comment on the different phenotypic results of polyploidy in plants and animals.

9. Recombinant DNA techniques permit the transfer of DNA between species. Does this new technology change the potential commercial value of the millions of non-domesticated species of plants and animals? How?

10. If 84% of a random-mating population express the dominant phenotype of being able to taste the bitter compound PTC (taster phenotype), what is the frequency of the recessive allele, which when homozygous makes its owner unable to taste this compound (a nontaster)?

11. If the frequency of a recessive allele is 30%, in a population of one thousand persons how many would you predict would be carriers of this allele, but would not express the recessive phenotype?

12. Of what significance is the fact that the proteins of humans and chimpanzees are so alike that any two homologous proteins chose at random have a 99% chance of being identical in their sequences?

13. Distinguish between euphenics and eugenics. Which program has the potential for effecting the greatest genetic change in the human population?

14. What immediate measures for the genetic health of the human population can you think of that could be implemented now without raising serious moral issues? Are there associated economic costs?

15. How many specific human traits can you think of that are the likely result of natural selection?

16. Compare genetic variability in the population of melting-pot countries such as Canada and the United States with genetic variation in countries that are less racially diverse.

For More Information

Cann, R. L., M. Stoneking, and A. C. Wilson. 1987. "Mitochondrial DNA and Human Evolution." *Nature,* 325:31–36.

Cohen, M. L., and R. V. Tauxe. 1986. "Drug-Resistant *Salmonella* in the United States: An Epidemiologic Perspective." *Science,* 234:964–69.

Daugherty, C. H., A. Cree, J. M. Hay, and M. B. Thompson. 1990. "Neglected Taxonomy and Continuing Extinction of Tuatara (*Sphenodon*)." *Nature,* 347:177–79.

Diamond, J. M., and J. I. Rotter. 1987. "Observing the Founder Effect in Evolution." *Nature,* 329:105–6.

Ehrlich, P. R. 1977. "Ecologists, Ethics, and the Environment." *Bioscience,* 27:239.

Elkington, J. 1985. *The Gene Factory: Inside the Genetic and Biotechnology Business Revolution.* Carroll & Graf Publishers, New York.

Gould, S. J. 1983. *The Mismeasure of Man.* W. W. Norton and Co., New York.

Hardin, G. 1974. "Living on a Lifeboat." *Bioscience,* 24:561–68.

Harris, M. 1973. "The Withering Green Revolution." *Natural History,* 82(3):20–22.

Hughes, R. 1988. *The Fatal Shore. A History of the Transportation of Convicts to Australia 1787–1868.* Pan Books Ltd., London. (A compelling story of the criminals transported to Australia, including the presumed inferior nature of the transportees.)

Johannessen, S., and C. A. Hastorf. 1989. "Corn and Culture in Central Andean Prehistory." *Science,* 244:690–92.

Jukes, T. H. 1986. "Frost Resistance and *Pseudomonas.*" *Nature,* 319:617.

Kevles, D. J. 1985. *In the Name of Eugenics. Genetics and the Uses of Human Heredity.* Alfred A. Knopf, New York.

King, M. C., and A. C. Wilson. 1975. "Evolution at Two Levels: Molecular Similarities and Biochemical Differences between Humans and Chimpanzees." *Science,* 188:107–17.

Knorr, D., and A. J. Sinskey. 1985. "Biotechnology in Food Production and Processing." *Science,* 229:1224–29.

Lewin, R. 1987. "The Unmasking of Mitochondrial Eve." *Science,* 238:24–26.

Lewin, R. 1987. "My Close Cousin the Chimpanzee." *Science,* 238:273–75.

May, R. 1990. "Taxonomy as Destiny." *Nature,* 347:129–30.

Miller, J. A. 1984. "The Mating Game." *Science News,* 126:232–35.

Miller, J. A. 1986. "Barnyard Biotech: Dissent on the Farm." *Science News,* 129:213.

Müller-Hill, B. 1988. *Murderous Science: Elimination by Scientific Selection of Jews, Gypsies, and Others, Germany 1933–1945.* Oxford University Press, Oxford, U.K.

Newhouse, J. R. 1990. "Chestnut Blight." *Scientific American,* 263:74–79.

Proctor, R. N. 1988. *Racial Hygiene. Medicine Under the Nazis.* Harvard University Press, Cambridge, Massachusetts.

Ruse, M. 1981. *Is Science Sexist? And Other Problems in the Biomedical Sciences.* D. Reidel Publishing Company, Boston, London.

Stepan, N. 1982. *The Idea of Race in Science.* Archon (Shoe String), Hamden, Connecticut.

Sun, M. 1986. "Engineering Crops to Resist Weed Killers." *Science,* 231:1360–61.

Tatum, L. A. 1971. "The Southern Corn Leaf Blight Epidemic." *Science,* 171:1113–16.

Tierney, J., L. Wright, and K. Springen. 1988. "The Search For Eve." *Newsweek,* January 11:46–52.

Wade, N. 1974. "Green Revolution, I and II." *Science,* 186:1093–96, 1186–92.

Wainscoat, J. 1987. "Out of the Garden of Eden." *Nature,* 325:17.

Classification of Life

Species are groups of organisms that can interbreed to produce new organisms of the same kind. Humans comprise one species, dogs another, and dandelions, onions, spruce budworms, brook trout, great white sharks, and starlings yet others. At least ten million and perhaps thirty million species now exist on earth.

Aristotle (384–322 B.C.) is often considered to be the first biologist, and among his many works is a classification of organisms known to the Greeks. Aristotle's system was used as the basis of biological classification by Western societies for almost 2,000 years, but most other societies had some form of classification. Perhaps the most common classification was to separate the living world into two categories of species: plants and animals. This was a commonsense approach. After all, plants are similar in being green, not ingesting other organisms, and not being able to move. Animals, on the other hand, are capable of movement and eat other organisms. Within the animal kingdom, smaller groups of organisms could be identified on the basis of increasingly precise characteristics.

In the mid-eighteenth century, a Swedish biologist, Carolus Linnaeus (1707–1778), invented the *binomial system,* which is the basis of modern biological nomenclature. In the Linnaean system, each species is assigned its own unique set of two latinized names. The first is the *generic* name, and the second is the *specific* name. (Genus is Latin for birth or race; species is Latin for kind.) For example, the scientific name of the domestic dog is *Canis familiaris.* The generic name (*Canis* meaning dog) is the more inclusive category and refers to a group of very similar (which Linnaeus knew) and also evolutionarily closely related (which Linnaeus did not know) species. For example, the genus *Canis* also includes *Canis latrans,* the coyote, and *Canis lupus,* the wolf.

A genus may include only one species, or it may include dozens or even hundreds of species. Species that are very closely related are assigned the same generic designation. On the other hand, the same specific epithet may be given to many different species, but only if they are in different genera. Figure A.1 gives the classification of two distantly related frogs with the same specific epithet, the spring peeper (*Pseudacris crucifer*) and the South American toad (*Bufo crucifer*). Both the peeper and the toad are assigned the descriptive name *crucifer* because each has a mark on its back resembling an X.

Crucifer means crossbearing in Latin. No two species are ever assigned the same generic and specific names together.

The binomial system permits all scientists to speak one language. A species may be called by many common names in different places. A tree frog in England is a grenouille in France and a rana in Spain, but scientists in all countries know it as *Hyla arborea.* A squirrel in the United States is an ardilla in Spain and an ecureuil in France, but it is *Sciurus niger* to all scientists. Groups of species are organized into hierarchical series of increasingly less-similar groups. Genera are combined into Families, Families into Orders, then Classes, Phyla, and finally Kingdoms. These major categories (or *taxa*) may be further divided if necessary. The biological system of classification is designed both to organize the immense

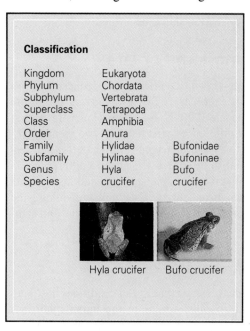

Classification

Kingdom	Eukaryota	
Phylum	Chordata	
Subphylum	Vertebrata	
Superclass	Tetrapoda	
Class	Amphibia	
Order	Anura	
Family	Hylidae	Bufonidae
Subfamily	Hylinae	Bufoninae
Genus	Hyla	Bufo
Species	crucifer	crucifer

Hyla crucifer   Bufo crucifer

Figure A.1

Example of classifying animals, using frog and toad having same specific name because both species have markings resembling a cross on their back.

© Thomas Gula/Visuals Unlimited; (b) Dr. J. P. Bogart, University of Guelph.

The common gut bacterium, *E. coli,* is a favorite organism of geneticists. The average cell is rod shaped and about 2u long by 1u in diameter. Under optimal conditions of food and temperature it reproduces by fission every twenty minutes to produce two identical daughter cells per division. In twenty-four hours the descendants of a single bacterium can number up to one billion.

DNA attached to cell membrane

Cell wall

Cell membrane

Polyribosomes   Free ribosomes

diversity of life into groupings of organisms according to similar or shared characteristics and to reflect the degree of evolutionary relationship among them.

Three Kingdoms of Life

A five-kingdom classification of life, introduced in 1969 by R. H. Whittaker, enjoyed widespread use until quite recently. In the late 1970s C. R. Woese demonstrated that major genetic differences exist between two groups comprising Whittaker's Prokaryotes and that these differences are equal to or exceed the genetic differentiation between the plants and animals. Woese and his students therefore recognize three kingdoms, dividing the Prokaryotes into the Archaea and the Bacteria, and combining all remaining taxa in the Eucarya.

Kingdom Archaea

Archaea are prokaryotes that live in conditions hostile to most life, but possibly similar to the harsh conditions under which life began. Although in external appearance they differ little from Bacteria, genetically they are extremely different.

Many of the Archaea are methanogenic (meaning methane producing). They live only where oxygen is absent, such as the decaying vegetation at the bottom of a swamp. A by-product of their metabolism is methane, a combustible gas that is known as swamp gas when seen burning in a swamp at night. Methanogens are also found in hot springs, stagnant water, sewage treatment plants, the intestinal tracts of animals, and the bottom of the

ocean. Other types of Archaea are halophiles (meaning salt loving), which inhabit highly saline waters such as the borders of evaporation ponds, the Dead Sea, and the Great Salt Lake, and the thermoacidophiles (meaning heat-acid loving), which occupy acidic hot springs, such as at Yellowstone National Park, and burning coal tailings.

Kingdom Bacteria

The other group of prokaryotes, the Bacteria, are ubiquitous and essential members of the living world. They occupy most habitats and have many different modes of life. Many are free living and function in the decay and breakdown of dead organic matter. If the Bacteria did not serve as recyclers to the world, we would soon be buried under the carcasses of plants and animals.

Other Bacteria occupy the bodies of living organisms. *Symbiotic* (meaning living together) bacteria benefit their host organisms: in return for a place to live, the gut bacterium of vertebrates, *Escherischia coli,* helps digest our food. *E. coli* is an important organism in genetic research because it is easy to culture in the lab and because it reproduces so rapidly—one generation every twenty minutes (figure A.2). Some bacteria are parasitic or pathogenic (meaning disease causing). The agent of common food poisoning is *Salmonella,* and *Staphylococcus* causes a variety of serious infections in humans.

Blue-green algae are another type of Bacteria. Most algae are eukaryotes, but blue-greens are clearly prokaryotic. They commonly inhabit stagnant water and many harsh environments, such as hot springs. Blue-green algae, as well as many bacteria, are capable of *photosynthesis*. They use the energy from sunlight to combine carbon dioxide and water into carbohydrates such as glucose, a simple sugar from which they then derive energy.

Kingdom Eucarya

Protists

Protistans are single-celled eukaryotes. They occur in a great variety of forms, but they are larger and more complex than any prokaryotes. The two basic types of protistans are the protozoa (meaning before animals), which exhibit many features of the multicelled animals, and the single-celled algae, which are much like the multicelled higher plants. The ancestors of the multicelled eukaryotic kingdoms probably originated from organisms much like living protistans.

Most protozoa are heterotrophic (meaning other food), i.e., they derive their energy from eating other organisms. Like the animals, many have a primitive ability to detect and respond to external stimuli. They occupy many habitats, and both free-living and parasitic, disease-causing species are known. Familiar protozoa include *Paramecium* and *Amoeba* (both found in freshwater ponds), *Trypanosoma* (the cause of African sleeping sickness), and *Plasmodium* (the cause of malaria).

One-celled algae are protistans that are primarily aquatic autotrophs (meaning self-feeding). They use photosynthesis to produce their own food, but, like multicelled plants and unlike photosynthetic bacteria, they possess specific organelles, *chloroplasts*, in which photosynthesis occurs. Photosynthesis requires the presence of one of a variety of pigments called *chlorophylls*. Common chlorophylls are green and cause the green color of most plants. In photosynthetic eukaryotes, chlorophyll occurs in chloroplasts.

Single-celled algae are extremely important because they form the primary source of food for many aquatic organisms. Common types of these algae are the diatoms (whose beautiful silicon shells have been used in toothpaste), dinoflagellates, and *Euglena*. One species of dinoflagellate causes the well-known *red tides*. Red tides occur when outbreaks of *Gonyaulax catanella* poison huge numbers of fish and other marine life.

With minor exceptions, the remaining members of the Eucarya are all multicellular.

Fungi

Toadstools and mushrooms are familiar to us all. Some exist as single-celled forms, but most are multicellular. Fungi used to be considered degenerate plants because they lacked chlorophyll, but the differences between fungi and all other organisms are so great that the fungi are now considered a major independent line of evolutionary descent.

Fungi are composed of a filamentous cell called a *hypha*. Often, hyphae form a multicelled structure called a *mycelium*, which is the body of the fungus. Fungal cells differ from plant cells in having neither chloroplasts nor chlorophyll (which is why they are often white in color) and in having a supportive cell wall of *chitin*, a polysaccharide that is also found in the skeletons of insects but never in plants. Unlike either plants or animals, fungi use extracellular digestion; digestive enzymes are secreted onto food, and the partially digested mixture is then absorbed by the fungal cells.

Fungi are important decomposer organisms. Like bacteria, they are one of the agents of decay of dead organic matter, particularly in forest communities. Many fungi such as molds and mildews are nuisance organisms, while others are serious disease organisms of crops (rusts and smuts of wheat and rye, for example) or humans (ringworm—not a worm at all—and thrush). Humans eat many species of fungi, but others such as the *Amanita* are highly toxic. Some are hallucinogenic. Fungi are the source of penicillin, an important antibiotic drug, and are used to produce Roquefort and other cheeses. Yeasts are a type of fungi whose metabolic waste products are used by humans: carbon dioxide from their respiration causes bread to rise, and fermentation involves the production of alcohol from the breakdown of sugars.

Plants

Plants are multicellular autotrophs. They range in size from unicellular algae to the largest organisms now on earth, the giant sequoias in California. The oldest living organisms are also trees. Some bristlecone pine trees in California and Nevada are almost 5,000 years old (figure A.3). They have been alive for almost the entire recorded history of human civilization.

Plants are the dominant organisms in most terrestrial ecosystems. They provide structure to the living community, a place for many organisms to live, and food for all. Organisms that do not eat plants nonetheless eat other organisms that do eat plants. Among the most important types of plants are mosses and liverworts, ferns, horsetails, gymnosperms (such as pine trees), and the familiar angiosperms or flowering plants.

Plants are able to attain great size because of their extraordinarily tough and strong cell walls. The cell walls of plants consist of cellulose, a polysaccharide. Cellulose is not only strong, but also resistant to decay. The cellulose trunks of some bristlecone pines that have been dead for four thousand years have yet to decay. Most heterotrophs are incapable of digesting cellulose. In a living tree, most of the mass consists of cellulose that once surrounded living cells. As the tree grew, the cells died, but the cellulose remained and continued to provide support for the surviving structure. Most of the living tissue of a tree consists of new leaves and twigs and branches and roots, plus a layer of growing tissue surrounding the trunk just under the bark.

The oldest bristlecone pine, from eastern Nevada, has been dated at forty-nine hundred years of age by analysis of its annual growth rings. These pines are among the world's oldest organisms and remain standing long after death.

USDA Forest Service.

Although plants now dominate terrestrial habitats, they originated in the oceans. Multicelled algae evolved from single-celled algal protistans, which then invaded dry land four hundred million to five hundred million years ago. Algae need not be small; many seaweeds like kelp are more than one hundred feet long.

Plants provide our primary source of food; civilization probably began as a result of the development of agriculture. But plants have a variety of less obvious but equally important uses to humans. Plants hold the soil in place, inhibiting erosion. Plants help cool the surface of the earth (compare a forested park to paved cities in summer). Plants are the source of hundreds of drugs. Plant products are used to build our homes and to furnish them, to write on, to write with, and to clothe us.

Plants also play a vital role in regulating the composition of our atmosphere. Plants consume carbon dioxide and release oxygen. As industrial pollution continues to pump carbon dioxide into the atmosphere, the capacity of plants to control levels of carbon dioxide is being strained. A particularly serious problem is the destruction by humans of the huge tropical forests, especially in the Amazon, which further reduces the mechanism for abating the levels of carbon dioxide. The potential buildup of carbon dioxide may induce a *greenhouse effect,* in which the surface temperatures of the earth will begin to increase. The effects of such an increase would be extremely serious, and there is good evidence that such an increase is underway.

Animals

Animals are multicellular heterotrophic eukaryotes. They differ from fungi (which are also multicellular and heterotrophic) in many ways: animal cells have no cell walls, digestion is usually within the body, animals usually are capable of active movement, and animals usually possess a nervous and sensory system.

The diversity of animal life is astonishing. Most of us are familiar with a few types of animals: mammals, birds, frogs, lizards, snakes, fish. All those are animals with backbones (or *vertebrates*), like humans. Over ten million species of animals are now alive, but only about 40,000 are vertebrates. There are probably at least one million species of insects alone, and three hundred thousand of those are beetles. Other types of *invertebrates* (animals without backbones) include sponges, corals and jellyfish, flatworms, roundworms, segmented worms (such as earthworms), ticks and spiders, crustaceans, sea squirts, and mollusks—by the tens of thousands.

Animals occur everywhere on the surface of the earth, beneath the surface, on the bottom of the ocean, and even in the air above the earth. Many species of insects and spiders have been caught in nets in wind currents 30,000 feet in the air, as well as at the top of Mt. Everest. Others have been brought up from wells in excess of 1,000 feet below the surface. Many species survive by parasitizing other animals and plants, and parasites often have their own parasites.

Plants cannot live deeper in the ocean than light penetrates, but animals survive on the rain of debris of all kinds that reaches many thousands of feet below the surface. In the 1970s, undersea expeditions discovered thriving invertebrate communities (including many species never before seen by humans) living around vents of hot water at depths of 1,500–2,000 meters in the eastern Pacific Ocean. These communities survive by eating the dense growths of Archaea surrounding the vents.

While plants usually determine the basic character of a terrestrial community, animals can also change the face of the earth massively. Humans have touched and changed nearly every spot on earth, but other animals can be equally formidable. Insect plagues can turn forests and croplands into deserts. A small colony of beavers can turn a valley into a series of lakes and meadows. (Some beaver dams are almost one kilometer long.) The largest individual organisms may be trees, but the largest structures are built by animals. Two are large enough to

(a)

(b)

be visible to astronauts (figure A.4): the Great Wall of
China, thirty-two hundred kilometers long and built by
thousands of humans; and the Great Barrier Reef, two
thousand kilometers long off the coast of Australia and
built from the bodies of billions of corals.

Human uses for animals are both mundane and
imaginative. We use animals for food, for clothing, for
shelter, and for transportation. They provide us with
companionship. Carrier pigeons communicate for us.
Mules pull our plows. And in every facet of biological,
medical, and psychological research, animals provide a
source of data through which we can learn about our
world and ourselves.

Viruses: What Are They?

Viruses are extremely small particles that possess some,
but not all, of the characteristics of normal living organisms. They seem to do little more than infect a living cell
and use the organelles of that cell to produce more viruses. Thus, they are capable of reproduction, but only
with the assistance (not freely given) of true cells. They
are not capable of other life processes such as metabolism. Because they must invade another organism to reproduce, they are parasitic, but their effects are not
always harmful to the host.

The structure of viruses is remarkably simple (figure
A.5). They consist of an outer protein coat surrounding
a segment of nucleic acid, either DNA or a related molecule, RNA. Often, a short tail is used to attach to the
surface of a host cell. Once attached, the virus injects its
DNA or RNA into the host. The virus nucleic acid then
commandeers the host system and causes the host cell to
begin manufacturing more viruses, at the expense of the
host cell's metabolic processes.

Thus, a virus may be viewed as a length of parasitic
DNA or RNA—a sort of free-lance genetic material. Its
only occupation is infection and self-reproduction. Biologists have not been able to decide if a virus is a living
organism or what its evolutionary origins might be. Is a
virus an extremely simple structure that might be like
the very first organisms or precursors of life on earth? Or
is a virus a degenerate bacterium, a piece of feral nucleic
acid?

Whatever the answers to such questions, viruses have
been of great importance to genetics. Because a virus is
little more than an independent piece of genetic material,
its function has taught us much about the ways in which
DNA and RNA function.

Evolutionary Relationships
of the Three Kingdoms

All species represent a branch tip of one large evolutionary tree. How do the branches fit together? Relationships among organisms are increasingly being
assessed on the basis of molecular differences. Because
genes carry encoded information, and because each gene
is a descendant of millions of generations of previous
genes, each gene carries information regarding its own
history. Thus, we can hope to understand the relationships of organisms by studying their molecular structure
and function.

(a)

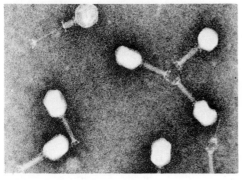

(b)

Figure A.5

The simplicity of viruses can be seen in these electron micrographs of T-2 and T-4 viruses. Viruses consist of an outer shell, with a head containing its genome and a tail for attaching to the bacterium it infects.

(a) E. Kellenberger; (b) Carl Zeiss, Inc.

Genetic evidence indicates that three primary kingdoms represent the fundamental types of life: the Archaea, the Bacteria, and Eucarya (figure A.6). These three groups are greatly but equally different from each other at the molecular level, and their ancestors probably diverged about 3.5 billion years ago. All three groups originated more or less directly from the first type of life on earth, the Progenotes. Progenotes appeared between 3.5–4.0 billion years ago, and it was in the Progenotes

that a simple genetic mechanism and metabolism developed. Perhaps while these complex processes were still in an experimental stage, the three primary kingdoms went their evolutionarily independent ways.

At first, the Archaea were probably the dominant group, as their metabolism is well suited to the conditions of primitive earth. As oxygen became a major component of the atmosphere and as temperatures decreased, they became restricted to their present habitats. The Bacteria probably became increasingly abundant as oxygen levels increased. The oldest known fossils are much like today's blue-green algae and are found in the Fig Tree Series of rocks in Africa (3 billion years old) and from rocks in Western Australia (3.5 billion years old).

Because eukaryotic cells are larger and more complex than prokaryotes, their origin has always been a puzzle. How did they get so large? How did their organelles originate? Recent hypotheses favor a symbiotic origin of the eukaryotes. Perhaps the primitive eukaryote was a slightly larger-than-normal cell capable of ingesting others. If this cell consumed cells that were resistant to digestion, sooner or later the consumed cell might establish a semi-independent existence within its predator. If the two cells survived better together than alone, the ancestor of today's eukaryote could have been formed.

Chloroplasts and mitochondria (chapter 2) are organelles that may have originated in just this way. Each is about the size of a bacterium. Both have their own DNA and apparatus for protein production independent from that of the rest of the cell. Both also reproduce independently from the rest of the cell. Great similarities have been found between chloroplast RNA and the RNA of blue-green algae, a group of the Bacteria. Some eukaryotic genes also appear to have originated from Archaeal genes. Whatever the origins of eukaryotes, they are certainly a very old group with an extremely complex origin. Archaea, Bacteria, and Progenote all may have participated in the formation of the first eukaryote.

The two prokaryotic types each represent a single kingdom, but the eukaryotes include five major lineages. The evolutionary divergences within the eukaryotes are certainly more recent than those between the primary kingdoms. The most ancient eukaryote fossils are protistans about 1.2–1.4 billion years old. Only a few fossils of multicelled organisms are older than 600 million years. Terrestrial life began in earnest about 400 million years ago, the date of the earliest plant fossils. Animals such as amphibians and arthropods soon followed the plants onto the land. Terrestrial habitats have been the site of the greatest diversity of both plant and animal life. The two most abundant terrestrial groups, the flowering plants and the insects, have experienced an immense diversification during the past 100 million years.

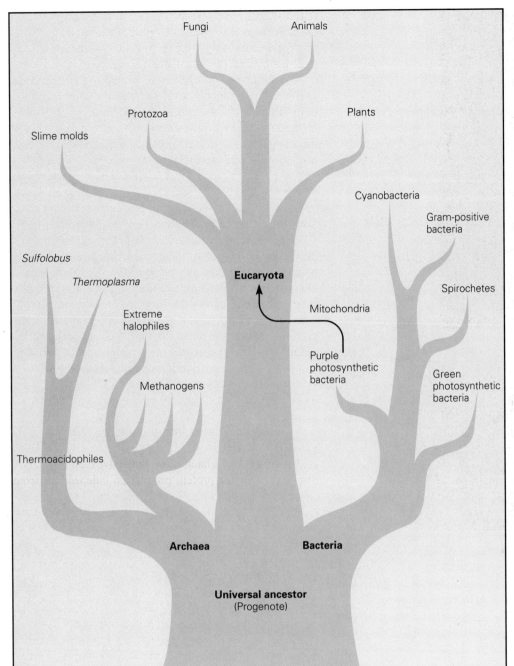

Three primary kingdoms proposed by Carl Woese. The former prokaryotes are grouped in the *Bacteria* and in the newly described *Archaea.* The Eucarya include the plants, animals, fungi, protozoans, and slime molds.

Fungi

Animals

Protozoa

Plants

Slime molds

Cyanobacteria

Gram-positive bacteria

Sulfolobus

Thermoplasma

Eucaryota

Spirochetes

Extreme halophiles

Mitochondria

Methanogens

Purple photosynthetic bacteria

Green photosynthetic bacteria

Thermoacidophiles

Archaea

Bacteria

Universal ancestor
(Progenote)

For More Information

Pool, R. 1990. "Pushing the Envelope of Life." *Science,* 247:158–60.

Whittaker, R. H. 1969. "New Concepts of Kingdoms of Organisms." *Science,* 193: 150–59.

Woese, C. R. 1981. "Archaebacteria." *Scientific American,* 244:98–122.

Woese, C. R., O. Kandler, and M. L. Wheelis. 1990. Towards a Natural System of Organisms: Proposal for the Domains Archaea, Bacteria, and Eucarya. *Proc. Natl. Acad. Sci. U.S.A.* 87:4576–79.

Probability

A major goal of science is *prediction*—the ability to foretell events. In genetics, one important type of prediction is the likely genotype of the offspring of specific matings. In some cases, prediction is easy, and the results are certain. For example, if homozygous round-seeded pea plants are bred (RR × RR) we know that, barring mutation, all offspring will be RR and produce round seeds. This result is certain because each parent has only R alleles to contribute to all offspring.

When heterozygotes are mated, however, several genotypes may occur in the offspring: RR, Rr (or its equivalent, rR), and rr. Some types of predictions cannot be made with certainty regarding the results of an Rr × Rr cross. For example, what will be the genotype of a particular offspring? Or, will there be more RR offspring than rr?

The mathematical science of *probability* has developed a language for expressing predictions regarding uncertain outcomes of particular events. A probability is a measure of the likelihood, the chance, that an event will have a particular outcome. Usually, a probability is expressed as a fraction between, and including, 0 and 1. An outcome with a probability of 1 is certain to occur; that with a probability of 0 is certain not to happen. For example, the probability that any one person will live forever is 0. We know that each of us, sooner or later, will die. The probability that a particular person will die is 1. The sum of all possible outcomes of a single event must always equal 1.

In all other cases, the chance of a particular outcome increases as the probability approaches 1, and decreases as 0 is approached. Thus, an outcome with a probability of 0.95 is highly likely, while that with a probability of 0.02 is very unlikely.

Many events in genetics are analogous to a series of coin-tossing experiments. A fair coin has two sides, heads (H) and tails (T), each of which has an equal chance of appearing on a single toss. Thus, the probability of getting heads is $\frac{1}{2}$, as is that of getting tails. The total probability is $\frac{1}{2} + \frac{1}{2} = 1.0$. In other words, the result of one toss must be either heads or tails. No other outcome can occur, and one of the two must occur. Furthermore, the outcome of a single coin toss is said to be *random,* because although each event has a specific probability, the outcome of the toss cannot be precisely predicted beforehand.

Note that the best prediction or educated guess that can be made regarding the outcome of a single coin toss is that heads or tails is equally likely, each with a probability of $\frac{1}{2}$. Knowledge of this single probability, however, makes possible other less-obvious predictions. For example, if the probability of H is $\frac{1}{2}$ for a single toss, what is the probability of getting heads on each of two consecutive tosses (H,H)? Because each toss is *independent* of (does not affect) the other, the latter probability is simply the product of the two individual probabilities: $\frac{1}{2} \times \frac{1}{2} = \frac{1}{4}$. The complete list of all possible outcomes shows that four equally likely outcomes may occur.

$$H,H = \tfrac{1}{2} \times \tfrac{1}{2} = \tfrac{1}{4}$$
$$H,T = \tfrac{1}{2} \times \tfrac{1}{2} = \tfrac{1}{4}$$
$$T,H = \tfrac{1}{2} \times \tfrac{1}{2} = \tfrac{1}{4}$$
$$T,T = \tfrac{1}{2} \times \tfrac{1}{2} = \tfrac{1}{4}$$
$$\text{Total} = 1.0$$

For three tosses, a similar list can be made, showing eight possible outcomes, each of equal probability.

$$H,H,H = \tfrac{1}{2} \times \tfrac{1}{2} \times \tfrac{1}{2} = \tfrac{1}{8}$$
$$H,H,T = \tfrac{1}{2} \times \tfrac{1}{2} \times \tfrac{1}{2} = \tfrac{1}{8}$$
$$H,T,H = \tfrac{1}{2} \times \tfrac{1}{2} \times \tfrac{1}{2} = \tfrac{1}{8}$$
$$H,T,T = \tfrac{1}{2} \times \tfrac{1}{2} \times \tfrac{1}{2} = \tfrac{1}{8}$$
$$T,T,H = \tfrac{1}{2} \times \tfrac{1}{2} \times \tfrac{1}{2} = \tfrac{1}{8}$$
$$T,H,T = \tfrac{1}{2} \times \tfrac{1}{2} \times \tfrac{1}{2} = \tfrac{1}{8}$$
$$T,H,H = \tfrac{1}{2} \times \tfrac{1}{2} \times \tfrac{1}{2} = \tfrac{1}{8}$$
$$T,T,T = \tfrac{1}{2} \times \tfrac{1}{2} \times \tfrac{1}{2} = \tfrac{1}{8}$$
$$\text{Total} = 1.0$$

Note that these sets of probabilities can be used to answer other questions. For example, considering two tosses, regardless of the order, what is the probability of obtaining one H and one T? This can occur from either result H,T or T,H. The sum of these two probabilities is $\frac{1}{4} + \frac{1}{4} = \frac{1}{2}$. In other words one-half the time a fair coin can be expected to produce one H and one T. For three tosses, the probability of obtaining at least one H and one T is $\frac{3}{4}$.

An offspring's genotype at one locus is analogous to a coin-toss experiment. An offspring's maternally derived allele may be regarded as the result of the toss of a coin, whose two possible outcomes represent the alleles possessed by the mother. The other allele represents a coin toss to choose among the two alleles of the father.

A homozygous parent is analogous to tossing a coin with two heads; only one result is possible. On the other hand, the meiotic process insures that heterozygous individuals produce gametes of two types (R and r, for example) in equal proportions. Precisely which gamete of the mother unites with which gamete of the father is random, or totally unpredictable.

Thus, if we wish to predict the genotype of a single offspring of monohybrid cross (Rr $\times$ Rr), we may apply the same logic as above. The probability of tossing an R for the first allele is $\frac{1}{2}$; of an r is also $\frac{1}{2}$, and so on. Four possible genotypes may occur: RR, Rr, rR, and rr. However, there is no order to the toss of the genetic coin, so Rr and rR are phenotypically and genotypically identical. Note that 50% of the time the result will be a heterozygote, just as 50% of all two-toss experiments produce one H and one T.

Two rules of probability need to be remembered: the rule of addition and the rule of multiplication. If we wish to calculate the probability of a particular event happening and there is *more than* one way in which it can occur, we *add* the independent probabilities. Thus the probability of obtaining a heterozygous offspring in the above monohybrid cross is the *sum* of the two probabilities, $\frac{1}{4}$ for Rr and $\frac{1}{4}$ for rR $= \frac{1}{2}$.

If, on the other hand, we wish to calculate the probability of two (or more) independent events occurring at the same time (or in succession), we then must *multiply* the probabilities of the independent events. The probability of the above monohybrid cross producing two offspring, one of which is RR and the other of which is rr, becomes $\frac{1}{4} \times \frac{1}{4} = \frac{1}{16}$.

A good rule of thumb in calculating probabilities of genetic events is first to calculate the probability of each independent event. Then state the question to determine whether you are after the probability of event X *or* event Y (add), or whether you are after the probability of event X *and* event Y (multiply).

For example, consider a family with six children. What is the probability that all six children will be girls? The probability of any child being a girl is $\frac{1}{2}$. Our question is concerned with the chance that child number one is a girl, *and* number two is a girl, *and* number three is a girl, and so on. Thus we must multiply $\frac{1}{2} \times \frac{1}{2} \times \frac{1}{2} \times \frac{1}{2} \times \frac{1}{2} \times \frac{1}{2} = \frac{1}{64}$. The probability that the six children will be all boys or all girls becomes the chance that all six are girls $(\frac{1}{64})$ *plus* the chance that all six are boys $(\frac{1}{64})$, or $\frac{1}{32}$. Table B.1 depicts a generalized way to calculate probabilities using the binomial expansion $(p + q)^n$ where p is the probability of one event, q is the probability of an alternative event, and n is the number of events considered.

Expanding the binomial

$(p + q)^n$

p and q represent the probabilities of occurrence of two alternative events (p = boys, q = girls) and n is the size of the family (group) involved.

$$4! = 4 \text{ factorial} = 4 \times 3 \times 2 \times 1 = 24$$

The probability of a particular combination can be calculated from the following formula:

$$(p + q)^n = \frac{n!}{s!t!} p^s q^t$$

Where n = the number in the group
Where s = the number in one class (i.e., boys)
Where t = the number in the other class (i.e., girls)

Example

If six babies are born in a given hospital on the same day, what is the probability that two will be boys and four will be girls?

$$p = \tfrac{1}{2}, q = \tfrac{1}{2}, n = 6, s = 2, t = 4$$

$$\frac{6!}{2!4!} p^2 q^4 = 15(\tfrac{1}{2})^2(\tfrac{1}{2})^4 = \frac{15}{64}$$

Example

If a couple has three children, what is the probability that two are girls?

$$p = \text{probability of a boy} = \tfrac{1}{2}$$

$$q = \text{probability of a girl} = \tfrac{1}{2}$$
$$n = 3, s = 1, t = 2$$
$$\frac{3!}{1!2!} pq^2 = 3(\tfrac{1}{2})(\tfrac{1}{2})^2 = \tfrac{3}{8}$$

Order of arrival: Different ways in which children may be born in a family of three.

Three boys
$$p\ p\ p = \tfrac{1}{8} = p^3$$

Two boys and one girl
$$\begin{array}{l} p\ p\ q = \tfrac{1}{8} \\ p\ q\ p = \tfrac{1}{8} \\ q\ p\ p = \tfrac{1}{8} \end{array} \quad = \tfrac{3}{8} = 3p^2q$$

Two girls and one boy
$$\begin{array}{l} q\ q\ p = \tfrac{1}{8} \\ q\ p\ q = \tfrac{1}{8} \\ p\ q\ q = \tfrac{1}{8} \end{array} \quad = \tfrac{3}{8} = 3pq^2$$

Three girls
$$q\ q\ q = \tfrac{1}{8} = q^3$$

$$(p + q)^3 = p^3 + 3p^2q + 3pq^2 + q^3$$

ABO blood group
A well-known blood group of humans, characterized by the presence or absence of A and B antigens on the surface of red blood cells.

absolutism
The belief that behavior should always follow a set of basic moral rules, regardless of the consequences of doing so.

acentric
A fragment of a chromosome lacking a centromere.

achondroplasia
A type of dwarfism caused by an autosomal dominant allele that produces abnormal growth in the long bones.

acquired immune deficiency syndrome (AIDS)
An infectious viral disease that destroys the immune system; first identified in 1981.

acquired immunity
Specific defense or resistance to a particular infective agent, which develops only after exposure to that agent.

acrocentric chromosome
A chromosome whose centromere is located very near one end.

active immunity
The production of antibodies by an individual's own immune system.

acute dose
A single massive dose of radiation.

AD.
See **Alzheimer's disease (AD)**

agammaglobulinemia
A sex-linked recessive condition in which individuals do not produce B cells, and therefore cannot produce antibodies.

agglutination
The clumping together of red blood cells in serum.

AIDS.
See **acquired immune deficiency syndrome (AIDS)**

albinism
Absence of normal pigmentation in skin, hair, and eyes.

allele
An alternate form of a gene; one of the differing sequences of nucleotide bases that occupy a specific genetic locus on a chromosome.

allergy
Immunological hypersensitivity to certain foreign antigens.

Alzheimer's disease (AD)
A form of senile dementia.

Ames test
A laboratory procedure that directly measures the mutagenicity of a chemical; because of the high correlation of mutagenesis and carcinogenesis, the Ames test is used as an indirect test for carcinogenicity.

amino acid
Small biological molecule; twenty common ones used in protein synthesis.

amniocentesis
A technique for sampling a small quantity of fetal cells from the amniotic fluid surrounding a fetus for tests.

amplification (gene amplification)
The production within the genome of many copies of a single gene.

anabolic steroids
A group of synthetic sex hormones whose use may produce increased growth of muscles, and which may also have unwanted masculinizing or feminizing side-effects, and even lead to serious disease and death.

anaphase
The third phase of mitosis, when sister chromatids separate forming two daughter groups of chromosomes.

anaphase I
The third phase of meiosis I, in which duplicated homologous chromosomes move towards opposite poles of the cell.

anaphase II
The third phase of meiosis II, in which sister chromatids move towards opposite poles of the cell.

aneuploid
A cell or individual varying from the normal diploid number; aneuploid individuals have one or more chromosomes too many or too few.

antibody
A large protein composed of four polypeptide chains produced by the body in response to exposure to a foreign antigen; antibodies protect the body from invading bacteria and other foreign substances.

anticodon
The sequence of three nucleotide bases in a tRNA molecule that is complementary to a codon of an mRNA molecule.

antigen
A molecule, often a protein, or a microorganism that elicits the production of antibodies in an individual.

antigenic determinant
The portion of an antigen that is recognized by the body's immune mechanisms and, thus, to which the immune mechanisms respond.

antigenic drift
A slight alteration in the antigenic determinants of influenza viruses.

antigenic shift
A major alteration in the antigenic determinants of influenza viruses.

anti-oncogenes
Genes whose products limit cell proliferation and control normal cell growth and differentiation; also called tumor suppressor genes. Loss or mutation of anti-oncogenes may lead to cancer.

Archaea (Archaebacteria)
One of the two major kingdoms of prokaryotes; a very unusual type of bacteria, including methanogens, halophiles, and thermoacidophiles.

artificial insemination
The collection (by agriculturalist, physician, or scientist) of male semen that after short-term or long-term storage, is placed in the reproductive tract of a female; used in animal husbandry and medicine.

artificial selection
The controlled breeding of domestic plants and animals so that desirable or valuable traits increase within a population or species.

asexual reproduction
Any form of reproduction such as fission, budding, or parthenogenesis, in which one parent produces offspring genetically identical to itself.

Australopithecus
A genus in the family Hominidae; a possible ancestor of the genus *Homo*.

autoantibody
An antibody produced to one's own tissues.

autoimmune disease
A disease in which the body produces antibodies against its own tissues.

autosomal chromosomes
All of the chromosomes in eukaryotes other than sex chromosomes.

auxin
A growth hormone in plants.

b

Bacteria (true bacteria)
One of two major kingdoms of prokaryotes; includes most common bacteria and blue-green algae.

bacteriophage
A virus that parasitizes bacteria.

Barr body
A darkly staining structure visible in the nucleus of cells of female mammals; an inactivated X chromosome; in individuals with more than one X chromosome all but one are inactivated, so an individual may have several Barr bodies.

base analogue
A chemical whose structure resembles that of a nucleotide base; sometimes base analogues are incorporated in a DNA molecule causing a mutation.

B cell
A type of lymphocyte that produces antibodies.

bioengineering
The ability to directly alter the genetic material of an organism; also *see* **biotechnology.**

bioethics
The study of moral decisions arising from new biological knowledge and technologies.

biological weapons
Microorganisms or toxins derived from living cells that are used to incapacitate, injure, or kill humans, or to damage or destroy agricultural crops and animals.

biotechnology
The industrial use of biological systems and processes.

bivalent
A synapsed pair of homologous chromosomes in meiosis I; each homologue consists of two sister chromatids. Thus, a bivalent consists of four chromatids.

blood group.
See **blood type (blood group)**

blood type (blood group)
A phenotypic trait defined by the type of antigens on the surface of red blood cells.

bovine growth hormone
A hormone produced in the pituitary of a cow that plays a role in controlling growth and milk production.

Burkitt's lymphoma
A cancer of the lymphatic tissue, associated with Epstein–Barr virus.

c

cancer
The uncontrolled growth of cells; cancer cells usually show no differentiation and invade tissues and metastasize to other parts of the body; untreated cancers are fatal.

carcinogenic
A cancer-causing substance.

carcinoma
One of the many types of cancer primarily affecting skin, glands, gastrointestinal and reproductive tracts, and urinary system.

cardiovascular system
The larger of the two circulatory systems of the body; including the heart, arteries, veins, and capillaries; providing circulation of the blood.

carrier
An individual heterozygous for a deleterious recessive allele.

C band
A darkened region on or near the centromere of a chromosome in a cell treated with stain.

cDNA
A copy of DNA made from RNA, using the enzyme reverse transcriptase; used extensively in recombinant DNA techniques.

cell
The basic structural and functional unit of life; consisting of cytoplasm, containing organelles and required molecules enclosed by a membrane.

cell cycle (mitotic cell cycle)
The complete life history of a single cell, from its origin by mitosis to its own subsequent mitotic division; divided into two major segments—interphase and mitosis.

cell theory
The concept that all organisms are composed of cells, and that all living cells arise from other living cells.

cellular immunity
A type of specific immunity resulting from the action of T cells, which directly attack foreign antigens.

central dogma
The concept describing the flow of hereditary information; within one generation information flows from DNA to RNA to protein, to produce the ultimate phenotype; between generations, DNA replicates itself precisely.

centriole
An organelle appearing to serve as an organizing structure for the formation and movement of the spindle during cell division in animal cells.

centromere
The primary constriction on a chromosome to which spindle fibers attach during cell division.

cerumen
Earwax; the yellow waxy secretion from the glands of the external ear.

Chargaff's rules
Within DNA, the amount of adenine equals the amount of thymine, and the amount of cytosine equals the amount of guanine.

chiasma
An X-shaped structure formed between chromatids of homologous chromosomes during synapsis; chiasmata are the mechanisms for recombination.

chimera
An individual consisting of two or more distinct cell lines; may result from the fusion, early in development, of two or more different zygotes.

chloroplast
A chlorophyll-containing organelle of plant cells and some protozoan cells; contains its own DNA; conducts photosynthesis.

chorion villus sampling
A rapid and painless way of sampling fetal tissue for detection of genetic disease during the first trimester of pregnancy. This methodology samples fetal tissue directly and results can be obtained with twenty-four to forty-eight hours of sampling.

chromatid
One of the two strands that comprise a single chromosome in the early stages of mitosis and meiosis; each chromatid, attached to its sister chromatid by a centromere, forms one-half of the chromosome. *See also* **sister chromatid exchange (SCE)**

chromatin
Chromosomes during interphase, the most common configuration of nuclear DNA, combined with histones and non-histone nuclear protein.

chromosomal *in situ* hybridization
The application of a DNA probe to intact chromosomes; the fluorescent or radioactive probe binds to a specific site on a chromosome, revealing the chromosomal location of a specific gene.

chromosome
In eukaryotes, one of many (usually ten to one hundred) rodlike structures, usually in pairs, formed by complexes of DNA, histones, and non-histone nuclear proteins.

chromosome puffs
Expanded regions on a polytene chromosome.

chromosome satellite
A small segment at the end of a chromosome separated by a secondary constriction.

chromosome set
Consists of one member of each chromosome pair; in humans there are twenty-three chromosomes in a set.

chronic dose
A series of small doses of radiation; may cause equal damage as an acute dose of radiation.

clonal selection
The selective growth of only those T cells specific to a single antigenic determinant on an invading organism or foreign substance; the result is efficient production a large clone of T cells that protects the body from the invader.

clone
A group of genetically identical genes, cells, or individuals; and, used as a verb, to produce a set of genetically identical genes, cells, or individuals.

cloning vehicle.
See **vector**

codominance
The expression of both alleles at a locus in an heterozygote; a type of allelic interaction at a locus.

codon
A sequence of three nucleotide bases in the sense strand of a DNA molecule that specifies which amino acid is to be added

to a polypeptide chain; also, the sequence of three nucleotide bases in a messenger RNA molecule that has been transcribed from and is complementary to the DNA codon; a few codons do not specify an amino acid, but signal the initiation or termination of protein synthesis.

coefficient of relationship (r)
The proportion of alleles in two individuals that are identical by descent; a measure of relationship.

colchicine
A chemical that induces polyploidy in a dividing cell; used for treatment of gout in humans.

colinear
The precise, one-to-one correspondence between the sequence of nucleotide triplets in a DNA molecule and the sequence of amino acids in the protein encoded by the DNA.

color blindness
The inability to distinguish certain colors.

colostrum
A secretion of a mother's mammary glands shortly after birth; colostrum is rich in antibodies and provides a nursing infant with short-term, passive specific immunity.

complementary (complementary base pairing)
The precise pattern of bonding of nitrogenous bases in opposing strands of nucleic acids; in DNA, adenine always pairs with thymine, and cytosine with guanine; when RNA is transcribed from DNA, uracil (rather than thymine) in the growing RNA chain pairs with adenine on the opposing DNA.

complete dominance.
See **dominance (complete dominance)**

consanguineous
"Of the same blood"; sharing genes derived from a common ancestor; related by descent.

contact inhibition
A phenomenon exhibited by cells grown in culture, limiting their growth to formation of a single layer within a culture dish; growth stops when a barrier, such as the wall of the dish or another cell, is encountered.

cri du chat syndrome
"Cry of the cat" syndrome; a serious genetic disease caused by a deletion in the short arm of chromosome 5.

crisscross inheritance
The pattern of transmission shown by X-linked alleles in a pedigree; an X-linked allele is transmitted from male-to-female, then female-to-male, in succeeding generations.

crossing-over (crossover)
The exchange of segments between homologous chromosomes during meiosis; usually is "reciprocal," involving the same set of homologous loci on each chromosome, but need not be; produces chromosomes with allelic combinations not found on either parental chromosome.

cryptic species
Genetically distinct species that are phenotypically identical.

cyclin
A cellular protein of eukaryotes whose concentration increases throughout the cell cycle and when in high concentration, activates the protein MPF.

cystic fibrosis
A disease caused by an autosomal recessive allele; most common genetic disease of Caucasians.

cytoplasm
The viscous fluid within a cell; in eukaryotes, cytoplasm is external to the nucleus and contains many organelles and molecules.

d

degenerate
With reference to the genetic code, redundant; the specification of some amino acids by more than one codon.

delayed onset
The absence of expression of a particular gene until later in life.

deletion
Loss of a segment of a gene or a chromosome.

deoxyribonucleic acid (DNA)
A long molecule comprised of two chains of nucleotides containing the sugar deoxyribose; the hereditary material in all organisms except some viruses.

deoxyribose
A five-carbon sugar; a chemical constituent of DNA.

development
The process by which a single unspecialized cell becomes a highly complex organism of many specialized cells; the result of growth and differentiation.

developmental cascade
A developmental sequence where completion of one stage activates genes that lead to initiation of the next stage, leading to a regular and predictable progression to the final phenotype.

diabetes
A disease resulting from an inability to metabolize carbohydrates properly, often due to a deficiency in insulin production.

differentiation
The structural and functional specialization of a cell or tissue.

dihybrid cross
Mating between individuals homozygous for alternate alleles at two loci.

dioxin
An extremely toxic mutagen and carcinogen that occurs as a contaminant of some herbicides and other chemicals.

diploid
Having two sets of chromosomes; one set derived from the mother, the other set from the father.

diploid number
The total number of chromosomes found in species in which chromosomes occur in pairs; often referred to as 2N, where N refers to the number of pairs of chromosomes; usually specific and constant for an eukaryotic species.

dispermy
The fertilization of one egg by two sperm.

dizygotic twins
Twins derived from the separate fertilization of two different eggs; fraternal twins; dizygotic twins share one half of all alleles, just as do normal siblings.

DNA.
See **deoxyribonucleic acid (DNA)**

DNA fingerprinting
RFLP analysis of highly variable segments of DNA, usually identifying unique patterns of individual variation.

DNA library
An extensive series of bacterial cultures that have incorporated different pieces of DNA for the purposes of gene cloning.

DNA sequencing
Determination of a sequence of bases in a strand of DNA.

dominance (complete dominance)
With reference to the interaction of two alleles at a locus, the expression by a heterozygote of a phenotype identical to that of one of its alleles (the dominant allele) when homozygous; the phenotype exhibited by the other allele (the recessive allele) when homozygous is suppressed.

dominant
Refers to an allele or trait that is expressed in a heterozygote.

dosage compensation
An adjustment (or compensation) in the number (dosage) of active alleles or chromosomes, usually by turning one chromosome off; usually refers to X inactivation.

double helix
The shape of the DNA molecule; often used as a synonym for DNA. *See also* **deoxyribonucleic acid (DNA)**

Down's syndrome
A set of phenotypic anomalies caused by trisomy of chromosome 21 in humans.

Drosophila melanogaster
The common fruit fly, used in many genetic experiments.

Duchenne muscular dystrophy.
See **muscular dystrophy (Duchenne muscular dystrophy)**

duplication
A type of mutation in which a nucleotide sequence that repeats (duplicates) a nucleotide sequence already present is added to a chromosome or genome.

dysgenic
Anything that acts to increase the frequency of deleterious genes in a population.

e

EBV.
See **Epstein-Barr virus (EBV)**

E. coli.
See ***Escherichia coli (E. coli)***

electrophoresis
A biochemical technique that uses differences in electrical charge, size, and shape to separate a mixture of proteins or nucleic acid fragments.

embryo
An early developmental stage of a plant or an animal; descendant of a zygote; in humans, usually refers to period between weeks two and eight of gestation.

embryo transfer
The removal of a developing embryo from its genetic mother and its placement for the remainder of gestation in a second female, a surrogate mother.

endoplasmic reticulum
A complex network of membranes within the cytoplasm of eukaryotes, on which are found ribosomes.

environment
The total of all conditions that impinge upon an individual cell or genome; genes and the environment interact to produce the phenotype of an individual; genes and gene products may comprise a genetic environment for other genes.

enzyme
A protein that catalyzes (increases the rate of) biochemical reactions.

epididymis
A long, convoluted tubule on the surface of each testis; the site of sperm storage and maturation.

epistasis
Interaction of alleles at different loci; the control or masking of the expression of alleles at one locus by alleles at a second locus.

Epstein–Barr virus (EBV)
A relatively common DNA virus producing cold-like symptoms, but also associated with Burkitt's lymphoma and infectious mononucleosis; one of the herpes viruses.

erythroblastosis fetalis
A type of hemolytic anemia in a fetus or newborn infant, resulting from Rh incompatibility of mother and fetus; potentially life-threatening; anti-Rh antibodies produced by an Rh-negative mother destroy red blood cells of an Rh-positive fetus; most serious for second and subsequent Rh-positive children.

Escherichia coli (E. coli)
The common gut bacterium of humans and other mammals, widely used in genetic experimentation.

estrogen
One of several types of female sex hormones.

ethics
The study of the standards of morality and human conduct.

Eucarya
The third kingdom of life, including all of the nucleated organisms.

euchromatin
Lightly staining chromatin; believed to be relatively uncoiled DNA and capable of transcription and other genetic activity.

eugenics
The study or practice of improvement in the genetic quality of a population.

eukaryotic cell (eukaryote)
One of two basic types of cells; about 1,000 times larger in volume than prokaryotic cells; well-defined nucleus with rod-like chromosomes; many organelles visible within extra-nuclear cytoplasm.

euphenics
The manipulation of the phenotype in an attempt to nullify an undesirable genetic trait.

evolution
Change with time; the biological concept that all life on earth changes through time; continuing genetic change may result in one species giving rise to new species.

exons
Those portions of the sense strand of DNA that are the template for protein synthesis; exons are interspersed with introns, both of which may be transcribed, but only exon sequences are translated.

expressivity
The extent to which an individual demonstrates particular phenotypic effects of a gene; a quantitative measure of expression of a gene.

f

F_1
The designation for progeny produced by the first mating in a pedigree; subsequent generations are labeled F_2, F_3, etc.; F refers to filial generation.

fallopian tube
The duct through which an ovum passes from the ovary to the uterus; oviduct.

fetal alcohol syndrome (FAS)
A set of teratologies produced in a fetus by maternal alcohol consumption during gestation.

fetal-cell surgery
The implantation of undifferentiated cells from an early fetus into a defective tissue or organ in another person; for example, into the pancreas of a person suffering from diabetes. The fetal cells do not induce an immune response in the recipient and may later differentiate into functional cells that produce a needed product—insulin in the case of diabetes—that cures the disease.

fetoscopy
A procedure for the visualizing of fetal blood vessels that permits removal of blood directly from a fetus.

fetus
The developing offspring of live-bearing animals; in humans, refers to developing individual from about age eight weeks until birth.

fixation
Where only one allele occurs at a particular locus in a population; that allele is said to be fixed, and all individuals are homozygous for it.

forensic
Relating to legal matters.

fragile site
A heritable constriction or gap in a chromosome of cells grown in highly specific culture conditions. These sites appear to be prone to breakage. About 40 different fragile sites are known in the human karyotype.

fragile X syndrome
An hereditary condition of variable phenotypic expression associated with a constriction (fragile site) near the tip of the X chromosome in mitotic white blood cells grown in culture. Fragile X syndrome is a major cause of mental retardation.

frameshift
A type of genetic mutation in which one or two bases are added or deleted to the nucleotide sequence, thus shifting the reading frame and changing all subsequent codons.

fraternal twins.
See **dizygotic twins**

g

G_0, G_1, G_2 phases
Three periods that occur within interphase of the cell cycle; G_0 is a quiescent period of variable duration; G_1 and G_2 are periods when the cell may be metabolically preparing for growth, but is still relatively quiescent. "G" is derived from the word "gap" between periods of growth.

gamete
A reproductive cell produced in the gonads; either spermatozoa or ova; gametes are usually haploid.

gametogenesis
Gamete formation; includes meiosis and the continued development of the haploid daughter products of meiosis to produce a mature gamete.

G band
A transverse band on a chromosome in a cell treated with Giemsa stain.

gel electrophoresis
A technique for separating molecules (often proteins and nucleic acid) on the basis of differences in their mobilities in an electric field.

gene
A specific location on a chromosome that encodes hereditary information in which sense it is synonymous with the terms *locus* or *genetic locus;* also used to refer to a specific molecular variant occupying a genetic locus, in which sense it is synonymous to *allele.*

gene amplification.
See **amplification (gene amplification)**

gene mutation (point mutation)
A type of mutation affecting only one or a few nucleotide bases; not microscopically visible.

gene pool
All the genes that occur within one species; all the genes possessed by all members of a species.

gene splicing
A phrase used to describe manipulation of the genome by insertion and/or deletion of DNA into or from a gene.

gene therapy
Replacement of disease-causing genes or nucleotide sequences with normal or non-pathological sequences; or inserting additional genetic material that can provide a normal phenotype.

gene transfer
Moving a gene from one species to another, often a virus or a plasmid is used to effect the transfer.

genetic bottleneck
A situation where only a few individuals establish a new population, thus representing only a small sample of the total genetic diversity of the species.

genetically significant dose (GSD)
The amount of radiation reaching the gonads.

genetic code
The triplet sequences of DNA bases that encode information for protein synthesis.

genetic counseling
Guidance provided by a physician or geneticist to a family or individual regarding genetic aspects of pathological conditions; genetic counselors may advise parents concerning the risks of producing genetically defective children or the treatment of individuals with genetic defects.

genetic distance
A measure of the average number of allelic differences per locus between populations; distances range from zero to infinity.

genetic engineering
Deliberate alterations of individual genotypes; any technique that creates new combinations of genetic material or that alters expression of preexisting combinations; recombinant DNA techniques.

genetic load
A measure of the proportion of deleterious alleles within a population.

genetics
The study of the structure and function of genes and their transmission between generations; the study of inheritance and sources of organismal variation.

genome
All the genes possessed by a cell or individual.

genotype
The genes possessed by an individual; may refer either to a pair of alleles at a specific locus, or to the entire genome.

germplasm bank
Repository for genetic material, cells, or tissues, where they are placed in a dormant but living state for long-term preservation and later use.

gestational surrogacy
A process used when a woman is unable to carry a fetus, but has functional ovaries. An egg is removed at the time of ovulation, subjected to *in vitro fertilization,* and the fertilized egg then implanted in a second woman for gestation.

gonad
Gamete-forming organ; testis in male and ovary in female.

grafting
Agricultural process of fusing cuttings from two (or more) different plants to grow as a single new (hybrid) individual.

Great Apes
Primates of the family Pongidae; the closest evolutionary relatives of humans; includes gorillas, chimpanzees, and orangutans.

Green Revolution
A series of international programs begun during the 1960s to increase agricultural productivity throughout the world, but especially in underdeveloped countries.

growth
With respect to a living organism, increase in size due to mitotic division of cells.

GSD.
See **genetically significant dose (GSD)**

guevodoces
Individuals from a village in the Dominican Republic born with a 46, XY karyotype, but as children, they possess ambiguous genitalia; at puberty, they undergo masculinization of the external secondary sex structures, becoming phenotypically normal males.

h

haplodiploid
A system of sex determination in bees, wasps, and termites, in which males are haploid and females are diploid.

haploid
Having only one chromosome of each homologous pair (a haploid set).

haploid number
The number of pairs of chromosomes in a cell; one-half the diploid number; often designated N; usually constant for a eukaryotic species.

Hardy–Weinberg equilibrium equations
A set of mathematical equations describing the relationship between allele and genotype frequencies in successive generations.

Hayflick Limit
The number of times normal cells will undergo mitosis in cell culture.

helper T cells
A type of T cell that participates with B cells in the immune response.

hemizygous
Genes or chromosomes that occur without a homologous partner; most commonly, refers to genes on the X and Y chromosomes of males.

hemoglobin
A protein consisting of four iron-bearing groups (heme) and four polypeptide chains (globin); found in vertebrate red blood cells and transports oxygen from the lungs to all body tissues.

hemophilia
A genetic disease characterized by the inability to coagulate blood; bleeder's disease; two forms are controlled by X-linked loci, one by an autosomal locus.

hepatitis B virus
The DNA virus causing hepatitis B.

heritability
The proportion of phenotypic variation within a population attributable to genetic rather than environmental variation.

hermaphrodite
An individual possessing both male and female gonads, testes and ovaries (true hermaphroditism); a pseudohermaphrodite possesses some secondary traits of both sexes, but gonadal tissue of only one sex (or none whatever).

herpes virus
One of many types of DNA viruses that infect humans and other vertebrates.

heterochromatin
Darkly staining chromatin; probably densely coiled and incapable of transcription.

heterogametic sex
The sex with only one sex chromosome, or two non-homologous sex chromosomes, which consequently produces two different types of gametes; in humans, the XY male is the heterogametic sex; XO males in some insects are also heterogametic.

heterogeneous nuclear RNA (HnRNA)
The initially transcribed product of the sense strand of DNA, found only in the nucleus; mRNA is produced by the excision of many segments of HnRNA.

heterozygous
Having two different alleles at a specific genetic locus.

hibakusha
A Japanese word meaning explosion-affected persons; refers to survivors of the atomic bomb explosions in Japan at the end of World War II (August 1945), who have not married or reproduced.

highly repetitive sequences
DNA sequences that have more than 100,000 copies in a genome.

histamine
A chemical released by mast cells that helps produce the allergic response.

histone
One of five types of nuclear proteins; they are involved in DNA coiling in chromosomes and regulation of gene activity.

HLA (HLA antigens)
Human leukocyte antigens; a set of antigens on the surface of white blood cells.

HnRNA.
See **heterogeneous nuclear RNA (Hn RNA)**

holandric
Y linked; a gene occurring on the Y chromosome and thus transmitted from father to each son.

homogametic sex
The sex having two identical sex chromosomes; in humans, the XX female is the homogametic sex.

homologous
Chromosomes that are similar in appearance and structure, possessing the same sequence of genetic loci; or, structures in different organisms that share a common evolutionary origin, such as the wing of a bat and the arm of a human.

homozygous
Having two identical alleles at a specific genetic locus.

hormone
A molecule that transmits information between cells; a chemical produced by one cell to regulate function of another cell.

HTLV.
See **human T-cell leukemia virus (HTLV)**

Human Genome Project
A project of the United States government, directed by James Watson, to produce maps and a sequence for the human genome by the year 2005.

Human Genome Organization (HUGO)
An organization directed by James Watson whose purpose is to coordinate international programs for mapping and sequencing the human genome.

human growth hormone
A pituitary hormone that plays a role in controlling growth and milk production.

human immunodeficiency virus (HIV)
The retrovirus associated with AIDS.

human papilloma virus
A DNA virus that can cause cervical and genital cancers.

human T-cell leukemia virus (HTLV)
A class of retroviruses that includes the causative agents of leukemia and AIDS.

human leukocyte antigen
See HLA (HLA antigens)

humoral immunity
A type of specific immunity resulting from the action of B cells, which produce antibodies, which in turn destroy foreign antigens.

Huntington's disease
A dominant genetic disease with delayed onset; the average age of onset of symptoms being twenty-five to forty-five years. The disease is characterized by progressive loss of motor control and mental deterioration. There is no known treatment or cure.

H-Y antigen
An antigenic protein found on all cells of males, but not females; the product of one or more Y-linked loci; essential for normal spermatogenesis.

hybrid
A cell or individual resulting from the mating of genetically different forms; a hybrid may be the result of a cross of genetically different individuals within a species, or the progeny of a cross of individuals of different species.

hybrid vigor
An increase in desirable phenotypic traits such as size, strength, and physiological health, which often occurs in hybrid individuals.

hypervariable minisatellite DNA
A highly variable segment of DNA in eukaryotes that can be used for identification of individuals by DNA fingerprinting.

i

identical by descent
Referring to two genes that are identical because they are derived from a recent common ancestor.

identical twins.
See monozygotic twins

immune response
The combined activity of T cells, B cells, and macrophages, in destroying or neutralizing the pathological effects of foreign antigens.

immunoglobulin
Antibody.

immunoglobulin E (IgE)
A type of antibody involved in the allergic response.

immunoglobulin G (IgG)
The most common type of antibody, providing protection against bacteria, viruses, and toxins.

immunologically privileged site
One of several structures or locations in the body, including the corneas, bone, cartilage, heart valves, and testes, which are isolated from the circulating T cells; these sites are thus not subject to normal immune surveillance and defense.

imprinting
A mechanism for marking a gene in order to regulate its expression in future generations—often accomplished by methylation of the DNA.

inborn error of metabolism
An inability to produce an enzyme needed for a particular metabolic pathway; many genetic diseases are due to inborn errors.

inbreeding
The mating of related individuals; consanguineous mating.

inbreeding depression
The reduction in phenotypic vigor, survival, and reproduction resulting from consanguineous mating.

incest taboo
Outlawing of consanguineous matings in tacit recognition of the harmful effects of inbreeding.

incomplete dominance (partial dominance)
With reference to two alleles at a locus, where a heterozygote exhibits a phenotype intermediate to the two homozygous types.

independent assortment.
See Mendel's Law of Independent Assortment

index case
An individual whose phenotype is the cause for the study of his or her pedigree.

inducible
With reference to an operon, a type of control in which protein (enzyme) synthesis occurs only when the substrate upon which those enzymes are active is present.

influenza
Viral disease caused by one of three different classes of RNA viruses; also called flu.

informational macromolecules
Proteins and nucleic acids.

insulin
A hormone produced in the pancreas that is essential for normal carbohydrate metabolism; in its absence, diabetes results.

intelligence quotient (IQ)
A standardized measure of some aspects of mental capacity, probably related, but in an unknown way, to intelligence.

interferon
A naturally occurring protein that has antiviral properties.

interleukins
Chemicals that stimulate the growth and division of B and T cells.

interphase
Usually the longest portion of the cell cycle, divided into the G_1 (Gap 1), S (synthesis), and G_2 (Gap 2) phases; a time of vegetative growth for most cells; DNA replication occurs during interphase; also refers to the brief pause between meiosis I and meiosis II.

introns
Segments of DNA interspersed between exons that do not contain information to be translated; introns are apparently transcribed to HnRNA, but the intron-coded portions are excised from HnRNA to produce mRNA.

inversion
A type of chromosomal mutation in which a nucleotide sequence is removed and then reinserted in reverse sequence to which it normally occurs.

in vitro
"In glass"; refers to experiments conducted in the laboratory, often in a flask or culture dish.

in vitro fertilization
Fertilization of an ovum in the laboratory rather than within the mother's body; a child produced by in vitro fertilization is misleadingly called "test tube baby" by the public news media. See also test tube baby

IQ.
See **intelligence quotient (IQ)**

k

karyotype
The total chromosomal complement of an individual; or, a formal arrangement of photographed chromosomes arrested at metaphase of mitosis from a single cell.

killer (cytotoxic) T cell
A type of T cell that kills invading cells by secreting toxic chemicals that kill such invaders.

Klinefelter's syndrome
A set of phenotypic anomalies associated with males having an extra X chromosome; the genotype is designated 47,XXY.

l

lac operon
The functional and structural unit of DNA of *E. coli* that controls the synthesis of lactose-metabolizing enzymes.

lampbrush chromosome
A type of chromosome found in oocytes of some species that has many loops of DNA extending out from the main axis.

latent period
The time interval when latent viruses are inactive and remain sequestered in the nervous system; characteristic of herpes viruses, AIDS virus, and others.

latent virus
One of many viruses that retreats to the tissues of the nervous system, where it may persist for years without producing pathological symptoms; may leave the latent stage at any time, causing the normal symptoms of disease.

LD50
"Lethal dose fifty percent"; the amount of a toxic substance or treatment which kills fifty percent of a test population.

leukemia
Cancer of the blood and blood-forming tissues, characterized by abnormally high production of white blood cells.

leukocyte
Another name for a white blood cell.

linkage
The occurrence of two or more genes on the same chromosome; such genes are said to be *linked* and are transmitted as a group to gametes; closely linked genes do not obey the law of independent assortment.

linkage group
All genes located on the same chromosome; humans have twenty-three linkage groups.

locus
A position or location on a chromosome; synonymous with *gene;* a locus is occupied by one of numerous molecular variants called *alleles.*

Lucy
An important hominid fossil, *Australopithecus afarensis,* found in Ethiopia.

lymphatic system
The smaller of the circulatory systems of the body; consists of a network of lymph vessels that collect lymph fluid from the tissues and transport it to the cardiovascular system, which it enters near the heart; the lymphatic system also includes lymph nodes, tonsils, spleen, and thymus gland, which participate in specific immunity.

lymphocyte
A type of white blood cell, which differentiates to become either a T cell or a B cell.

lymphokines
Chemicals released by T cells that have antiviral and antibacterial properties.

Lyon hypothesis
The concept that all but one of the X chromosomes are inactivated in each cell of placental mammals.

lysogenic infection
A type of viral infection in which viral DNA attaches to the DNA of a host cell, where it is replicated with every division of the host; this type of infection does not kill the host cell.

lytic infection
A pattern of viral life history in which the host cell ruptures following viral replication.

m

macrophage
A large phagocytic cell that attacks foreign antigens and also interacts with T cells in producing specific immunity.

major histocompatibility complex
A set of closely linked loci whose products act as self markers and play a role in the immune response.

mast cell
A type of cell responsible for histamine release in the allergic response.

meiosis
A process of cell division occurring only in the gonads; usually consists of two successive divisions of a diploid parent cell to form four haploid daughter cells.

meiosis I
The first meiotic division; a *reduction division,* in which the chromosomal complement is reduced from duplicated pairs of chromosomes to unpaired chromosomes, each still consisting of two sister chromatids.

meiosis II
The second meiotic division; an equational division in which each chromosome (consisting of two sister chromatids) splits equally through the centromere; meiosis II begins with a cell containing one haploid set of duplicated chromosomes, which divides to produce two daughter cells, each containing one haploid set of non-duplicated chromosomes.

membrane
A thin layer of lipid and proteins that encloses a cell and also penetrates the interior of the cell, providing internal structure.

membrane budding virus
A virus that reproduces by packaging viral RNA or viral DNA into the viral coat at the cell wall and occasionally including a piece of host cell membrane in the mature virus.

memory cell
A type of T cell or B cell that provides long-term recognition of an antigen to which the body has previously been exposed.

Mendel's Law of Dominance
A law of inheritance, articulated by Mendel, which describes one type of relationship between homologous alleles in a heterozygote; traits governed by dominant alleles are expressed in heterozygotes, while those governed by recessive alleles are expressed only in homozygotes.

Mendel's Law of Independent Assortment
A law of genetics, first described by Mendel, which states that genes on different chromosomes are distributed to gametes and occur in subsequent generations independently of one another.

Mendel's Law of Parental Equivalence
A law of inheritance, first described by Mendel, which recognizes that the expression of an allele is not dependent upon which parent transmitted that allele.

Mendel's Law of Segregation

The concept, first described by Mendel, that the two hereditary particles (alleles) controlling a phenotypic trait separate (segregate) into different gametes during meiosis; each allele is transmitted unchanged to an offspring.

messenger RNA (mRNA)

A type of RNA, complementary to the protein-encoding segments of the sense strand of DNA, which carries the genetic information to be translated into a polypeptide chain.

metacentric chromosome

A chromosome with the centromere in the middle.

metaphase

The second phase of mitotic division, when the duplicated chromosomes are arranged along the metaphase plate of the spindle, towards the middle of the cell.

metaphase I

The second phase of meiosis I; paired duplicated homologous chromosomes (bivalents) line up at the metaphase plate.

metaphase II

The second phase of meiosis II; unpaired duplicated chromosomes line up at the metaphase plate.

metastasize

With reference to all diseases but especially cancer, to spread from a point of origin to other locations in the body.

methylation

Regulation of gene activity by the attachment of methyl groups (CH_3) to bases in DNA. Adding methyl groups blocks gene transcription; removal of methyl groups "turns on" gene activity.

middle repetitive sequences

DNA sequences that occur 10–100,000 times in a genome.

missense mutation

A substitution mutation that produces a triplet encoding a different amino acid than was encoded prior to the mutation.

mitochondrion

Membrane-bound organelle in the cytoplasm of eukaryotes; mitochondria produce energy for the cell and have their own DNA, separate from that of the nucleus. Mitochondria usually are inherited only from the female.

mitosis

A type of cell division in which a diploid cell divides once to produce two diploid daughter cells, each identical to the parent; the type of cell division found in all non-germinal cells.

mitotic cell cycle.

See **cell cycle (mitotic cell cycle)**

MN blood group

A blood group of humans characterized by the presence or absence of M and N antigens on red blood cells.

mobile genes

Segments of DNA that move between different locations in the genome.

monoculture

The cultivation of a single genetic strain over large geographic areas.

monohybrid cross

Mating between individuals homozygous for different alleles at one locus.

monomorphic

Having only one form; having only one kind of allele at a locus.

mononucleosis

A viral disease considered a self-limiting leukemia, caused by EBV.

monozygotic twins

Twins derived from the splitting of a single zygote shortly after fertilization; identical twins.

mosaic

An individual having two or more genetically different cell lines within the body.

MPF

Maturation promoting factor; a cellular protein, activated by cyclin, that both initiates mitosis and degrades cyclin, thus playing a key role in controlling the cell cycle in eukaryotic cells.

mRNA.

See **messenger RNA (mRNA)**

multiple alleles

The occurrence in a population of more than two alleles for a single gene.

multiple sclerosis

An autoimmune disease in which an individual produces antibodies to the myelin tissue of his or her nervous system.

muscular dystrophy (Duchenne muscular dystrophy)

An X-linked recessive genetic disease causing severe muscular deterioration and early death; primarily affects young males.

mutagen

Any substance or process that causes a genetic mutation.

mutation

Any heritable change in the sequence of nucleotide bases in DNA; may affect only one or two bases (point mutation) or long segments (chromosomal mutation).

n

nail-patella syndrome

An autosomal dominant trait, producing absent or malformed fingernails, toenails, and kneecaps.

natural killer (NK) cells

A type of lymphocyte that specializes in destroying viruses and cancer cells.

natural resistance

A set of innate, nonspecific defense mechanisms that protect an individual against a broad spectrum of toxic and pathogenic agents.

natural selection

Non-random reproduction; differential survival and reproduction on the basis of genotype, often popularly designated "survival of the fittest."

negative eugenics

The application of programs that reduce the number of deleterious or undesirable genes in a population, often by prohibiting individuals carrying such genes from breeding.

negative feedback

A type of system control in which there is an inverse relationship between present and future levels of function of the system; in other words, present high levels of production result in decreased future levels of production, and vice versa.

neoplasm

A cancerous tumor.

newborn screening

The testing of individuals at birth for genetic disorders.

nitrogenous base

One of four molecules (adenine, cytosine, guanine, and thymine) that link together the two chains of a DNA molecule; in RNA, uracil instead of thymine comprises the fourth base.

nondisjunction

The failure of homologous chromosomes to separate during meiotic division, producing gametes with one extra chromosome, or one chromosome too few.

nonoverlapping

A feature of the genetic code; the genetic code is read in consecutive groups of three nucleotide bases (a codon).

nonsense codon (terminator codon)

A codon that does not specify the addition of any amino acid to a polypeptide chain; therefore, it terminates the reading of the genetic message and ends the production of that protein.

normal distribution

A pattern of quantitative variation in which a set of phenotypic values shows a "bell-shaped" pattern of dispersion about a mean value; most individual values are close to the mean, while decreasingly small numbers show large variation from the mean; an example would be the distribution of heights of corn plants in a field.

norm of reaction

The range of all possible phenotypic responses of a particular gene or set of genes over all environments.

nuclear transplantation

The replacement of the nucleus of one cell with the nucleus from a second cell.

nucleic acid

DNA and RNA; long molecules composed of a chain of nucleotides; the genetic material.

nucleolus

"Little nucleus"; a dark-staining body within the nucleus.

nucleotide

A fundamental building block of DNA or RNA, consisting of a five-carbon sugar, a phosphate, and a nitrogenous base.

nucleus

The largest organelle within an eukaryotic cell; usually appears as a darkened, central area; contains chromosomes.

o

oncogenes

Mutational forms of genes that lead to the development of cancers.

one gene-one polypeptide hypothesis

The concept that each gene encodes information for the production of a single polypeptide chain; the single polypeptide chain may be either a complete protein itself or an entire chain that combines with other chains to form a protein.

oogenesis

The development of ova.

oogonium

A precursor cell within an ovarian follicle, which develops into a primary oocyte.

operator gene

A regulatory gene in an operon that exerts control over transcription of adjacent structural genes.

operon

A model of a structural and functional unit of the DNA of prokaryotes; includes a set of adjacent structural and regulatory genes involved in a particular metabolic pathway.

organelle

"Little organ"; one of many structural and functional sub-units within a cell.

organoids

A small growth of cells produced by recombinant DNA methods that are implanted in a specific site in the body for therapeutic purposes; the cells contain one gene that directs the development of blood vessels linking the organoid to the body's circulatory system and a second gene that synthesizes a chemical that the body requires but does not produce.

ovary

Female gonad; produces mature sex cells (ova).

overlapping genes

A phenomenon whereby a segment of DNA is transcribed twice, but with the reading frame in one transcription offset from that in the other; thus, one nitrogenous base participates in two codons, but occupies a different position (i.e., first, second, or third); the two codons often encode different amino acids.

oviduct

The duct through which an ovum passes from the ovary to the uterus.

ovulate

To release a secondary oocyte from the ovary.

ovum

Egg.

p

P

A designation for the original or first generation in a pedigree.

palindrome

Genes that encode the same message regardless of the direction of reading, e.g., *ATCCTA*.

parthenogenesis

A type of reproduction in which an egg undergoes complete development without fertilization.

partial dominance.

See incomplete dominance (partial dominance)

partially X linked (pseudoautosomal)

Homologous genes that occur on the X and Y chromosomes; their pattern of inheritance is the same as that of autosomal genes; see figure 5.2b.

passive immunity

Immune protection that results from the introduction of antibodies from one person into another.

pattern baldness

A phenotype showing sex-influenced inheritance, which produces early baldness; the individual retains a distinct rim of hair around the head.

PCR.

See polymerase chain reaction

pedigree

A diagram representing a set of familial relationships; a genealogy.

penetrance

The expression of a single gene in the phenotype—with respect to a single individual a gene may be either zero or 100% penetrant—expressed or not dependent upon the individuals total genotype.

penetrant

Expressed; a gene showing *any* phenotypic manifestation is said to be penetrant; with respect to an individual, a gene is either penetrant or not; a qualitative measure of gene expression in an individual.

peptide bond

A linkage between adjacent amino acids in a polypeptide.

pesticide treadmill

The necessity, when using pesticides to control insect and other crop pests, to increase continually the dose of pesticides applied to maintain the same level of pest control.

phage.

See bacteriophage

phenocopy

A phenotype caused by an environmental factor that resembles a phenotype normally caused by a genetic factor; for example, sun-bleached hair (environmentally caused) resembles blond (genetically caused) hair.

phenotype

Any expression of genes; phenotype may refer to any trait of an organism, such as its structure, physiology, biochemistry, behavior, or to all traits collectively; the phenotype is determined by the interaction of genes and environment.

phenylketonuria (PKU)
A genetic defect characterized by inability to produce the enzyme phenylalanine hydroxylase; phenylketonurics are unable to metabolize phenylalanine properly and without a controlled diet become mentally retarded.

pheromone
A type of hormone that is released into the environment as a mechanism of chemical communication between individuals.

Philadelphia (Ph1) chromosome
A shortened chromosome 22 of humans resulting from translocation of the long arm, usually to the long arm of chromosome 9; usually found in persons with leukemia.

placenta
A structure formed by the complex growth and intertwining of adjacent tissue from mother and fetus; provides a means for the passage of nutrients from mother to fetus and waste products from fetus to the mother; also secretes some hormones.

plasma
The liquid portion of the blood.

plasma cell
A type of cell formed by the differentiation of B cells; produces antibodies.

plasmid
A small ring of DNA found in the cytoplasm of bacteria.

pleiotropic
The multiple phenotypic effects of a single allele.

ploidy
The number of sets of chromosomes possessed by a cell or individual.

point mutation.
See **gene mutation (point mutation)**

polar body
A daughter cell of either the first or second meiotic division in females that has a normal chromosomal complement, but almost no cytoplasm.

polygenic inheritance
The control of a single phenotypic trait by many different genes.

polymerase chain reaction (PCR)
A technique for amplifying a specific segment of DNA.

polymorphic
Having many forms; having two or more alleles at a locus.

polypeptide
A sequence of three or more amino acids linked by peptide bonds; may be a complete protein.

polyploid
A cell or individual possessing entire extra *sets* of chromosomes; designated with reference to the number of haploid sets present, such as triploid (3N), tetraploid (4N), etc.

polyribosome
A structure formed by the attachment of many ribosomes to a single mRNA molecule during translation.

polytene chromosome
A giant chromosome consisting of many identical DNA molecules, arranged side by side, resulting from chromosome replication without subsequent cell division.

position effect
Where the expression of a gene depends upon its physical location within the genome; a change in the chromosome position may change the phenotypic effect.

positive eugenics
The application of techniques that selectively increase or otherwise favor beneficial genes in a population.

prenatal diagnosis
Identification of genetic diseases of fetuses from tissue samples taken *in utero.*

prenatal screening
Biochemical tests for genetic diseases conducted on tissue samples taken from a fetus or its surrounding amniotic fluid *in utero.*

primary nondisjunction
Nondisjunction of homologous chromosomes during meiosis I.

primary oocyte
A cell within an ovarian follicle, derived from an oogonium, that enters into the first meiotic division, giving rise to a secondary oocyte and the first polar body.

primary sex characters
The gonads.

primary sex ratio
The sex ratio at conception.

primary spermatocyte
A cell within the seminiferous tubules derived from a spermatogonium; divides in meiosis I to produce two secondary spermatocytes.

primary structure
Of a protein referring to the sequence of amino acids in the protein.

primates
The order of mammals including monkeys, Great Apes, and humans.

probability
A measure of the uncertainty of occurrence of a specific event in the future; usually expressed as a number between zero (the event is impossible) and one (the event is certain to occur); likelihood.

probe
A molecule (usually a nucleic acid) that has been labeled with a radioisotope, dye, or enzyme and is used to locate a particular gene or DNA sequence on a DNA molecule or chromosome.

progeria
An autosomal dominant genetic disease characterized by extremely rapid aging and a very short life span.

progesterone
A female sex hormone.

prokaryotic cell (prokaryote)
One of two basic types of cells; smaller and simpler in appearance than eukaryotic cells; lacks a nucleus and has few distinguishable features under a light microscope; has only one circular DNA molecule, not a true chromosome.

proliferate
In reference to living cells, to increase in numbers as a result of cell division.

promoter
That region of the bacterial operon to which RNA polymerase attaches to initiate gene transcription.

prophase
The first stage of mitosis, beginning when chromosomes become visible under the microscope.

prophase I
The first of the four phases of meiosis I.

prophase II
The first phase of meiosis II.

protein
A major category of molecules required by living organisms; consists of one or more chains of amino acids, sometimes combined with other types of molecules.

proto-oncogenes
Genes that in their normal condition promote cellular growth, but that sometimes undergo mutation to become oncogenes that cause cancer.

protoplast fusion
The production of hybrid cells in tissue culture, as a result of the union of cells from different species or different cell lines.

pseudoautosomal.
See **partially X linked**

pseudohermaphrodite
An individual possessing only one type of gonad (or none at all), but externally showing some secondary sex characteristics of both sexes.

pure-breeding
A genetic lineage in which all members are homozygous for a particular gene or set of genes and, hence, show little or no phenotypic variation for the trait or traits controlled by that gene or genes; such a lineage is said to "breed true."

q

quantitative variation
A type of phenotypic variation in which individual phenotypes differ on a finely graded, or continuous scale.

r

r.
See **coefficient of relationship (r)**

race
In biological terms, almost any population or subgroup of a species that can be distinguished from other sub-groups (a definition so broad as to be almost meaningless); in genetics, race often refers to a population differing in the frequency of certain alleles, chromosomes, or phenotypic features from other populations.

racism
The belief that one race or ethnic type is inferior to another (usually one's own) race or ethnic type, almost always associated with stigmatization of the type believed inferior.

rad
A measure of the amount of radiation energy absorbed by living tissues.

random mating
In a genetic sense, mating without respect to genotype.

recessive
An allele whose phenotypic effects are suppressed by its homologous partner in a heterozygote; may also refer to the phenotypic trait that is suppressed.

reciprocal cross
A hybrid mating involving the same genetic traits as in a previous cross, but in which the sexes of the individuals of each type are reversed; for example, a mating of a short female pea plant and a tall male pea plant is the reciprocal cross of a tall female pea plant and a short male pea plant.

reciprocal translocation
A type of chromosomal mutation in which acentric portions are exchanged by non-homologous chromosomes.

recombinant DNA technologies
Genetic engineering techniques; artificial alteration of DNA sequences or individual genotypes; may include transfer of genetic material from one individual to another or insertion of artificially constructed DNA sequences; or combination in hybrid progeny of genetic material from two or more parents of different genetic types; or any other technique that produces new genetic combinations or altered expression of preexisting combinations.

recombination
The exchange of alleles among homologous chromosomes as a result of crossing over in meiosis I; produces new allelic combinations on chromosomes; may also refer to or include the results of independent assortment, which have the similar consequence of producing gametes with allelic combinations found in neither parent.

reduction division
The first meiotic division, which produces two daughter cells, each containing one duplicated member of each pair of chromosomes.

regulator (regulator gene)
A gene in an operon that encodes a repressor protein and thus controls transcription.

regulatory gene
A gene that controls the expression of structural or other regulatory genes.

rem
Roentgen equivalent man; the amount of biological damage done by radiation; one rem = 10 millisieverts.

replicate
To duplicate precisely, often in reference to chromosomes.

repressor
In an operon, a protein produced by a regulator gene that, when bound to the operator, inhibits transcription.

restriction endonucleases
Bacterial enzymes that each cut DNA at specific base sequences. These enzymes are used in many modern genetic engineering methodologies.

restriction fragment length polymorphism (RFLP)
A large variety of patterns of the DNA after it has been incubated with a series of bacterial enzymes that cut the DNA at specific base sequences. Each individual has his or her own unique RFLP pattern, which can be used for both identification of genetic disease and for a variety of other medical and forensic purposes.

retrovirus
One of a group of RNA viruses that causes tumors in vertebrates.

reverse transcriptase
An enzyme produced by RNA viruses; used to construct a complementary DNA copy of the viral RNA for insertion into the host's genome.

reverse transcription
Synthesis of a DNA molecule from an RNA template. Viral RNA lysogenically infects cells in this manner, often resulting in cancer.

RFLP.
See **restriction fragment length polymorphism (RFLP)**

Rh blood group
A blood group of humans characterized by the presence or absence of the Rh antigen on red blood cells.

Rh incompatibility
Where mother and fetus have different Rh blood types; incompatibility between an Rh-negative mother and Rh-positive fetus often causes erythroblastosis fetalis.

ribonucleic acid (RNA)
A molecule comprised of a chain of nucleotides containing the sugar ribose; RNA is the hereditary material in a few viruses, and is involved in protein synthesis in all organisms.

ribose
A five-carbon sugar; one of the structural components of RNA.

ribosomal RNA (rRNA)
A type of RNA that combines with proteins in the cytoplasm to form ribosomes.

ribosome
A cytoplasmic organelle formed by the association of rRNA and proteins; the site of translation.

ribozymes
Ribosomal RNA that functions as an enzyme to catalyze chemical reactions; ribozymes catalyze the excision of introns from HnRNA.

ring chromosome
A circular chromosome in eukaryotes; formed after both ends of a chromosome are lost by terminal deletion, and the free ends then join.

RNA.
See **ribonucleic acid (RNA)**

RNA polymerase
An enzyme that catalyzes the formation of RNA (polymerization); facilitates the union of ribonucleotides in a complementary sequence to that of a DNA molecule.

rRNA.
See **ribosomal RNA (rRNA)**

S

S phase
The synthesis period of interphase, immediately following G_1, during which the chromosomes are replicated.

same sense (same sense mutation)
A type of point mutation in which one nucleotide base replaces another, but because of the redundancy of the genetic code, the new codon encodes the same amino acid as the old codon; therefore, same sense mutations do not change the protein encoded by that gene.

sarcoma
A type of cancer affecting bones and muscles.

satellite DNA
A highly variable region of the human genome that can be isolated by a cDNA probe and then RFLP mapped for individual human "genetic fingerprinting."

SCID
Severe combined immune deficiency, a sex-linked recessive disease leaving a victim without any immune capability—generally lethal.

secondary nondisjunction
Nondisjunction of sister chromatids of a duplicated chromosome during meiosis II.

secondary oocyte
A cell within an ovarian follicle resulting from the first meiotic division of a primary oocyte; if fertilized it divides to produce the ovum and a second polar body.

secondary sex characters
Traits, other than gonads, that show sexual dimorphism; i.e., breasts in females, beards in males.

secondary sex ratio
The sex ratio at birth.

secondary spermatocyte
Either of two cells produced during meiosis I by the division of a primary spermatocyte; subsequently divides in meiosis II to produce two spermatids.

semi-conservative replication
The type of replication found in DNA, in which each daughter double helix receives one complete parental strand of DNA, which in turn serves as a template for the construction of the complementary strand.

seminiferous tubule
Extremely long and convoluted tubule within the testis; the site of spermatogenesis.

sense strand
The strand of DNA that is transcribed to mRNA; the complementary strand is not transcribed.

sex chromosomes
In eukaryotes, an X or Y chromosome, variation in which is related to sex determination; usually, in one sex both members of the pair are homologous, while in the other sex they differ in structure; in some species, one sex has two sex chromosomes, the other sex has only one.

sex-influenced inheritance
A pattern of inheritance in which the expression of alleles is different in each sex; usually, an allele is dominant in one sex, recessive in the other; refers only to autosomal alleles.

sexism
The belief that one sex is inferior to another, and that the inferior sex should be stigmatized; most commonly, women are the victims of sexism.

sex-limited inheritance
A pattern of inheritance in which autosomal genes are expressed only in one sex, but not the other.

sex-linked genes
Genes located on a sex chromosome, usually the X.

sex ratio
The proportion of males to females in a population; often expressed as the number of males per 100 females.

sexual dimorphism
Phenotypic differences between males and females.

sexual reproduction
Reproduction usually involving the fusion of gametes from two (or more) parents to produce offspring; the key aspect is reorganization in the progeny of genetic material derived from the parents.

sickle-cell anemia
Hereditary disease caused by an autosomal recessive allele; causes hemoglobin to crystallize, deforming red blood cells and producing numerous phenotypic abnormalities.

sickle-cell trait
The phenotype of heterozygotes for sickle-cell anemia who produce two types of hemoglobin.

sister chromatid exchange (SCE)
The exchange of identical segments between sister chromatids during cell division; does not produce genetic recombination such as occurs due to crossover between non-sister homologues in meiosis. *See also* **chromatid**

SLE.
See **Systemic lupus erythematosus (SLE)**

somatic
"Of the body"; sometimes refers to all bodily tissues, and sometimes used to distinguish between germinal and all other bodily (i.e., somatic) tissues.

somatic hypermutation
The process where each B cell produces a gene for a specific type of antibody by selecting segments for the variable portions of the antibody from hundreds of possible sequences contained in the nuclear genome. Each B cell produces a unique antibody, and the result is an enormous variety of antibody types, from which those specific to a particular disease organism or toxin can subsequently be amplified.

speciation
The process of evolutionary change in which one species evolves into one or more new species.

species
A group of individuals of one type that can freely interbreed in nature and produce fertile offspring.

specific immunity
Resistance of an individual to a particular microorganism or pathogen.

spermatid
Either of two cells that are the result of division by a secondary spermatocyte in meiosis II; spermatids subsequently develop into spermatozoa.

spermatogenesis
The development of spermatozoa.

spermatogonium
A precursor cell within the seminiferous tubules that initiates spermatogenesis by dividing mitotically, becoming two primary spermatocytes.

spermatozoon (sperm)
Male gamete; sex cell.

sperm banking
The collection and preservation of sperm in a dormant but living condition for later use.

spermiogenesis
The final stage of spermatogenesis, during which spermatids undergo maturation to become functional spermatozoa.

spindle apparatus
A set of fibers, shaped like a spool, that become visible in the cytoplasm during cell division; in animal cells, the spindle extends between two centrioles; in plants, the spindle is similar to that in animals, but no centrioles are present.

spleen
An amorphous abdominal organ, part of the lymphatic system, that serves as a reservoir for blood and functions in the immune response.

SRY gene
A gene located on the short arm of the Y chromosome. The SRY gene product directs differentiation of the embryonic gonad to a testis. The name *SRY* refers to the "sex-determining region of the Y."

sterile male technique
The release of large numbers of sterilized males of a specific insect pest species for the limitation or control of that species; the sterile males mate with fertile females, who subsequently produce no offspring.

structural gene
A gene encoding the amino acid sequence of a polypeptide.

submetacentric chromosome
A chromosome with the centromere located slightly off-center.

substitution (substitution mutation)
A type of point mutation in which one nucleotide base replaces another; affects only the codon in which substitution occurs.

superfecundation
The fertilization of two ova by sperm from different males during the same menstrual cycle; twins resulting from superfecundation are only half siblings, in spite of sharing a uterus during gestation.

superovulation
The ovulation of more eggs than normal; induced artificially by injecting a female with gonadotropic hormones.

surrogate mother
A female who receives an implanted conceptus or embryo from another female (the genetic mother) and who then carries that embryo for the remainder of its gestational period.

symbiosis
"Living together"; any biological relationship or interaction involving two or more species; symbionts may be benefited, harmed, or not affected by the relationship.

synapsis
The close side-by-side pairing of duplicated homologous chromosomes in meiosis I; does not normally occur in mitosis.

syndrome
A set of symptoms associated with one disease or pathological condition.

systemic lupus erythrematosus (SLE)
An autoimmune condition resulting from the production of antibodies to one's own DNA.

t

T cell
A type of lymphocyte responsible for cellular immunity.

TDF.
See **testis determining factor (TDF)**

telocentric chromosome
A chromosome whose centromere is located at the very end.

telophase
The fourth and last stage of mitosis; nuclear membranes enclose each set of daughter chromosomes, and the cytoplasm of the parent cell divides to form two complete daughter cells.

telophase I
The fourth and last phase of meiosis I.

telophase II
The fourth and last phase of meiosis II.

teratogen
Any chemical, radiation, or other factor that causes physical deformity.

terminator codon.
See **nonsense codon (terminator codon)**

tertiary sex ratio
Sex ratio measured at any specified time after birth.

testcross
The mating of a suspected heterozygote to a recessive homozygote.

testicular feminization
A syndrome probably due to an X-linked recessive mutation, in which individuals with a normal male karyotype (46, XY) possess small internal testes, but no other internal reproductive structures; externally, they have an apparently normal female phenotype.

testis
Male gonad; produces mature sex cells, spermatozoa; the primary sexual character of males; usually paired in vertebrates.

testis determining factor (TDF)
A long-hypothesized gene and its product, thought to be responsible for the induction of testis formation. Assumed to be located on the Y chromosome, this gene is probably identical to the SRY gene.

testosterone
The male sex hormone.

test tube baby
An evocative but misleading phrase popularized by the public news media for *in vitro* fertilization; freshly ovulated eggs are collected from a female and these eggs are then fertilized *in vitro* (usually in a petri dish, *not* a test tube) and subsequently reintroduced into the mother's uterus, where normal gestation occurs; a child produced in this manner differs from another child only in that his or her conception occurs outside the mother's body.

thalassemia
A group of diseases in which affected individuals produce a reduced amount of the alpha chain or the beta chain of hemoglobin.

thymine dimer
The bonding to one another of adjacent thymine molecules on a single strand of DNA, rather than to the adenines on the complementary strand.

thymus gland
An amorphous gland, part of the lymphatic system, located in the upper chest; the site of maturation of T cells.

tissue culture
The growth of selected cells or tissues in defined culture medium in the laboratory.

totipotent
A cell with the capacity to develop into a complete organism.

transcription
The production of mRNA from a DNA template.

transduction
Transfer of a segment of DNA from one cell to another by a virus; the DNA from the first cell can be incorporated into the genome of the second.

transfer RNA (tRNA)
The smallest type of RNA, only about eighty nucleotides long, which transports an amino acid to a ribosome for incorporation into the growing protein.

transformation
A cell incorporating free DNA from the surrounding medium into its own genome; also refers to the change of normal cells into cancerous ones.

translation
The reading of mRNA by ribosomes, resulting in the production of a polypeptide chain; the sequence of nitrogenous bases in the mRNA specifies the sequence of amino acids in the polypeptide.

translocation
A type of chromosomal mutation in which a segment of DNA is detached from its normal chromosome and attached to a non-homologous chromosome.

translocation Down's syndrome
A type of Down's syndrome caused by the translocation of the long arm of chromosome 21 to one of the other autosomes, most frequently to chromosome 14.

transposon
A gene that moves between locations on one or more chromosomes, also called transposable elements.

triplet
A codon.

trisomy
The presence in a cell of three homologues of a particular chromosome; produces a diploid number of $2N+1$.

tRNA.
See **transfer RNA (tRNA)**

tumor suppressor genes
Genes whose products limit cell proliferation and control normal cell growth and differentiation; also called anti-oncogenes. Loss or mutation of tumor suppressor genes may lead to cancer.

Turner's syndrome
A set of phenotypic anomalies associated with females having only one X chromosome; the genotype is designated 45, XO.

U

ultraviolet radiation (UV)
Radiation with wavelengths shorter than visible light and longer than X rays.

unconventional viruses
A group of infectious agents, extremely small in size, but bearing few similarities to normal viruses; they produce diseases involving slow degeneration of nervous tissue and ultimately death.

unique sequences
DNA sequences that occur only once in a genome.

utilitarian
The belief that behavior should be determined by assessing the consequences of alternative actions and choosing the one that has the most desirable consequences.

V

vector or **cloning vehicle**
Plasmids or phages used in molecular cloning to move genes and/or pieces of DNA from one cell to another.

vertebrates
Animals with backbones; includes fish, sharks, amphibians, reptiles, birds, and mammals (including humans).

virus
A structure with some but not all features normally attributed to living organisms; consists of nucleic acid, either RNA or DNA, enclosed in a protein coat; reproduces only when parasitizing living cells.

W

Wilms tumor
Cancer of the urinary bladder; associated with a deletion on chromosome 11.

X

X chromosome
The sex chromosome found in both sexes in species with two sex chromosomes and in which the male is the heterogametic sex; the larger of the two sex chromosomes in humans; females have two X chromosomes, males only one.

xeroderma pigmentosum
A disease resulting in numerous skin cancers caused by a recessive autosomal allele.

X inactivation
The "turning-off," or inactivation, of an X chromosome during development.

X-linked genes
Any or all genes on the X chromosome; often used synonymously with *sex-linked*.

X rays
High enery radiation capable of causing damage to DNA and chromosomes, resulting in mutations.

XX-XO system
The chromosomal system of sex determination, found in some insects, in which the female has two sex chromosomes (XX) and the male has only one (XO).

XX-XY system
A chromosomal system of sex determination, in which females are homogametic (XX) and males are heterogametic (XY).

Y

Y chromosome
The sex chromosome found only in males in species with two sex chromosomes and in which the males are the heterogametic sex; the smaller of the sex chromosomes in humans; males have one Y chromosome, females none.

Y-linked genes
Any or all genes on the Y chromosome.

Z

zygote
A fertilized egg; also refers to any developmental stage of an organism prior to hatching or birth; in humans, most commonly refers to period from conception to end of second week of gestation.

ZZ-ZW system
The chromosomal system of sex determination found in birds and some insects, in which the male is the homogametic sex (ZZ) and the female heterogametic (ZW).

hemoglobin, 200–201
sexual phenotype, 131–34
in eukaryotes
DNA structural modification,
206–7
nucleus organization, 201
regulatory molecules, 206
variation in DNA amounts, 208
in prokaryotes
lac operon, 197–99
operon and genetic engineering,
248, 253
operon defined, 197
operon hypothesis, 197–99
regulatory genes, functions of,
197–200
repressor, 198–99
transcription inhibition, 198–99
X inactivation, 116–18
Gene splicing, 236
Gene therapy, 262–66, 268, 363
Genetically significant dose (GSD),
radiation, 308–9
Genetic bottleneck, 369
Genetic code, 179–82
Genetic counseling, 85–86
Genetic dictionary, 180–81
Genetic diseases, 78, 90, 109, 154
Genetic distance, 383–84
Genetic diversity, 356, 380
Genetic engineering
gene pool manipulation, 236
operon as tool of, 248, 253
and viral transduction, 340
Genetic load, 307
Genetic pharmacology, 247
Genetic probes, 236, 240
Genetics
defined, 6, 8, 22
developmental, 196
ecological, 359
and evolution, 10, 14, 357–59
genetic code, 179–81
genetic dictionary, 180–81
genetic diversity, 356, 380
population, 356–59
Genetics and health, 11
genetic counseling
Down's syndrome, 277–79
sickle-cell anemia, 86, 242
therapeutic abortion, 86, 159,
161
genetic techniques
amniocentesis, 154
chorion villus sampling, 157
fetoscopy and fetal surgery, 157
newborn screening, 88–89
prenatal diagnosis, 156–59
new techniques
cell fusion, 363
cloning, 236–39
recombinant DNA, 236
transduction, 340
transformation, 171–73
pest control, 365

reproductive techniques
artificial insemination, 146–47
in vitro fertilization (IVF), 11,
147–56, 159
sperm banks, 146–47
surrogate motherhood, 147–51
Genetic sexual disorders, 131–40
Genetics of sex
sex determination
chromosomal mechanisms of,
103
environmental factors, 103
sex-influenced inheritance, 119
sex linkage
X, 109–15
X inactivation, 116–18
Y, 109
ZZ-ZW, 105
sex ratios, 119–20
Gene transfer, 363–64
Gene transformation, 171–73
Genome, 356
Genotype, 37, 51
Genus, classification, 395
Germinal tissues, 39
Glossary, 407–21
Glucose-6-phosphate dehydrogenase
deficiency, 109
Gonads, 39, 105, 124
Gould, S. J., 217, 226
Grafting, 362
Great Apes, 15–16, 23, 40–41,
378–79
Green Revolution, 12, 14, 367–68
Griffith, F., 171–72
Growth, 45, 249–54
Guanine, 174–75
Guevodoces, 135
Guthrie, Woody, 12

h

Haplodiploid, 105–6
Haploid, 37, 105
Hardy, G. H., 357
Hardy-Weinberg equilibrium
equations, 357–59
Hardy-Weinberg law, 357–59
Hayflick Limit, 293
Helper T-cells, 327–28
Hemizygousity, 109–10, 281
Hemoglobin, 79, 83, 169, 188, 196,
255
development of, 200
polypeptide chains of, 200
Hemophilia, 109, 111–13
Hemophilia B (Christmas disease),
109, 111
Heritability, 222–23
Hermaphrodite, 102, 135
Herpes, 343
Herpes virus, 338, 343
Hershey, A. D., 172–73
Heterochromatin, 34
Heterochromatinization, 203
Heterogametic sex, 105

Heterogeneous nuclear RNA,
(HnRNA), 183–84
Heterosexual, 142
Heterozygote, 55
Heterozygous, 37, 55
Hierarchical body organization, 31
Highly repetive sequences, DNA, 208
Hiroshima and Nagasaki studies,
radiation, 310, 312–13
Histamine, function of, 337
Histones, 35, 323
Holandric inheritance, 109, 115–16
Homeobox genes, 196
Hominids, 16
Homogametic sex, 105
Homologous, chromosomes, 37
Homosexual, 142
Homozygote, 55
Homozygous, 37, 55
Hong Kong flu, 342
Hopi Indians, and albinism, 80–81
Hormones, 134, 150, 168, 206,
249–50, 366
Human Genome Organization,
(HUGO), 261–62, 268
Human Genome Project, 259–62, 268
Human immunodeficiency virus
(HIV), 346–50
Human leukocyte antigens (HLA)
and disease, 95–96, 345
and forensic medicine, 95–97
and insulin dependent diabetes, 96,
345
and tissue rejection, 332
Human origins, 17, 378–83
Human sexuality
female reproductive anatomy,
124–26, 133
gametogenesis, 128–31
gender, 140–42
genetic sexual disorders
chimeras and mosaics, 136
chromosomal, 136–40
single gene, 134
male reproductive anatomy,
126–28, 133
oogenesis, 130
sexual phenotype development,
131–32, 134
spermatogenesis, 129
Human T-cell leukemia virus
(HTLV-1), 297
Humoral immunity, 323
Huntington's disease, 12, 90, 211–12,
251, 254
H-Y antigen, protein, 133
Hybridization, 39, 241–44, 362–63
Hybrid vigor, 362
Hydatidiform moles, 207
Hydrocephalus, 149
Hypotrichosis, 78

i

Ichthyosis, 109, 113–15
Identical twins, 76, 222

Mitosis, 39–40, 43–45
Mitotic cell cycle, 41–44
Mitotic nondisjunction, 137
MN blood group, 93–94, 219, 332
Modifying loci, 216
Molecules
    described, 32
    informational macromolecules, 178
    nucleotide, 175
    organic, 32
    protein, 168–70
    regulatory, 205–6
Monoculture, and agricultural
        genetics, 363, 365
Monod, Jacques, 197
Monohybrid crosses, 52, 56
Mononucleosis, 341
Monosomy, 137, 277
Monozygotic twins, 76
Mosaic, 118, 136, 277
Muller, H. J., 147, 290, 306–7, 365
Müllerian ducts, 131–32, 135
Müllerian inhibiting substance, 132,
        135
Multiple alleles, 62, 95
Multiple sclerosis (MS), 96, 343–44
Mumps, 344
Mutagenic chemicals, 294–95
Mutation, 64, 204, 272, 316
    and artificial selection, 361–62
    chromosome, 273–85
        aneuploidy, 273, 316
        deletions, 280–82
        duplications, 273–77, 282–83
        inversions, 283–84
        polyploidy, 273–77, 316, 362
        translocations, 284–85
    defined, 10, 64, 204, 272
    gene
        causes of, 286–92, 361
        DNA repair, 292
        frameshift, 181, 286
        missense, 181, 286
        point, 286
        randomness of, 289
        same sense, 181, 286
        spontaneous, rate of, 287–88
        substitution, 181, 286, 289
        thymine dimers and, 291–92
        types of, 181
    germinal, 272
    and Mendel's laws, 64
    mutation theory of aging, 292–93
    and radiation, 290–92, 310–14
    somatic, 272
Mutation theory of aging, 292–93

Negative eugenics, 390
Negative feedback control, and lac
        operon, 199
Neoplasms, and cancer, 295
Neurofibromatosis, 87, 90
Newborn screening, 88–89
Nicotine, 126
Nirenberg, M. W., 180
Nitrogenous base, nucleotide, 175
Nondisjunction, 136–37, 139, 277–78
Nonoverlapping, and genetic code,
        181
Non-repetitive sequences, DNA, 208
Nonsense codons, and genetic code,
        181
Nonspecific defense mechanisms, and
        body, 322, 323
Normal distribution, 216
Norm of reaction, 209–10
Nuclear transplantation, and cloning,
        247
Nucleic acids, function of
    central dogma, 178, 188–89
    DNA replication, 10, 179
    genetic code, 179–82
    as hereditary material, 168
    protein synthesis, 186–88
    RNA structures and functions,
        177, 182–85
    unified view of gene, 187
Nuclein, 170
Nucleolus, 34, 36
Nucleotide molecule, 175
Nucleus, cell, 34
Nucleus organization, and gene
        regulation, 201
Nullisomic, chromosomes, 277
Nutrition, and IQ, 226

described, 172–73, 177, 399
DNA, 172–73
human immunodeficiency virus
  (HIV), 146, 346–47
latent
  described, 338
  insulin dependent diabetes, 344
  multiple sclerosis, 343
life cycle of
  described, 172–73, 338–39
  reverse transcriptase, 339–40
  transduction, 340
lytic, 338
membrane budding, 343–44
  mumps, 344
  RNA, 177
viral disease
  AIDS, 345–50
  herpes, 298, 338, 343
  influenza, 338, 342
  leukemia, 297, 338
  measles, 343–44

mumps, 344
polio, 338
rubella, 338
Vitamin D resistant rickets, 109, 113
Vries, Hugo de, 8